MTP International Review of Science

Lanthanides and Actinides

MTP International Review of Science

Publisher's Note

The MTP International Review of Science is an important new venture in scientific publishing, which we present in association with MTP Medical and Technical Publishing Co. Ltd. and University Park Press, Baltimore. The basic concept of the Review is to provide regular authoritative reviews of entire disciplines. We are starting with chemistry because the problems of literature survey are probably more acute in this subject than in any other. As a matter of policy, the authorship of the MTP Review of Chemistry is international and distinguished; the subject coverage is extensive, systematic and critical; and most important of all, the new issues of the Review will be published every two years.

In the MTP Review of Chemistry (Series One), Inorganic, Physical and Organic Chemistry are comprehensively reviewed in 33 text volumes and 3 index volumes, details of which are shown opposite. In general, the reviews cover the period 1967 to 1971. In 1974, it is planned to issue the MTP Review of Chemistry (Series Two), consisting of a similar set of volumes covering the period 1971 to 1973. Series Three is planned for 1976, and so on.

The MTP Review of Chemistry has been conceived within a carefully organised editorial framework. The over-all plan was drawn up, and the volume editors were appointed, by three consultant editors. In turn, each volume editor planned the coverage of his field and appointed authors to write on subjects which were within the area of their own research experience. No geographical restriction was imposed. Hence, the 300 or so contributions to the MTP Review of Chemistry come from many countries of the world and provide an authoritative account of progress in chemistry.

To facilitate rapid production, individual volumes do not have an index. Instead, each chapter has been prefaced with a detailed list of contents, and an index to the 10 volumes of the MTP Review of Inorganic Chemistry (Series One) will appear, as a separate volume, after publication of the final volume. Similar arrangements will apply to the MTP Review of Physical Chemistry (Series One) and to subsequent series.

Butterworth & Co. (Publishers) Ltd.

Inorganic Chemistry Series One

Consultant Editor
H. J. Emeléus, F.R.S.
Department of Chemistry
University of Cambridge

Volume titles and Editors

1 **MAIN GROUP ELEMENTS— HYDROGEN AND GROUPS I–IV**
Professor M. F. Lappert, *University of Sussex*

2 **MAIN GROUP ELEMENTS— GROUPS V AND VI**
Professor C. C. Addison, F.R.S. and Dr. D. B. Sowerby, *University of Nottingham*

3 **MAIN GROUP ELEMENTS— GROUP VII AND NOBLE GASES**
Professor Viktor Gutmann, *Technical University of Vienna*

4 **ORGANOMETALLIC DERIVATIVES OF THE MAIN GROUP ELEMENTS**
Dr. B. J. Aylett, *Westfield College, University of London*

5 **TRANSITION METALS—PART 1**
Professor D. W. A. Sharp, *University of Glasgow*

6 **TRANSITION METALS—PART 2**
Dr. M. J. Mays, *University of Cambridge*

7 **LANTHANIDES AND ACTINIDES**
Professor K. W. Bagnall, *University of Manchester*

8 **RADIOCHEMISTRY**
Dr. A. G. Maddock, *University of Cambridge*

9 **REACTION MECHANISMS IN INORGANIC CHEMISTRY**
Professor M. L. Tobe, *University College, University of London*

10 **SOLID STATE CHEMISTRY**
Dr. L. E. J. Roberts, *Atomic Energy Research Establishment, Harwell*

INDEX VOLUME

Inorganic Chemistry Series One

Consultant Editor
H. J. Eméleus, F.R.S.

MTP International Review of Science

Volume 7
Lanthanides and Actinides

Edited by **K. W. Bagnall**
University of Manchester

Butterworths · London
University Park Press · Baltimore

THE BUTTERWORTH GROUP

ENGLAND
Butterworth & Co (Publishers) Ltd
London: 88 Kingsway, WC2B 6AB

AUSTRALIA
Butterworth & Co (Australia) Ltd
Sydney: 586 Pacific Highway 2067
Melbourne: 343 Little Collins Street, 3000
Brisbane: 240 Queen Street, 4000

NEW ZEALAND
Butterworth & Co (New Zealand) Ltd
Wellington: 26–28 Waring Taylor Street, 1

SOUTH AFRICA
Butterworth & Co (South Africa) (Pty) Ltd
Durban: 152–154 Gale Street

ISBN 0 408 70221 4

UNIVERSITY PARK PRESS

U.S.A. and CANADA
University Park Press Inc
Chamber of Commerce Building
Baltimore, Maryland, 21202

Library of Congress Cataloging in Publication Data
Bagnall, K   W
    Lanthanides and actinides

    (Inorganic chemistry, series one, v. 7) (MTP
international review of science)
    Includes bibliographies
    1. Rare earth metals. 2. Actinide elements.
I. Title
QD151.2.15 vol. 7 [QD172.R2] 546'.4 75-160326
ISBN 0-8391-1007-3

First Published 1972 and © 1972
MTP MEDICAL AND TECHNICAL PUBLISHING CO. LTD
Seacourt Tower
West Way
Oxford, OX2 OJW
and
BUTTERWORTH & CO. (PUBLISHERS) LTD

Filmset by Photoprint Plates Ltd., Rayleigh, Essex
Printed in England by Redwood Press Ltd., Trowbridge, Wilts
and bound by R. J. Acford Ltd., Chichester, Sussex

Consultant Editor's Note

The problem of keeping abreast of research literature on as broad a front as possible is one that confronts all chemists. In the past this difficulty has been met, in the main, by literature surveys and by several uncorrelated reviews of progress in certain subject areas. There are obvious inadequacies in this approach, which have become increasingly apparent in recent years. I was, therefore, grateful for the opportunity of helping to plan this new series, which has been designed to provide a comprehensive, critical survey of each of the main branches of chemistry.

This section of the MTP International Review of Science deals with progress in Inorganic Chemistry. The subject is developing at an astonishing rate and in many directions. Fortunately, however, it lends itself to a systematic treatment. Ten volumes have been prepared, three dealing with the main group elements and two with the general chemistry of the transition metals. Organometallic derivatives of the main group elements and lanthanides and actinides are covered separately, as is the subject of reaction mechanisms. The two remaining volumes on radiochemistry and solid state chemistry have been planned to avoid, as far as possible, overlap with those that have gone before.

It is a pleasure to thank the many experts who have collaborated as authors and volume editors in making this publication possible. While working to a pre-arranged over-all plan, they have been able to assess and interpret the literature in terms of their own experience in specialised fields. I believe that in this way they will not only provide a record of what has been done, but will stimulate further exploration in this fascinating branch of chemistry.

Cambridge   H. J. Eméleus

Preface

Although some of the lanthanide and actinide elements were discovered almost 200 years ago, their chemistry has only been investigated in depth in the 30 years since the discovery of the first of the synthetic transuranium elements. The stimulus for the very considerable growth in interest in these two f-transition series has been twofold. First, the discovery of nuclear fission, leading to the development of nuclear power, indicated a need for a better understanding of the chemistry of the lanthanides—the major fission products produced in a nuclear reactor—and for research into the chemistry of the transuranium part of the actinide series, produced at the same time. Secondly, the lanthanide elements have become increasingly important in industry, so that research on their production and properties has increased very substantially.

Because of these factors, research on the separation chemistry of the elements in the two series has been of particular importance, while the mixed oxide, the chalcogenide and the pnictide systems have been extensively investigated because of their industrial applications. At the same time, there has been a notable increase in research on the fundamental chemistry of the halide and pseudohalide complexes, organometallic compounds, the interpretation of electronic spectra and on the measurement of the thermodynamic properties of compounds of both series of f-transition elements, which together make up approximately 28% of the periodic table.

These major growth areas are reviewed, some for the first time, in this volume. It is hoped that the contents will provide the kind of up-to-date account of the present state of knowledge which will stimulate further research in these areas and in other fields of f-transition-element chemistry which have not yet received such detailed treatment.

Manchester   K. W. Bagnall

Contents

1
Separation Chemistry of the Lanthanides and Transplutonium Actinides

E. K. HULET and D. D. BODÉ
Lawrence Radiation Laboratory, Livermore, California

1.1   INTRODUCTION

Some of the most elegant chemistry and techniques yet devised have been introduced during the past 25 years for the mutual separation of the lanthanide and actinide elements in the common tripositive state. The laborious process of fractional crystallisation has disappeared, and the scientist now looks to advanced methods of ion exchange and solvent extraction for fast, effective separations. This chapter deals primarily with these modern methods of separating lanthanides (Ln) and actinides (An).

The chemical principles underlying these processes are reasonably well understood and progress today is mainly in the direction of improving the techniques or introducing newer ones such as extraction chromatography. Successful new techniques have, in turn, created opportunities for investigating the chemistry of the Ln and An further than was possible before their innovation. Because advances in techniques are contributing so abundantly

	2+	3+	4+		2+	3+	4+	5+	6+
La		X		Ac		X			
Ce		X	X	Th			X		
Pr		X		Pa			X	X	
Nd		X		U		⊗	X	⊗	X
Pm		X	⊗	Np		X	X	X	X
Sm	⊗	X		Pu		X	X	X	X
Eu	X	X		Am		X	⊗	X	X
Gd		X		Cm		X			
Tb		X		Bk		X	X		
Dy		X		Cf		X			
Ho		X		Es		X			
Er		X		Fm		X			
Tm		X		Md	X	X			
Yb	⊗	X		No	X	X			
Lu		X		Lr		X			

Figure 1.1   Oxidation states of the actinides and lanthanides in solution. Very unstable species are denoted by shaded circles

to these investigations, this review will not be confined to purely chemical information, but will include progress in the technology of performing Ln and An separations.

Separation methods for 24 elements, comprising nearly one-quarter of those in the Periodic Table, are presented here. The chemical behaviour of the lighter An elements (Th to Pu) is complex, and it seemed appropriate that their separation chemistry should be discussed apart from the heavier group at some later time. However, the chemistry of a majority of the An and Ln is remarkably similar and there is no reason not to treat these family members as a single class when describing separation methods. Accordingly, this review is organised into three sections for discussing advances, primarily within the last 5 years, in separating all Ln and the transplutonium elements (Am to Lr).

Although a large portion of this chapter (Section 1.2) is devoted to reviewing the separation of the lanthanide and the transplutonium groups as tripositive ions, it is significant that one-third of these elements can be oxidised or reduced from the $3+$ state (see Figure 1.1), resulting in greatly simplified and specific separations. A discussion of separations that are based on oxidation or reduction can be found in Section 1.3 for the following: Eu, Sm, Yb, Md, and No as dipositive ions or amalgams, Ce^{4+} and Bk^{4+}, and Am in the penta- or hexapositive oxidation state.

The separation chemistry of any element is usually examined for some ultimate application to a practical problem. Investigations of the Ln and An are no exception, so when the eventual application is a large-scale separation process, we normally find that the research and testing have been more thorough than usual. It is instructive to review a selection of the large-scale processes since they represent the state-of-the-art in the field of chemical separations. Therefore, major processes are described and illustrated in Section 1.4 in the form of flow-diagrams. It was necessary to abbreviate the description of each process and, regrettably, not all large-scale methods could be included.

In recent years there has been a vast increase in the production of rare earths and transplutonium elements and, as part of this growth, the literature dealing with their chemical separation has proliferated tremendously. Earlier books and reviews[1-23] have consolidated a part of this information, but many new reports appear every month. It is not possible in a review of this size to examine all of these references; therefore, we will generally emphasise the more effective chemistry and techniques along with newer developments that in our opinion may be of eventual service. Not all separation methods that have been studied extensively have found significant use, either because they are less effective than other methods or because they leave the product in a medium where recovery becomes an added burden. Among those methods which are not covered in this review are thin-layer and paper chromatography, electrophoresis, and the ring-oven technique.

1.2   SEPARATION OF THE TRIVALENT LANTHANIDES AND ACTINIDES

Separating individual Ln and An is no longer viewed as a difficult procedure in the light of the high discrimination afforded by modern extraction and ion-exchange methods. Exploiting very small chemical differences between adjacent elements, these methods easily resolve complex mixtures of the tripositive ions. Such separations are founded upon variations in the strength of complexes formed between the metal cations and diverse anionic ligands. Among the factors involved in the formation of these complexes are steric hindrance, rearrangement of water molecules in the hydration spheres of the cation and the ligand, and electrostatic attraction between the metal ions and ligand. In the vast majority of cases, the latter two are decisive factors regulating the relative complex strengths among members of the Ln and An series.

It would be difficult indeed to devise ways for separating these elements from one another were it not for a contraction of the size of the aqueous ion with increasing nuclear charge. This results in slight incremental increases in charge density of the metal ions as the atomic number increases through either series of elements. Due to greater electrostatic attraction, the smaller ion will form the stronger ionic complex. This small but significant effect furnishes the grounds for most of the Ln≕An separation chemistry known today. Of course, hydration or dehydration of the metal ion, ligand, or chelate can greatly modify and, in many cases, invert the ordering of complex stability within a series. An inversion in ionic size occurs in outer-sphere complexes because water molecules attracted to the smaller uncoordinated metal ion produce the larger hydrated ion.

Much of the ion-exchange and extraction chemistry discussed here is based on these simple principles, although a detailed knowledge of these systems is still very imperfect.

1.2.1   Ion-exchange chromatography

With the mounting variety of separation techniques based on an ion-exchange mechanism (liquid extraction, extraction chromatography, etc.), one of the original definitions of the term, ion exchange, is becoming slightly blurred. We would hope that it continues to describe a separation method performed on column beds of organic and inorganic exchange materials, as well as to define a reaction mechanism.

The tripositive Ln–An are most often separated from one another by elution from columns filled with cation-exchange resin using complexing agents as eluants. These separations depend more on differences in the complexing powers of the eluants toward the metal ions than on any selectivity by the resin. The heavier metals with the smaller ionic radii form the stronger complexes and, as would be expected, they are eluted first. The metal cation is attracted by both the exchange sites in the cation resin and the competing ligands in the eluant.

Cation exchange is by now a rather old and familiar method of separating individual Ln or An. Significant advances in recent years are restricted mainly to enhancing the performance of exchange columns by bettering the kinetic processes. Completely new developments, which are described later, concern inventions dealing with ion-exchange separations of Ln on standing columns of foam or by electrochromatography in cation-exchange resin beds. Otherwise, progress is represented by improvements in the theories governing the displacement mode of operation in column chromatography[24]. The displacement method is associated with processing multi-gramme amounts of Ln and An, in contrast to the familiar elution technique used for separating lesser quantities in the laboratory.

Complexing agents that most effectively resolve mixtures of either the Ln or An (but not members of both series simultaneously) using cation exchange are α-hydroxyisobutyric acid (HIBUT) and ethylenediaminetetra-acetic acid (EDTA). At high pH values, chelates formed with EDTA are exceedingly stable, leaving insignificant concentrations of uncomplexed metal ions in the aqueous eluate. Unfortunately, the solubility of the free acid, $H_4(EDTA)$, is low and it tends to precipitate in the resin bed ahead of the Ln during elution. This can be circumvented in displacement chromatography by saturating the cation-resin bed with retaining ions such as Cu^{2+} or even lanthanides.

Kinetics for the overall ion-exchange process are exceptionally slow with EDTA and other amino-polycarboxylic acids, apparently due to the great stability of the chelates.

Actinide separations by cation exchange are usually made with HIBUT, whereas EDTA is employed chiefly for technical-scale separations of the

Table 1.1 Separation factors of lanthanides and transplutonium elements obtained by elution with several complexing agents using Dowex 50 ion-exchange resin

Elements	HIBUT* (at 87 °C) S	HIBUT* (at 25 °C) S	HIMBUT† (at 87 °C) S	EDTA‡ S§
Md				
Fm	1.4||			
Es	1.7			
Cf	1.5			
Bk	2.2			(2.0)¶
Cm	1.7			(3.1)
Am	1.4			(2.0)
Lu				
Yb	1.3**	1.4††	2.4‡‡	(1.9)§§
Tm	1.4	1.5	1.4	(1.8)
Er	1.4	1.6	1.7	2.0||||
Ho	1.3	1.4	1.3	2.0¶¶
Dy	1.6	2.0	1.8	1.9***
Tb	1.8	1.8	1.8	(2.3)
Gd	1.9	1.9	1.9	(4.2)
Eu	1.4	1.5	1.4	1.4†††
Sm	1.7	1.9	2.0	(1.5)
Pm	1.7	1.7	1.8	1.5‡‡‡
Nd	1.5	1.6	1.5	2.2§§§
Pr	1.5	1.4	1.4	(1.8)
Ce	1.7	1.7	1.6	(2.5)
La	2.1	2.1	3.0	(3.7)

* α-hydroxyisobutyric acid (2-hydroxy-2-methylpropanoic acid)
† α-hydroxy-α-methylbutyric acid (2-hydroxy-2-methylbutanoic acid)
‡ ethylenediaminetetra-acetic acid
§ Values in () were derived from static equilibrium measurements or the ratios of the 1:1 stability constants
|| Avg. S from elution data; Ref. 25, 26, 58
¶ Ref. 27
** Avg. S from elution data; Ref. 25, 28
†† Avg. S from elution data; Ref. 29, 30
‡‡ Ref. 31
§§ Ref. 32
|||| Avg. S from elution data; Ref. 33–35
¶¶ Ref. 34
*** Avg. S from elution data; 33, 34
††† Avg. S from elution data; Ref. 33, 34, 36
‡‡‡ Ref. 37
§§§ Calculated from Ref. 32, 37

Ln. The comparative performance of EDTA, HIBUT and HIMBUT is rated in Table 1.1. HIMBUT, a hydroxycarboxylic acid very similar to HIBUT, might be the slightly more selective of these two, although it has not been examined for An separations[31, 38].

Separation factors (S) listed in Table 1.1 for adjacent pairs of tripositive elements are defined for two-phase systems, such as ion exchange and extraction, by the ratio of distribution coefficients of the two metals, i.e.,

$$S = K_{d(Z)}/K_{d(Z \pm 1)}$$

Occasionally, separation factors for non-adjacent pairs of elements will be mentioned in this report, and these are indicated by, for example, La/Pr, $S = 3.4$. So long as the concentration units used in expressing the distribution coefficients are consistent, it makes little difference whether values of S are measured by static equilibrations or by column elutions with tracer amounts of Ln and An.

In addition to separation factors, the quality of the separation is measured by peak resolution, which is a parameter governed largely by kinetic processes within the exchanger. For the most part, diffusion of ions inside the resin particle is the rate limiting step. The exchange rates are indeed very small inside the tight network of a highly cross-linked resin, which is the type most often used for Ln–An separations. Improving these rates of diffusion continues to attract the attention of a number of investigators[39–43]. Raising the operating temperature of the exchange column improves the resolution (peak widths) somewhat[40] by increasing the diffusion coefficients. Highly cross-linked resins (12% divinylbenzene) have an adverse effect on the resolution[44] but, by increasing the temperature or column length and by lowering the rate of flow of the eluant, it can be compensated[40]. Without question, the most direct way of improving the resolution is to reduce the resin particle size (to 5–20 μm), since the diffusion period is directly proportional to the square of the particle diameter[40–43]. This has an important practical consequence; well resolved peaks are obtainable from columns packed with a finely divided, homogeneous resin at much higher flow rates than could be tolerated by columns filled with coarser-sized particles[42]. There may be a limit as to how far one can reduce the particle size before processes other than particle diffusion control the exchange rate. In this connection, evidence now exists[39] that diffusion from the eluant to the resin boundary (film diffusion) is becoming an important rate step which should, however, be improved by increasing the temperature and by creating turbulence in the solution phase through higher flow rates.

A high-pressure method of cation exchange, based on the kinetic concepts discussed above, is a very important new development in separating Ln and An[45, 46]. Derived from a similar method used in biochemistry for the analysis of body fluids and amino acids, the technique was modified by Campbell and Burton specifically for separating milligramme amounts of transcurium elements. Operating at pressure drops up to 2500 lb in^{-2} across the resin bed, long columns packed with finely-sized (10–20 μm) Dowex 50×12 resin readily separate adjacent pairs of Ln or An in substantially less than 1 h. Flow rates of the eluant, HIBUT, can be made amazingly high without affecting the separations undesirably. In one elution, the actual velocity of the eluant down the column was 60 cm/min, which corresponds to a flow rate of 10 ml min^{-1}. Many Ln–An separations were made with flow rates near 24 ml cm^{-2}min^{-1}, which is ~25 times the rate normally found satisfactory in ion exchange.

The elution curves obtained by the high-pressure technique are unsymmetrical. They show a slow increase in metal ion concentration upon approaching the peak maximum and a very rapid decrease in concentration behind the peak. This shape is just opposite to the commonly encountered steep rise followed by tailing. Lowering the temperature from 80 °C to 25 °C had little effect upon the resolution of Nd and Pr elution peaks, but loading the resin bed to more than 4% of its capacity caused some overlapping of elution bands. Separations of the transplutonium elements are now made routinely by high-pressure cation exchange at the Oak Ridge and Savannah River laboratories[47]. Damage caused by exposure of the resin to intense radiation is reduced by the high flow rates and, because of the high pressures, radiolytic gases are either dissolved or swept out of the column before gas pockets can form in the resin bed.

More in the class of a novelty at the moment, foam fractionation of the Ln is being evaluated for large-scale separations[48]. The separation mechanism is clearly ion exchange, but the process is extraction at surfaces of thin films. In a long glass column, stable foams are created from a cationic surfactant by bubbling nitrogen through the column base into an aqueous solution containing the Ln and a chelating agent, EDTA. Partitioning many times on the surface of the bubbles, the individual Ln separate into distinct bands that move upward with the foam and eventually emerge from the top of the column. Even though this technique is at a very early stage in its development, a respectable separation factor of ~ 2 for Nd/Ce was obtained.

A d.c. voltage applied through annular rings in a bed of Dowex 50 resin produces a transverse migration of Ln ions, which is in addition to their movement in the direction of eluant flow. In this unusual system[49, 50], the Ln are fed to the centre area at the top of the resin bed and an eluant (EDTA) is forced evenly (laminar flow) through the column. Due to the electric-potential field between the rings and a central electrode, uncomplexed Ln ions migrate outward toward the walls, while the complexed ions stay nearer the centre axis of the column. Concentric ports at the column base divide the output stream containing the radially separated Ln. The resulting separation depends on solution- and resin-phase mobilities of the ions and on the selectivity of the complexing ligand. Currently, a unit 22.5 cm in diameter and 120 cm high is operating, but the Ln separation factors are unreported.

Intra-Ln or An separations by anion exchange have never gained wide acceptance, principally because the formation of anionic complexes required high concentrations of inorganic salts. Besides increasing the problem of recovering the metals, high ion concentrations caused poor resolution of the elution bands. At present, investigations of mixed-media eluants consisting of mineral acids diluted with polar solvents are being made in order to overcome these objections[51-57].

1.2.2 Solvent extraction

With the use of the most selective extractants and simple new methods of multiple extraction, solvent extraction has proved a viable alternate to ion exchange for the separation of adjacent An and Ln. Similarly, extraction systems, comparable to ion exchange are successfully used for the group separation of Ln from the An. The raising of solvent extraction to a position competitive with ion exchange for such separations is mainly due to three important advances that began 11–15 years ago. These may be listed as the studies by Peppard of acidic phosphorus-based extractants[59], the development by Moore of long-chain tertiary and quaternary amines as anion extractants[60, 61], and the development of the new technique of extraction chromatography, first successfully applied by Siekierski and Kotlinska[62]. Because these developments form the foundation for the most effective separations of tripositive An and Ln elements, we shall concentrate upon their features, by passing a discussion of some of the older extractants such as tributyl phosphate (TBP) and thenoyltrifluoroacetone (TTA).

In technical terms, solvent extraction is defined as the distribution of a dissolved metal salt, ion-association complex or chelate between two immiscible liquids that are in contact. Because of the very marked ionic

Table 1.2   Abbreviations for some common extractants and their nomenclature. First group extracts cations and the second extracts anions

HDEHP	bis(2-ethylhexyl)phosphoric acid
HEHϕP	2-ethylhexyl hydrogen phenyl phosphonic acid
OPPA	bis(2-ethylhexyl)pyrophosphoric acid*
HDOP	bis(n-octyl)phosphoric acid
HD(DIBM)P	bis(2-6-dimethyl-4-heptyl)phosphoric acid
DDCP	dibutyl-N,N-diethylcarbamyl phosphonate
TLMANO$_3$	tridodecylmethylammonium nitrate (trilaurylmethylammonium nitrate)
Alamine 336 Adogen-364-HP	trioctyl- and tridecylamines*
Aliquat 336-S-X	trioctyl- and tridecylmethylammonium salt†

* A commercial mixture consisting of $\sim 50\%$ OPPA with mono- and dialkyl esters of phosphoric acid
† S and HP denote high purity products; X denotes anion

character of the Ln and An, one of the immiscible liquids is normally an aqueous solution; the other is an organic phase containing the extractant. At equilibrium, the metal salt is found distributed (partitioned) between the two liquids in a manner governed by its solubility in each phase. The equilibrium distribution of the metal is most commonly given as a ratio of the metal concentration in the organic phase to its concentration in the aqueous solution: i.e.

$$K_d = C_{org}/C_{aq} = \text{'distribution coefficient'} \qquad (1.1)$$

Occasionally, a simple mass distribution ratio is used to express the results of an extraction equilibrium. However, results expressed by such a ratio are sometimes confused with 'distribution coefficient', and in addition, the information content is less than that of K_d.

A list of the extractants that are reviewed in this section and referred to in subsequent sections is introduced in Table 1.2. Several of those listed were only recently investigated, whereas most have already proven useful for effective separations. The list in Table 1.2 is also intended to guide the reader unfamiliar with these compounds through the thicket of abbreviations and synonyms spread throughout the literature.

The organic extractants identified in the Table are divided into two categories, depending on whether they behave as extractants for cations or for anions. Inasmuch as the extraction mechanisms are quite different, each group is discussed separately in the following subsections.

1.2.2.1 Cation extraction

A group of acidic compounds that are discussed in this section form the major source of extractants for the intra-series separation of An and Ln. The extraction of metal cations, at least when in low concentrations, follows a simple ion-exchange reaction, and for this reason they are often termed liquid cation-exchangers. The most useful and widely studied extractants in this class are the monoacidic organophosphorus compounds. Similar diacidic compounds are not considered here because their selectivity is inadequate[63, 64], and because they are highly polymerised in the diluent[65]. We include, however, several bidentate phosphorus compounds that have seen recent applications in the bulk separation of the transplutonium and Ln elements from highly acidic solutions.

Among the many cation extractants tested, HDEHP and HEHϕP are the more selective for the separation of individual tripositive Ln and An ions. A variety of other dialkyl-[66] and diaryl-esters[67] of phosphoric acid have been evaluated for intra-lanthanide separations, but none offered advantages over HDEHP and HEHϕP. Within the range of 0.1–2.0 M $[H^+]$, extraction of tripositive ions by these monoacidic derivatives of phosphoric and phosphonic acids may be represented by the following equation[65]:

$$M^{3+}_{aq} + 3(HA)_{2,org} \rightleftharpoons M(HA_2)_{3,org} + 3H^+_{aq} \tag{1.2}$$

In this reaction, hydrogen is replaced in the weakly acidic extractant by a metal ion, followed by solvation of the metal complex in the organic phase. It should be noted that HDEHP and HEHϕP exist as dimers in most diluents[16]. In principle, the reaction is mainly ionic; thus the strength of the complex consisting of ions of the metal and extractant is dependent mainly on the size and electrostatic charge of the metal ion. This, of course, implies that in a regular series of elements with constant ionic charge, such as the An and Ln, the order of extraction is dependent upon the contraction of the ionic radii with increasing atomic number. The smaller metal ion is bound tighter to the extractant and will be more favourably recovered in the organic phase (Table 1.3).

Separation factors (defined in the ion-exchange section) that were measured for HDEHP and HEHϕP are compared in Table 1.4 with the separation factors found for one of the most selective ion-exchange systems. The separation factors form an interesting grouping described by Peppard et al.[75] as the tetrad effect. A plot of the logarithm of distribution coefficients or separation factors vs. Z displays four smooth curves, with Gd and Cm, the elements possessing half-filled f electron shells, each providing a point common to the second and third tetrad of their respective curves. A quantum mechanical explanation derived from symmetry in interelectron repulsion energies has been proposed by Nugent[77]. A less generalised statement of these regularities was made earlier by Fidelis and Siekierski[78], which prompted Rowlands[79] to review the effect and find that there was no correlation with many complexing agents. Nevertheless a variety of solvent extraction systems provide examples of this effect[76]; an especially clear case is shown in Figure 1.2 for the HEHϕP–HCl system. This figure further illustrates the selectivity of

Table 1.3   Distribution coefficients of lanthanides and transplutonium elements in HDEHP and HEHϕP

Elements	HDEHP 0.4F in heptane			HEHϕP	
	(0.4 M HCl) K_d*	(0.4 M HNO$_3$) K_d*	(1.0 M HCl) K_d†	HNO$_3$ (M)	K_d‡
Am	0.10	0.11	0.37		
Cm	0.145	0.144	0.52		
Bk	2.23	1.77	14.0		
Cf	7.1	5.7	45.0		
Es	7.0	5.8	58.0		
Fm	15.0	13.9	146.0		
La				0.21	0.123
Ce					0.543
Pr					0.750
Nd					0.980
Pm					2.39
Sm					8.54
Eu					21.7
Gd					37.7
Tb				1.0	0.919
Dy					3.01
Ho					6.41
Er					17.8
Tm					70.8
Yb				4.5	2.21
Lu					4.36

* Liquid-liquid extraction at 25 °C; derived from Ref. 68 and 69.
† 1.0 F HEHϕP in heptane; room temperature; derived from Ref. 70
‡0.038 F HEHϕP in heptane; 25°C; Ref. 71

HEHϕP for separating many individual tripositive ions of the Ln and transplutonium elements, although the separation of certain elemental pairs, Am–Cm, Cf–Es, Ce–Pr, and Pr–Nd, is not outstanding. Taking the Ln series as a whole, the separation factor of Lu/La reaches the remarkably large value of nearly 10^6. The unusual break between Cm and Bk or Cm and Cf has been extensively exploited in practical applications[70].

Under no circumstances can the solvent used in diluting these monoacidic extractants be considered as inert[80]. Gureev and co-workers[66] found that in extracting Am^{3+} with HDEHP the value of K_d fell by a factor of almost 1000 upon changing the diluent from iso-octane to chloroform. The distribution coefficient decreased with the following order of diluents: Iso-octane > Cyclohexane > CCl_4 > Toluene > Benzene > $CHCl_3$. Baybarz[70] found a factor of 15 decrease in the distribution of Am^{3+} and Cf^{3+} between 1.0 M HCl and

1.0 M HEHϕP by changing the diluent from heptane to toluene. There appeared to be a direct correlation between the distribution coefficients and the dielectric constant of the diluents. From this, it seems that the more polar diluents suppress the formation of an extractable complex, resulting in less extraction of the metal.

Deviations from the rather simple reaction given by equation (1.2) begin when competing complexes form with anions in the aqueous phase[81], when HNO_3 and HCl concentrations are greater[82, 83] than 4 M, and when the metal concentration becomes so large as to saturate the extractant appreciably[84]. In the HDEHP system, the distribution coefficients of the Ln pass through

Table 1.4 Separation factors of adjacent pairs of lanthanides and trans-plutonium elements obtainable with acidic extractants and by cation-exchange

Elements	HIBUT	HDEHP		HEHϕP	
	(at 87 °C)	(HCl)	(HNO$_3$)	(HCl)	(HNO$_3$)
	S	*S*	*S*	*S*	*S*
Am	1.4††	1.26*	1.24*	1.4§	
Cm	1.7	9.4	8.3	30.0	
Bk	2.2	2.67	2.7	3.3	
Cf	1.5	0.993†	1.02†	1.3	
Es	1.7	2.04	2.20	2.5	
Fm	1.4	4.0‡	4.4‡		
Md					
La	2.0	2.8‖	2.7‖	3.3¶	4.31**
Ce	1.7	1.5		1.5	1.41
Pr	1.5	1.3	}1.55	1.3	1.19
Nd	1.5	2.7	2.1	2.8	2.66
Pm	1.7	3.2	2.7	3.6	3.45
Sm	1.7	2.2	2.1	2.3	2.57
Eu	1.4	1.5	1.7	1.6	1.71
Gd	1.9	5.0	5.5	5.4	6.96
Tb	1.8	2.6	3.0	2.1	3.22
Dy	1.6	2.1	2.2	1.9	2.19
Ho	1.3	2.8	2.7	2.9	2.74
Er	1.3	3.4	3.5	3.8	3.96
Tm	1.4	2.8	3.1	3.2	3.76
Yb	1.3	1.9	1.9	3.0	1.99
Lu					

* Extraction chromatography at 60 °C; Ref. 69
† Same conditions as *; Ref. 68
‡ Same conditions as *; Ref. 72
§ Liquid-liquid extractions at room temperature; Ref. 70
‖ Extraction chromatography at room temperature; Ref. 73
¶ Same conditions as ‖; Ref. 74
** Liquid-liquid extractions at 25 °C; Ref. 71
†† Reference sources listed in Table 1.1

a minimum at HNO_3 concentrations of 4–6 M and then strongly increase up to 16 M HNO_3. A similar sharp minimum is found at an HCl concentration of 6 M. The separation factors also decrease somewhat in either acid system at these higher aqueous acidities[83]. Since the extraction mechanism is apparently changing with the kind of mineral acids tested, it is not surprising that the separation factors also vary with the acid ligand in the aqueous phase. Recent studies suggest that separation factors of Ln, in high concentrations, are somewhat greater in aqueous HCl than in HNO_3 and $HClO_4$ [84].

A continuing interest in the use of alkyl phosphates for purifying the transplutonium elements and for separating the lanthanides is demonstrated by reports from throughout the world[85-89]. Extraction chromatography,

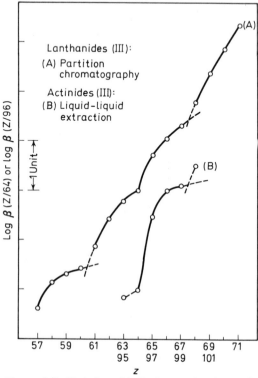

Figure 1.2 Variation with Z of separation factors for Ln^{III} and An^{III} in HEHϕP *vs.* aqueous HCl systems; Ln^{III}: (A) extraction chromatography (Ref. 74); An^{III}: (B) liquid-liquid extraction (diethyl benzene diluent) (Ref. 70). (Reprinted with permission from Peppard *et al.* (Ref. 75) and Pergamon Press)

which is described in a later section, has been widely used for investigating many specific separations, temperature dependencies, and the effect of strongly complexing ligands in the aqueous phase.

Organo-carboxylic[90-94] and -sulphonic[95] acids are among the newer extractants being considered for the extraction and separation of the Ln. These weak acids are effective extractants only at very low $[H^+]$, and so far they have given poorer Ln separations than the alkyl phosphoric acids. Similarly, separations with the β-diketone 2,2,6,6-tetramethyl-3,5-heptanedione (Hthd) are unattractive[96, 97].

In addition to the need for separating individual elements, there is a desire to find extractants for the recovery and concentration of transplutonium elements that have been mixed with vast quantities of other material. These bulk separations are useful mainly in the assay of biological and environmental samples and for recovery of the valuable heavy elements produced in underground nuclear explosions. Concentrations of the actinides in many instances range downward from 10^{-13} g/g of soil.

Essential features required of a suitable extractant are: (1) exceptionally high extractive powers when the aqueous phase is strongly acidic and (2) high discrimination against extraction of the common elements, particularly Ca, Al, Mg, and Fe. An ideal extractant has not yet been found but two

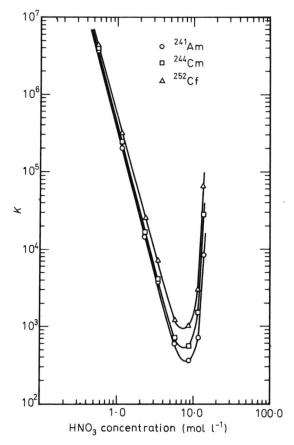

Figure 1.3 Distribution coefficients of representative An^{III} in 8 wt % bis(2-ethylhexyl) pyrophosphoric acid (OPPA) diluted with kerosene. (Ref. 100)

bidentate organophosphorus compounds have been recently developed to a point of practical application.

Butler and Hall[98] described the separation of nine actinides from biological samples using DDCP (formula in Table 1.2) as the primary extractant. This extractant is one of several evaluated earlier by Siddall[99]. Over 90%

of the transplutonium elements are distributed into undiluted DDCP from 12 M HNO_3 at an aqueous to organic volume ratio of 50:1. Other actinides (Th, U, Np, Pu), Fe, and Ln are as well extracted, but Ca and Cs are left in the HNO_3 solution. Back extraction into 2 M HNO_3 or HCl after dilution of the DDCP with toluene completes the extraction cycle. Siddall's studies show that considerable concentrations of HNO_3 and H_2O appear in the organic phase which implies an ion-association mechanism of extraction in addition to chelation. Among the drawbacks of this separation system are the appearance of phosphate residues in the back-extract and the high cost of DDCP.

An octyl pyrophosphate (OPPA) appears especially suited for concentrating heavy actinides in soil and rock samples[100]. The distribution coefficient

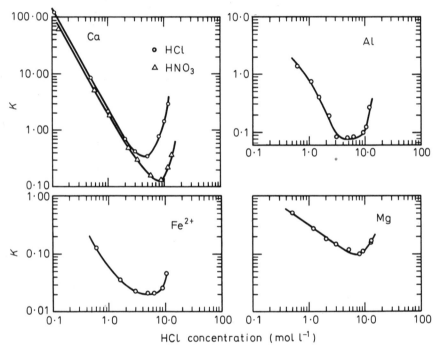

Figure 1.4 Distribution of common elements into 8 wt % bis(2-ethylhexyl) pyrophosphoric acid (OPPA) (Ref. 100)

for these elements are immense over all HCl and HNO_3 molarities. As shown in Figure 1.3, the distribution coefficients are never lower than 100 and approach 10^7 in equilibrations between 0.5 M HNO_3 and 8 wt % OPPA dissolved in kerosene. Fortunately, Ca, Al, Mg and Fe^{2+} are barely extractable over a useful range of acid concentrations (see Figure 1.4). Presumably a chelation mechanism predominates, thus the tetrapositive metals (Pu, Np, Zr, Hf, etc.) are quantitatively recovered in the organic phase, whereas most fission products with low ionic charge are not noticeably extracted. Extraction coefficients of the Ln and transplutonium elements are so large that appreciable concentrations of aqueous F^-, SO_4^{2-}, and HSO_4^- offer no serious hindrance to heavy-element extraction.

The advantages of OPPA are flawed by an inherent property of pyrophosphates, namely, their breakdown into *ortho*-phosphates by the hydrolysis reaction indicated below:

$$RO-\overset{\overset{\displaystyle O}{\|}}{\underset{\underset{\displaystyle OH}{|}}{P}}-O-\overset{\overset{\displaystyle O}{\|}}{\underset{\underset{\displaystyle OH}{|}}{P}}-OR+H_2O \longrightarrow 2 \left[RO-\overset{\overset{\displaystyle O}{\|}}{\underset{\underset{\displaystyle OH}{|}}{P}}-OH\right]$$

The rate of hydrolysis of OPPA in contact with 0.5 M HCl is a sensitive function of temperature as seen in Figure 1.5. Surprisingly, a degradation of the extractive properties of OPPA with time also occurs after dilution with octane carefully dried with sodium metal. The degradation rate is

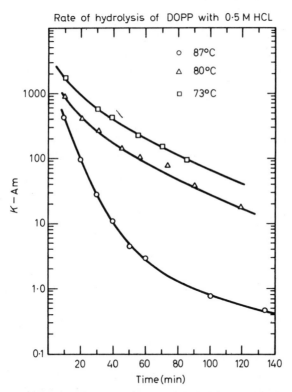

Figure 1.5 Hydrolysis rate of 8 wt% bis(2-ethylhexyl) pyrophosphoric acid (OPPA) stirred with 0.5 M HCl (Ref. 100)

negligible within the first 25 days after dilution, but at the end of 5 months, K_d for Am^{3+} has fallen by a factor of 2000.

Once the heavy elements are drawn into this very powerful extractant, back extraction into an aqueous solution is somewhat challenging. A workable method involves hydrolysing the extractant with 0.5 M HCl at 90 °C for 1 h, draining the HCl solution, and forcing the tripositive actinides into a fresh 4–6 M HCl solution by the mass-action effect of saturating the extractant with Fe^{3+}.

In a later section, a pilot-plant process using OPPA for recovering the heaviest elements from rock debris is described. Prior information about OPPA is a result of its development for the recovery of uranium from phosphate ores[101, 102].

1.2.2.2 Anion extraction

The extraction systems discussed up to now were based on the formation, by cation-exchange, of complexes between the metal cation and extractant. In this section, we describe systems where an anion-exchange reaction prevails. The primary extractants in this category are long-chain tertiary and quaternary alkylamines, the latter being obtained by reaction of alkylhalides with a tertiary amine.

Ammonium salts of the amines are made upon neutralising the amines with aqueous solutions of strong acids as described by the equilibrium,

$$(R_3N)_{org} + H^+_{aq} + X^-_{aq} \rightleftharpoons (R_3NHX)_{org}$$

The ammonium salts mainly exist as ion-pairs, i.e. $R_3NH^+ \cdots X^-$, where the radical R_3NH^+ can operate freely as an attractant for anionic metal complexes formed in the aqueous solution. The exchange reaction between a tertiary amine and anionic complexes of the Ln or An is then,

$$(n-3)(R_3NH^+ \cdots X^-)_{org} + MX^{-(n-3)}_{n,\,aq} \rightleftharpoons [(RNH^+)_{n-3} \cdots MX_n^{-(n-3)}]_{org} + (n-3) X^-_{aq},$$

where n is equal to the number of singly-charged ligands attached to the complex. The equilibrium is driven forward, toward stronger extraction, if the association of the ammonium salt is weaker than association of the metal–amine complex. Factors governing these anion–cation interactions are aqueous acidity and ligand concentrations, hydration equilibria in the aqueous phase, polar qualities of the diluent, and length and branching of alkyl groups in the amine. These aspects as well as actinide separations were recently reviewed thoroughly by Müller[103].

It should be understood that for amine extraction to succeed, a reasonably stable anionic form of the metal must be produced in the aqueous solution. Unlike the d electrons in the first and second transition series metals, f electrons seldom enter into covalent bonding, so that most complexing in solution is identified as a hard ionic type. Consequently there is a limited number of ligands suitable for forming extractable anionic complexes of the tripositive Ln and An elements. The ligands most commonly used are chloride, nitrate, thiocyanate, amino-acid salts and hydroxycarboxylates. Separations by amine extraction using these complexing agents are entirely analogous to, and were mainly derived from similar separations with anion-exchange resin.

There are considerably fewer separations with basic amine extractants than those employing acidic organophosphates. We mention several that are notable because of their superiority over nearly any other method devised for the same purpose. First among these is the separation of the trans-plutonium elements, as a group, from tripositive Ln by tertiary amine extraction. Studied first by Moore[104] and later fully developed by Baybarz and co-workers[105], Alamine 336 preferentially extracts the heavy actinides from slightly acidic 11 M LiCl solutions. Separation factors between any member of the Ln series and Cm, the least extractable actinide, range from 100 upward to 600. A slightly modified version for separating Bk from Ce was reported by Moore[106]. Additional details are not presented here since the system has been reviewed before [16, 107, 108, 103]. A complete process for isolating transplutonium elements that incorporates this amine extraction system is described in a later section.

The higher stability of An thiocyanate complexes compared to those of the lanthanides affords another approach to separating transplutonium elements from Ln fission products. The An complexes, being stronger, are extracted more favourably into amine salts than the Ln. A quaternary ammonium salt, Aliquat 336-S-SCN, offered greater selectivity and larger distribution coefficients than primary, secondary and tertiary amine salts when tested for the extraction of actinide–thiocyanate complexes[109]. Xylene was used to dilute the amine to a concentration of 30 wt./vol.%. After several trial experiments, the aqueous conditions for separating Am from Eu were fixed at a NH_4SCN concentration of 0.6 M in solutions of 0.2 M H_2SO_4. Because the thiocyanate ion concentration changes with acidity, extraction by the amine is very sensitive to the H_2SO_4 (or HCl) concentration. Separations of the An from the Ln are very good; we have calculated from the equilibrium data of Moore[109] a distribution coefficient of 10.4 for Am and a separation factor of 130 in relation to the extraction of Eu. The Am/Yb separation factor is about three times smaller than that for Eu. The extracted species are easily recovered through decomplexing and back extraction, which occurs upon increasing the acidity above 1 M $[H^+]$. Further investigations of this system have been made by Barbano and Rigali, who, additionally, separated Am as well as La, Ce, Pm, Eu, and Tm by a gradient elution technique using extraction chromatography[110, 111].

Most recently, a study identifying the thiocyanate complexes of selected Ln in a molten eutectic mixture of KSCN–NaSCN was made by extracting the complexes with dioctylamine thiocyanate[112]. The predominant extractable complex was a singly charged anion, implying the association of four linear thiocyanate ions per metal ion. It was possible to separate the few lanthanides examined in this research, although molten-salt baths are rather unattractive for many applications. As an example, S for the Ho/Tb pair was given as 10.9.

After comparing the NH_4SCN and LiCl systems for separating heavy elements from Ln fission products, we prefer the LiCl method. Occasionally, gramme quantities of americium have accelerated a disastrous polymerisation of NH_4SCN left in 0.2–1.0 M acid solutions. In spite of the rarity of these events, they are worth avoiding, as the recovery problems are considerable.

Complexation in a nitrate system affords a particularly good separation of Am^{III} from Cm^{III}. Only recently investigated, the extraction of these elements by quaternary ammonium nitrates from high-nitrate solutions yields a larger separation factor than any other method[113-115]. Horwitz and co-workers[113] evaluated a number of possible systems for the amine separation of Eu, Am, Cm, Cf, and Es. Essentially, their measurements show that a quaternary amine (Aliquat 336) is superior to a tertiary amine (Alamine 336), that $LiNO_3$ and $Mg(NO_3)_2$ are more active nitrating agents than are $Al(NO_3)_3$ and $Ca(NO_3)_2$, and that dilution of the quaternary amine with an aromatic solvent gives greater An extraction than dilution with cyclohexane, whereas the opposite is true for the tertiary amine. Separation

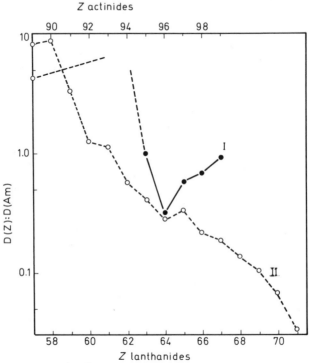

Figure 1.6 Distribution of tripositive actinides (I) and lanthanides (II) as a function of atomic number. The distribution coefficients are normalised to $K_d(Am)$. Organic phase: 0.1 M TLMANO₃ in o-xylene; aqueous phase: 5 M $LiNO_3$ adjusted with HNO_3 to pH 2.0. (By courtesy of van Ooyen (Ref. 115))

factors for Am/Cm and Es/Cf are a little larger in Aliquat 336 compared to Alamine 336, but the dependence upon type and concentration of the nitrate salt is slight.

The value of S for Es/Cf is 1.4, roughly comparable to the separation achieved with a cation-exchange resin using HIBUT. However, an unusually large separation factor of 2.4 was found for Am/Cm distributions between 0.39 F Aliquat 336 nitrate and 4.0 M $LiNO_3$. This separation factor is nearly twice the value of 1.3 found in HDEHP extractions[85] and of 1.4 from HIBUT-cation-exchange.

In a thorough study of a quaternary amine (TLMANO$_3$) as an extractant for tripositive actinides, van Ooyen[115] determined several relevant physical and thermodynamic properties of the extractant in equilibrium with concentrated LiNO$_3$ solutions. There are only a few previous studies accurately establishing the state of aggregation of the extractant molecules in a diluent that compare with this work. Aggregates of 2, 3, and 5 molecules in *ortho*-xylene are the most favoured species.

The major advance arising from this research, in addition to refining our knowledge of the extractant, is the increase in the Am/Cm separation factor from 2.7 to a value of 3.5. The use of high purity TLMANO$_3$ was apparently responsible for raising the factor to 3.1, and the separation factor became even larger upon cooling the solution from room temperature to 10 °C.

For the purpose of indicating all possible separations with LiNO$_3$–TLMANO$_3$, normalised distribution coefficients of every Ln and tripositive An are shown in Figure 1.6. Plutonium(III) measurements were attempted, but Pu could not be held in the 3 + state during the extractions. It is apparent from the distribution data that the transplutonium elements can be adequately separated from both the lightest and heaviest lanthanides. One may view this as a bonus when making Am–Cm separations; otherwise, the LiCl or NH$_4$SCN systems are preferable for group separations.

1.2.3   Extraction chromatography

In our discussion of extraction chemistry, we have written encouragingly whenever the separation factor for adjacent elements is equal to or exceeds a value of 2. A few may wonder at this optimism when it is clear that many extraction stages are required to separate efficiently elements with such small differences in their extractability. The answer is that hundreds of stages are made possible by the simple method of extraction chromatography. In this technique, the extractant is adsorbed on the surfaces of a fine porous powder, which, after being slurried and transferred to a glass tube, forms the bed of a chromatographic column. Loading of the column with the metal ions to be separated, elution with a suitable aqueous solution, collecting fractions, and other operations are exactly analogous to ion-exchange chromatography. Similarly, the metal ions, on passing through the column bed, are partitioned many times between the mobile aqueous and stationary organic phases, hence the large number of extraction stages.

The technique has been developed into a useful tool for making inorganic separations and for investigating new extraction chemistry. The earliest work was aimed mainly at demonstrating extraction separations of the Ln and An, while in recent years the emphasis has been on improving the column performance and in applying this technique to analytical and radiochemical separations. At present, extraction chromatography is accepted for about a third of the separation requirements in transplutonium work, while ion exchange is still preferred for Ln separations. In this section, we have tried to create an understanding of the operating principles and summarise those separations that are based upon the most selective extractants. Several other methods of partition chromatography that depend on fixing the extractant in a stationary position, such as thin-layer or paper chromatography, have been omitted. These techniques are more restrictive than the column method in that their capacity is limited to separating extremely small quantities of pure materials. For the same reason, but on a different scale, the column method of extraction chromatography will probably never gain acceptance by the rare earth industry[13]. Capacities of extraction columns are approximately four to ten times less than those of solid ion-exchange columns of the same size.

A new review by Cerrai and Ghersini[116] is essential reading for those wishing to know nearly all that is known about extraction chromatography. An additional source, compiled by Eschrich and Drent[117], is an indexed bibliography and abstracts, listing all publications pertaining to the subject until mid-1967.

1.2.3.1 Performance of extraction columns

The following are some desirable properties of the powder supporting the extractant. Primarily, the support must be wettable by the organic phase (organophilic) while repelling aqueous solutions (hydrophobic). Other necessary features include an inertness to all chemicals contacting it and a lack of any secondary sorption of the metal ions. As shown later, the particles should be very small and of a uniform size for maximum resolution of the elution bands.

Table 1.5 summarises the support materials most commonly used in extraction chromatography. Attempts at producing a better support by grafting vinyl monomers to a hydrophilic polymer or by copolymerising a polyethylene glycol and styrene graft to the surface of Kel–F powder have been only partially successful[125, 126]. The modified Kel–F was promising; however, all other grafted copolymers lacked chemical stability.

The efficiency of a chromatographic column is characterised by the number of theoretical plates, or better still, the height of a single plate (HETP). The terminology dates back to the plate-equilibrium theory derived for pulsed-extraction columns. Since then the theory has been modified to fit ion exchange, and with no further changes, it now includes extraction chromatography. Most results are calculated from the Glueckauf equation,

$$N = \frac{8 V_r^2}{W^2} = \frac{L}{\text{HETP}} \qquad 1.3)$$

where N is the total number of plates, V_r is the volume of eluant to peak-maximum, W is the width of the elution peak at $1/e$ times the peak-height, and L is the length of the bed.

The importance of small values of HETP is being stressed because an excellent An or Ln separation may become marginal if no care is taken in controlling elution peak-widths. A few specific evaluations have been made of the influence of support materials, temperature, flow-rates, surplus extractant, and metal loadings upon the height of a plate[126, 208, 121, 73, 127]. In every case, data were gathered from the elution of Ln from column beds

Table 1.5 Common support materials used in extraction chromatography and some of their performance qualities

Chemical compound	Examples (Trade-names, etc.)	Quantity of extractant supported (ml HDEHP/g support)	Particle size (μm)	HETP (mm)	References
Polytrifluoro-chloroethylene	Kel-F 300 LD	⎰0.75	62–88*	0.6*	118
	Plaskon CTFE 2.300	⎱0.05	57–88*		119
	Hostaflon C2				
	Voltalef 300 LD				
Polytetrafluoroethylene	Teflon-6	0.25	178–210*	3.9*	120
	Fluorplast 4				
	Fluon				
Poly(vinylchloride-vinylacetate) copolymer	Corvic	0.2	104–152*	1.5*	121, 122
SiO₂	Kieselguhr	0.12	37–74*	0.4	123
	Celite 545	0.09	35–42*	0.29–0.33	68
	Hiflo Super-Cel	0.102	30–35*	0.2–0.33	73
SiO₂-gel (silica gel)	KSS-3	0.8	15	0.15–0.2	208
	KSK-2	0.6	40–71	2.9	124

* Particle sizes or HETP were estimated from data in reference

supporting HDEHP. After sorting out the kinetics of chemical reactions from these data, we find that diffusion of the metal-HDEHP complex from within the particle out to the aqueous-extractant interface is a slow and important rate-determining step. This conclusion is not surprising, since diffusion to and from the exchange sites is often the slowest kinetic process in ion exchange. It then follows that the diffusion rate can be increased by:

  (a) increasing the temperature[127, 121, 69, 68, 128];
  (b) preparing the column bed with small (10–30 µm) and uniformly sized particles[208, 73, 127, 68]; and
  (c) holding the amount of extractant below the quantity necessary to saturate the support[208, 68, 73]—otherwise excess extractant on the particles adds to the thickness of the diffusion layer.

Concerning item (c), Herrmann[208] found that HETP is virtually independent of extractant loadings *below* the level needed to saturate the support. His data are directly opposed to data taken with diatomaceous earth[68, 73]

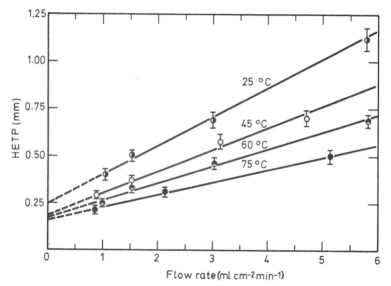

Figure 1.7 The effect of flow rate and temperature on the HETP for the elution of Cm^{III} from HDEHP on Celite. Eluant: 0.50 M HNO$_3$. (Reprinted with permission from Horwitz *et al.* (Ref. 69) and Pergamon Press)

in which HETP began climbing rapidly when concentrations of HDEHP in the support exceeded 20% of saturation. The data from all groups appear valid, so it can only be assumed that the KSK silica-gel and the column loading techniques used by Herrmann were responsible for his unique results. Additional studies by Herrmann suggested that pore radii of his silica-gel supports should be ⩾ 35 Å for minimising HETP.

Slowing the flow of the eluant is an obvious way of reducing peak-widths, since this will allow more residence time for the metal complex to complete any kinetically slow process. The relation between HETP and flow-rate is linear[127, 121, 69], as is clearly evident in Figure 1.7. We can easily deduce from this information and equation (1.3) that the peak-width is proportional to the square root of the flow-rate, providing other factors remain constant. In the report by Horwitz et al.[69], the data show an increase in peak-widths with increasing atomic number. We are somewhat skeptical of their interpretation, which suggests that rates of sorption-desorption are dependent on the atomic number. In their experiments, peak-widths as a function of the flow-rate were measured for Am–Cm, Bk, and Cf–Es in separate experiments using 0.168 M, 0.301 M, and 0.408 M HNO_3, respectively. We prefer not to assume that column kinetics are related to the An ion extracted, but rather that the extraction mechanism is changing with the increasing concentrations of HNO_3. This, in turn, alters the reaction kinetics. Lenz[84] has shown that mixed complexes containing nitrate and metal ions are extracted by HDEHP from moderate concentrations of HNO_3, which led him to suppose a neutral species TBP-type extraction mechanism. He suggested that an ion-exchange reaction is most important at low ionic strengths, however.

The operating temperature of a chromatographic extraction column is another issue relevant to column performance. Increasing the temperature reduces the peak-widths (illustrated in Figure 1.7), thus aiding in the separation of the individual elements. Offsetting this benefit, the higher operating temperatures can easily degrade the basic separation factors, S. The extent of this effect over the 15–50 °C temperature range was recently determined for all Ln by Fidelis[71]. A considerable amount of very accurate tabular and graphic data are supplied in her report, but briefly, more separation factors (10) decrease with increasing temperature than are improved (4). The extremes in either direction represent 21–25 % changes of S over the rather short temperature range of her measurements. In sum, each separation should be evaluated to determine if narrower peak-widths at higher operating temperatures are overshadowed by smaller separation factors.

1.2.3.2 Applications to An and Ln separations

Extraction chromatography was first applied to fractionating the lanthanides[129]. From a selection of potential extractants, chromatographers early centred upon HDEHP as the extractant of choice and its use continues even though HEHϕP is known to be more selective[74]. Preference for the type of support has been far more random with, perhaps, the silica-based and polytrifluoro-chloroethylene products showing the greatest popularity. Excellent separations of adjacent pairs of Ln have been obtained, especially with tracer or microgramme quantities of these elements. Because of small separation factors, several pairs, namely, Ce–Pr, Pr–Nd, Eu–Gd, and Yb–Lu, offer more resistance than most Ln to separation by this and any other method (see Table 1.4).

The choice of acid used as the eluant is crucial for the more exacting separations. At room temperature, separation factors within the group of light Ln are greater in HCl than in HNO_3, but they are comparable with either acid for heavy Ln [73]. Nevertheless, peak-widths of the heavier group are several times smaller with HNO_3 than with HCl as an eluant[124, 73], which indicates HNO_3 should be used for separating the latter half of the Ln series. A general approach to separating more than 6–7 consecutive Ln, and one having the advantages of conserving time and eluate volumes, is the gradient elution method by which the eluant acid concentration is continuously increasing during the elution[118, 122, 130].

Separations of tracer and microgramme quantities of the Ln have been successful, but what if columns containing HDEHP were burdened with larger amounts of material? Capacities of the principal support materials are limited to ~0.8 ml of HDEHP for each gramme of support (Table 1.5) which represents 1.2–1.3 milliequivalents per gramme of support, assuming HDEHP is dimeric. Then, at most, the exchange capacity is about one-fourth of a column containing an equivalent amount of Dowex 50 ion-exchange resin. Still, there are a number of reported separations of milligramme to gramme quantities of Ln elements. Sochacka and Siekierski[73] examined the separation of increasing amounts of Er from tracer Tb in column beds containing 40 wt % HDEHP. Elution peaks were fully resolved when 110 mg of Er was eluted from a 7 mm diameter by 100 mm long bed holding 2.469 g of Kieselguhr and HDEHP. An exceptionally small value of 3.32 was furnished as the molar ratio of HDEHP to Er in this column. One must presume that HDEHP is no longer dimeric at these extractant loadings; otherwise this molar ratio should correspond to six formula weights of HDEHP for each mole of metal. Essentially the same ratio (3:1) was found by Mikhailichenko and Pimenova[124] from column elutions of macro quantities of Y, Tb, and Yb. The extraction of a 3:1 complex occurred only when large amounts of Ln were present. Resolution of their peaks is noticeably poorer than that of other workers, especially with metal loadings greater than 10% of bed capacity. Even so, a large column (3.3 × 99 cm) filled with 60 wt % HDEHP-silica gel was used for separating gramme amounts of Y from Gd, Tb and Dy.

Watanabe[131] tested the separation of 1 mg of Sm from micro-amounts of Eu, Pm, Am, and Cm on a column bed 3 × 70 mm containing 50 vol/wt % HDEHP. Valleys between all but the Am–Cm band were over a factor of 1000 lower than the peak heights. Using the same sized column, he also found good separations, except between Am and Pr, when 5 mg La, 1 mg Pr, and tracer Am and Pm were eluted. To demonstrate the separation of 110 mg of La, which had been added in co-precipitating Pm and Eu hydroxides, he used a 6 × 140 mm bed to obtain sufficient metal capacity. Elution peaks were broader than before but tailing from the peaks was not obvious nor were separations impaired.

Extraction chromatography is now being developed for rapidly separating short-lived isotopes of the lanthanides. In their studies at Mainz, Riccato and Herrmann[132] attempted to balance flow-rate, column length, and HCl concentration in order to optimise the separation of Tb from Eu in the shortest time span. Admittedly they were willing to take a 50% loss of Eu, provided the other 50% contained $\leqslant 1\%$ Tb. Separations meeting this criterion were performed in 1.5 min using 4 × 140 mm column beds of Hostaflon C2 holding 10 vol/wt % HDEHP. An effective separation in such a short time interval hinged principally on obtaining high flow-rates (2.8–4.6 ml/min) and, without equilibrium conditions, tailing of the Eu peak and early breakthrough of Tb was very bad. These unfavourable features might have been improved by changing the support to a fine grade of Kieselguhr and increasing the temperature from 50 °C to 80 °C. Similar work on fast separations are continuing at Mainz in the hope of applying them to many areas of nuclear chemistry, including the separation of the heaviest actinides.

Extraction chromatography has already added a new dimension to An separations in spite of a long delay by heavy-element chemists in systematically studying this field. The earliest chromatographic separations of tripositive transplutonium ions with HDEHP were made by Kooi et al.[123], Gavrilov et al.[133] and by Moore and Jurriaanse[120]. With the exception of Gavrilov and co-workers they reported neither distribution coefficients nor separation factors. To repair this oversight, Horwitz et al. furnished, in three publications appearing in 1969 and 1970[68, 69, 72], definitive data on column performance, separation factors and distribution coefficients for the elements Am through Md. The dependence of the distribution coefficients upon acid concentration, acid ligand, dilution of HDEHP and temperature was also reported. They left little unexplored though it should prove even more rewarding to substitute HEHϕP for HDEHP, judging from the larger values of S found by Baybarz[70].

In appraising the HDEHP results by Horwitz et al., we find that the Bk–Cm, Cf–Bk, and Md–Fm separations are outstanding, whereas the Fm–Es separation is of marginal usefulness compared to ion exchange. Fs, Cf and Cm–Am are virtually inseparable by this method. Considering only the better separations, an increase in temperature caused a slight decrease in separation factors and distribution coefficients, although K_d's for Cm and Bk were almost unchanged over the temperature range of 25–75 °C. The influence of temperature was much more pronounced in their liquid–liquid extractions. Separation factors for Bk/Cm were a little larger for elutions with 0.1–0.6 M HCl than with HNO_3 elutions. Either acid gave the same results for Cf–Bk separations, but for the Md–Fm pair, S rose from 4.0 to 4.4 upon changing the eluant from HCl to HNO_3. The effect on K_d and peak shapes of increasing the mass of Eu^{III} and Bk^{III} was also investigated. Horwitz et al.[69] observed that 'as the mass of Eu^{III} loaded on the column was increased, the number of free column volumes to peak maximum decreased, the front of the elution curve became sharper, the width broader, and the tail longer.' These comments furnish a clear picture of the outcome whenever HDEHP-Celite columns are loaded to more than a few percent of their capacity.

Following their tracer-scale studies, Horwitz and co-workers applied HDEHP-extraction chromatography to the separation of valuable daughters from the nuclear decay of Bk, Cf, or Es. In one case[134], the columns were operated in a shielded cell for recovering ^{248}Cm from 0.25 mg of its parent, ^{252}Cf. Most recently, they quickly separated Md^{III} from Es targets after bombarding the Es with helium ions[135].

In the application of extraction chromatography to transplutonium separations, we find an advantage in that elution with common mineral acids furnishes a pure product, immediately ready for the next separation step or for preparing samples. The elution sequence (peak position with ascending Z) can be either an asset or a liability, depending on the separations problem. In the synthesis of heavy artificial elements with neutrons, the quantities produced drop sharply with increasing A and Z. Amounts measured in atoms of such isotopes as ^{257}Fm must then be separated from grammes of ^{244}Cm target material. The elution order presented by cation-extraction systems is undesirable in this situation, for the reason that tailing from the elution band of the most abundant species is in the direction of the least abundant elements. On the other hand, the order is beneficial for separating α or electron-capture decay daughters of elements of higher Z. For the latter class of separations, we believe extraction chromatography offers an excellent alternative to cation-exchange, wherein elution is in order of descending atomic numbers.

Recalling that Am and Cm are well separated by extraction with quaternary alkylamines, we now consider experiments where this separation was practised with the amines adsorbed on chromatographic columns. Shortly after their original research on Am–Cm separations with amines using liquid–liquid extraction[113], Horwitz and co-workers[136, 137] employed extraction chromatography to demonstrate very successful tracer- and macroscale separations. The column beds were filled with Aliquat 336 adsorbed on 200–400 mesh Celite. In their tracer experiments, the concentration of the amine in the form of its chloride salt was 0.17 g per gramme of dry support; this concentration was reduced about 30% for milligramme separations after it was found that HETP was increasing with higher amine loadings. In either case, near optimum column performance was observed at flow-rates of 1.0–1.1 ml $cm^{-2}min^{-1}$. Single plate heights (HETP) were within 0.45–0.52 mm in the tracer experiments.

Room temperature elutions with slightly acidified 3.5–3.6 M $LiNO_3$ produced Am/Cm separation factors of 2.7–2.8 and fully separated greater than 99% of tracer Am and Cm with less than 0.1% cross-examination. Up to 20 mg of ^{244}Cm was well separated in 2 h from 2.7 mg of ^{243}Am on 8×140 mm column beds. Inasmuch as Am suffers from contamination by tailing of the Cm band, the purity of the Cm product is much the higher. As an example, results of an elution of 20 mg of Cm indicated 7% Cm in 99% of the Am and 5 p.p.m. Am in 99% of the Cm. In the few years since these trial separations, steadily increasing quantities of Cm (as the intensely α-active isotope, ^{242}Cm) were purified from Am by this technique[138, 139].

Research of a more basic nature was carried on concurrently with that of Horwitz et $al.$ by van Ooyen[115]. The major conclusions from his work were reported in the section on extraction. In conjunction with his studies of $TLMANO_3$, he performed numerous chromatographic separations of Am–Cm mixtures using $LiNO_3$–$TLMANO_3$ that are noteworthy for improving the separation factors. Several examples of his elution curves are presented in Figure 1.8 and Figure 1.9. A different type of support material was used in obtaining each chromatogram, so that it is possible to compare their relative effectiveness from the curves. The broader peaks obtained in elutions from the Kel–F–supported extractant can be accounted for by the somewhat larger particle sizes of the Kel–F powder (110–120 mesh $v.$ 200–280 mesh Celite 535). On balance, Kel–F seems the better support material, as noticeably less tailing is evident in the chromatogram shown in Figure 1.8.

Concluding the discussion of extraction chromatography, it should be

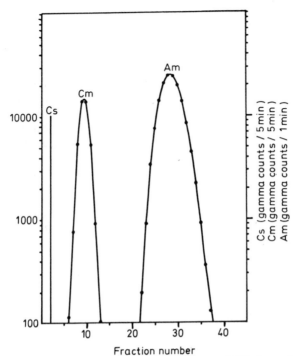

Figure 1.8 Separation of Am and Cm at 10 °C. Support material Kel-F, 0.2 M TLMANO$_3$. Eluant: 4 M LiNO$_3$ adjusted to pH 2.0. (By courtesy of van Ooyen (Ref. 115))

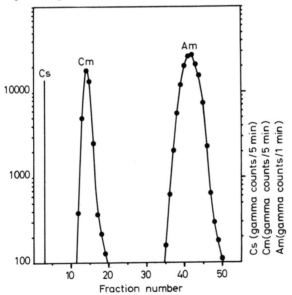

Figure 1.9 Separation of Am and Cm at 10 °C. Support material Celite 535; 0.2 M TLMANO$_3$. Eluant: 4 M LiNO$_3$ adjusted to pH 2.0. (By courtesy of van Ooyen (Ref. 115))

noted that Stronski[140], in a continuing investigation, has used $LiNO_3$-Aliquat 336 to separate some five pairs of Ln and An. Separation factors derived from his chromatographic elutions are compared in Table 1.6 with those coming from an HEHϕP system. Only the separation of Yb from Tm is obviously enhanced by amine extraction from $LiNO_3$ solutions, but further research might find separations of untested pairs are also improved by this method.

Table 1.6   Separation coefficients from the $LiNO_3$-Aliquat 336 and HNO_3-HEHϕP systems

Element pairs	$LiNO_3$ (M)	S ($LiNO_3$-Aliquat 336)	S (HNO_3-HEHϕP)
Eu-Am	4.2	2.50	
Gd-Eu	5.25	1.71	1.7
Er-Ho	5.8	1.34	2.74
Yb-Tm	6.0	4.90	3.76
Lu-Tm	6.0	5.00	7.5

1.3 SEPARATIONS BASED ON OXIDATION AND REDUCTION

From the previous sections on ion exchange and extraction, it can be seen that the majority of the described separations are dependent on properties associated with filling of the f orbitals, i.e. the An and Ln contraction. These separations deal with ions in the tripositive state. As a consequence of the stability of specific f orbitals, other oxidation states are observed for ions of the metals at the beginning of the series and for those adjacent to the half-filled or fully-filled f shells. Elements easily qualifying under this simplifying principle are Ce, Eu, Yb, Th, Bk and No. Americium behaves counter to expectations by exhibiting higher oxidation states rather than a lower dipositive state, which should be stabilised by a half-filled 5f shell. Despite the Am exception, the existence of Sm and Md as 2+ ions would appear to reinforce a corollary of the above proposition concerning enhanced stability of dipositive states upon *approaching* half- or completely-filled f shells.

In this section, we are considering several diverse separations derived from either reducing tripositive ions to the dipositive state or oxidation to states higher than tripositive. We are treating amalgam formation as a reduction separation even though we are unable to identify through the literature the oxidation state of many An and Ln absorbed into Hg. Categorising the few known separations of these elements in the metallic state presents a less troublesome distinction, since we are not considering such methods in any case.

1.3.1   Dipositive Ln and An

Standard oxidation potentials of the easily reduced Ln and An are presented in Table 1.7.

In the dipositive state, the metals listed in Table 1.7 behave similarly to the alkaline earths such as Ba^{2+} and Sr^{2+} because of analogous valence-electron configurations. Properties classically associated with these dipositive species are the solubility of the hydroxides, insolubility of sulphates and — for those 2+ ions that are not oxidised in concentrated HCl — insolubility

Table 1.7   Oxidation potentials of dipositive actinide and lanthanide ions

Reaction	Oxidation potential (V)
$Sm^{2+} \rightarrow Sm^{3+} + e^-$	1.55*
$Eu^{2+} \rightarrow Eu^{3+} + e^-$	0.43*
$Yb^{2+} \rightarrow Yb^{3+} + e^-$	1.15*
$Md^{2+} \rightarrow Md^{3+} + e^-$	0.1 to 0.2†
$No^{2+} \rightarrow No^{3+} + e^-$	−1.4 to −1.5‡

* From Ref. 141
† From Ref. 142 and 143
‡ From Ref. 144

of the chloride salts[207]. These precipitations are not nearly as specific as one would desire for efficient separations; therefore the newer methods depend on the lower charge density of the dipositive ions compared to tripositive ions. Liquid and solid cation exchangers provide the high discrimination needed to separate ions in these charge states effectively. At low acid and metal concentrations, several cation extractants are known to give separation factors approaching 10^6 for separating tripositive Ln from Eu^{II} [145]. Although the literature is rather sparse in this subject area, we shall mention several extraction experiments and then discuss the chemistry of amalgam separations.

1.3.1.1   Extraction

Solvent extraction separations have been applied to Eu^{II} [145, 146], Md^{II} [142] and No^{II} [144], whereas a cation-exchange resin was used to identify the stable dipositive state of No^{II} [147]. The investigation supplying the greatest detail compared the extraction of Eu^{II} and Eu^{III} at macro- and tracer-concentrations, using several phosphoric and phosphonic acid esters[145]. The acidity of the aqueous solution is so chosen as to favour >99% extraction of the tripositive ions by the organic extractant while barely removing any Eu^{II} from the aqueous phase. A separation factor for Eu^{III}/Eu^{II} of 7.5×10^5, obtained with HEHϕP, affords a very definite separation in a single liquid–liquid partitioning. Using similar chemistry, Moskvin[146] separated macro-amounts of Eu^{II} from the light Ln by extraction chromatography. Maintaining Eu in the 2+ state is the principal problem in any separation method based on reduction. Moskvin was unsuccessful at tracer levels, but Peppard et al.[145] found that Cr^{II} is partially extracted into the organic phase and would thus serve as a reducing scavenger for minute amounts of oxidising impurities in the extractant. Tracer amounts of Eu^{II} were then stabilised by mixing a three-phase system consisting of solid Zn(Hg) amalgam, the organic extractant, and an aqueous phase containing 0.001–0.01 M Cr^{II}.

The strong preference shown by HDEHP for tripositive ions over those of lower charge states played an important role in discovering the dipositive oxidation state of Md[142, 148] and in estimating the oxidation potential of No^{II}[144]. There had been no hint of divalency from any previous chemical experience with Md[149, 150, 133]. In these recent experiments, extraction chromatography was used to distinguish the portion of Md or No in the 2+ state after treating Md with various reductants, or No with oxidants. The extraction beds were prepared from 1.5 M HDEHP in n-heptane supported on Kel-F powder. The most efficient separations of dipositive Md from tracer quantities of Es-Fm were obtained by eluting with 0.1 M HCl and 0.6 M Cr^{2+}, the latter acting as a reductant. The Md yields were 99% with 0.6% Es-Fm contamination in the Md fraction. We believe that the Es-Fm retention by HDEHP could be greatly improved by decreasing the Cr^{2+} concentration, since Cr appeared to saturate the extractant and lower the Es-Fm distribution coefficients by mass-action. At lower Cr^{2+} concentrations it is particularly important to displace oxygen above the extraction bed with pure CO_2 in order to prevent air oxidation of Cr^{2+} and Md^{2+}.

Existing as a stable dipositive ion in solution, No occupies a singular position among all An and Ln. The separation chemistry of this element is much more closely associated with Ba^{2+} and Sr^{2+} than it is with the other actinides. Nobelium tends to coprecipitate with BaF_2 rather than LaF_3; it is complexed by HIBUT to a lesser extent than the other An and forms less soluble sulphates than the other heavy elements[147]. Only after oxidation does No follow the conventional behaviour of other members in the series. Oxidising agents that have given nearly complete oxidation of No^{2+} to No^{3+} are Ce^{4+}, H_5IO_6 and $(NH_4)_2S_2O_8$. The latter two reagents were tested along with $HBrO_3$, CrO_3 and HIO_3 in determining the approximate oxidation potential of No by means of extraction chromatography[144]. Paraperiodic acid is the preferred oxidant, in so far as the percentage of No retained in an HDEHP extraction-bed was large and comparable to the percentage of Cm and Cf co-extracted (90%).

1.3.1.2 Amalgamation

Amalgamation with mercury is a somewhat different approach to separating the individual An and Ln. Under carefully controlled experimental conditions, this method is capable of separating Sm, Eu, Yb, and Md from other Ln or An, and in some instances, from one another. The usual procedure is to shake an alkali-metal amalgam(Hg) with a buffered aqueous solution containing Ln or An ions, allow the phases to separate, and withdraw the mercury which now carries the reduced elements. A sophistication is possible by applying an electrical potential between the alkali-metal amalgam (cathode) and a platinum anode. Complexing agents such as citric or tartaric acids are often added to prevent hydrolysis and precipitation of the Ln and An at the necessarily low H^+ concentrations needed for electrolysis or amalgam reductions. These agents also increase the selectivity of reduction and, by lowering the electronic charge of the tripositive ion, they aid the ion in crossing the electrode double layer at the Hg-electrolyte interface.

The chemical reactions occurring at the surface of alkali-metal amalgams are the reduction of hydrogen ions, the reduction of Ln or An ions, and the oxidation of the alkali metal. At pH values lower than about 4, the liberation of hydrogen is a more favourable reaction than reduction of the Ln ions[151, 152]. From a thermodynamic viewpoint, the force impelling the Ln and An

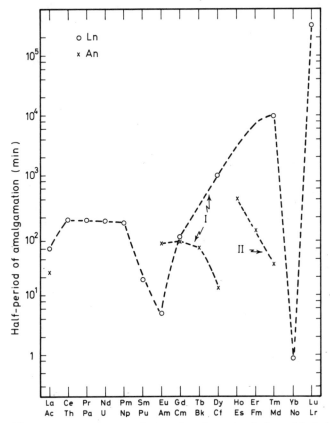

Figure 1.10 Relative half-periods of amalgamation for Ln and An in lithium citrate. Corrections were made for complexing in the aqueous phase from stabilities of the citrate complex relative to Eu(III). I: Curves derived from the data of David and Bouissiéres (Ref. 157). II: Curve derived from the data of Malý (Ref. 143)

towards reduction and amalgamation with the Hg originates from the energy difference between the Hg-cathode potential and the potentials of amalgamation of the Ln or An species[153]. The cathode potential of a Li(Hg) amalgam[154] is approximately 2.4 V, and it is usually implied but not proven that the potential of the cell

$$Li(Hg) \,|\, Ln^{3+} \,|\, Li^+ \,|\, Ln(Hg)$$

is always positive. With complexing agents present, the difficulty of reducing the tripositive ions increases due to the shift in potential caused by the enchanced stability of Ln^{III} complexes over the free Ln^{III} ions[153].

During electrolysis or amalgam reductions, there is a diffusion of alkali-metal ions into the depth of the solution and a counter current of Ln ions across the amalgam-solution boundary toward the Hg. Diffusion rates and rates of reaction at the amalgam surface are slow for most Ln and An[155–158] and equilibrium is seldom reached except for Eu^{2+} ions[154]. Indeed, a substantial number of electrolysis experiments suggest that separations of the Ln by this technique are largely dependent upon relative differences in the rates of amalgamation offered by each Ln. Examples are given in Figure 1.10 where half-periods of amalgamation, corrected for citrate complexing, are presented for many tripositive elements. Most of these data were obtained by David and Bouissières[157] with electrolytic cells of Li(Hg) and 0.1 M lithium citrate solutions. Why the rates vary in the manner shown is undetermined.

The workings of amalgam systems are complex and an understanding of their basic chemistry is far from complete. For instance, no measurements exist which identify the oxidation state of the Ln or An alloyed in the amalgam, but it is commonly *assumed* that they are metals or metal hydrides[151, 156, 159–162]. Further, there is no logic to the assumption that only Ln and An with quasi-stable dipositive states are predisposed toward amalgamation, since all Ln and most transplutonium elements are known to form amalgams[143, 158, 162, 163] particularly if time is allowed to reach equilibrium. A correlation between amalgamation rates and oxidation potentials of dipositive ions may eventually develop, but it is difficult for us to see a close relationship after noticing that the amalgamation periods of La and Ac are roughly comparable to those of Sm, Eu, and Yb (Figure 1.10). Therefore, we feel that recent reports[157, 164] equating the ease of amalgamation of Ac, Cf, Es and Fm with appreciable stability of dipositive states are premature.

The separation of Eu is probably the most important application of this technique. Recent emphasis has been placed on attaining rapid separations[165–167] and recovering Eu without appreciable Sm contamination[154, 160, 168]. With recoveries of Eu in the range 95–98 %, a separation factor Eu(Hg)/Sm(Hg) of 20–40 can be obtained[154], but if the Ln amalgam is separated from the aqueous phase and decomposed with cold, concentrated HCl, Eu^{2+} is precipitated while Sm^{2+} is rapidly oxidised to the soluble 3 + state[167]. With the latter procedure, Eu is separated by about a factor of 1000 from Sm, and with respect to other Ln, one to several orders of magnitude greater. Malý[143], in similar experiments using Eu carrier, separated Md from Am, Cm, Cf, Es and Fm. The yields of Md approached 97 %, and the Md enrichment in relation to Fm was ~25.

1.3.2 Ce^{IV}, Bk^{IV}, and $Am^{V \text{ and } VI}$

Cerium and Bk are normally separated from other Ln and An after having been oxidised to the tetrapositive state. On the other hand, separating Am in oxidation states higher than 3 + is seldom worthwhile unless the quantities exceed several milligrammes. Perhaps as a result of two very new developments in separating $Am^{V \text{ and } VI}$, this viewpoint will change.

1.3.2.1 Ce^{IV} and Bk^{IV}

The chemistries of Ce and Bk in their tetrapositive states are so very much alike that separating one from another is troublesome. Their standard oxidation potentials lie close together, near -1.7 V, with Ce^{4+} being a slightly stronger oxidant by ~ 0.08 V in HNO_3 or $HClO_4$ at concentrations below 7.5 M [169]. The potentials of both elements are considerably more positive in H_2SO_4 and H_3PO_4. A number of oxidants are capable of producing tetrapositive ions in HNO_3 and H_2SO_4, but the following are known to be effective: BrO_3^-, $Cr_2O_7^{2-}$, MnO_4^-, $S_2O_8^{2-}$, $NaBiO_3$, PbO_2, also, electrolysis. Reducing with H_2O_2 in acid solutions quickly regenerates the tripositive ions, although this process occurs spontaneously with Bk via the slow radio-lytic formation of H_2O_2 [170, 171].

On a technical scale, $Ce(OH)_4$ is selectively precipitated from pH 6–7 solutions containing other Ln while oxidising with either H_2O_2 [172, 173], O_3 [174], or air [175, 176]. Greater enrichments of Ce or Bk are easily achieved, however, by solvent extraction or by precipitation of the iodates. The preferential extraction of Bk^{4+} and Ce^{4+} by HDEHP is the leading method used for purifying these elements from tripositive Ln and An [177, 178]. Nitric acid solutions containing the An or Ln are heated for a short period with BrO_3^- to oxidise Ce and Bk and are then equilibrated with HDEHP diluted with an alkyl solvent. To oxidise Ce and Bk completely [179, 180] the nitric acid concentration should be 5.0 M or greater and BrO_3^- concentrations should be above 0.4 M. The distribution coefficients of the 4+ ions are larger in HDOP than in HDEHP [180], but since in 0.15 F HDEHP they are approaching 2000 anyway, the choice of extractant would seem to make little difference. Separations from tripositive species are excellent with factors in the range of 10^4 to 10^6. Ce^{IV} and Bk^{IV} are reduced and completely back-extracted [179] into an aqueous phase of 4 M HCl containing 0.01 to 1.0 M H_2O_2.

The solubility of $Ce(IO_3)_4$ in 2 M HNO_3 containing 0.057 M HIO_3 is 6.8 µg/1, and the solubility of $Bk(IO_3)_4$ is even lower under the same con-ditions [181, 182]. Again these elements are oxidised with 0.6 M $NaBrO_3$, and complete precipitation is obtained after digesting for about 1 h at 60–80 °C. This method is very satisfactory for the final preparation of very pure samples of Bk or Ce, although Zr, Hf and Th are co-precipitated if present initially.

Prior to the advent of an effective extraction technique, Ce^{III} was separated from Bk^{III} by elution from cation resin with 13 M HCl containing C_2H_5OH. Moore [183] has now developed an amine extraction system whereby Ce^{IV} is selectively extracted, leaving Bk^{IV} in the aqueous solution. At HNO_3 con-centrations near 2 M, Ce^{IV} as a hexanitrato complex is quantitatively ex-tracted into the quaternary amine, Aliquat 336-S-NO_3 (30 vol/wt % in xylene). The amine was substituted for Dowex 1 × 4 anion-exchange resin which had been examined earlier and found to differentiate strongly between tetrapositive Ce and Bk [184]. For maximum effectiveness, the amine system also should be adapted to a chromatographic technique. The poor extracta-bility of Bk^{IV} ($K_d \approx 0.02$), seemingly implies a wide divergence between Bk^{IV} and Ce^{IV} with respect to nitrate complexing. This is unexpected, inas-much as other tetrapositive ions of the An (Pu, Np, Th) form strong anionic nitrate complexes which are extensively extracted by alkyl amines.

1.3.2.2 **Am** $^{V and VI}$

There are relatively few methods of separating Am that are based on oxidation to the penta- or hexapositive states. The double potassium carbonate salt of AmO_2^+ is insoluble in basic solutions of K_2CO_3 [185]; in acid solutions, the fluoride[186] and oxalate of AmO_2^{2+} are soluble. By contrast, the corresponding compounds of tripositive species have reciprocal solubilities, thus affording a separation of oxidised Am from other Ln and An. Aside from these precipitation schemes, which are practical for weighable amounts of Am, there are several not-so-useful methods that depend on the selectivity of cation-exchange resins or extractants for ions of different charge[178, 187–189]. They fail, principally because americyl compounds are extremely prone to reduction by unstable impurities in organic extractants and ion-exchange resins. The half-cell potential[190] for oxidising Am^{3+} to AmO_2^{2+} in 1.0 M $HClO_4$ is -1.69 V, which classifies AmO_2^{2+} as a strong oxidant capable of reacting with all but the most inert organic substances.

By introducing an inorganic cation exchanger, zirconium phosphate, Moore[191] and Horwitz[192] may have overcome the problems of separating $Am^{V, VI}$ caused by the sensitivity of organic exchangers to oxidation. Ions in the 3+ state are held strongly by this exchanger at low acid concentrations, whereas Am oxidised in very dilute HNO_3 is but slightly retained in the column bed. Americium in about 90% yield is completely separated from Eu, Cm, and Cf with decontamination factors exceeding 2.5×10^5. The tripositive ions are later released from the exchanger by increasing the HNO_3 concentration to 10 M.

A standard method[190], used by Moore, of oxidising Am^{III} to Am^{VI} is by heating 0.01–0.1 M HNO_3 solutions of Am with 0.1 M $(NH_4)_2S_2O_8$ for ~ 10 min at 80–90 °C. Normally, a trace of Ag^+ is added to accelerate oxidation reactions with $S_2O_8^{2-}$. This functions as a catalyst by producing Ag^{2+} which acts as a single electron-carrier performing the actual oxidation of Am^{3+}. Moore, by omitting the Ag^+, presumed that Am would be stabilised in the pentapositive state after having been oxidised to the hexapositive state, although we know of no evidence supporting the assumption that Am^{VI} would reduce to Am^V while $S_2O_8^{2-}$ was still present. In any event, he proposes Am^V as the non-adsorbable species in the zirconium phosphate exchanger.

A close relative of HDEHP has been synthesised recently by Peppard for preferentially extracting Am^{VI} [193]. The extraction of Ln^{III} and An^{III} ions is depressed, presumably because of steric hindrance within the extractant molecule bis(2,6-dimethyl-4-heptyl) phosphoric acid, HD(DIBM)P. The difference between the extractability of Am^{VI} and Cm^{III} is enormous, amounting to 10^7. In a system of 0.6 F HD(DIBM)P (heptane) *v.* 0.025 M HNO_3 containing Ag^+ and $K_2S_2O_8$, the K_d of Am^{VI} was provisionally about 50, depending on whether Am^{3+} was oxidised entirely or not. Were it not for complexing of the americyl ions (and of 3+ ions, for that matter) by HSO_4^- or SO_4^{2-} formed during the reduction of persulphate, the distribution coefficient would be two to three orders of magnitude larger. As an example[194], the K_d of U^{VI} obtained in nearly the same system, but without persulphate, is 10^5. In view of the adverse action of persulphate upon the K_d of Am^{VI}, it would be interesting to try other oxidants.

As already suggested by Peppard, his system should be adapted to extraction chromatography so as to obtain the highest performance. Several mildly unfavourable features are associated with his extraction system as well as with the one using zirconium phosphate exchanger. For one, Ce and Bk accompany Am; for another, some artfulness is necessary to oxidise tracer Am. This means using, literally, pure reagents and solutions of tracer, and pretreating all extractants, column beds, and glassware with an oxidising solution of HNO_3 and persulphate.

1.4. SELECTED LARGE-SCALE PROCESSES

Detailed descriptions of individual separation methods are discussed in the earlier sections of this report. We will now illustrate how these were united to form complete processes for accomplishing important objectives. Only large-scale processes were selected as examples because, by economic necessity, such processes must embrace the most modern and effective separation methods. By 'large-scale' we mean only the largest separation projects carried on today, for surely, with the passing of time, larger facilities will be needed to fill new and pressing demands for these valuable elements.

1.4.1   Rare earth processing

Industrial processing is introduced with some trepidation, owing to the scant details available for proprietary processes. Spokesmen for the rare earth industries are unwilling to talk about their methods of separating Eu and Y, currently the two products with the highest commercial value. Fortunately, several process descriptions have been given for the world's largest rare earth concentrating facility at Mountain Pass, California[13, 195–197]. From their adjacent open mine, the Molybdenum Corporation of America is taking about 1200 tons of bastnasite ore ($LnFCO_3$) a day for concentration and separation of the light rare earths. Assayed as oxides, bastnasite contains $\sim 8\%$ rare earths, largely La, Ce, Pr, and Nd but also with Sm through Lu present as 5% of the total rare earth abundance. The extraction of these elements by the process shown diagrammatically in Figure 1.11 represents a major advance over older methods.

Solvent extraction was used for the first time for large-scale fractionation of the lanthanides. Counter-current extraction in mixer-settler stages is carried out in large plywood boxes coated with epoxy-fibreglass and polyvinylchloride for protection against HCl. A technical grade of La is recovered from the extraction cycle as are concentrates of Pr-Nd, Sm-Gd, and heavier Ln.

Europium, the most profitable product, is reportedly separated from Sm and Gd by precipitation of $EuSO_4$ after passage through a reduction column of Zn-amalgam[196]. Following an early 1969 visit to the Mountain Pass plant where the europium recovery area was classified, we suspected that this part of the process had been modified. Without considering the technological aspects of application on a large scale, we believe the most attractive tech-

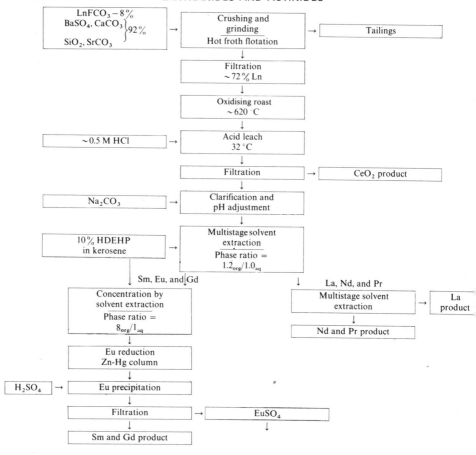

Figure 1.11 Process for concentrating rare earths used by the Molybdenum Corporation of America

nique to be extracting the tripositive Ln with HDEHP after reducing Eu^{3+} to Eu^{2+}. In this way, a separation factor for Ln^{III}/Eu^{II} of about 10^5 can be utilised[145] and this, at least, is a thousand times larger than a comparable factor for sulphate precipitations.

Europium is eventually recovered as a 99.99%-pure oxide at the Mountain Pass plant. The Pr-Nd, La, and Sm-Gd products are refined later to high purities by a licensed process at Louviers, Colorado.

1.4.2   Recovery of transplutonium elements

At one time each of the major US A.E.C. research laboratories manufactured and chemically recovered their own supply of transplutonium elements. Since these independent programmes were inefficient compared to a single production source which would supply all of the A.E.C. and university needs,

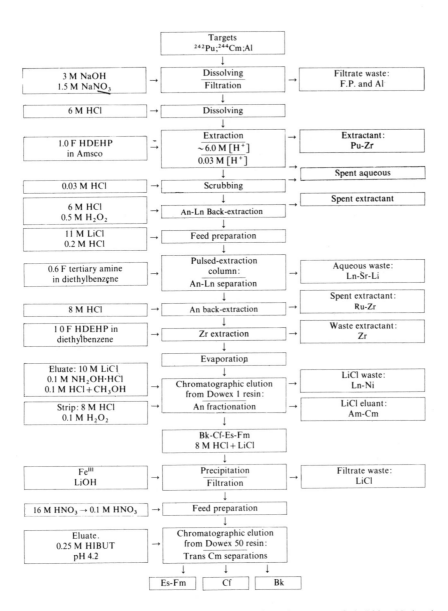

Figure 1.12 Process for the recovery of transplutonium elements at Oak Ridge National Laboratory

the production of these rare elements was eventually consolidated and made a national programme. The programme was enlarged and a special high-flux reactor (HFIR) and associated processing plant (TRU) were constructed at the Oak Ridge National Laboratory. After becoming fully operational in 1966, the TRU facility has seen increasingly larger quantities of heavy actinides channelled through its remotely operated hot-cells. Most recently, 120 mg of ^{252}Cf was isolated in a single processing.

The separation process used at TRU[107, 108, 198], outlined in Figure 1.12, is a refinement based on 6 years of development work and 4 years of operating experience by the Oak Ridge chemists. The obvious purpose is to separate Bk, Cf, Es, and Fm free of each other, fission products, and the starting target isotopes — ^{242}Pu or a mixture of ^{243}Am and ^{244}Cm. This kind of processing is tremendously difficult, as many extra complications in the separation chemistry are created by the extremely high levels of radioactivity. Hydrogen peroxide evolution from radiolysis produces gas blockages in ion-exchange resin beds, frothing of liquids, and unwanted reducing or oxidising conditions in solution. Reagents are destroyed, glass vessels become opaque and fragile, plastics soften and decompose, and the corrosion rate of metals is accelerated by the intense radioactivity. In reality, these problems are never truly solved — only minimised by compensating efforts on the part of the chemists and engineers.

Several critical separation steps require more explanation than is given in the flow diagram of Figure 1.12. The two extractions with HDEHP, following dissolution of the target, are for removing Pu, numerous fission products, and extraneous metal ions. In many ways, this extraction fulfils the function of a fluoride precipitation. Plutonium is first separated from the tripositive heavy elements by extraction from 6 M HCl; the aqueous acidity is then reduced to 0.03–0.06 M [H^+] where the Ln^{III} and An^{III} are easily extractable into fresh HDEHP[198]. Hydrogen peroxide is added to the stripping solution to reduce Bk^{IV} and prevent its retention by the extractant. The Ln fission products are separated in the next two steps. Most are removed by counter-current extraction using the high purity tertiary amine, Adogen 364-HP. Final Ln^{III} rejection and the separation of Am-Cm from transcurium elements takes place during a LiCl elution from a heated bed of Dowex 1 × 10 (200–300 mesh) resin. After 90–95 % of the curium has been eluted, the heavier elements remaining on the column are quickly eluted with 8 M HCl. The last strip solution, containing some LiCl salt, is treated by co-precipitating the An^{III} with $Fe(OH)_3$. After dissolving the precipitate in 12.5 M HCl, $(FeCl_4)^-$ is adsorbed by a small column of Dowex 1 resin.

The transcurium elements are separated from each other in high-pressure (1000 lb in^{-2}) ion-exchange columns filled with very fine Dowex 50 × 8 resin. The 120 cm long columns fractionate the transcurium elements in about 2.5 h operating at 80 °C. The Cf fraction is normally quite pure at this point and, after recovery from HIBUT, it is ready for distribution or further neutron irradiations to produce Es. However, Bk is contaminated with Am and Cm and requires additional purification by extraction of Bk^{IV}. The Es and Fm are collected in a single fraction from the first HIBUT-cation-exchange column, and later they are cleanly separated after passage through another high-pressure column.

Regarding improvements to the overall process, we believe in removing Bk in the tetrapositive state at an early stage in the processing. This eliminates the need for holding reductants throughout the rest of the separations as well as leaving three, rather than four, transcurium elements to be separated with the final HIBUT-column. By adding the oxidant $NaBrO_3$ to a HNO_3 solution resulting from the initial dissolution of the An, Bk^{III} is oxidised to Bk^{IV} and is then quantitatively extracted into HDEHP. Plutonium, Ru, Zr and Ce are also extracted in this step, but Bk and Ce are selectively back extracted into 8 M $HNO_3 + 0.5$ M H_2O_2.

Many of the separation steps that have been combined into the TRU process have seen wide usage for 10–20 years, either in processing reactor targets or in laboratory separations. Other chemists, trying to avoid the handling of corrosive HCl and LiCl in their equipment, have developed processes using only HNO_3 [199–201]. The separation of pure Am and Cm from irradiated Pu by Berger and co-workers[199], using nitrate-based systems, worked very well. On the other hand, Am and Cm were incompletely separated from one another by a cation-exchange purification developed by Wheelwright et al.[200, 201].

1.4.3 Recovery of transplutonium elements from underground nuclear explosions

Specially designed thermonuclear devices have produced high yields of the transplutonium elements[202]. Isotopes of these elements are created during the underground explosion when a uranium target undergoes an instantaneous sequence of multiple neutron captures, and subsequently β-decays to isotopes of elements with higher atomic numbers. The synthesised elements are dispersed in thousands of tons of explosion-fused rock and rubble. Their recovery depends upon drilling back into the explosion zone, bringing kilogrammes of An-rich rock to the surface, and chemically isolating the minute amounts of heavy elements from the mass of siliceous material. The concentration of An is low even in debris from the best of these experiments. As an example, the atom ratio of SiO_2 to ^{257}Fm was 10^{16} in prime rock debris taken from the recent 'Hutch' test[202]. To recover useful amounts of these rare isotopes, it is imperative to rapidly process kilogramme quantities of rock.

The concentration procedure outlined in Figure 1.13 was developed especially for treating alluvium, tuff, and dolomite on a large scale[100]. At the moment, a pilot plant capable of handling two 5-kg batches in parallel process streams every 16 h has been installed and was placed in operation after each underground experiment[203, 204]. The practical capacity of this shielded facility is about 50 kg per week. The process chemistry is almost totally based upon the powerful extracting properties of OPPA (abbreviated as DOPP in this flow-diagram and described in the extraction section). Once a batch is started, step-by-step progression requires little, if any, intermediate preparation such as frequent adjustments of pH or chemical concentrations. Output yields of the transplutonium elements vary from 70–100%, depending more upon mechanical handling of flow-streams than on chemical conditions.

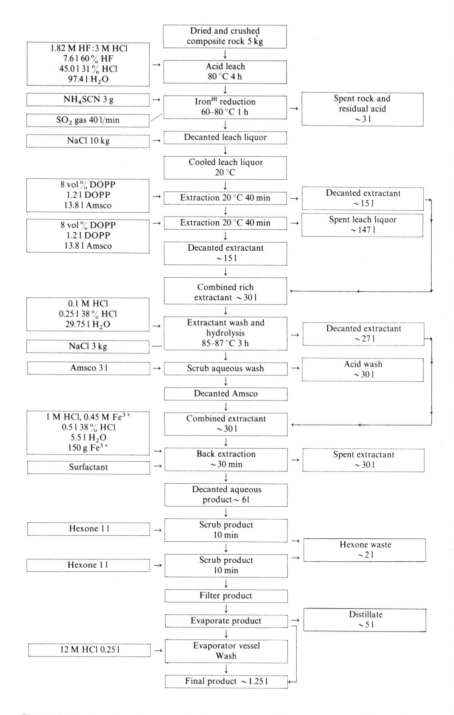

Figure 1.13 Transplutonium concentration process used for the recovery of these elements from crushed rock (Ref. 204)

The product solutions, containing appreciable concentrations of Fe, Al, and Ln, are contacted several times with hexone to extract Fe. The iron-free products from up to ten 5 kg batches are later combined and evaporated preparatory to separating the Ln with a large LiCl-anion-exchange column. The bulk of the Ln occurs naturally in the rock while much smaller amounts were formed as fission products during the explosion. After the Ln are removed, the remaining separation steps are normal ion-exchange procedures, i.e., C_2H_5OH–HCl and HIBUT elutions from cation-exchange resins.

Other groups have developed processes on a smaller scale which are directed toward the recovery of actinides from these special nuclear explosions[205, 206]. The initial steps in the process by Horwitz et al.[206] involve fluoride and hydroxide precipitations. The Los Alamos procedure, developed by Wolfsberg and co-workers[205], begins with dissolving the finely-ground rock with HF–$HClO_4$; subsequent steps include precipitating a fluoride and extracting the An and Ln with TBP from a solution heavily salted with $Al(NO_3)_3$. An upward scaling of the size of these two processes would be somewhat difficult due to the mechanical awkwardness of handling voluminous precipitates or treating them with fuming $HClO_4$.

1.5 ACKNOWLEDGEMENTS

We wish to thank Dr. John F. Wild for reviewing the manuscript and Mr. Lee O. Mead for his literature search. We also thank Mrs. B. G. Hulet for typing and Mrs. Carol E. Patrick for obtaining reprints of articles and publications.

References

1. Katz, J. J. and Seaborg, G. T. (1957). *The Chemistry of the Actinide Elements*, (London: Methuen and Co.)
2. Pascal, P., ed. (1970). *Nouveau Traité de Chimie Minérale*, **15**, (Paris: Masson et Cie)
3. Ryabchikov, D. I. and Ryabukhin, V. A. (1970). *Analytical Chemistry of Yttrium and the Lanthanide Elements*, (London: Ann Arbor)
4. Korkisch, J. (1969). *Modern Methods for the Separation of Rarer Metal Ions*, (New York: Pergamon Press)
5. Moeller, T. (1963). *Chemistry of the Lanthanides*, (New York: Reinhold)
6. Starý, J. (1964). *Solvent Extraction of Metal Chelates*, (New York: Macmillan)
7. Penneman, R. A. and Keenan, T. K. (1960). *National Academy of Sciences, Washington, D.C.*, NAS-NS-3006
8. Stevenson, P. C. and Nervik, W. E. (1961). *National Academy of Sciences, Washington, D.C.*, NAS-NS-3020
9. Starý, J. (1966). *Talanta*, **13**, 421
10. Starý, J. and Caletka, R. (1966). *Chemicke Listy*, **60**, 1267
11. Hafez, M. B. and Patti, F. (1970). *Centre d'Etudes Nucléaires de Fontenay-aux-Roses Rept.* CEA-BIB-171
12. Cunningham, B. B. (1964). *Annual Reviews of Nuclear Science*, eds., Segre, E., Friedlander, G., and Noyes, H. P., **14**, 323, (Palo Alto: Annual Reviews, Inc.)
13. Pings, W. B. (1969). *Colo. Sch. Mines, Miner. Ind. Bull.*, **12**, 1
14. Asprèy, L. B. and Penneman, R. A. (1967). *Chem. Eng. News*, **45**, 75
15. Bagnall, K. W. (1964). *Sci. Prog. (London)*, **52**, 66
16. Peppard, D. F. (1966). *Advances in Inorganic and Radiochemistry*, ed. Eméleus, H. J. and Sharpe, A. G., **9**, 1, (New York: Academic Press)

17. Peppard, D. F. and Mason, G. W. (1969). *Coordination Chemistry*, ed. Kirschner, S., 289, (New York: Plenum Press)
18. Peppard, D. F. (1963). *Conferencia Interamericana de Radioquimica, 1°, Montevedio*, 63, (Washington, D.C.: Union Panamerican)
19. Weaver, B. (1965). *Proceedings of the Conference on Rare Earths, 4th, Phoenix, Arizona*, 415, (New York: Gordon and Breach)
20. Ulstrup, J. (1966). *At. Energ. Rev.*, **4,** 3
21. Transplutonium Elements: A Bibliography (1968). *U.S. At. Energ. Comm.*, Oak Ridge, TID-3317; (1969) TID-3317 (Suppl. 1); (1970) TID-3317 (Suppl. 2)
22. Miller, H. W. (1968). *Dow Chemical Co., Rocky Flats Div. Rept.* RFP-1081
23. Kleinberg, T., compiled by (1967). *Los Alamos Scientific Laboratory Rept.* LA-1721
24. Helfferich, F. G. and James, D. B. (1970). *J. Chromatogr.*, **46,** 1
25. Brandstetr, J., Zvarova, T. S., Krivanek, M. and Malý, J. (1963). *Radiokhimiya*, **5,** 694
26. Lougheed, R. W., Evans, J. E. and Hulet, E. K., unpublished data
27. Fuger, J. (1961). *J. Inorg. Nucl. Chem.*, **18,** 263
28. Choppin, G. R. and Silva, R. J. (1956). *J. Inorg. Nucl. Chem.*, **3,** 153
29. Deelstra, H. and Verbeek, F. (1965). *J. Chromatogr.*, **17,** 558
30. Vobecky, M. and Mastalka, M. (1963). *Collect. Czech. Chem. Commun.*, **28,** 709
31. Nishi, T. and Fujiwara, I. (1964). *Nippon Genshiryoku Gakkaishi*, **6,** 15; *Lawrence Rad. Lab. Trans.* UCRL-Trans-1118(L)
32. Powell, J. E. (1964), in *Progress in the Science and Technology of the Rare Earths*, **1,** 62, ed. by Eyring, L. (New York: Macmillan)
33. Noguchi, M., Yoshifugi, A. and Hagiwara, Z. (1969). *Bull. Chem. Soc. Jap.*, **42,** 2286
34. Hagiwara, Z. and Noguchi, M. (1970). *Bull. Chem. Soc. Jap.*, **43,** 401
35. Hagiwara, Z., Banno, A. and Kamei, A. (1969). *J. Inorg. Nucl. Chem.*, **3,** 3295
36. Powell, J. E. and Burkholder, H. R. (1967). *J. Chromatogr.*, **29,** 210
37. Safonova, N. D., Matorina, N. N. and Chmutov, K. V. (1969). *Russ. J. Phys. Chem.*, **43,** 1619
38. Bleyl, H. J. and Münzel, H. (1968). *Radiochim. Acta*, **9,** 149
39. Nazarov, P. P. and Chmutov, K. V. (1967). *Zh. Fiz. Khim.*, **41,** 2985
40. Aly, H. F., Latimer, R. M. and Abdel-Rassoul, A. A. (1970). *Talanta*, **17,** 265
41. Aubouin, G. and Laverlochere, J. (1968). *J. Radioanal. Chem.*, **1,** 123
42. Beranova, H. and Pettuzila, V. (1964). *Collect. Czech. Chem. Commun.*, **29,** 500
43. Hoehlein, G., Voeller, H. and Weinlaender, W. (1969). *Radiochim. Acta*, **11,** 172
44. Mikler, J. (1967). *Monatsh. Chem.*, **98,** 1899
45. Campbell, D. O. and Buxton, S. R. (1970). *Ind. Eng. Chem., Proc. Des. Develop.*, **9,** 89
46. Campbell, D. O. (1970). *Ind. Eng. Chem., Proc. Des. Develop.*, **9,** 95
47. Hale, W. H. and Lower, J. T. (1969). *Inorg. Nucl. Chem. Lett.*, **5,** 363
48. Robertson, G. H. (1970). *Dissertation* (Univ. of Calif., Berkeley)
49. Nady, L. and Vermuelen, T. (1970). *Lawrence Radiation Lab., Berkeley, Rept.* UCRL-19526 Abst.
50. Vermuelen, T., Nady, L., Krochta, J. M., Ravoo, E. and Howery, D. (1971). *Ind. Eng. Chem., Proc. Des. Develop.*, **10,** 91
51. Orlandini, K. A. and Korkisch, J. (1968). *Argonne Nat. Lab. Rept.* ANL-7415
52. Bochkarev, V. A. and Voevodin, E. N. (1965). *Radiokhimiya*, **7,** 461
53. Morrow, R. J. (1966). *Talanta*, **13,** 1265
54. Ross, R. and Roemer, J. (1967). *Isotopenpraxis*, **3,** 197
55. Stewart, D. C., Bloomquist, C. A. A. and Faris, J. P. (1965). *Argonne Nat. Lab. Rept.* ANL-6999
56. Greene, R. G. and Fritz, J. S. (1965). *Iowa State Univ. Rept.* IS-1153
57. Faris, J. P. (1967). *J. Chromatogr.*, **26,** 232
58. Choppin, G. R., Harvey, B. G. and Thompson, S. G. (1955). *J. Inorg. Nucl. Chem.*, **2,** 66
59. Peppard, D. F., Mason, G. W., Maier, J. L. and Driscoll, W. J. (1957). *J. Inorg. Nucl. Chem.*, **4,** 334
60. Moore, F. L. (1957). *Anal. Chem.*, **29,** 1660
61. Leddicotte, G. W. and Moore, F. L. (1952). *J. Amer. Chem. Soc.*, **74,** 1618
62. Siekierski, S. and Kotlinska, B. (1959). *Atomnaya Energ.*, **7,** 160
63. Mason, G. W., McCarty, S. and Peppard, D. F. (1962). *J. Inorg. Nucl. Chem.*, **24,** 967
64. Lewey, S., Mason, G. W. and Peppard, D. F. (1971). *J. Inorg. Nucl. Chem.*, in press
65. Peppard, D. F., Mason, G. W., Driscoll, W. J. and Sironen, R. J. (1958). *J. Inorg. Nucl. Chem.*, **7,** 276

66. Gureev, E. S., Kosyakov, V. N. and Yakovlev, G. N. (1964). *Radiokhimiya*, **6**, 655
67. Krasovec, F. and Klofutar, C. (1964). *Nukl. Inst. "Josef Stefan", Ljubljana, Yugoslavia Rept.* NIJS-R-440
68. Horwitz, E. P., Bloomquist, C. A. A. and Henderson, D. J. (1969). *J. Inorg. Nucl. Chem.*, **31**, 1149
69. Horwitz, E. P., Bloomquist, C. A. A., Henderson, D. J. and Nelson, D. E. (1969). *J. Inorg. Nucl. Chem.*, **31**, 3255
70. Baybarz, R. D. (1963). *Nucl. Sci. Eng.*, **17**, 463
71. Fidelis, I. (1967). *Nukleonika*, **12**, 477
72. Horwitz, E. P. and Bloomquist, C. A. A. (1969). *Inorg. Nucl. Chem. Lett.*, **5**, 753
73. Sochacka, R. J. and Siekierski, S. (1964). *J. Chromatogr.*, **16**, 376
74. Fidelis, I. and Siekierski, S. (1965). *J. Chromatogr.*, **17**, 542
75. Peppard, D. F., Bloomquist, C. A. A., Horwitz, E. P., Lewey, S. and Mason, G. W. (1970). *J. Inorg. Nucl. Chem.*, **32**, 339
76. Peppard, D. F., Mason, G. W. and Lewey, S. (1969). *Solvent Extraction Research*, ed. by Kertes, A. S. and Marcus, Y. (New York: John Wiley and Sons, Inc.)
77. Nugent, L. J. (1970). *J. Inorg. Nucl. Chem.*, **32**, 3485
78. Fidelis, I. and Siekierski, S. (1966). *J. Inorg. Nucl. Chem.*, **28**, 185
79. Rowlands, D. L. G. (1967). *J. Inorg. Nucl. Chem.*, **29**, 809
80. Mason, G. W., Schofer, N. L. and Peppard, D. F. (1971). *J. Inorg. Nucl. Chem.*, in press
81. Goto, T. and Smutz, M. (1965). *J. Inorg. Nucl. Chem.*, **27**, 1369
82. Michelsen, O. B. and Smutz, M. (1971). *J. Inorg. Nucl. Chem.*, **33**, 265
83. Harada, T. and Smutz, M. (1970). *J. Inorg. Nucl. Chem.*, **32**, 649
84. Lenz, T. G. (1967). *Ames Lab., Iowa State Univ. Rept.* IS-T-138
85. Rhee, C. T. (1969). *Daehan Hwahak Hwoejee*, **13**, 205
86. Gusmini, S. and Nonnenmacher, R. (1970). *Centre d'Etude Nucléaires de Fontenay-aux-Roses Rept.* CEA-R-4004
87. Galaud, G. (1967). *Communaute Eur. Energ. At.-EURATOM* EUR-3277.f
88. Thomas, N. E. (1970). *Thesis, Iowa State Univ.* IS-T-370
89. Kotlinska-Filipek, B. and Siekierski, S. (1963). *Nukleonika*, **8**, 607
90. Alekperov, R. A. and Geibatova, S. S. (1968). *Dokl. Akad. Nauk SSSR*, **178**, 349
91. Alekperov, R. A. (1969), *Issled. Obl. Neorg. Fiz. Khim. Ikh Rol Khim. Prom., Mater. Nauch. Konf.*, ed. by Sogomonyan, M. S. (Baku, USSR: AzINTI)
92. Sherrington, L. G. and Kemp, W. P. (1970). *Brit. Pat.*, 1 180 922
93. Schweitzer, G. K. and Sanghvi, S. M. (1969). *Anal. Chim. Acta*, **47**, 19
94. Bauer, D. J. and Lindstrom, R. E. (1964). *U.S. Bur. Mines Rept. Inv.* 6396
95. Khopkar, P. K. and Narayanankutty, P. (1968). *J. Inorg. Nucl. Chem.*, **30**, 1957
96. Sweet, T. R. and Parlett, H. W. (1968). *Anal. Chem.*, **40**, 1885
97. Sweet, T. R., Parlett, H. W., Porell, A. L., Kahn, W. and Blair, S. D. (1968). *U.S. Clearinghouse Fed. Sci. Tech. Inform.* AD-685238
98. Butler, F. E. and Hall, R. M. (1970). *Anal. Chem.*, **42**, 1073
99. Siddall, T. H. (1963). *J. Inorg. Nucl. Chem.*, **25**, 883
100. Hulet, E. K., Evans, J. E., Quong, R. and Qualheim, B. J. (April 1968). *Abst. 131, ACS Symp. Macroscopic Studies of the Actinides, San Francisco, Calif.*
101. Ellis, D. A. (1952). *Dow Chemical Co. Rept.* 81
102. Long, R. S., Ellis, D. A. and Bailes, R. H. (1956). *Proceedings of the International Conference on the Peaceful Uses of Atomic Energy, Geneva, 1955*, **8**, 77, (United Nations)
103. Müller, W. (1967). *Actinides Rev.*, **1**, 71
104. Moore, F. L. (1961). *Anal. Chem.*, **33**, 748
105. Baybarz, R. D., Weaver, B. S. and Kinser, H. B. (1963). *Nucl. Sci. Eng.*, **17**, 457
106. Moore, F. L. and Mullins, W. T. (1965). *Anal. Chem.*, **37**, 687
107. Leuze, R. E. and Lloyd, M. H. (1970). *Progress in Nuclear Energy, Series III, Process Chemistry*, **4**, 549, ed. by Stevenson, C. E., Mason, E. A. and Gresky, A. T. (New York: Pergamon Press)
108. Baybarz, R. D. (1970). *At. Energ. Rev.*, **8**, 327
109. Moore, F. L. (1964). *Anal. Chem.*, **36**, 2158
110. Gerontopulos, P. Th., Rigali, L. and Barbano, P. G. (1965). *Radiochim. Acta*, **4**, 75
111. Barbano, P. G. and Rigali, L. (1967). *J. Chromatogr.*, **29**, 309
112. David, J. and Zangen, M. (1969). *5th Solvent Extr. Res., Proc. Int. Conf. Solvent Extr. Chem.* (1968). 219, ed. by Kertes, A. S. (New York: Wiley-Interscience)

113. Horwitz, E. P., Bloomquist, C. A. A., Sauro, L. J. and Henderson, O. J. (1966). *J. Inorg. Nucl. Chem.*, **28**, 2313
114. van Ooyen, J. (1970). *Atoomenergie–Haar toepassingen,* **12**, 263
115. van Ooyen, J. (1970). *Reactor Cent. Ned. Rept.* RCN-113
116. Cerrai, E. and Ghersini, G. (1970). *Advances in Chromatography,* **9**, ed. by Giddings, J. C. and Keller, R. A. (New York: Marcel Dekker, Inc.)
117. Eschrich, H. and Drent, W. (1967). *Eurochemic Rept.* ETR 211
118. Cerrai, E. and Testa, C. (1963). *J. Inorg. Nucl. Chem.*, **25**, 1045
119. Stronski, I. (1970). *Radiochim. Acta,* **13**, 25
120. Moore, F. L. and Jurriaanse, A. (1967). *Anal. Chem.*, **39**, 733
121. Pierce, T. B. and Hobbs, R. S. (1963). *J. Chromatogr.*, **12**, 74
122. Pierce, T. B. and Peck, P. F. (1962). *Nature (London)*, **194**, 84
123. Kooi, J. and Boden, R. (1964). *Radiochim. Acta,* **3**, 226
124. Mikhailiehenko, A. I. and Pimenova, R. M. (1969). *Zh. Prikl. Khim. (Leningrad)*, **42**, 1010
125. Hornbeck, R. F. (1967). *J. Chromatogr.*, **30**, 438
126. Hornbeck, R. F. (1967). *J. Chromatogr.*, **30**, 447
127. Siekierski, S. and Sochacka, R. J. (1964). *J. Chromatogr.*, **16**, 385
128. Cerrai, E., Testa, C. and Triulzi, C. (1962). *Energia Nucl.*, **9**, 377
129. Winchester, J. W. (1958). *U.S. At. Energ. Comm. Rept.* CF-58-12-43
130. Bosholm, J. and Grosse-Ruyken, H. (1964). *J. Prakt. Chem.*, **26**, 83
131. Watanabe, K. (1965). *J. Nucl. Sci. Technol.*, **2**, 45
132. Riccato, T. M. and Herrman, G. (1970). *Radiochim. Acta,* **14**, 107
133. Gavrilov, E. G., Starý, J. and Wang, T. S. (1966). *Talanta,* **13**, 471
134. Horwitz, E. P., Bloomquist, C. A. A., Buzzell, J. A. and Harvey, H. W. (1969). *Argonne Nat. Lab. Rept.* ANL-7546
135. Fields, P. R., Ahmad, I., Barnes, R. F., Sjoblom, R. K. and Horwitz, E. P. (1970). *Nucl. Phys.,* **A154**, 407
136. Horwitz, E. P., Orlandini, K. A. and Bloomquist, C. A. A. (1966). *Inorg. Nucl. Chem. Lett.*, **2**, 87
137. Horwitz, E. P., Bloomquist, C. A. A., Orlandini, K. A. and Henderson, D. J. (1967). *Radiochim. Acta,* **8**, 127
138. Horwitz, E. P., Bloomquist, C. A. A. and Griffen, H. E. (1969). *Argonne Nat. Lab. Rept.* ANL-7569
139. Stewart, D. C., Horwitz, E. P., Youngquist, C. H. and Wahlgren, M. A. (1970). *Nucl. App. Tech.,* **9**, 875
140. Stronski, I. (1969). *Chromatographia,* **2**, 285
141. Cotton, F. A. and Wilkinson, G. (1962). *Advanced Inorganic Chemistry,* 888, (New York: Interscience)
142. Hulet, E. K., Lougheed, R. W., Brady, J. D., Stone, R. E. and Coops, M. S. (1967). *Science,* **158**, 486
143. Malý, J. (1969). *J. Inorg. Nucl. Chem.*, **31**, 741
144. Silva, R. J., Sikkeland, T., Nurmia, M., Ghiorso, A. and Hulet, E. K. (1969). *J. Inorg. Nucl. Chem.*, **31**, 3405
145. Peppard, D. F., Horwitz, E. P. and Mason, G. W. (1962). *Rare Earth Research,* 15, (New York: Gordon and Breach)
146. Moskvin, L. N. (1963). *Radiokhimiya,* **5**, 747
147. Malý, J., Sikkeland, T., Silva, R. J. and Ghiorso, A. (1968). *Science,* **160**, 1114
148. Malý, J. and Cunningham, B. B. (1967). *Inorg. Nucl. Chem. Lett.*, **3**, 445
149. Choppin, G. R., Harvey, B. G. and Thompson, S. G. (1956). *J. Inorg. Nucl. Chem.*, **2**, 66
150. Beranova, H., Brandstetr, J., Druin, V., Ermakov, V., Zvarova, T., Krzywanek, M., Malý, J., Polikanov, S. and Su, H. K. (1962). *Nukleonika,* **7**, 465
151. Barrett, M. F., Sweasey, D. and Topp, N. E. (1962). *J. Inorg. Nucl. Chem.*, **24**, 571
152. Shvedov, V. P. and Frolkov, A. Z. (1968). *Radiokhimiya,* **10**, 482
153. Onstott, E. I. (1956). *J. Amer. Chem. Soc.*, **78**, 2070
154. Onstott, E. I. (1955). *J. Amer. Chem. Soc.*, **77**, 2129
155. Onstott, E. I. (1961). *Anal. Chem.*, **33**, 1470
156. Onstott, E. I. (1963). *Inorg. Chem.*, **2**, 967
157. David, F. and Bouissières, G. (1968). *Inorg. Nucl. Chem. Lett.*, **4**, 153
158. Shvedov, V. P. and Antonov, P. G. (1967). *Radiokhimiya,* **9**, 478
159. McCoy, H. N. (1941). *J. Amer. Chem. Soc.*, **63**, 1622

160. Sayun, M. G. and Timofeeva, T. G. (1967). *Radiokhimiya*, **9**, 261
161. Marsh, J. K. (1943). *J. Chem. Soc.*, 531
162. Shvedov, V. P. and I-Bei. F. (1960). *Radiokhimiya*, **2**, 231
163. Antonov, P. G. and Shvedov, V. P. (1969). *Radiokhimiya*, **11**, 311
164. Malý, J. (1967). *Inorg. Nucl. Chem. Lett.*, **3**, 373
165. Ross, R. (1969). *Isotopenpraxis*, **5**, 233
166. Okashita, H. (1967). *Radiochim. Acta*, **8**, 85
167. Malan, H. P. and Muenzel, H. (1966). *Radiochim. Acta*, **5**, 20
168. Pankratova, L. N. (1968). *Radiokhimiya*, **10**, 122
169. Weaver, B., Coleman, C. F. and Stevenson, T. N. (April 1971). *161st National Meeting of the American Chemical Society, Los Angeles, Calif.*
170. Propst, R. C. and Hyder, M. L. (1970). *J. Inorg. Nucl. Chem.*, **32**, 2205
171. Gutmacher, R. G., Bodé, D. D., Lougheed, R. W. and Hulet, E. K. (1970). To be published.
172. Golinski, M. and Korpak, W. (1967). *Przem. Chem.*, **46**, 328
173. Facchini, A., Teatini, A. and Verganti, G. (1969). *Chim. Ind. (Milan)*, **51**, 970
174. Bauer, D. J. and Lindstrom, R. E. (1968). *U.S. Bur. Mines, Rep. Invest.* 7123
175. Richter, H., Koenig, O., Schmitt, A. and Schwerdt, G. (1965). *Pat. Ger. (East)*, 37 258
176. Zelikman, A. N. and Lyapina, Z. N. (1962). *Izv. Vysshikh Uchebn. Zavedenii, Tsvetn. Met.*, **5**, 115
177. Peppard, D. F., Moline, S. W. and Mason, G. W. (1957). *J. Inorg. Nucl. Chem.*, **4**, 344
178. Hulet, E. K. (1964). *J. Inorg. Nucl. Chem.*, **26**, 1721
179. Knauer, J. B. and Weaver, B. (1969). *Oak Ridge Nat. Lab. Rept.* ORNL-TM-242
180. Krtil, J. and Bezdek, M. (1969). *Collect. Czech. Chem. Commun.*, **34**, 1406
181. Weaver, B. (1968). *Anal. Chem.*, **40**, 1894
182. Fardy, J. J. and Weaver, B. (1969). *Anal. Chem.*, **41**, 1299
183. Moore, F. L. (1969). *Anal. Chem.*, **41**, 1658
184. Moore, F. L. (1967). *Anal. Chem.*, **39**, 1874
185. Burney, G. A. (1968). *Nucl. Appl.*, **4**, 217
186. Moore, F. L. (1963). *Anal. Chem.*, **35**, 715
187. Hara, M. (1970). *Bull. Chem. Soc. Jap.*, **43**, 89
188. Stokely, J. R. and Moore, F. L. (1967). *Anal. Chem.*, **39**, 994
189. Moore, F. L. (1968). *Anal. Chem.*, **40**, 2130
190. Penneman, R. A. and Asprey, L. B. (1955). *Proceedings of the International Conference on the Peaceful Uses of Atomic Energy, Geneva*, **7**, 355, (United Nations; 1956)
191. Moore, F. L. (1971). *Anal. Chem.*, **43**, 487
192. Horwitz, E. P. (1966). *J. Inorg. Nucl. Chem.*, **28**, 1469
193. Mason, G. W., Bollmier, A. F. and Peppard, D. F. (1970). *J. Inorg. Nucl. Chem.*, **32**, 1011
194. Peppard, D. F., Mason, G. W., Bollmier, A. F. and Lewey, S. (1971). *J. Inorg. Nucl. Chem.*, in press
195. McKinley, J. R. and Kruesi, P. R. (1967). *Fr. Pat.* 1 503 042
196. 'Moly Corp uses solvent extraction to separate europium from rare earths,' (March 1966). *World Mining*, 16
197. DeBruyne, P. (1969). *Fermentatio*, **65**, 274
198. Burch, W. D., Biglow, J. E. and King, T. J. (1970). *Oak Ridge Nat. Lab. Rept.* ORNL-4540
199. Berger, R., Koehly, G., Musikas, C., Pottier, R. and Sontag, R. (1970). *Nucl. Appl. Technol.*, **8**, 371
200. Wheelwright, E. J., Roberts, F. P. and Bray, L. A. (1969). *Battelle-Northwest Lab. Rept.* BNWL-SA-1492
201. Wheelwright, E. J. and Roberts, F. P. (1969). *Battelle-Northwest Lab. Rept.* BNWL-1072
202. Eccles, S. F. and Hulet, E. K. (1969). *Lawrence Rad. Lab., Livermore, Rept.* UCRL-50767, unpublished
203. Quong, R. and McNabb, J. R. (1968). *Lawrence Rad. Lab., Livermore, Rept.* UCRL-50499, unpublished
204. Quong, R. and Cowles, J. O. (1970). *Lawrence Rad. Lab., Livermore, Rept.* UCRL-50847, unpublished
205. Wolfsberg, K., Daniels, W. R., Ford, G. P. and Hitchcock, E. T. (1967). *Nucl. Appl.*, **3**, 568
206. Horwitz, E. P., Bloomquist, C. A. A., Harvey, H. W. and Hoh, J. C. (1966). *Argonne Nat. Lab. Rept.* ANL-7134, unpublished
207. McCoy, H. N. (1935). *J. Amer. Chem. Soc.*, **57**, 1756
208. Herrmann, E. (1968). *J. Chromatogr.*, **38**, 498

2
Lanthanide and Actinide Mixed Oxide Systems with Alkali and Alkaline Earth Metals

CORNELIUS KELLER

Institute for Radiochemistry, Karlsruhe

2.1 INTRODUCTION

In the last decade the solid state chemistry of the lanthanide and actinide mixed oxides has received considerable attention. This growing interest is partly due to the application of special oxides and oxide phases in inorganic and nuclear technology, examples being the use of some rare earth compounds in colour television and of $(U,Pu)O_2$ mixed oxides as a nuclear fuel for fast breeder reactors. The purely scientific interest in this area is, however, increasing.

The divalent to heptavalent lanthanide and actinide metals are a special class of ions inasmuch as they are multivalent ions with large ionic radii. In this respect the actinides are much more interesting than the lanthanides, having a greater range of valencies, from $+2$ (e.g. No^{II}) to $+7$ (Np,Pu, Am(?)) and perhaps to $+8$ (Pu(?)). The trivalent lanthanide metal ions, however, are also very interesting, because they form a series of fifteen ions (including La), which makes it possible to study the influence of the ionic radius on the crystal structures. However, there are only a few systems in which a series is isostructural from lanthanum to lutetium, examples being the vanadates $REVO_4$ and niobates $RENbO_4$, the oxides $KREO_2$, the oxides $(RE^{III}_{0.5},M^V_{0.5})O_2$ with $M^V = Pa$, U, Np and the $UO_3(NpO_3)\cdot 6REO_{1.5}$ ternary oxides (RE represents a rare earth element, including La).

In most of these isostructural series a so-called gadolinium point (or curium point in the actinide series), that is a discontinuity of physical or chemical data as a function of the atomic number, is observed when going from Gd^{3+} to Tb^{3+} (or from Cm^{3+} to Bk^{3+}). This effect is ascribed to the special stability of the half-filled $4f^7$ $(5f^7)$ electron shell. Two further smaller discontinuities, the tetrad effect[1], at present known only for solution chemistry data, perhaps may also be observed in solid state chemistry if more accurate data are obtained.

In the following chapters two areas of actinide and lanthanide mixed oxide chemistry are discussed, the reactions with alkali metal and alkaline earth oxides. In the last decade these reactions have received very much attention, especially from structural and crystal–chemical points of view. Using such reactions it is also possible to prepare in a very simple manner ternary oxides of actinide and lanthanide elements in their highest oxidation states, for example:

$$Np^{VII} \text{ and } Pu^{VII} \text{ in } Li_5MO_6 \quad \text{and } Ba_2LiMO_6,$$
$$Am^{VI} \qquad\qquad \text{in } Li_6AmO_6 \text{ and } Na_4AmO_5,$$
$$Pr^{IV} \text{ and } Cm^{IV} \text{ in } Li_8MO_6$$

These compounds, which are extraordinarily stable towards self-irradiation effects, give us a convenient starting point in the search for physico-chemical data of these oxidation states. Another very broad field of research in lanthanide (actinide)–alkali metal (alkaline earth) mixed oxides is devoted to answering the question of whether it is possible to substitute a medium valent metal ion by corresponding amounts of higher valent and lower valent metal ions and if so, to what extent ordering of the substituted metal ions occurs. The following series serves as an example:

$$BaNpO_3 - Ba(La^{III}_{0.5},Np^V_{0.5})O_3 - Ba(Sr^{II}_{0.5},Np^{VI}_{0.5})O_3 - Ba(Li^I_{0.5},Np^{VII}_{0.5})O_3$$

Having written this short review, the author considers that, in the future, the following points should attract much more scientific interest, resulting in the filling of the large gaps which still exist in the knowledge of the actinide and lanthanide mixed oxides:

(a) Although very many compounds and some structures are known, in most cases no study concerning the stoicheiometry or the nonstoicheiometry of the compounds, and the influence of the latter on the crystal structure or on other physico-chemical data, is recorded.

(b) Much more precise physical data, especially lattice constants to the third decimal place, are necessary to establish inconstancies in the homologous ternary oxide series which result from small changes in the electronic configuration.

(c) Spectroscopic (i.r., e.s.r., nuclear γ-resonance), and especially magnetic data down to at least liquid helium temperature, are necessary to establish the electronic configuration and its influence on or by the other constituent atoms in the lattice in question.

(d) There are practically *no* thermodynamic data, such as heats of formation or heats of solution of oxygen in stoicheiometric and non-stoicheiometric compounds. Besides applying normal calorimetric or high-temperature galvanic cell methods, the development and use of new methods, such as fluorine bomb calorimetry, seem necessary to obtain reliable thermodynamic data.

(e) The study of self-irradiation effects in highly radioactive actinide compounds should be pursued to obtain a better understanding of the formation of metamict minerals.

(f) The evaluation of phase diagrams at high temperatures should attract more interest.

2.2   REACTIONS WITH MONOVALENT METAL OXIDES

2.2.1   Compounds with heptavalent actinide elements

The reaction of Li_2O with $NpO_2(PuO_2)$ in a molar ratio of about (2.7–2.9):1 and 400–420 °C in a flow of oxygen leads to the formation of the penta-lithiumactinide(VII)-hexoxide[2, 3], the first compounds of these unusual oxidation states:

$$2.5\,Li_2O + NpO_2(PuO_2) \xrightarrow[\;400-420\,°C\;]{O_2} Li_5NpO_6(Li_5PuO_6)$$

(An excess of about 0.2–0.4 moles of Li_2O is necessary to achieve complete oxidation.)

By using a thermobalance, installed in a glove-box, the weight gain of the reaction mixture gives a direct indication of the mean valency state in the reaction product, oxidation states of 6.97 ± 0.04 being normally obtained without any difficulty. Li_5NpO_6 may also be prepared from Li_2O_2 and $NpO_3·H_2O$ as starting materials[4]. Li_5NpO_6 and Li_5PuO_6 do not decompose on storing for some weeks if CO_2 and humidity are strictly excluded. Long-term storage, however, leads to partial radiation chemical decomposition as shown, for example, by changes in the Mössbauer spectrum of Np(VII).

These compounds are isomorphous with Li_5ReO_6 (and $Li_5TcO_6 + Li_5IO_6$)[5]; they crystallise in a hexagonal structure (space group $P3_112$) with lattice constants of $a = 5.21$ Å and $c = 14.61$ Å for Li_5NpO_6, and $a = 5.19$ Å and $c = 14.48$ Å for Li_5PuO_6. In contrast to the products isolated from aqueous solution, these solid compounds contain isolated MO_6 octahedra and not MO_2^{3+} or MO_5^{3-} groups.

The heptavalency of Np and Pu in these compounds has been demonstrated by physico-chemical methods. In the nuclear-γ-resonance spectrum of Li_5NpO_6 (Figure 2.1) an isomeric shift of -6.84 cm s^{-1} (with respect to NpO_2) is characteristic of Np^{VII} whilst the small quadrupole-splitting at 4.2 K indicates a deviation from O_h-symmetry around the Np^{7+} ion, that is, there is a slightly distorted octahedral oxygen environment[6]. The following Mössbauer data were derived from the spectra in Figure 2.1:

Oxidation state	Isomeric shift [mm s^{-1}]	Quadrupole coupling constant 1/4 eq Q [mm s^{-1}]	Asymmetry parameter	Half width $\frac{1}{2}\Gamma$exp
Np(VII)	-68.4 ± 2	10.0 ± 1	0.045 ± 0.05	1.6 ± 0.2
Np(VI)	-48.4 ± 2	10.0 ± 2	0	1.8 ± 0.3

The Np^{VI} present in the spectrum of Li_5NpO_6 is probably due to the presence of a small amount of Li_6NpO_6.

The very small magnetic susceptibility ($\mu_{eff} = 0.34\ \mu_B$), which does not follow the Curie–Weiss relationship, is an indication of the 5f^1 system of Pu^{VII}[7]. As expected, Np^{VII} is diamagnetic [7,8].

Li_5NpO_6 and Li_5PuO_6 are soluble in water and alkaline solutions to give

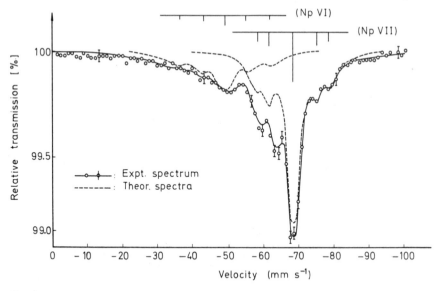

Figure 2.1 Mössbauer spectrum of $Li_5Np^{VII}O_6$ at 4.2 K (the sample contains small amounts of $Li_6Np^{VI}O_6$ as an impurity) (source: Am(Th) at 4.2 K)

the fairly well-known dark green solutions of the MO_5^{3-} ions. From these solutions the following compounds can be precipitated: $[Co(NH_3)_6]$ $NpO_5\cdot$aq, $[Co(en)_3]NpO_5\cdot$aq, $[Cr(NH_3)_6]NpO_5\cdot$aq, $[Pt(NH_3)_3Cl]NpO_5\cdot$aq, $M_3(NpO_5)_2\cdot$aq (M = Ba,Sr,Ca) and $Ba_3(PuO_5)_2\cdot$aq [9-13]. These compounds are relatively unstable, decomposition occurring within a short time. The

Mössbauer-spectrum of $[Co(NH_3)_6]NpO_3 \cdot aq$ (isomeric shift: $-6.28\,cm\,s^{-1}$) shows that the strong central resonance has satellite peaks typical of quadrupole splitting. Their presence is attributed to two non-equivalent Np^{VII} species or sites in the sample; both sites are highly asymmetric (asymmetry parameter $\eta = 0.83$ and $\eta = 0.69$, respectively)[14].

Attempts to oxidise americium to the heptavalent state in the presence of Li_2O, in order to obtain Li_5AmO_6, were unsuccessful. The oxidation only leads to Am^{VI} [2]. This observation agrees with Zaitseva's finding that Am^{VII} can only be prepared by disproportionation of Am^{VI} and not by direct oxidation in solution[15].

2.2.2 Compounds with hexavalent actinide elements

Whereas the systems of hexavalent actinide oxides with lithium and sodium oxides are well known, there is less information on compounds with the heavier alkali metals, potassium, rubidium, and caesium.

The best method of preparing the different types of ternary oxide is to heat the corresponding actinide oxide (U_3O_8, NpO_2, PuO_2, AmO_2) with the oxide, peroxide, hydroxide, or carbonate of the appropriate alkali metal at temperatures between 400 °C and 1300 °C, depending on the thermal stability of the compound. The preparation of ternary oxides with a high alkali metal: actinide ratio (e.g. of Li_6UO_6, Na_4NpO_5, or Na_6AmO_6) cannot be carried out at high temperatures. Consequently, alkali metal oxides must be used instead of carbonates or hydroxides.

The thermal stability of the alkali metal–actinide(VI) oxides strongly depends on:
(a) the alkali metal to actinide ratio, the stability decreasing in the order[16]:

$$M_2U_6O_{19} > M_2U_3O_{10} > M_2U_2O_7 > M_2UO_4 > M_4UO_5,$$

(i.e. $Na_2U_6O_{19}$ is thermally more stable than $Na_2U_3O_{10}$);
(b) the atomic number of the alkali metal, that is, upon the volatility and the polarisation behaviour. The following order of decreasing thermal stability is observed

$$Li_2UO_4 > Na_2UO_4 > K_2UO_4 > Rb_2UO_4 > Cs_2UO_4.$$

Heating these compounds for 6 h at 1100 °C results in the reduction of the M(I) : U ratio of Cs_2UO_4 to 0.32 and of K_2UO_4 to 0.93, whereas Li_2UO_4 is stable[17];
(c) the atomic number of the actinide element. Here there is a decrease in thermal stability in the order $U > Np > Pu > Am$, the uranates(VI) being the most stable compounds and the americates(VI) the least stable ones[18]. Attempts to prepare Cm(VI) by oxidising $^{244}CmO_2-$ Li_2O and $^{244}Cm_2O_3-Li_2O$ mixtures failed even at very low reaction temperatures[19].

The thermal stability and preparative conditions of the ternary oxides in the Li–Am–O system are summarised in Figure 2.2. By choosing proper reaction conditions all compounds can be very easily prepared in a pure form.

In addition to the lithium and sodium uranates(VI) and trans-uranates(VI), mentioned in Table 2.1, several other polyuranates with a ratio M(I) : U of less than 1 : 3 exist. A polyuranate phase with the approximate composition $Na_2O \cdot 13UO_3$ was identified by Carnall et al.[20], with cell parameters closely related to those for U_3O_8 and $\alpha\text{-}UO_3$. Kovba reports structural data for two potassium polyuranates $K_8U_{16}O_{52}$ and $K_2U_7O_{22}$[21]. In addition to $Li_2U_3O_{10}$ and the orthorhombic phase $Li_2O \cdot (1.60\text{–}1.64)UO_3$ ($\sim Li_{22}U_{18}O_{65}$) a very uranium rich phase $Li_2O \cdot 6UO_3$ also seems to exist.

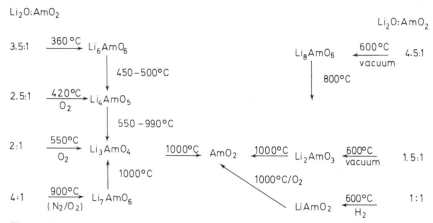

Figure 2.2 Preparative conditions and thermal stability of compounds in the Li_2O–$AmO_{1.5}$-O_2 system

Table 2.1   Schematic representation of lithium and sodium uranates(VI) and transuranates(VI)

Composition	U Li	Na	Np Li	Na	Pu Li	Na	Am Li	Na
$Me_2X_3O_{10}$	+	−	−	−	−	−	−	−
$Me_2X_2O_7$	+	+	−	+	−	−	−	−
$\alpha\text{-}Me_2XO_4$	+	+	−	+	−	−	−	−
$\beta\text{-}Me_2XO_4$	+	+	−	+	−	−	−	−
$\alpha\text{-}Me_4XO_5$	−	+	−	+	−	+	−	+
$\beta\text{-}Me_4XO_5$	+	+	+	+	+	+	+	−
Me_6XO_6	+	−	+	+	+	+	+	+

+ = existent
− = non-existent

Investigations by Anderson[22] on potassium polyuranates do not agree with the data of Kovba[21]. Anderson observed only two polyuranate phases of compositions near $K_2O \cdot 3UO_3$ and $K_2O : UO_3$ 1 : 8, the structural arrangements of these phases being pseudohexagonal layers. From the data on precipitated uranates there seems to be some confusion concerning the composition of polyuranates, so that a new and detailed study of this problem is desirable. Silver uranate, Ag_2UO_4, and several uranates of thallium(I) have also been prepared, namely Tl_2UO_4, $Tl_2U_2O_7$, $Tl_2U_3O_{10}$, and a solid solution between $Tl_2O \cdot 3UO_3$ and $Tl_2O \cdot 6UO_3$ [23]. These compounds have only been characterised by their formulae.

In the alkali metal (Li,Na) oxide-transuranium(VI) oxide systems only compounds with $M^I : M^{VI}$ ratios of $\geqslant 4$ are known, with the exception of the Na_2O–NpO_3 system in which the mononeptunate(VI), α-(β-)Na_2NpO_4, and the dineptunate, $Na_2Np_2O_7$, have been described[18, 24]. Although the preparation of $Li_2Np_2O_7$ failed, it is possible, however, to replace the U^{VI} in $(Li,Na)_2U_2O_7$ by Np^{VI} to give $(Li,Na)_2Np_2O_7$ [25]. Such products may be obtained from molten $LiNO_3$–$NaNO_3$ eutectics containing BrO_3^-.

The structures of the diuranate phases $M_2^I U_2(Np_2)O_7$ (Na,K,Rb) share essentially the same basic hexagonal network of uranium atoms lying in the a–b plane (Table 2.2). This network is slightly contracted as compared

Table 2.2   Unit cell dimensions of alkali metal diuranates

| Compound | Lattice parameters | | | Number of molecules per unit cell |
	a [Å]	b [Å]	c [Å]	
$Na_2U_2O_7$	6.812 ($\sqrt{3} \times 3.930$)	11.790 (3×3.930)	17.742 (3×5.914)	9
$K_2U_2O_7$	6.902 ($\sqrt{3} \times 3.985$)	7.971 (2×3.985)	19.643 (3×6.548)	3
$Rb_2U_2O_7$	24.0 (6×4.00)	13.86 ($2\sqrt{3} \times 4.00$)	20.57 (3×6.86)	36

to the basic network of α-UO_3 and U_3O_8. The sodium and potassium diuranates are a superstructure of this type attributable to a slight displacement of the uranium atoms from the ideal hexagonal positions of the basic unit. $Na_2U_2O_7$ loses oxygen on heating to give a series of oxygen-deficient compounds $Na_2U_2O_{7-x}(x < 0.5)$ without changing the structure.

The crystal structure of the monouranates is known (Table 2.3)[26]. In all the compounds $M_2U(Np)O_4$, except α-$Na_2U(Np)O_4$, the uranium atoms are octahedrally bonded to six oxygen atoms, two of which (parallel to the z-axis, the uranyl group) are more strongly bonded than the remaining four oxygen atoms lying in the xy plane. The crystal lattice consists of somewhat flattened U—O octahedra, joined by sharing the four secondary oxygen atoms to form infinite $[(UO_2)O_2]^{2-}$ layers. The structure of the uranates can be regarded as a stack of such layers. The low temperature (α-) form of $Na_2U(Np)O_4$ is reported to share two opposite edges of the UO_6 octahedra, thus forming infinite $[(UO_2)O_2]^{2-}$ chains, as described for $MgUO_4$. Its structure may be regarded as a superlattice of the NaCl-type, three-quarters of the octahedral holes being filled with sodium and uranium atoms in an ordered fashion[27].

The structures of Li_4MO_5 and β-Na_4MO_5 (M = U—Am) are interesting in that they represent a M^{VI}—O configuration which does not contain the actinyl group, that is, $(MO_4)^{2-}$ rather than $(MO_2)^{2+}$ ions characterise the bonding[18, 24, 28, 29]. From neutron diffraction studies the following atomic positions have been assigned (space group $I4/m$):

Li_4UO_5		β-Na_4UO_5
	2U in (a)	
	$2O_{II}$ in (b)	
$x = 0.280$		$x = 0.244$
$y = 0.097$	$8O_I$ in (h)	$y = 0.097$
$x = 0.197$		$x = 0.180$
$y = 0.383$	$8M(I)$ in (h)	$y = 0.419.$

This structure may be considered as a NaCl-type lattice with every fifth alkali metal atom being replaced by a uranium atom in an ordered form.

Table 2.3   Structural data for alkali metal uranates(VI) and transuranates(VI)

Compound	Symmetry	Space group	Lattice parameters [Å]			M(VI—O distance [Å]
			a	b	c	
$Li_2U_3O_{10}$	monoclinic	$P2_1/c$	6.821	18.91	7.300 $\beta = 121.56$	
$Li_2O \cdot 6UO_3$	orthorhombic		6.701	4.01	4.148	
$Na_2O \cdot 13UO_3$	orthorhombic		6.807	15.934	8.254	
$Na_2Np_2O_7$	orthorhombic		3.91	6.77	17.11	
$K_8U_{16}O_{52}$	tetragonal	$P6_3$	14.290		14.014	
$K_2U_7O_{22}$	orthorhombic	$Pbam$	6.945	19.533	7.215	
Li_2UO_4	orthorhombic	$Fmmm$	6.06	5.13	10.52	1.89(2×), 1.98(4×)
α-Na_2UO_4	orthorhombic	$Cmmm$	9.74	5.72	3.49	1.90(2×), 2.24(4×)
β-Na_2UO_4	orthorhombic	$Fmmm$	5.97	5.795	11.68	1.93(2×), 2.08(4×)
α-Na_2NpO_4	orthorhombic	$Cmmm$	9.685	5.705	3.455	1.89(2×), 2.22(4×)
β-Na_2NpO_4	orthorhombic	$Fmmm$	5.936	5.785	11.652	1.92(2×), 2.07(4×)
K_2UO_4	tetragonal	$I4/mmm$	4.335		13.10	1.90(2×), 2.17(4×)
Rb_2UO_4	tetragonal	$I4/mmm$	4.345		13.85	1.91(2×), 2.17(4×)
Cs_2UO_4	tetragonal	$I4/mmm$	4.38		14.79	1.91(2×), 2.19(4×)
Li_4UO_5	tetragonal	$I4/m$	6.721		4.451	2.00(4×), 2.22(2×)
Li_4AmO_5	tetragonal	$I4/m$	6.666		4.410	1.98(4×), 2.21(2×)
α-Na_4NpO_5	cubic		4.379			2.37
α-Na_4PuO_5	cubic		4.718			2.36
β-Na_4NpO_5	tetragonal	$I4/m$	7.515		4.597	2.10(4×), 2.30(2×)
β-Na_4PuO_5	tetragonal	$I4/m$	7.449		4.590	2.09(4×), 2.30(2×)
Li_6NpO_6	hexagonal	$R\bar{3}$	5.217		14.70	
Li_6AmO_6	hexagonal	$R\bar{3}$	5.174		14.59	
Na_6NpO_6	hexagonal	$R\bar{3}$ (?)	5.78		16.0	
Na_6PuO_6	hexagonal	$R\bar{3}$ (?)	5.76		15.9	

A disordered replacement of Na by U can be postulated for the low-temperature cubic α-Na_4MO_5. The absorption spectrum of solid Li_4NpO_5 (Teflon powder pellet) is shown in Figure 2.3.

Li_6MO_6(U—Am) and, probably, Na_6MO_6 are isostructural with hexagonal Li_6ReO_6 and Li_5ReO_6 (space group $P3_112$), having cubic closed packing of oxygen atoms[30]. Regular or slightly distorted MO_6 octahedra, therefore, can be assumed to be present in these compounds.

Several publications on the infrared spectra of alkali metal actinide oxides are available[29, 31, 32]. A full normal coordinate treatment of the optically active U—O lattice vibrations of some uranates, based on a simple method of infinite $[(UO_2)O_2]^{2-}$ layers, is given by Ohwada[30]. The force constants obtained from these calculations are summarised in Table 2.4. These values agree well with those calculated by using Badger's rule:

$$r_{U-O}[\text{Å}] = 1.08 \, K^{-1}[\text{mdyn/Å}] + 1.17.$$

Table 2.4 Force constants for alkali and alkaline earth metal uranates
(From Ohwada[32], by courtesy of Pergamon Press)

Compound	Force constant [mdyn/Å]*				
	K_1	K_2	H_2/r_0^2	$(H_2+H_3)/r_1^2$	$(H_1+H_3)/r_1^2$
Li_2UO_4	1.18	4.84	0.290	0.101	0.0884
β-Na_2UO_4	1.19	4.61	0.316	0.101	0.0884
K_2UO_4	1.19	4.44	0.292	0.101	0.0884
Rb_2UO_4	1.06	4.37	0.304	0.101	0.0884
Cs_2UO_4	0.835	4.38	0.331	0.101	0.0884
β-$SrUO_4$	1.24	4.79	0.240	0.101	0.0884

* K_1, K_2:   U—O stretching force constants
H_1, H_2, H_3:   bending force constants around the uranium and oxygen atoms
r_1, r_2:   equilibrium distances of the primary U—O$_I$ and the secondary U—O$_{II}$ bonds

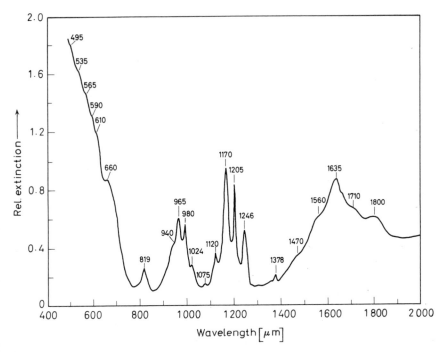

Figure 2.3 Absorption spectrum of Li_4NpO_5 (15 mg Li_4NpO_5 + 45 mg Teflon powder measured against 20 mg Al_2O_3 + 40 mg Teflon powder)

Observed and calculated frequencies for M_2UO_4 (K–Cs) in the 200 cm^{-1} to 700 cm^{-1} region are reproduced in Table 2.5.

A somewhat complicated problem is the so-called ammonium diuranate (ADU), an orange-yellow precipitate which is obtained by adding ammonia to a U(VI) solution. It now seems clear that the precipitate has not the

Table 2.5 Observed and calculated frequencies [cm^{-1}] for K_2UO_4, Rb_2UO_4 and Cs_2UO_4 (From Ohwada[32], by courtesy of Pergamon Press)

Species		K_2UO_4 Observed	Calculated	Rb_2UO_4 Observed	Calculated	Cs_2UO_4 Observed	Calculated
A_{1g}	ν_1		686		681	(680)	682
A_{2u}	ν_2	732	732	726	726	727	727
	ν_3		220		220		220
B_{2u}	ν_4		207		207		207
E_g	ν_5		249		254		265
E_u	ν_6	520	520	490	490	435	435
	ν_7	265	265	270	270	280	280
	ν_8		200		200		200

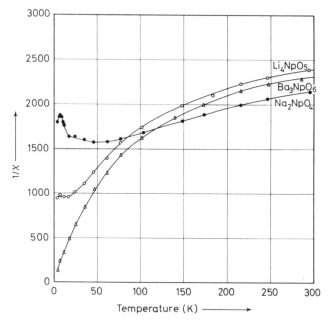

Figure 2.4 Magnetic susceptibility of some ternary Np(VI) oxides between room temperature and 4.2 K

composition $(NH_4)_2U_2O_7$ but consists of several compounds in varying amounts: $UO_3 \cdot 2H_2O$, $UO_3 \cdot 0.33NH_3 \cdot 1.67H_2O$, $UO_3 \cdot 0.5NH_3 \cdot 1.5H_2O$ and $UO_3 \cdot 0.67NH_3 \cdot 1.33H_2O$[33]. Stuart and Whateley, however, find no indications for distinct compounds, but rather a single phase in which the $NH_3 : U$ ratio varies continuously[34].

The compound precipitated from Pu^{VI} solutions by NH_4OH is not ammonium diplutonate, as previously reported[35], but rather is $PuO_2(OH)_2 \cdot H_2O$ or its stoicheiometrically equivalent $PuO_3 \cdot 2H_2O$ [36].

The magnetic susceptibilities of some Np^{VI} ternary oxides are shown in Figure 2.4.

2.2.3 Compounds with pentavalent actinide elements

Whereas most pentavalent oxide–alkali metal oxide systems are very complicated and characterised by the existence of several ternary oxides, the corresponding systems with pentavalent actinides (Pa–Am) are simple. Only three types of compounds are known, $M^IM^VO_3$, $M_3^IM^VO_4$, and $M_7^IM^VO_6$.

The pure compounds are normally prepared as follows:

(a) symproportionation of $M^{IV} + M^{VI}$, e.g.
$$Na_2UO_4 + UO_2 \rightarrow 2NaUO_3;$$

(b) thermal decomposition or reduction of actinides(VI), e.g.
$$Na_4AmO_5 \rightarrow Na_3AmO_4,$$
$$Na_2U_2O_7 \xrightarrow{H_2} 2NaUO_3;$$

(c) volatilisation of alkali metal oxides from compounds having higher $M^I : M^V$ ratios, e.g.
$$Li_7AmO_6 \rightarrow Li_3AmO_4 + Li_2O;$$

(d) direct reaction of M_2O with ternary actinide(V)oxides which have a lower $M^I : M^V$ ratio:
$$Li_3NpO_4 + 2Li_2O \xrightarrow{\text{(closed system)}} Li_7NpO_6;$$

(e) oxidation of ternary actinide(IV)oxides in a flow of oxygen, or of MO_2 in the presence of alkali metal peroxide, using a closed system:
$$Li_2O + AmO_2 \xrightarrow{O_2} Li_3AmO_4,$$
$$2UO_2 + Na_2O_2 \xrightarrow{\text{(closed system)}} 2NaUO_3;$$

(f) reduction of uranium(VI) compounds with alkali metals or azides:
$$UO_3 + KN_3 \xrightarrow{800\,°C,\,65\,kbar} KUO_3 + 1\tfrac{1}{2}N_2.$$

At 400 °C, liquid sodium reduces U_3O_8 to the dioxide, no double oxide being present in the product[11]. At 600 °C, the sodium oxide formed reacts with UO_2 according to the equation:

$$2Na_2O + UO_2 \rightarrow Na_3UO_4 + Na,$$

a reaction also observed in the attempted preparation of alkali metal uranates(IV) and neptunates(IV).

The crystal structures of most of these compounds are known[37–46] (Table 2.6). A cubic pervoskite type of structure only is reported for KM^VO_3 and RbM^VO_3, whereas the sodium compounds crystallise in the orthorhombic deformed perovskite lattice of the $GdFeO_3$-type and $LiUO_3$ has been shown by single crystal data to crystallise in the rhombohedral $LiNbO_3$ lattice. The $Li_3M^VO_4$ ternary oxides have a structure based on a

$3:1$ order of univalent and pentavalent cations in the rocksalt lattice. The powder patterns of Na_3UO_4 and Na_3AmO_4 were indexed on the basis of a cubic NaCl lattice, but a detailed structural analysis is lacking. The crystal structure of Li_7SbO_6 and of the isostructural actinide(V) compounds is closely related to the Li_8SnO_6-type with its hexagonal closed-packing of oxygen atoms, but with lower symmetry[47].

Chemical, spectroscopic and magnetic data on some of the uranium compounds have shown clearly that these compounds contain U^V and

Table 2.6 Structural data for compounds of alkali metal oxides and pentavalent actinide oxides

Compound	Symmetry	Space group	Lattice parameters			
			a [Å]	b [Å]	c [Å]	α (β) [°]
$LiUO_3$	rhombohedral	R3c	5.901			54.60
$NaPaO_3$	orthorhombic	Pbnm	5.82	5.97	8.36	
$NaUO_3$	orthorhombic	Pbnm	5.775	5.905	8.25	
$KPaO_3$	cubic	Pm3n	4.341			
KUO_3	cubic	Pm3n	4.290			
$RbPaO_3$	cubic	Pm3n	4.368			
$RbUO_3$	cubic	Pm3n	4.323			
$TiUO_3$	cubic	Pm3n	11.28			
Li_3PaO_4	tetragonal		4.52		8.48	
Li_3UO_4	tetragonal		4.49		8.46	
Li_3NpO_4	tetragonal		4.485		8.390	
Li_3PuO_4	tetragonal		4.464		8.367	
Li_3AmO_4	tetragonal		4.459		8.355	
Na_3PaO_4	tetragonal		6.865		9.598	
Na_3UO_4	cubic (?)		4.77			
Na_3AmO_4	cubic (?)		4.75			
Li_7PaO_6	hexagonal-orthorhombic	R3	5.55		15.84	
Li_7UO_6	hexagonal-orthorhombic	R3	5.52		15.80	
Li_7NpO_6	hexagonal-orthorhombic	R3	5.54		15.74	
Li_7AmO_6	hexagonal-orthorhombic	R3	5.54		15.65	

not a mixture of $U^{IV}+U^{IV}$ [48-52]. The magnetic susceptibility measurements of some ternary U^V oxides by the Tübingen research group was recently extended to liquid helium temperature[52] (Figure 2.5). The ternary oxides with a $5f^1$ electronic configuration show a magnetic behaviour which depends strongly on the oxygen environment of the metal atom, i.e. on the ligand field and its symmetry. The reciprocal susceptibility $1/\chi$ is highest for Li_7UO_6 (CN = 6) and lowest, e.g. for $(U_{0.5},Th_{0.5})O_{2.25}$ (CN = 8). Table 2.7 shows that the magnetic moment of U^V and other $5f^1$ ternary oxides is very low. The μ_{eff} values listed in Table 2.7 are obtained from the linear part of the $\chi = f(1/T)$ curves.

$LiUO_3$ and $NaUO_3$ show a very unusual behaviour as the reciprocal susceptibility drops rapidly at about 35 K ($NaUO_3$) and 19 K ($LiUO_3$); at lower temperatures the susceptibility depends on the strength of the magnetic field (Figure 2.6). The same effect is observed for Li_3UO_4 at <7 K.

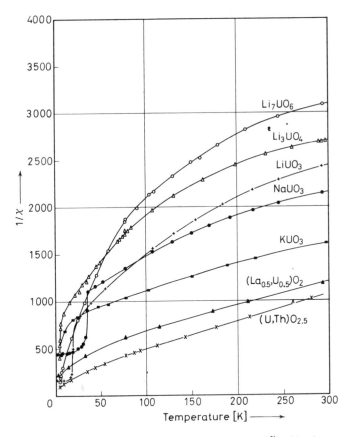

Figure 2.5 Magnetic susceptibility of some ternary U^V oxides between room temperature and 4.2 K, demonstrating clearly the ligand field effects

Table 2.7 Magnetic moments for several ternary oxides with a 5f^1 electronic configuration

Compound	μ_{eff} [μ_B]	Region of linearity of the magnetic susceptibility [K]
LiUO$_3$	0.57	100–300
NaUO$_3$	0.54	150–300
KUO$_3$	0.66	130–300
Li$_3$UO$_4$	0.44	70–300
Li$_7$UO$_6$	0.45	4–300
(La$_{0.5}$,U$_{0.5}$)O$_2$	0.39(4.2 K) to 1.41 (R.T.)*	none
(U$_{0.5}$,Th$_{0.5}$)O$_{2.25}$	0.59(4.2 K) to 1.46 (R.T.)*	none
Li$_5$PuO$_6$	0.34	4–300
α-Na$_2$NpO$_4$	0.48	200–300
Li$_4$NpO$_5$	0.46	100–300
Ba$_3$NpO$_6$	0.49	4–300

*R.T. = room temperature

A low temperature x-ray study, currently in progress[53], will show if structural changes occur in these compounds.

The reflectance spectra of some of the ternary uranium oxides also show that the optical properties are strongly influenced by the symmetry of the ligand field. The spectra of, for example, KUO_3 and $RbUO_3$ have been

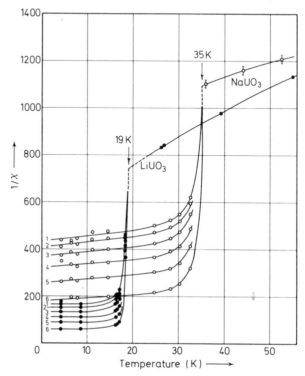

Figure 2.6 Low temperature region of the magnetic susceptibility of $LiUO_3$ and $NaUO_3$, showing the dependence of the reciprocal susceptibility on the strength of the magnetic field (1:3.75 kG, 2:5.73 kG, 3:7.65 kG, 4:9.55 kG, 5:11.3 kG, 6:12.5 kG)

analysed successfully on the basis of the $5f^1$ configuration for the U(V) ion in an octahedral ligand field[49].

2.2.4 Compounds with tetravalent actinide and lanthanide elements

In the alkali metal oxide–actinide(IV)/lanthanide(IV) oxide systems only ternary oxides of the compositions $M_2^I M^{IV} O_3$ (M^{IV} = Th,Am,Ce,Pr,Tb) and $Li_8 M^{IV} O_6$ (Pu,Am,Ce,Pr,Tb) have been described. The preparation of these compounds is very simple; a mixture of the alkali metal oxide (or peroxide) is heated with the actinide or lanthanide oxide to temperatures of about 600–800 °C, either in an oxygen flow (Th,Ce,Pr,Tb) or in a closed system $(Am,Pu)^{54-62}$. The reaction of UO_2 and NpO_2 with Li_2O or Na_2O in a closed system always results in the formation of uranates(V) and neptunates(V) with the liberation of alkali metals, which sublime to the colder parts of the reaction system. A similar reaction—formation of uranate(V)—was also found in liquid sodium (see preceding section). It therefore seems very improbable that alkali metal uranates(IV) and neptunates(IV) exist.

The crystal structure of most of these $M_2^I M^{IV}O_3$ compounds, with the exception of Na_2AmO_3, was determined by Hoppe et al. at Giessen. In the americium compound the high radioactivity of $Am(^{241}Am : 3.2$ Ci $g^{-1})$ and the self-radiation damage prevent the study of single crystals. Most of the observed types of structure (Table 2.8) may be compared with either the α-NaFeO$_2$-lattice $(\beta$-Na$_2$TbO$_3 =$ Na(Na$_{1/3}$, Tb$_{2/3}$)O$_2$) or the NaCl-lattice (e.g. α-Na$_2$TbO$_3 =$ (Na$_{2/3}$, Tb$_{1/3}$)O) or the Li$_2$SnO$_3$-lattice (e.g. β-K$_2$TbO$_3$) or the Rb$_2$PbO$_3$-lattice (e.g. Rb$_2$ThO$_3$).

Magnetic data for some ternary lanthanide oxides have been measured but not interpreted. The susceptibilities follow the Curie–Weiss relationship with relatively high Weiss constants. The measured values of the magnetic moment of Pr^{4+} and Tb^{4+} are remarkably smaller than the calculated values of 2.56 $\mu_B(Pr^{4+})$ and 7.9 $\mu_B(Tb^{4+})$:

	μ_{eff} [μ_B]	Weiss constant Δ
K$_2$PrO$_3$	2.40	-140
Li$_2$TbO$_3$	7.6	-95
Na$_2$TbO$_3$	7.6 (7.8)	-68 (-62)
β-K$_2$TbO$_3$*	8.5	-47
Rb$_2$TbO$_3$	7.9	-57
Cs$_2$TbO$_3$	7.3	-64

*value too high; the sample probably contained Tb^{3+} ($\mu_{eff} = 9.7$ μ_B)

The Li$_8$M^{IV}O$_6$ ternary oxides are isostructural with Li$_8$SnO$_6$ (space group $R\bar{3}$), the structure of which consists of isolated M^{IV}O$_6$ octahedra and a slightly distorted hexagonal closed packing of oxygen atoms[63].

2.2.5 Compounds with trivalent actinide and lanthanide elements

The alkali metal oxide-trivalent actinide (lanthanide) oxide systems are very simple, for only ternary oxides of the composition $M^I M^{III}O_2$ are known; however, these have different crystal structures[64-72]. To prepare these compounds, the alkali metal oxides (or peroxides) are usually heated with the corresponding heavy metal oxides to temperatures of 400–800 °C. Sometimes (e.g. in the cases of Pr,Tb,Am), the use of an inert or reducing atmosphere is necessary to prevent oxidation.

In the LiM^{III}O$_2$ series three types of structure have been found (Table 2.9); the ternary oxides of lithium with the light lanthanides up to europium and with americium are isostructural with monoclinic α-LiEuO$_2$. The

Table 2.8 Ternary oxides of tetravalent actinides and lanthanides with alkali metals

Compound	Symmetry	Lattice parameters			
		a [Å]	b [Å]	c [Å]	$\alpha(\beta)$[°]
Na_2ThO_3	monoclinic	6.16	10.23	6.07	112.4
K_2ThO_3	hexagonal (monoclinic)	6.41	11.09	12.72	99.40
Rb_2ThO_3	hexagonal	3.75		19.7	
Na_2AmO_3	monoclinic	5.92	10.26	11.23	100.12
α-Na_2CeO_3	cubic	4.82			
β-Na_2CeO_3	hexagonal (monoclinic)	3.448		16.49	
K_2CeO_3	hexagonal (monoclinic)	3.59		18.68	
α-Na_2TbO_3	cubic	4.740			
β-Na_2TbO_3	hexagonal	3.35		16.41	
α-K_2TbO_3	cubic	5.11			
β-K_2TbO_3	hexagonal (monoclinic)	3.48		18.62	
Rb_2TbO_3	orthorhombic	10.9	7.29	6.13	
α-Na_2PrO_3	cubic	4.84			
β-Na_2PrO_3	hexagonal (monoclinic)	3.406		16.44	
K_2CeO_3	hexagonal (monoclinic)	3.59		18.86	
Li_8PuO_6	hexagonal	5.64		15.95	
Li_8AmO_6	hexagonal	5.62		15.96	

Table 2.9 Structural data for some ternary oxides of alkali metals with trivalent lanthanides

Compound	Symmetry	Space group	Lattice parameters			
			a [Å]	b [Å]	c [Å]	β [°]
α-$LiEuO_2$	monoclinic	$P2_1/c$	5.6815	5.9885	5.6221	103.17
β-$LiEuO_2$	orthorhombic	$Pnam$	11.405	5.3353	3.4711	
$LiLuO_2$	tetragonal		4.37		9.95	
$NaLaO_2$	tetragonal		4.80		11.29	
$NaGdO_2$	tetragonal		4.66		10.53	
$NaLuO_2$	hexagonal	$R\bar{3}m$	3.32		16.52	
$KLaO_2$	hexagonal	$R\bar{3}m$	3.70		18.71	
$KPrO_2$	hexagonal	$R\bar{3}m$	3.64		18.65	
$KYbO_2$	hexagonal	$R\bar{3}m$	3.39		18.40	
$RbDyO_2$	hexagonal	$R\bar{3}m$	3.48		19.4	
$RbLuO_2$	hexagonal	$R\bar{3}m$	3.40		19.15	
$LiEu_3O_4$	orthorhombic	$Pbnm$	11.565	11.535	3.480	

orthorhombic structure of what is probably the high-temperature modification of β-LiEuO$_2$ is also observed for LiTbO$_2$. LiGdO$_2$, as well as the series
LiErO$_2$ to LiLuO$_2$ and NaLaO$_2$ to NaGdO$_2$, crystallise in the tetragonal
α-LiFeO$_2$ type, and NaTbO$_2$ to NaErO$_2$ in the β-LiFeO$_2$ lattice. NaTmO$_2$
to NaLuO$_2$ (including the Y, Sc, and In ternary compounds), all the potassium (KREO$_2$) (Lu(?)) and all the rubidium (RbREO$_2$) ternary oxides,
which are known at present, possess the hexagonal α-NaFeO$_2$ type of
structure. The arrangement of metal and oxygen atoms in KLaO$_2$ is shown

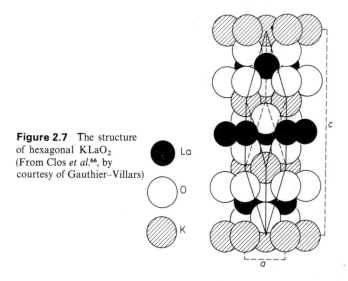

Figure 2.7 The structure
of hexagonal KLaO$_2$
(From Clos et al.[66], by
courtesy of Gauthier-Villars)

La

O

K

in Figure 2.7. The oxygen atoms are arranged in the plane perpendicular to
the c-axis with the sequence ABC–ABC. The potassium and lanthanide
atoms occupy the octahedral holes between the oxygen planes. In the series
of KM^{III}O$_2$ compounds, the c-parameter changes very little compared to the
a-parameter. The lattice constant a is the distance between two lanthanide
atoms in the same plane and changes with the ionic radius of the M^{III}
atom.

KPrO$_2$ forms solid solutions with K$_2$PrO$_3$, K$_{1+x/3}$Pr$^{III}_{1-x}$Pr$^{IV}_{2x/3}$O$_2$ with
$0 \leqslant x \leqslant 0.33$, which may be prepared by reacting K$_2$O with a mixture of
Pr$_2$O$_3$ and Pr$_6$O$_{11}$. In these solid solutions the substitution $3Pr^{3+} = 2Pr^{4+}$
$+ 1K^+$ occurs.

Another interesting compound is LiEu$_3$O$_4$, isostructural with orthorhombic LiSr$_2$EuO$_4$[73]. LiEu$_3$O$_4$ undergoes a topotactical transformation to
Eu$_3$O$_4$ when heated in a high vacuum. There is a very close structural
relationship between LiEu$_3$O$_4$ and Yb$_3$S$_4$; except for lithium, all the atoms
are correlated as for isostructural compounds. As in Eu$_3$O$_4$ it is possible
to distinguish between Eu^{2+} and Eu^{3+} in LiEu$_3$O$_4$.

Only a few magnetic data are known. LiGdO$_2$ and NaGdO$_2$ remain
paramagnetic down to 4 K, the Curie temperature being -5 K (LiGdO$_2$)
and -15 K (NaGdO$_2$) with a magnetic moment of $\mu_{eff} = 7.7$ μ_B (spin-only
value, 7.94 μ_B).

Recently Peppard *et al.* observed a so-called 'tetrad effect'[1], that is, in addition to the gadolinium point, two further discontinuities between the elements with $Z = 60–61(Nd–Pm)$ and $Z = 68–69(Er–Tm)$ or $100–101(Fm–Md)$, respectively, in a detailed study of distribution data and stability constants of the lanthanide and actinide series. To observe a similar effect in solid state chemistry, it is necessary to have very accurate and reliable lattice constants of the series in question, at least up to the third decimal place. These data can usually be obtained without difficulty in cubic, tetragonal, hexagonal and orthorhombic systems by using existing computer programs. A very simple system which would be adequate for the study of the tetrad effect in solids is the $KM^{III}O_2$ series in which all compounds with $M^{III} = $ La–Yb(Lu?) have the same type of structure[66]. However, the lattice constants are not known as exactly as would be necessary (and as is possible to determine). A notable exception from this general trend, namely that structural chemists determine adequate structures but mostly report inadequate lattice constants, is to be seen in Reference 73.

Because of the lack of accurate data for a series of isostructural compounds ranging from La to Lu with a high-symmetry type of structure and not too low a M^{III}:other atoms ratio, an investigation of the solid-state tetrad effect is not at present possible.

2.3 REACTIONS WITH DIVALENT METAL OXIDES

2.3.1 Compounds with heptavalent actinide elements

The reaction of BaO_2 with $NpO_3 \cdot H_2O$ at 400–500 °C leads to the formation of $Ba_3(NpO_5)_2$, the diffraction pattern of which is very similar to that of $Ba_3(TcO_5)_2$[4]. However, it is not known whether the thermally prepared and the precipitated $Ba_3(NpO_5)_2$ compounds are identical.

The cubic ordered perovskite compounds Ba_2LiNpO_6 and Ba_2NaNpO_6[4] are obtained by heating BaO_2 with $Li_2O_2(Na_2O_2)$ and $NpO_3 \cdot H_2O$ at 450 °C. These are isostructural with the corresponding compounds of rhenium and technetium:

M =	Tc^{VII}	Re^{VII}	Np^{VII}
Ba_2LiMO_6	$a = 8.092$ Å	8.118 Å	8.367 Å
Ba_2NaMO_6	$a = 8.292$ Å	8.296 Å	8.590 Å

2.3.2 Compounds with hexavalent actinide elements

In the alkaline earth oxide–uranium trioxide systems, ternary oxides of the formula M_3UO_6 (M = Ba,Sr,Ca), M_2UO_5 (M = Sr,Ca), and MUO_4 (M = Ba,Sr,Ca,Mg), as well as several so-called polyuranates, such as $Ba_2U_3O_{11}$ and MgU_3O_{10}, are known. No compounds with beryllium oxide have been found, but similar ternary oxides with other divalent metal atoms (e.g. Pb, Co, and Cu) have been described. Exact preparative conditions and crystal structures of most of these compounds have been established, but as for most other oxide compounds, thermodynamic data, oxygen pressures, phase widths, etc. are almost completely lacking.

The best-known compounds are the alkaline earth monouranates, MUO_4. Three different types of crystal structure have been found. represented by the compounds $BaUO_4$, $CaUO_4$, and $MgUO_4$. Each of these ternary oxides possesses a uranyl group with two short (covalent) uranium—oxygen bonds with bond lengths of about 1.90 Å. The uranium atoms in all alkaline earth uranates are 6-coordinated except in $CaUO_4$, in which the U-atom coordination is eight. The lattice constants, space groups, and some bond lengths of the compounds which have been structurally investigated are summarised in Table 2.10.

The crystal structure of orthorhombic $Ba(UO_2)O_2$ was determined by Samson and Sillén[74], and a refinement of this structure, as well as of some other types of uranate structures, was reported by Loopstra and Rietvield, using neutron diffraction[75]. In this type of structure there are infinite UO_4 layers, the oxygen atoms forming distorted $(UO_2)O_4$ octahedra. The atoms are in a slightly distorted face-centred-cubic array. The transition from the three-dimensional U—O linkage in UO_2 to the two-dimensional layered ternary uranium(VI) oxide phases is caused primarily by the formation and the manner of the arrangement of the uranyl ions. The same structure is found for β-$SrUO_4$, $PbUO_4$, and $BaNpO_4$. Force constants of the infinite $[(UO_2)O_2]^{2-}$ layers assumed in β-$SrUO_4$ are $K_1 = 1.24$ and $K_2 = 4.79$ mdyn/Å [32].

Barium neptunate(VI), as well as the corresponding strontium and calcium compounds with the $CaUO_4$-type of structure, are best prepared by prolonged heating at 1050 °C of mixtures of BaO and NpO_2 in an oxygen flow with repeated mixing of the material during the reaction[76]. $PbUO_4$ can be prepared by thermal reaction methods and also by heating U_3O_8 with sheet lead and Pb_3O_8 or PbO in water at 280 °C [77]. The oxygen deficient composition $PbUO_{4-}$ is obtained either as a rhombohedral phase—up to a composition of about $PbUO_{3.6}$—or as a *fcc* phase $PbU_x^{IV}U_{1-x}^{VI}O_{4-x}$ ($0 < x < 0.51$), depending on the method of preparation and the reaction temperature[78].

The structure of $CaUO_4$, and of the isostructural α-$SrUO_4$, $SrNpO_4$, $SrPuO_4$, and $CaNpO_4$, is quite different from that of $BaUO_4$. According to Zachariasen[79] the rhombohedral $CaUO_4$ type of structure may be considered as being a slightly deformed fluorite lattice. Each uranium atom is surrounded by eight oxygen atoms, the geometric arrangement being a puckered hexagonal bipyramidal unit, as found in several other complexes of hexavalent uranium (e.g., UO_2F_2). A schematic representation of the hexagonal $(UO_2)O_6$- and the tetragonal $(UO_2)O_4$-layers is shown in Figure 2.8.

The structure of orthorhombic magnesium uranate, $MgUO_4$, contains endless UO_2O_2 chains and—in contrast to the uranates of the heavy alkaline earth metals—not endless layers[80]. The chains are linked together by the magnesium atoms, each magnesium atom being bonded to six

Table 2.10 Structural data for ternary oxides of hexavalent actinides with alkaline earth oxides

Substance	Symmetry	Space group	Lattice parameters					U—O bond lengths [Å]
			a [Å]	b [Å]	c [Å]	α [°]	β [°]	
BaU_2O_7	tetragonal	$I4_1/amd$	7.128		11.95			1.84(2×), 2.12(2×), 2.325(2×)
$BaUO_4$	orthorhombic	$Pbcm$	5.7553	8.1411	8.2335			1.887(2×), 2.214(2×), 2.187(2×)
Ba_3UO_6	(pseudo-) cubic		8.89					
$BaNpO_4$	orthorhombic	$Pbcm$	5.730	8.080	8.167			
Ba_3NpO_6	(pseudo-) cubic		8.860					
Ba_3PuO_6	(pseudo-) cubic		8.844					
Ba_3AmO_6	(pseudo-) cubic		8.81					
α-$SrUO_4$	rhombohedral	$R\bar{3}m$	6.54			35.53		1.91(2×), 2.33(6×)
β-$SrUO_4$	orthorhombic	$Pbcm$	5.4896	7.9770	8.1297			1.886(2×), 2.208(2×), 2.186(2×)
Sr_2UO_5	monoclinic	$P2_1/c$	8.1043	5.6614	11.9185		108.985	for U(1): 2.005(2×), 2.012(2×), 2.217(2×); for U(2): 2.118(2×), 1.986(2×), 2.189(2×)
Sr_3UO_6	monoclinic	$P2_1$	5.9588	6.1795	8.5535		90.192	2.104, 1.975, 2.157, 2.091, 2.077, 2.047 (each 1×)
$SrNpO_4$	rhombohedral	$R\bar{3}m$	6.522			35.66		
$SrPuO_4$	rhombohedral	$R\bar{3}m$	6.51			35.68		
$CaUO_4$	rhombohedral	$R\bar{3}m$	6.2683			36.040		1.963(2×), 2.298(6×)
Ca_2UO_5	monoclinic	$P2_1/c$	7.9137	5.4409	11.4482		108.803	for U(1): 2.022(2×), 2.027(2×), 2.247(2×); for U(2): 2.128(2×), 1.953(2×), 2.206(2×)
Ca_3UO_6	monoclinic	$P2_1$	5.7275	5.9564	8.2982		90.568	2.102, 2.034, 2.023, 2.112, 2.026, 2.181 (each 1×)
$CaNpO_4$	rhombohedral	$R\bar{3}m$	6.245			35.68		
$MgUO_4$	orthorhombic	$Imma$	6.520	6.924	6.595			1.92(2×), 2.16(2×), 2.20(2×)

oxygen atoms belonging to four different chains. The six oxygen atoms about each uranium atom form a distorted UO_2O_4 octahedron. According to Haag and Muncy[81] $MgUO_4$ also may exist as an oxygen deficient compound up to a limiting composition of $MgUO_{3.75}$. The heat of oxidation of fcc $MgUO_{3.5}$ to orthorhombic $MgUO_{3.75}$.:

$$MgUO_{3.5} + 0.125\ O_2 \rightarrow MgUO_{3.75}$$

is 83.3 kcal $mol^{-1}(O_2)$. $MgUO_4$ appears to be soluble in excess U_3O_8 without changing the U_3O_8 unit cell dimensions[82].

There is another proposed structure for $MgUO_4$ (monoclinic, $a = 9.34$ Å, $b = 7.00$ Å, $c = 9.30$ Å, $\beta = 89.47$ degrees, space group $C2/c$) but full details of this are lacking[83]. The structure, however, seems to be very similar to that of the $MgUO_4$ reported by Zachariasen ($a \simeq a\sqrt{2}$,

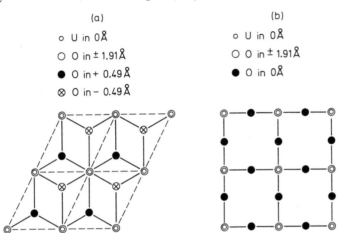

Figure 2.8 Schematic representation of hexagonal (a) and tetragonal (b) uranium—oxygen layers in some alkaline earth uranates (From *Acta Crystallographica*[79], by courtesy of Munksgaard)

$b \simeq c$, $c \simeq b\sqrt{2}$, $\alpha \simeq 90$ degrees) and may be the structure of oxygen deficient $MgUO_{4-x}$, for the single crystal used for the structural studies was obtained from a U_3O_8–MgO melt. A detailed investigation would certainly be of interest.

All monouranates, MUO_4, and diuranates, MU_2O_7 (M = Ba,Sr,Ca, Mg), show a weak paramagnetism which may be attributed to the linear UO_2^{2+} group[84]. The low magnetic moment of Np^{VI} in $BaNpO_4$ and Ba_3NpO_6 is typical of a $5f^1$ system.

The $MgUO_4$ type of structure is also found for some transition metal uranates (e.g. $CoUO_4$, β-$NiUO_4$, $ZnUO_4$ and $MnUO_4$ (Table 2.11)). $CoUO_4$ decomposes at $900\ °C$ in a high vacuum to give $CoUO_{3.5}$ without changing the structure. Treatment at $1000\ °C$ yields $CoUO_3$ (tetravalent uranium) but other information on this compound is lacking[85].

The magnetic structure of $CoUO_4$ at liquid helium temperature requires doubling of the a and c axis[86, 87]. The spin orientation is either along a or c.

Table 2.11   Structural data for ternary oxides of hexavalent actinides with divalent transition metal oxides

Substance	Symmetry	Space group	Lattice parameters					U–O bond lengths [Å]
			a [Å]	b [Å]	c [Å]	α [°]	β [°]	
$PbUO_4$	orthorhombic	Pbcm	5.528	7.952	8.180			
Pb_3UO_6	orthorhombic	Pnam	13.71	12.36	8.21			2.15, 2.19, 2.03, 2.05, 2.04, 1.90 (each 1×)
α-$CdUO_4$	orthorhombic base-centred		7.01	6.836	3.519			
β-$CdUO_4$	rhombohedral (hexagonal)	$R\bar{3}m$	3.587		17.41			
Cd_2UO_5	monoclinic	$P2_1/c$	8.074	5.312	11.52		110.52	1.91(4×), 2.07(2×)
$Cr^{III}U^VO_4$	orthorhombic	Pbcn	4.871	11.787	5.053			2.06(2×), 2.16(2×), 2.22(2×)
$MnUO_4$	orthorhombic	Imma	6.645	6.983	6.749			
$FeUO_4$	orthorhombic	Pbcn	4.888	11.937	5.110			2.05(2×), 2.18(2×), 2.22(2×)
$CoUO_4$	orthorhombic	Imma	6.497	6.952	6.497			1.92(2×), 2.17(4×)
α-$NiUO_4$	orthorhombic	Pbcn	4.820	11.627	5.188			
β-$NiUO_4$	orthorhombic	Imma	6.472	6.870	6.472			
$CuUO_4$	monoclinic	$P2_1/n$	5.475	4.957	6.569		118.87	1.90(2×), 2.15(2×), 2.24(2×)
$ZnUO_4$	orthorhombic	Imma	6.492	6.994	6.574			

The moment of cobalt, determined by susceptibility measurements in the paramagnetic region, is near to $S = 4/2$. The Néel temperature of $CoUO_4$ and $MnUO_4$ is 12 K with paramagnetic Curie temperatures of -23 K and -8 K respectively. The Néel temperature of both compounds are weak because the significant interactions are of the super-exchange type at right angles and of the type super-super-exchange[88].

Two modifications of $CdUO_4$ are known[89]. The orthorhombic α-$CdUO_4$ is stable up to about 720 °C, when it transforms to rhombohedral β-$CdUO_4$, with the $CaUO_4$ type of structure. At about 925 °C β-$CdUO_4$ decomposes to β-$CdUO_{4-x}$, which has a defect structure of the $CaUO_4$ type. The loss of oxygen is accompanied by a change of colour from yellowish-orange to black. The fcc $(Cd,U)O_{2-x}$ phase is formed at 1000 °C.

$HgUO_4$ and $CuUO_4$ have been prepared by heating the appropriate nitrates in an oxygen flow[90]. Copper monouranate crystallises in the monoclinic system (space group $P2_1/n$)[91]. In this type of structure, each uranyl group is surrounded by four secondary oxygen atoms in a planar array. The oxygen array surrounding the copper atoms is formed of two uranyl and two secondary oxygen atoms in a square planar configuration, with Cu—O bond lengths of 1.95 Å and 1.96 Å. Two very long copper—oxygen bonds, of lengths 2.59 Å each, are observed above and below the copper-oxygen plane.

$CuUO_4$ is not stable above 910 °C, decomposing as follows:

$$6CuUO_4 \rightarrow 3Cu_2UO_4 + U_3O_8 + O_2$$

The structure of Cu_2UO_4 is not known[92]. Cu_2UO_4 is also obtained by thermal decomposition of copper triuranate:

$$6CuU_3O_{10} \rightarrow 3Cu_2UO_4 + 5U_3O_8 + 4O_2$$

α-$NiUO_4$, $FeUO_4$ and $CrUO_4$ are orthorhombic, space group $Pbcn$, and are believed[93, 94] to be isostructural with $BiVO_4$.

Iron monouranate is obtained by heating mixtures of the calculated amounts of iron and uranium oxides containing the required oxygen composition in evacuated silica tubes for 2 weeks at 1050 °C:

$$FeO + UO_3 \rightarrow FeUO_4 \; or$$
$$2Fe_2O_3 + U_3O_8 + UO_2 \rightarrow 4FeUO_4$$

No compound formation was observed when mixtures of iron and uranium oxides or ferric and uranyl nitrates were heated in air to temperatures as high as 1200 °C. Magnetic measurements of $FeUO_4$ show two discontinuities near 54 and 42 K due, respectively, to the two transitions:

$$\text{paramagnetic} \rightarrow \text{antiferromagnetic} \rightarrow \text{ferromagnetic state.}$$

The ferromagnetic moment is $\mu_{eff} = 3.08 \; \mu_B$ at 4.2 K[95]. $FeUO_4$ is one of the rare examples of an oxide which is a ferromagnetic insulator.

$CrUO_4$ contains pentavalent uranium[96, 97], the alternative—the presence of equal amounts of U^{IV} and U^{VI}—being incompatible with the infrared data[93]. $CrUO_4$ volatilises at elevated temperatures, probably dissociating into UO_3 and CrO_3. The activation energy for this process is 97 kcal mol^{-1}.

$CrUO_4$ shows an antiferromagnetic arrangement at 4.2 K, as shown by neutron diffraction.

The thermal stability of the transition metal uranates appears to vary directly with the characteristic transition element oxidation state; $Cr^{III} > Mn$, $Co^{III,II} > Ni$, $Zn^{II} > Cu^{II,I}$. The iron compound, which decomposes at about 820 °C, constitutes a definite exception to this pattern. The Fe, Ni, and Zn uranates tend to decompose directly into their constituent oxides, while the Mn, Co, and Cu compounds decompose to other double oxides.

The yellow dialkaline earth uranates, Sr_2UO_5, and Ca_2UO_5, are prepared by the reactions:

$$CaUO_4 + Ca_3UO_6 \xrightarrow[\text{air}]{975\,°C} 2Ca_2UO_5$$

$$SrUO_4 + Sr_3UO_6 \xrightarrow[\text{air}]{1000\,°C} 2Sr_2UO_5$$

or by calcination of a $CaO(SrO)–UO_3$ mixture at 900 °C [98,99].

The first structural investigation of the monoclinic compounds by Sawyer[100] was extended by Loopstra and Rietvield[75]. Each uranium atom is surrounded by six oxygen atoms forming a distorted octahedron. The U—O bond lengths, 2.005 Å up to 2.217 Å (Sr_2UO_5), indicate the absence of a distinct uranyl group. Cd_2UO_5 has the same structure[101], which is shown in Figure 2.9.

Three polyalkaline earth uranates, Ba_3UO_6, Sr_3UO_6, and Ca_3UO_6, are known. According to the formula $M(M^{II}_{0.5}, U^{VI}_{0.5})O_3$, these compounds can be regarded as deformed substituted perovskite structures. Ba_3UO_6 has been found to possess a slightly deformed cubic structure of this substituted perovskite type[102], whilst the strontium and calcium compounds are monoclinic[103] in contrast to the orthorhombic structure, suggested earlier[99, 102].

In Sr_3UO_6 and Ca_3UO_6 each uranium atom is surrounded by six oxygen atoms, but the bond lengths are not equal, ranging from 1.85 Å (U—O^{II}) to 2.42 Å (U—O^V). There are also two different alkaline earth metal atom sites. The Ca(Sr) atoms at the cube corners are coordinated by twelve oxygen atoms, whilst the atoms at the centre of the perovskite lattice—alternately occupied by uranium and alkaline earth metal atoms—are surrounded by six oxygen atoms.

The real structure of $Ca_3UO_6(Sr_3UO_6)$ is markedly different from an idealised substituted perovskite structure, but the essential features of the idealised structure are retained. In particular, the U—O—Ca(Sr) bonds are no longer collinear, but make an average angle of 140 degrees. This is caused by the covalent character of these bonds which, by using the noncollinear $2p$ orbitals of the oxygen atoms, are prevented from taking up an angle of 180 degrees to each other.

Similar compounds for the transuranium elements Np, Pu, and Am are also known. The x-ray powder patterns of the barium compounds have been indexed on the basis of a cubic structure[76, 104], but a slight deformation of the cubic lattice cannot be excluded. The best way to prepare these transuranium compounds is to heat the appropriate mixtures of alkaline earth oxides and actinide dioxides at 1000–1200 °C in an oxygen flow. The alkaline earth transuranates(VI) are also much more stable to radiolytic decomposition than the corresponding aqueous solutions.

The structure of the dark red Pb_3UO_6 differs radically from those of other known M_3UO_6 uranates[105]. The uranium atoms of the orthorhombic unit cell are octahedrally coordinated to six oxygen atoms, and the slightly

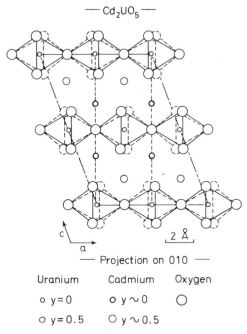

— Cd_2UO_5 —

— Projection on 010 —

Uranium	Cadmium	Oxygen
o y = 0	o y ~ 0	O
o y = 0.5	O y ~ 0.5	

Figure 2.9 The structural arrangement of Cd_2UO_5
(From Sterns and Sawyer[101], by courtesy of Pergamon Press)

distorted octahedra, by sharing apices, form infinite puckered UO_5-chains in the z direction, the lead atoms and the remaining oxygen atoms being distributed in a rather irregular array between the chains.

In the polyalkaline earth compounds one or two of the alkaline earth metal atoms may be substituted by other divalent metal atoms (e.g., Mn, Fe,Mg,Ni,Co,Hg,Pb,Cd, and Cu[102, 106]). Using a tolerance factor t

$$t = (r_A + r_O)/\sqrt{2}(r_{B^{2+}} + + r'_{M^{6+}}).$$

Sleight and Ward calculated for the $A(B_{0.5},U_{0.5})O_3$ type perovskites that a number of compounds containing barium in the A position should form the cubic ordered perovskite structure, and observed that they did form this structure. All compounds containing Sr in the A position have distorted structures. The unit cells are pseudomonoclinic, but the powder pattern may be indexed on a smaller orthorhombic unit cell. Compounds like Ba_2SrUO_6 ($a = 8.84$ Å), Ba_2MnUO_6 ($a = 8.52$ Å), Ba_2HgUO_6 ($a = 8.83$ Å), and Ba_2ZnUO_6 ($a = 8.397$ Å) possess this cubic or slightly distorted cubic structure, whilst for other polynary oxides, such as Sr_2GdUO_6 ($a = 6.03$ Å, $b = 8.42$ Å, $c = 5.91$ Å) or $BaSr_2UO_6$ ($a =$

6.26 Å, $b = 8.76$ Å, $c = 6.13$ Å), an orthorhombic deformed lattice is suggested. Ba_2CuUO_6 is one of the very rare perovskite oxides containing copper. The tetragonally distorted lattice ($a = 8.18$ Å, $c = 8.84$ Å) is to be expected from the Jahn–Teller effect if the copper is in the octahedral sites. It has also been reported that the oxygen deficient phase $Ba_3UO_{5.2}$ has the same structure as Ba_3UO_6, but with a larger lattice parameter[83].

Corresponding mixed oxides are also known for neptunium and plutonium, (e.g. Ba_2SrNpO_6 ($a = 8.799$ Å), $BaSr_2NpO_6$ ($a = 8.735$ Å), and Ba_2PbPuO_6 ($a = 8.58$ Å) [76, 104, 107].

Another type of structure, which may be related to the perovskite one, is given by the formula $M_3^{II}UM^{III}O_9 = M^{II}(M_{2/3}^{III}, U_{1/3}^{VI})O_3$ [102, 108, 109]. Compounds such as $Ba_3Y_2UO_9$ ($a = 8.70$ Å), $Ba_3Fe_2UO_9$ ($a = 8.232$ Å), $Pb_3Fe_2UO_9$ ($a = 8.140$ Å) and $Sr_3Cr_2UO_9$ ($a = 8.00$ Å) all have a cubic structure. The distribution of the iron and uranium atoms in the mixed oxide phase $Sr_3UFe_2O_{9-x}$ ($a = 8.066$ Å, space group $Fm3m$) is given best by the formula $Sr_8[(Fe_4)(Fe_{4/3},U_{8/3})]O_{24-y}$ ($0 \leqslant y \leqslant 0.3$). This compound shows ferri- and ferromagnetism with a Curie point of about 38 °C [108]. The substitution of uranium by tungsten in $Sr_3UFe_2O_9$, which leads to $Sr_3(U_x,W_{1-x}Fe_2)O_9$, has a remarkable effect; for $0.54 \leqslant x \leqslant 1$ there is an ordering of the metal atoms leading to a lattice constant of, for example, 8.016 Å for $x = 0.7$. For $0 \leqslant x \leqslant 0.54$ the ordering seems to disappear and the lattice constants of the normal perovskite lattice are observed, thus $a = 3.980$ Å for $x = 0.40$ [109].

There is much less information concerning the polyuranates of alkaline earth metals, the general formula of which is $Me_xU_yO_{3y+x}$. Most of these compounds are characterised only by their chemical formula. The single phases BaU_2O_7 and $Ba_2U_3O_{11}$ are prepared by reaction of $BaUO_4$ with U_3O_8 at 900 °C in air[110]:

$$6BaUO_4 + U_3O_8 \xrightarrow{O_2} 3Ba_2U_3O_{11}$$

$$3BaUO_4 + U_3O_8 \xrightarrow{O_2} 3BaU_2O_7$$

Both compounds are also formed by the thermal decomposition of $Ba(UO_2)_2(CH_3COO)_6 \cdot 2H_2O$:

$$2Ba(UO_2)_2(CH_3COO)_6 \cdot 2H_2O \xrightarrow[\text{fast}]{500-800\,°C} Ba_2U_3O_{11} + UO_3$$

$$800-900\,°C,\ \text{slow} \quad | \quad 900-1000\,°C,\ \text{fast}$$

$$2BaU_2O_7 \xrightleftharpoons[\text{slow}]{900\,°C} Ba_2U_3O_{11} + \tfrac{1}{3}U_3O_8$$

$Ba_2U_3O_{11}$ is stable to at least 1500 °C.

The uranium atoms in the tetragonal BaU_2O_7 ($a = 7.128$ Å, $c = 11.95$ Å, space group $I4_1/amd$) are in octahedral coordination, and a linear uranyl group with a U—O distance of 1.84 Å is present[111]. The octahedra are joined by sharing edges, and form endless chains in the a and b directions (Figure 2.10). The structure can be derived from the cubic close-packed BaO, with half of the Ba atoms replaced by O^{III} atoms, and the lattice expanded, particularly in the a and b directions, because of the presence of U atoms. This chain structure differs considerably from the monouranate structures. The strong absorption in the i.r.-spectrum of BaU_2O_7 at 765 cm^{-1} is the antisymmetric stretching frequency v_3 of the uranyl group; this leads to a value of 4.8 mdyn/Å for the force constant of the uranyl bond. Another polyuranate phase, $BaO \cdot (2.2-2.5)UO_3$, is mentioned by Ippolitova et al.[112], but no details are given.

The Sr- and Ca-polyuranate phases, CaU_2O_7, $Me_2U_3O_{11}$, and MeU_4O_{13} have been reported by Cordfunke and Loopstra[113], but Brisi et al.,[114] in an extensive study of the calcium uranate system, only found evidence for

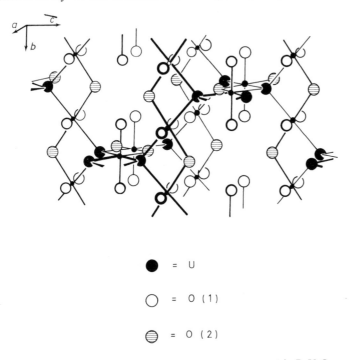

$$\bullet \ = \ U$$

$$\bigcirc \ = \ O \ (1)$$

$$\ominus \ = \ O \ (2)$$

Figure 2.10 The uranium—oxygen arrangement in BaU_2O_7
(From Allpress[111], by courtesy of Pergamon Press)

CaU_2O_7, $Ca_2U_3O_{11}$ and $CaU_5O_{15.4}$. They also published a phase diagram which represents the phase equilibria in the $CaO-UO_2-O_2$-system in air.

The structures of these compounds are not known. It has been mentioned that the subcells of CaU_4O_{13} and SrU_4O_{13} are related to the subcell of U_3O_8, and a b-face centred subcell with monoclinic or orthorhombic symmetry is therefore suggested[113]:

SrU_4O_{13}: $a = 6.734$ Å $\qquad\qquad$ CaU_4O_{13}: $a = 6.656$ Å
$\phantom{SrU_4O_{13}:}$ $b = 4.193$ Å $\qquad\qquad$ $\phantom{CaU_4O_{13}:}$ $b = 4.161$ Å
$\phantom{SrU_4O_{13}:}$ $c = 4.065$ Å $\qquad\qquad$ $\phantom{CaU_4O_{13}:}$ $c = 4.030$ Å.
$\phantom{SrU_4O_{13}:}$ $\beta = 90.16°$

Similarly, the structures of the isomorphous compounds $Ca_2U_3O_{11}$ and $Sr_2U_3O_{11}$ are related to rhombohedral $CaUO_4$; the following lattice constants for the triclinic pseudo-rhombohedral subcell have been calculated:

$Sr_2U_3O_{11}$: $a = c =$ 6.484 Å
$b =$ 6.523 Å
$\alpha = \gamma =$ 37.44°
$\beta =$ 36.10°

$Ca_2U_3O_{11}$: $a = c =$ 6.186 Å
$b =$ 6.212 Å
$\alpha = \gamma =$ 37.12°
$\beta =$ 37.56°

The various polyuranates show a marked difference in thermal stability. $Sr_2U_3O_{11}$ appears to be stable to at least 1300 °C, whilst the calcium compound decomposes in air at about 900 °C; CaU_4O_{13} decomposes to CaU_2O_7 above 1060 °C, which in turn decomposes to the *fcc* CaU_2O_6-fluorite phase at only slightly higher temperatures. SrU_4O_{11} is stable to about 1130 °C. The following thermal decomposition steps have been observed:

$$6SrU_4O_{13} \xrightarrow[\text{air}]{1130\,°C} 3Sr_2U_3O_{11} + 5U_3O_8 + 2\tfrac{1}{2}\,O_2,$$

$$3Ca_2U_3O_{11} \xrightarrow{900\,°C} 3CaU_2O_7 + 3CaUO_4$$

$$\Big\downarrow \; 1075\,°C$$

$$3CaU_2O_6 + 1\,1/2\,O_2$$

$$\Big\uparrow \; 1075\,°C$$

$$3CaU_4O_{13} \xrightarrow{1060\,°C} 3CaU_2O_7 + 2U_3O_8 + O_2$$

When mixtures of U_3O_8 and MgO are heated, the first compound to be observed, at about 600°C, is the yellow-brown MgU_3O_{10}[115, 116], which seems to have a structure analogous to α-UO_3, but details of this structure are lacking. It decomposes at 1250 °C to $MgUO_4$ and U_3O_8.

Similar isostructural compounds are also known for some transition elements[117-119]. The following hexagonal lattice constants have been reported on the basis of powder diffraction data and a close relationship to the α-UO_3 structure:

MgU_3O_{10} $a =$ 7.57 Å, $c =$ 16.32 Å [120]
NiU_3O_{10} $a =$ 3.78 Å, $c =$ 4.04 Å [117]
MnU_3O_{10} $a =$ 3.79 Å, $c =$ 4.14 Å [117]
CoU_3O_{10} $a =$ 3.79 Å, $c =$ 4.08 Å [117]
CuU_3O_{10} $a =$ 3.77 Å, $c =$ 4.17 Å [117]
ZnU_3O_{10} $a =$ 7.56 Å, $c =$ 16.38 Å [119].

The lattice constants of the transition metal triuranates are probably those of a subcell. Preliminary single crystal data reported by Hoekstra and Marshall[93] indicate, however, that the true structure of MU_3O_{10} is monoclinic (e.g. $a =$ 7.575 Å, $b =$ 6.473 Å, $c =$ 16.679 Å, $\beta =$ 91 degrees for CuU_3O_{10}). These data also indicate that the CuU_3O_{10} lattice contains oxygen vacancies in the sheets, rather than the chains as reported by Ippolitova *et al.*[186].

According to the i.r.-spectrum of MgU_3O_{10}, the presence of absorption bands at 733 cm^{-1} and 842 cm^{-1} indicates a uranyl group[122]. Compounds of composition $NiO \cdot 3UO_3$ and $CuO \cdot 3UO_3$ are also obtained in the aqueous systems UO_3–$NiO(CuO)$–SO_3–H_2O at high pressures[123, 124]. It is not known if these compounds are isostructural with those prepared by thermal reactions.

There has been only one thermodynamic investigation of alkaline earth uranates. Cordfunke and Loopstra[113] determined the heat of solution and the heat of formation of some Sr-uranates (± 1 kcal mol^{-1}):

	Heat of formation $-\Delta H_{298 \text{ K}}$ kcal mol^{-1}
$Sr_2U_3O_{11}$	1242
α-$SrUO_4$	469.6
β-$SrUO_4$	469.9
Sr_2UO_5	617.2
Sr_3UO_6	760.2

2.3.3 Compounds with pentavalent actinide elements

Only very few stoicheiometric or nearly stoicheiometric ternary oxides with divalent metals are known. Compounds like $Mg(UO_3)_2$, $Ca(UO_3)_2$ and $Sr(UO_3)_2$ are members of the $Me(II)$–UO_{2+x}-fcc series. It has been deduced from magnetic measurements that $Ca_2U_2O_7$ and the 'fluorite compound' $Ca(UO_3)_2$ possess a uranyl group, UO_2^+ [125], but new and more reliable (e.g. spectroscopic) measurements are necessary to prove this. The only possible arrangement of anions on the equatorial plane of the uranyl group should lower the symmetry very much, so that an undisturbed fcc arrangement would, therefore, be quite unusual and is not to be expected. The red-brown compound $Ba_2U_2O_7$ probably crystallises in a monoclinic deformed structure of the pyrochlore type with lattice constants of $a = b = 11.56$ Å, $c = 11.31$ Å, $\beta \approx 90$ degrees[126]. Spectroscopic and magnetic measurements indicate an octahedral environment of the U^{5+} ion.

The brown ternary oxides CoU_2O_6 and NiU_2O_6 are prepared by heating UO_2 with $MeUO_4$ (or $MeO + UO_2 + UO_3$) in a sealed system at 800 °C [121, 127]. They crystallise in a hexagonal type of structure (space group $P321$) with the lattice constants $a = 9.095$ Å, $c = 4.990$ Å (CoU_2O_6) and $a = 9.015$ Å, $c = 5.013$ Å (NiU_2O_6), respectively. In the system PbO–U_2O_5 a phase of variable composition, ranging from $Pb_{1.5}U_2O_{6.5}$ up to $Pb_{2.5}U_2O_{7.5}$ is found[128]. With increasing PbO content, a steady conversion from a cubic ($Pb_{1.5}U_2O_{6.5}$: $a = 11.16$ Å) to a distorted rhombohedral structure ($Pb_{2.5}U_2O_{7.5}$: $a = 11.23$ Å, $\alpha = 90.25$ degrees) occurs. This phase may be deduced from the intermediate pyrochlore type $Pb_2U_2O_7$ by the addition or subtraction of PbO up to 0.5 mol. $Pb_2U_2O_7$ forms a solid solution with $NaUO_3F$ up to a 1 : 1 composition[129].

The interesting class of III–V-ordered perovskites, $Ba_2Me^VMe^{III}O_6$, is obtained by replacing the +4 cations in $Me^{II}Me^{IV}O_3$-perovskite-compounds with +3 and +5 cations. As would be expected from calculations of the Madelung energy of the lattice[130], all these compounds possess an ordering of the metal cations. The barium compounds are cubic or hexagonal, but most strontium compounds are pseudocubic, the x-ray powder patterns being indexable as orthorhombic. Typical compounds are Ba_2YUO_6 ($a = 8.69$ Å), Ba_2ScUO_6 ($a = 8.49$ Å), Ba_2RhUO_6 ($a = 5.84$ Å, $c = 14.9$ Å), Ba_2LaPaO_6 ($a = 8.885$ Å), Ba_2AmPaO_6 ($a = 8.793$ Å), and Ba_2NdPuO_6 ($a = 8.50$ Å). Ba_2CrUO_6 can be obtained either in the hexagonal barium titanate structure ($a = 5.83$ Å and $c = 14.4$ Å) or in the cubic form ($a = 8.297$ Å). The mode of preparation of the two modifications suggests that the achievement of the particular crystal form depends to some extent on the kinetics of the reaction. The compound $Ba_3PaO_{5.5}$ ($= Ba(Ba_{0.5}^{II}, Pa_{0.5}^{V})O_{2.75}$) also crystallises in the ordered perovskite structure ($a = 8.932$ Å), but with vacancies in the anion sublattice. Similar ternary oxides should also be obtained for Np^V, but the very low thermal stability of Am^V makes it unlikely that the analogous americium compounds can be prepared.

There exist several combinations where different oxidation states of the B-cations are possible, for example, $Ba_2Mn^{II}U^{VI}O_6$ or $Ba_2Mn^{III}U^VO_6$ and $Ba_2Ce^{IV}Pa^{IV}O_6$ or $Ba_2Ce^{III}Pa^VO_6$. The very small lattice constant of Ba_2CePaO_6, as compared with those of the neighbouring polynary La- and Nd-oxides, suggests that Ba_2CePaO_6 is a IV–IV-compound that is, a 1:1 mixed crystal of $BaCeO_3$ and $BaPaO_3$, but with ordering of the Me^{IV} cations[131]. According to the results of spectroscopic and magnetic measurements, the charge distribution in the Ba_2MnUO_6 perovskite is $Ba_2(Mn_{0.4}^{III},Mn_{0.6}^{II})(U_{0.6}^V,U_{0.4}^{VI})O_6$, whilst Ba_2FeUO_6 is a pure III–V compound and $Ba_2CoUO_6 + Ba_2NiUO_6$ are pure II–VI-compounds. As expected, Ba_2FeUO_6 and $Ba_2Fe_{1-x}Me_xUO_6$ (Me = Ga,In,Mg,Zn) are ferrimagnetic at low temperatures, for example, below 160 K for the pure iron compound[126, 132].

Another perovskite structure with cation ordering is found for the type $Ba_3M^{II}U_2O_9$ (M = Mg ($a = 8.526$ Å), Fe ($a = 8.485$ Å), Zn ($a = 8.542$ Å), Mn, Co, Ni); the ordering corresponds to the formula $Ba_2(M_{0.67}^{II},U_{0.33})UO_9$. $Ba_3FeU_2O_9$ is ferrimagnetic below 123 K[133].

Kemmler-Sack[49] has reported a detailed study of the spectroscopic behaviour of the U^V-ion in ternary oxides by measuring the reflectance spectra from 4000–40 000 cm^{-1}. It was found that the optical properties are strongly influenced by the symmetry of the ligand field. For compounds such as $KSrU_2O_6F$ and $RbBaU_2O_6F$ an octahedral environment was observed for the U^V ion, for which a $5f^1$ configuration is deduced.

2.3.4 Compounds with tetravalent actinide and lanthanide elements

The best known ternary oxides of tetravalent actinides and lanthanides are the perovskite compounds $M^{II}M^{IV}O_3$. Most of these compounds possess cubic structures with lattice constants between 4 Å and 5 Å, examples being $BaPaO_3$ ($a = 4.45$ Å), $BaNpO_3$ ($a = 4.357$ Å), $BaAmO_3$ ($a = 4.35$ Å) and $BaTbO_3$ ($a = 4.285$ Å)[61]. The symmetry of $BaThO_3$ is not known exactly, but the powder pattern has been indexed on the basis of a larger

($\sim 2a$) unit cell with $a = 8.985$ Å (space group $P2_13$ or $P4_232$)[134].

The most extensively investigated compound is $BaUO_3$ ($a = 4.411$ Å), which may take into solid solution up to 2 moles of BaO. On storing in air the compound slowly oxidises up to a limiting composition $BaUO_{3.3}$ and the lattice constant decreases to 4.392 Å. From density measurements it has been shown that the superstoicheiometric oxygen is interstitial. Fujino and Naito investigated the solubility of KUO_3 in $BaUO_3$, both compounds being pure perovskites[135]. The reaction begins at 550 °C, and the solubility of KUO_3 exceeds 30 mol% KUO_3 in $BaUO_3$ at 750 °C. The electrical conductivities of the samples varied with composition and showed a distinct maximum at 34 mol% KUO_3.

Compounds with the ordered perovskite structure, such as Ba_2UZrO_6 ($a = 8.35$ Å), Ba_2UGeO_6 ($a = 8.56$ Å), Ba_2PuTiO_6 ($a = 8.06$ Å), and Ba_2CePuO_6 ($a = 8.72$ Å)[106, 107], may be considered as a 1:1 mixed crystal of two simple perovskites, for example $BaUO_3 + BaZrO_3$, the M^{IV} atoms being in an ordered arrangement. A more detailed investigation would be of particular interest, enabling the degree of solid solution in the normal and in the ordered perovskite state to be ascertained. In addition, there is the question of whether there is a steady transition from the $BaMO_3$ to the $Ba_2MM'O_6$ structure or whether a two-phase region exists. These ternary oxides may also be used to solve the problem of whether the ordered perovskite compounds take up excess oxygen like $BaUO_3$.

The Sr-compounds $SrMO_3$ (M = Pa,U,Np,Pu) are not cubic and the x-ray powder patterns are normally indexed as pseudocubic. The powder pattern of $SrPuO_3$ has been indexed on the basis of a monoclinic unit cell with $a = 5.980$ Å, $b = 4.276$ Å, and $c = 6.114$ Å[136]. The magnetism of $BaTbO_3$ and pseudocubic $SrTbO_3$ follow the Curie-Weiss relationship with $\Delta = -36$ K and $\mu_{eff} = 7.75 \mu_B$[61].

Perovskite compounds of M^{IV} are only known with Ba^{II} and Sr^{II}, but not with the light alkaline earth elements calcium, magnesium, and beryllium. These systems are simple eutectic ones. The $BeO-ThO_2$ system, for example, has a eutectic at 70 mol% BeO ($= 18.1$ wt %) and 2155 °C. The solid solubility of ThO_2 in BeO below the eutectic temperature is less than 0.016 mol% ThO_2 and that of BeO in ThO_2 less than 0.1 mol% BeO. ThO_2 has no effect on the $\alpha \rightleftharpoons \beta$ BeO phase inversion[137].

2.3.5 Compounds with trivalent actinide and lanthanide elements

Very few data on the reaction of trivalent actinide oxides with the divalent metal oxides are available. The reaction of Pu_2O_3 and Am_2O_3 with BaO at about 1250 °C in an evacuated quartz ampoule yields the compounds $BaO·Pu_2O_3$ and $BaO·Am_2O_3$ which are isomorphous with the corresponding ternary lanthanum oxides.

Many more data are available for the rare earth(III) oxide-alkaline earth oxide systems[138-150]. Several types of ternary oxides are known, represented by the general formulae: $MO·RE_2O_3$ ($= M(REO_2)_2$), $MO·2RE_2O_3$ ($= MRE_4O_7$), $2MO·RE_2O_3$ ($= M_2RE_2O_5$), $3MO·RE_2O_3$ ($= M_3(REO_3)_2$), $3MO·2RE_2O_3$ ($= M_3RE_4O_9$), $5MO·4RE_2O_3$ ($= M_5RE_8O_{17}$), but, with the exception of some $MO \cdot RE_2O_3$ compounds which are isomorphous with the orthorhombic $CaFe_2O_4$, no structural data are known.

The solubility of CaO, SrO, and BaO in La_2O_3 increases with temperature[151] and attains values of 16, 18, and 14 mol% for CaO, SrO, and BaO, respectively, for molten and rapidly quenched specimens. According to an observation by Queyroux[139], a monoclinic, type B, solid solution is found in some $SrO-Re_2O_3$ systems, even when the pure rare earth oxide possesses the hexagonal La_2O_3 structure. Near the melting point up to 20, 22, and 30 mol% of SrO dissolve in the monoclinic oxides Dy_2O_3, Gd_2O_3, and Sm_2O_3, respectively. SrO does not dissolve in cubic rare earth oxides, nor does CaO in Yb_2O_3 (Gd_2O_3,La_2O_3) below 1850 °C, but in the $CaO-Yb_2O_3$ system at temperatures above 1950 °C, Yb_2O_3 dissolves considerable amounts of CaO with the formation of a monoclinic Yb_2O_3 solid solution. At about 1850 °C there is no solubility, in accordance with an observation by Barry[152], who did not mention the monoclinic Yb_2O_3–CaO solid solution in his extensive study of the Yb_2O_3-CaO system. There are many open and contradicting points in the description of solid solutions of MO in RE_2O_3, so that a new and very exact investigation of this problem would be of interest.

In contrast, the solubility of La_2O_3 in alkaline earth oxides is very small, ranging from 1 mol% for BaO to 2.4 mol% for SrO. No solubility is found for CaO in La_2O_3 at 1000 °C, but at 1850 °C a small solubility of 0.4 mol% is observed. The solubility, however, seems to increase with decreasing ionic radius of the rare earth ions, being slightly less than 6 mol% for Gd_2O_3 in CaO and about 8 mol% for Yb_2O_3 in CaO. The defect type produced when trivalent rare earth ions are introduced into the NaCl structure has been shown from precise density measurements to consist of cation vacancies.

The decomposition of these solid solutions, that is the precipitation of, for example, Yb_2O_3 from CaO crystalline solutions, has been shown to be diffusion-controlled, at least in the initial stages[152].

The most thoroughly investigated class of compounds in these systems has the composition $MO \cdot RE_2O_3$. These compounds are usually obtained by heating mixtures of the appropriate binary oxides at high temperatures. The calcium compounds are only stable at high temperatures (e.g. $CaYb_2O_4$ at >1850 °C and $CaDy_2O_4$ at >1700 °C) and undergo subsolidus dissociation to CaO and M_2O_3 on cooling. Rapid quenching is therefore necessary to get these compounds metastable at room temperature. The decomposition of CaY_2O_4 was shown to proceed according to 'zero order' kinetics[153]. No phase width has been reported for the compounds in the various $MO-Re_2O_3$ systems.

The ternary oxides $BaO \cdot RE_2O_3$ (La–Ho) and $SrO \cdot Re_2O_3$ (Nd–Lu) crystallise in the orthorhombic $CaFe_2O_4$ type structure (space group $Pnam$, $Z = 4$), as shown by x-ray powder photography and subsequently confirmed by single crystal investigations of $CaSc_2O_4$, $SrTb_2O_4$, and Eu_3O_4 ($= Eu^{II}Eu_2^{III}O_4$)[142, 147, 149]. In the $MO \cdot RE_2O_3$ type of structure each RE(III) atom is surrounded by a distorted octahedron of oxygen atoms with bond lengths of 2.27 Å ($2\times$), 2.34 Å ($2\times$), 2.40 Å ($1\times$), and 2.53 Å ($1\times$) for Tb_1 and 2.19 Å ($1\times$), 2.23 Å ($1\times$), 2.33 Å ($2\times$) and 2.54 Å

$(2 \times)$ for Tb_{II} in $SrO \cdot Tb_2O_3$. The lattice constants of some of these compounds are:

$BaO \cdot La_2O_3$	$a = 10.675$ Å	$b = 12.645$ Å	$c = 3.702$ Å
$BaO \cdot Dy_2O_3$	$a = 10.415$ Å	$b = 12.147$ Å	$c = 3.476$ Å
$SrO \cdot Eu_2O_3$	$a = 10.133$ Å	$b = 12.081$ Å	$c = 3.498$ Å
$SrO \cdot Tb_2O_3$	$a = 10.10$ Å	$b = 11.98$ Å	$c = 3.45$ Å.

In Eu_3O_4 the trivalent europium atom may be replaced by other trivalent rare earth elements, leading to compounds such as $EuGd_2O_4$ or $EuDy_2O_4$, which also crystallise in the $CaFe_2O_4$ type of structure and which have interesting co-operative magnetic transitions[154]. Other ternary oxides of EuO with rare earth sesquioxides include $EuO \cdot 2RE_2O_3$ and $3EuO \cdot 2RE_2O_3$ (RE = Nd,Gd) [155].

The replacement in MRE_2O_4 of one RE^{3+}-ion by a smaller trivalent ion, (e.g. Fe^{3+}, Cr^{3+}, Rh^{3+} or Ga^{3+}) leads to the formation of compounds with the tetragonal K_2NiF_4 type of structure (space group $I4/mmm$) [156, 157], examples being $BaLaFeO_4$ ($a = 3.885$ Å, $c = 12.73$ Å) and $SrLaRhO_4$ ($a = 3.92$ Å, c = 12.78 Å). The compound $La_2Li_{0.5}Co_{0.5}O_4$ ($a = 3.77$ Å, $c = 12.58$ Å, $c/a = 3.34$), obtained by replacing Ba^{2+} by $1/2$ Li^+ and $1/2$ Co^{3+}, is of special interest, for the energy difference between the high-spin and low-spin states of Co^{3+} is very small[158]. In the analogous polynary oxide $La_2Li_{0.5}Ni_{0.5}O_4$ ($a = 3.75$ Å, $c = 12.89$, $c/a = 3.44$) the Ni^{3+} ion is in the low-spin state as shown by magnetic measurements. The large value of c/a is ascribed to the Jahn-Teller effect acting on the Ni^{3+} ion.

Ignition of basic cupric carbonate with rare earth oxides in air at 1000 °C leads to the formation of $CuO \cdot RE_2O_3$-compounds (RE = La to Gd, except Ce) [159]. Except for $CuLa_2O_4$, these compounds are isostructural with $NiLa_2O_4$ [160] and have the K_2NiF_4 structure (e.g. $CuNd_2O_4$: $a = 3.94$ Å, $c = 12.15$ Å). $CuLa_2O_4$ decomposes above 1200 °C to Cu_2O and La_2O_3. Whereas the congruently melting $CaSc_2O_4$ (m.p. 2110 °C) [161] also has the orthorhombic $CaFe_2O_4$ structure ($a = 9.65$ Å, $b = 11.20$, $c = 3.16$ Å), a monoclinic distortion of this structure is reported for $CaYb_2O_4$ ($a = 9.729$ Å, $b = 3.322$ Å, $c = 11.567$ Å, $\beta = 89.8$ °). The number of compounds in the $CaO–RE_2O_3$ systems increases with decreasing ionic radius of RE^{III}. While the $CaO–La_2O_3$ system is a simple eutectic one, one compound ($CaO \cdot 2Gd_2O_3$) has been detected in the $CaO–Gd_2O_3$ system. In the $Yb_2O_3–CaO$ system, however, four crystalline phases are observed $Y_2O_3 \cdot nCaO$ with $n = 1/2$, 1, 2, and 3 [152].

In contrast to the behaviour with tetra-, penta- and hexavalent actinide oxides, beryllium oxide reacts with rare earth oxides to give ternary oxides of different compositions[148, 162]. The crystal structures of two of these compounds have been solved. In monoclinic $La_2Be_2O_5$ ($a = 7.536$ Å, $b = 7.348$ Å, $c = 7.439$ Å, $\beta = 91.55$ degrees, space group $C2/c$ or Cc) BeO_4 tetrahedra (bond distances 1.602–1.678 Å) form a unique three dimensional framework of corner-sharing tetrahedra with cage-like sites for the large lanthanum cations. The lanthanum—oxygen co-ordination polyhedron is rather irregular, with five close La—O contacts (2.415–2.556 Å), two longer contacts (2.719 Å and 2.755 Å) and three still longer contacts (2.903–2.999 Å) in which interactions must be relatively weak.

In the metastable compound Y_2BeO_4 (orthorhombic, $a = 3.5315$ Å, $b = 9.8989$ Å, $c = 10.400$ Å, space group $Pmcn$, $Z = 4$) an unusual trigonal Be—O co-ordination with Be—O bond lengths of 1.549 Å to 1.564 Å is present, whereas the yttrium atoms are surrounded by distorted octahedra of oxygen atoms.

No compound formation between MgO and RE_2O_3 is observed at 1650 °C. In the binary system PbO–La_2O_3 only one compound, $4PbO·La_2O_3$, with a congruent melting point of 1225 °C exists (Figure 2.11); it is flanked

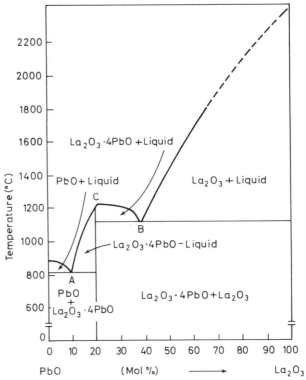

Figure 2.11 The La_2O_3—PbO phase diagram

(From Warzec *et al.*[163], by courtesy of the American Ceramic Society)

by two eutectics[163]. The structure of this compound, although of lower symmetry, is intimately related to the C modification of the rare earth oxides. Solid solutions existing between the compositions $La_2O_3·2PbO$ and pure La_2O_3 have a *bcc* lattice. The systems of PbO with Sm_2O_3 and Gd_2O_3 are quite similar to that with La_2O_3. The compound $4PbO·Sm_2O_3$ decomposes at 1000 °C with evaporation of PbO.

Palladium oxide, PdO, which is stable in air up to 800 °C, reacts with the light rare earth oxides to form binary oxides of differing compositions, the structures of which are unknown[164]. No compound formation or reaction in the solid state was observed for the heavier rare earth oxides, Ho to Lu and including Y. In the most thoroughly investigated system,

Nd_2O_3–PdO three compounds, $PdO \cdot 2Nd_2O_3$, $PdO \cdot Nd_2O_3$ (metastable), and $2PdO \cdot Nd_2O_3$ occur which decompose at 1135, 860 and 1085 °C respectively. Above 1135 °C the system corresponds to the Nd_2O_3–Pd binary system. Three compounds of the same compositions also occur in the Sm_2O_3–PdO and Eu_2O_3–PdO systems. Two compounds, 2:1 and (Figure 2.12) 1:2, occur in the La_2O_3–PdO system; other ternary oxides include the

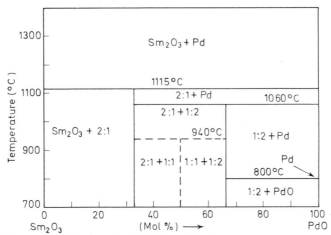

Figure 2.12 The Sm_2O_3–PdO phase diagram
(From McDaniel and Schneider[164], by courtesy of the U.S. Government Printing Office)

2:1 and 1:1 compositions in the Gd_2O_3–PdO system and the 1:1 composition in the Dy_2O_3–PdO system. The most stable compound is $2PdO \cdot Nd_2O_3$, which dissociates at 1190 °C. The decomposition temperatures of the 2:1 and 1:2 compounds decrease with decreasing radius of the RE(III) ion, whilst that of the 1:1 compound increases from 860 °C for $PdO \cdot Nd_2O_3$ to 1030 °C for $PdO \cdot Dy_2O_3$.

References

1. Peppard, D. F., Bloomquist, C. A. A., Horwitz, E. P., Lewey, S. and Mason, G. W. (1970). *J. Inorg. Nucl. Chem.*, **32**, 339
2. Keller, C. and Seiffert, H. (1969). *Angew. Chem.* **81**, 294; (1969) *Angew. Chem. Intern. Ed.*, **8**, 279
3. Keller, C. and Seiffert, H. (1969). *Inorg. Nucl. Chem. Lett.*, **5**, 51
4. Awasthi, S. K., Martinot, L., Fuger, J. and Duyckaerts, G. *Inorg. Nucl. Chem. Lett.*, (in press)
5. Hauck, J. (1968). *Z. Naturforsch.*, **23b**, 1603
6. Fröhlich, K., Gütlich, P., Keller, C. and Gross, J. (Unpublished results)
7. Kanellakopulos, B., Fischer, R. D., Henrich, E. and Keller, C. (Unpublished results)
8. Cajkhorskij, A. A. and Romanov, G. A. (1969). *Radiokhimiya*, **11**, 599; (1969) *Radiochimie* (French translation), **11**, 567
9. Gross, J., Keller, C. (unpublished results)
10. Krot, N. N., Mefodeva, M. P., Zakharova, F. A., Smirnova, T. V. and Gelman, A. D. (1968). *Radiokhimiya*, **10**, 638; (1968) *Radiochimie* (French translation), **10**, 690
11. Krot, N. N., Mefodeva, M. P., Smirnovo, T. V. and Gelman, A. D. (1968). *Radiokhimiya*, **10**, 412; (1968) *Radiochimie* (French translation), **10**, 452
12. Mefodeva, M. P., Krot, N. N. and Gelman, A. D. (1970). *Radiokhimiya*, **12**, 232

13. Komkov, J. A., Krot, N. N. and Gelman, A. D. (1968). *Radiokhimiya*, **10**, 625; (1968) *Radiochimie* (French translation), **10**, 685
14. Stone, J. A., Pillinger, W. L. and Karraker, D. G. (1969). *Inorg. Chem.*, **8**, 2519
15. Zaitseva, V. P. (1969). *Dokl. Akad. Nauk. SSSR*, **188**, 826
16. Spitsyn, V. I., Ippolitova, E. A., Efremova, K. M. and Simanov, Y. P. (1961). *Investigations in the Field of Uranium Chemistry*, English translation of *Issledovaniya v. oblasti khimii urana*, Moscow University, ANL-trans-33, US-Atomic Energy Commission 1964, p. 142
17. Efremova, K. M., Ippolitova, E. A., Simanov, Y. P. and Spitsyn, V. I. (1959). *Dokl. Akad. Nauk. SSSR*, **124**, 1057
18. Keller, C., Koch, L. and Walter, K. H. (1965). *J. Inorg. Nucl. Chem.*, **27**, 1205
19. Seiffert, H., Toussaint, J. C., Keller, C. and Müller, W. (Unpublished results)
20. Carnall, W. T., Walker, A. and Neufeldt, S. J. (1966). *Inorg. Chem.*, **5**, 2135
21. Kovba, L. M. (1970). *Radiokhimiya*, **12**, 522
22. Anderson, J. S. (1969). *Chimia*, **23**, 438
23. Prigent, J., Grislain, B. (1964). *Compt. Rend.*, **258**, 1814
24. Keller, C. (1967). 'The Solid-State Chemistry of Americium Oxides' *Lanthanide/Actinide Chemistry, Advan. Chem. Ser.*, **71**, 228. (Washington, D.C.: American Chemical Society)
25. Carnall, W. T., Neufeldt, S. J. and Walker, A. (1965) *Inorg. Chem.*, **4**, 1808
26. Kovba, L. M., Ippolitova, E. A., Simanov, Y. P. and Spitsyn, V. I. (1958). *Dokl. Akad. Nauk. SSSR*, **120**, 1042
27. Kovba, L. M., Polunina, G. P., Ippolitova, E. A., Simanov, Y. P. and Spitsyn, V. I. (1961). *Zh. Fiz. Khim.*, **35**. 719
28. Kovba, L. M. (1962). *Zh. Strukt. Khim.*, **3**, 159
29. Hoekstra, H. and Siegel, S. (1964). *J. Inorg. Nucl. Chem.*, **26**, 693
30. Hauck, J. (1969). *Z. Naturforsch.*, **24b**, 455
31. Hoekstra, H. R. (1965). *J. Inorg. Nucl. Chem.*, **27**, 801
32. Ohwada, K. (1970). *J. Inorg. Nucl. Chem.*, **32**, 1209; (1970) *Spectrochim. Acta*, **26A**, 1723
33. Cordfunke, E. H. P. (1962). *J. Inorg. Nucl. Chem.*, **24**, 303; (1970), **32**, 3129
34. Stuart, W. I. and Whateley, T. L. (1969). *J. Inorg. Nucl. Chem.*, **31**, 1639
35. Gelman, A. D., Moskvin, A. I., Zaitsev, L. M. and Mefodeva, M. P. (1961). *The Complex Chemistry of Transuranium Elements*, English translation FTD-TT-61-246
36. Cleveland, J. M. (1970). *Inorg. Nucl. Chem. Lett.*, **6**, 535
37. Kemmler, S. (1965). *Z. Anorg. Allgem. Chem.*, **338**, 9
38. Keller, C., Koch, L. and Walter, K. H. (1965). *J. Inorg. Nucl. Chem.*, **27**, 1225
39. Scholder, R. and Gläser, H. (1964). *Z. Anorg. Allgem. Chem.*, **327**, 15
40. Rüdorff, W., Kemmler, S. and Leutner, H. (1962). *Angew. Chem.*, **74**, 429
41. Keller, C. (1965). *Physico-Chimie du Protactinium*, 73. (Paris: Colloq. Intern. Centre Nat. Rech. Sci.)
42. Chamberland, B. L. (1969). *Inorg. Chem.* **8**, 1183
43. Iyer, P. N. and Smith, A. J. (1965). *Physico-Chimie du Protactinium*, 81. (Paris: Colloq. Intern. Centre Nat. Rech. Sci.)
44. Blasse, G. (1964). *Z. Anorg. Allgem. Chem.*, **331**, 44
45. Ippolitova, E. A. and Kovba, L. M. (1961). *Dokl. Akad. Nauk. SSSR*, **138**, 605
46. Bartram, S. F. (1969). Report *GEMP*-1013
47. Hauck, J. (1969). *Z. Naturforsch.*, **24b**, 252
48. Addison, C. C., Barker, M. G., Lintonbon, R. M. and Pulham, R. J. (1969A). *J. Chem. Soc.*, 2457
49. Kemmler-Sack, S. (1968). *Z. Anorg. Allgem. Chem.*, **363**, 295
50. Rüdorff, W. and Menzer, W. (1957). *Z. Anorg. Allgem. Chem.*, **292**, 197
51. Kemmler-Sack, S., Stumpp, E., Rüdorff, W. and Erfurth, H. (1967). *Z. Anorg. Allgem. Chem.*, **354**, 287
52. Henrich, E. and Kanellakopulos, B. (Unpublished results)
53. Keller, C. *et al.* (Unpublished results)
54. Hoppe, R. and Seeger, K. (1968). *Naturwiss.*, **55**, 297
55. Hoppe, R. and Lidecke, W. (1962). *Naturwiss.*, **49**, 255
56. Hagenmuller, P., Devalette, M. and Claverie, J. (1966). *Bull. Soc. Chim. France*, 581
57. Hoppe, R. (1965). *Bull. Soc. Chim. France*, 1115
58. Keller, C. and Walter, K. H. (Unpublished results)
59. Walter, K. H. (1965). Dissertation T. H. Karlsruhe, *Ber. KFK*-280

60. Scholder, R. (1958). *Angew. Chem.,* **70,** 583
61. Paletta, E. and Hoppe, R. (1966). *Naturwiss.,* **53,** 611
61a. Scholder, R., Räde, D. and Schwarz, H. (1968). *Z. Anorg. Allgem. Chem.,* **362,** 149
62. Hoppe, R. and Seeger, K. (1970). *Z. Anorg. Allgem. Chem.,* **375,** 264
63. Trömel, M. and Hauck, J. (1969). *Z. Anorg. Allgem. Chem.,* **368,** 248
64. Seeger, K. and Hoppe, R. (1969). *Z. Anorg. Allgem. Chem.,* **365,** 22
65. Blasse, G. (1964). *J. Inorg. Nucl. Chem.,* **26,** 901; (1966), **28,** 2444
66. Clos, R., Devalette, M., Hagenmuller, P., Hoppe, R. and Paletta, E. (1967). *Compt. Rend.,* **265C,** 801
67. Bärnighausen, H. (1963). *Acta Cryst.,* **16,** 1073; (1965), **19,** 1048
68. Clos, R., Devalette, M., Fouassier, Cl. and Hagenmuller, P. (1970). *Mater. Res. Bull.,* **5,** 179
69. Sevostyana, N. I., Muraveva, I. A., Kovba, L. M., Martynenko, L. I. and Spitsyn, V. I. (1965). *Dokl. Akad. Nauk. SSSR,* **161,** 1359
70. Gonrand, M. and Bertaut, E. F. (1962). *Bull. Soc. Franc. Minéral. Crist.,* **86,** 301
71. Foex, M. (1961). *Bull. Soc. Chim. France,* 109
72. Bertaut, F. and Gonrand, M. (1962). *Compt. Rend.,* **255,** 1135
73. Bärnighausen, H. (1967). *Z. Anorg. Allgem. Chem.,* **349,** 280; (1970), **374,** 201
74. Samson, S. and Sillén, L. G. (1947). *Ark. Kemi Min. Geol.,* **251,** No. 21
75. Loopstra, B. O. and Rietveld, H. M. (1969). *Acta Cryst.,* **B25,** 787
76. Keller, C. (1963). *Nukleonik,* **5,** 89
77. Frondel, C. and Barnes, I. (1958). *Acta Cryst.,* **11,** 562
78. Kuznetsov, L. M., Kirkinsky, V. A. and Makarov, E. S. (1964). *Zhur. Neorgan. Khim.,* **9,** 1187
79. Zachariasen, W. H. (1948). *Acta Cryst.,* **1,** 281
80. Zachariasen, W. H. (1954). *Acta Cryst.,* **7,** 788
81. Haag, R. M. and Muncy, C. R. (1964). *J. Am. Ceram. Soc.,* **47,** 34
82. Amato, I. and Negro, A. (1970). *J. Less-Common Metals,* **20,** 37
83. Rüdorff, W. and Pfitzer, F. (1954). *Z. Naturforsch.,* **9b,** 568
84. Brochu, R. and Lucas, J. (1967). *Bull. Soc. Chim. France,* 4764
85. Bobo, J. C. (1966). *Compt. Rend.,* **262C,** 553
86. Bacman, M. and Bertaut, E. F. (1969). *J. Physique,* **30,** 949
87. Bertaut, E. F., Delapalme, A., Forrat, F., Roult, G., de Bergevin, F. and Pauthenet, R. (1962). *J. Appl. Phys. Suppl.,* **33,** 1123
88. Bertaut, E. F., Delapalme, A., Forrat, F. and Pauthenet, R. (1962). *J. Phys. Radium.,* **23,** 477
89. Ippolitova, E. A., Polunina, G. P., Kovba, L. M. and Simanov, Y. P. (1961). *Investigations in the Field of Uranium Chemistry,* English translation of *Issledovaniya v. oblasti khimii urana,* Moscow University: (1964), ANL-trans-33, US-Atomic Energy Commission, p. 213
90. Weigel, F. and Neufeldt, S. (1961). *Angew. Chem.,* **73,** 468
91. Siegel, S. and Hoekstra, H. R. (1968). *Acta Cryst.,* **B24,** 967
92. Montorsi, M. A. (1969). *Atti. Accad. Sci. Torino, Classe. Sci. Fis. Mat. Nat.,* **103,** 689
93. Hoekstra, H. R. and Marshall, R. H. (1967). 'Lanthanide/Actinide Chemistry', *Advan. Chem. Ser.,* **71,** 211. (Washington, D.C.: American Chemical Society).
94. Bacman, M. and Bertaut, E. F. (1967). *Bull. Soc. Franc. Minéral. Crist.,* **90,** 257
95. Bacman, M., Bertaut, E. F., Blaise, A., Chevalier, R. and Roult, G. (1969). *J. Appl. Phys.,* **40,** 1131
96. Bacman, M., Bertaut, E. F. and Bassi, G. (1965). *Bull. Soc. Franc. Minéral. Crist.,* **88,** 24
97. Smith, D. K., Cline, C. F. and Sands, D. E. (1961). *Nature,* **192,** 867
98. Scholder, R. and Brixner, L. (1955). *Z. Naturforsch.,* **10b,** 178
99. Bereznikova, I. A., Ippolitova, E. A., Simanov, Y. P. and Kovba, L. M. (1961). *Investigations in the Field of Uranium Chemistry,* English translation of *Issledovaniya v. oblasti khimii urana,* Moscow University; (1964) ANL-trans-33, US-Atomic Energy Commission, p. 176
100. Sawyer, J. O. (1963). *J. Inorg. Nucl. Chem.,* **25,** 899
101. Sterns, M. and Sawyer, J. O. (1964). *J. Inorg. Nucl. Chem.,* **26,** 2291
102. Sleight, A. W. and Ward, R. (1962). *Inorg. Chem.,* **1,** 790
103. Rietveld, H. M. (1966). *Acta Cryst.,* **20,** 508
104. Keller, C. (1962). *Nukleonik,* **4,** 271

84 LANTHANIDES AND ACTINIDES

105. Sterns, M. (1967). *Acta Cryst.*, **23**, 264
106. Awasthi, S. K., Chakraburtty, D. M. and Tondon, V. K. (1968). *J. Inorg. Nucl. Chem.*, **30**, 819
107. Awasthi, S. K., Chakraburtty, D. M. and Tondon, V. K. (1967). *J. Inorg. Nucl. Chem.*, **29**, 1225
108. Berthon, J., Ropars, C., Bernier, J. C. and Poix, P. (1966). *Compt. Rend.*, **263C**, 1304
109. Berthon, J. and Poix, P. (1968). *Compt. Rend.*, **267C**, 1585
110. Allpress, J. G. (1964). *J. Inorg. Nucl. Chem.*, **26**, 1847
111. Allpress, J. G. (1965). *J. Inorg. Nucl. Chem.*, **27**, 1521
112. Ippolitova, E. A., Bereznikova, I. A., Leonidov, V. Y. and Kovba, L. M. (1961). *Investigations in the Field of Uranium Chemistry*, English translation of *Issledovaniya v. oblasti khimii urana*, Moscow University; (1964) ANL-trans-33, US-Atomic Energy Commission, p. 186
113. Cordfunke, E. H. P. and Loopstra, B. O. (1967). *J. Inorg. Nucl. Chem.*, **29**, 51
114. Brisi, C. and Montorsi Appendino, M. (1969). *Ann. Chim. (Rome)*, **59**, 400
115. Negro, A. and Amato, I. (1969). *J. Less-Common Metals*, **19**, 159
116. Hoekstra, H. R. and Katz, J. J. (1952). *J. Am. Chem. Soc.*, **74**, 1683
117. Brisi. C. (1963). *Ann. Chim. (Rome)*, **53**, 325
118. Jakeš, D., Sedláková, L. N., Moravec, J. and Germanić, J. (1968). *J. Inorg. Nucl. Chem.*, **30**, 525
119. Polunina, G. P., Kovba, L. M. and Ippolitova, E. A. (1961). *Investigations in the Field of Uranium Chemistry*, English translation of *Issledovaniya v. oblasti khimii urana*, Moscow University; (1964) ANL-trans-33, US-Atomic Energy Commission, p. 220
120. Klima, J., Jakeš, D. and Moravec, J. (1966), *J. Inorg. Nucl. Chem.*, **28**, 1861
121. Kemmler-Sack, S. (1968). *Z. Anorg. Allgem. Chem.*, **358**, 226
122. Marshall, W. L. and Gill, J. S. (1963). *J. Inorg. Nucl. Chem.*, **25**, 1033
123. Marshall, W. L. and Gill, J. S. (1964). *J. Inorg. Nucl. Chem.*, **26**, 277
124. Gill, J. S. and Marshall, W. L. (1961). *J. Inorg. Nucl. Chem.*, **20**, 85
125. Leroy, J. M. and Tridot, G. (1966). *Compt. Rend.*, **262C**, 1376
126. Kemmler-Sack, S. (1968). *Z. Naturforsch.*, **23b**, 1260
127. Kemmler-Sack, S. and Rüdorff, W. (1967). *Z. Anorg. Allgem. Chem.*, **354**, 255; (1968) **358**, 226
128. Kemmler-Sack, S. and Rüdorff, W. (1966). *Z. Anorg. Allgem. Chem.*, **344**, 23
129. Kemmler-Sack, S. (1969). *Z. Anorg. Allgem. Chem.*, **364**, 135
130. Rosenstein, R. D. and Schor, R. (1963). *J. Chem. Phys.*, **38**, 1789
131. Keller, C. (1965). *J. Inorg. Nucl. Chem.*, **27**, 321
132. Dianoux, A. J. and Poix, P. (1968). *Compt. Rend.*, **266C**, 283
133. Kemmler-Sack, S. (1969). *Z. Naturforsch.*, **24b**, 1398
134. Smith, A. J. and Welch, A. J. E. (1960). *Acta Cryst.*, **13**, 653
135. Fujino, T. and Naito, K. (1969). *J. Am. Ceram. Soc.*, **52**, 574
136. Chackraburtty, D. M. and Jayadevan, N. C. (1964). *Indian J. Phys.*, **38**, 585
137. Otto, H. E. (1965). Report *DRI*-1092-219
138. Schwarz, H. and Bommert, D. (1964). *Z. Naturforsch.*, **19b**, 955
139. Queyroux, F. (1965). *Rev. Hautes Temt. Réfractaires*, **2**, 307; (1969), **6**, 111
140. Barry, T. L. and Roy, R. (1967). *J. Inorg. Nucl. Chem.*, **29** 1243
141. Barry, T. L., Stubican, V. S. and Roy, R. (1966). *J. Am. Ceram. Soc.*, **49**, 667
141a. Lopato, L. M., Lugin, L. I. and Shevchenko, A. V. (1970). *Dopovidi. Akad. Nauk. Ukr. RSR., Ser. B.*, **6**, 535
142. Müller-Buschbaum, H. (1964). *Naturwiss.*, **21**, 508
143. Bchargava, K. D., Ovba, L. M., Martinenko, L. S. and Spitsyn, V. I. (1963). *Dokl. Akad. Nauk. SSSR*, **153**, 1318; (1965), **161**, 594
144. Purt, G. and Modern, E. (1968). *Mh. Chem.*, **99**, 2171
145. Andreeva, A. B. and Keler, E. K. (1965). *Zhur. Prikl. Khim.*, **38**, 2166; (1965), *J. Appl. Chem. USSR* (English translation), **38**, 2128
146. Tresvyatskii, S. G., Pavlikov, V. N., Lopato, L. M. and Lugin, L. I. (1970). *Izv. Akad. Nauk. SSSR Neorg. Materialy.*, **6**, 41
147. Bärnighausen, H. and Brauer, G. (1962). *Acta Cryst.*, **15**, 1059
148. Harris, L. A. and Yakel, H. L. (1967). *Acta Cryst.*, **22**, 354; (1968), **B24**, 672
149. Paletta, E. and Müller-Buschbaum, H. (1968). *J. Inorg. Nucl. Chem.*, **30**, 1425
150. Lopato, L. M., Yaremenko, Z. A. and Tresvyatskii, S. G. (1966). *Ukr. Khim. Zh.*, **32**, 437

151. Foex, M. (1961). *Bull. Soc. Chim. France,* 109
152. Barry, T. L. and Roy, R. (1967). *J. Am. Ceram. Soc.,* **50,** 105
153. Barry, T. L., Stubican, V. S. and Roy, R. (1967). *J. Am. Ceram. Soc.,* **50,** 375
154. Wickman, H. H. and Catalano, E. (1968). *J. Appl. Phys.,* **39,** 1248; UCRL-12046 (1970)
155. Lopato, L. M. and Shevchenko, A. V. (1970). *Zh. Neorgan. Khim.,* **15,** 1995
156. Blasse, G. (1965). *J. Inorg. Nucl. Chem.,* **27,** 2683
157. Hennig, H., Sieler, J., Papstein, H. and Holzapfel, H. (1968). *Z. Chem.,* **8,** 188
158. Blasse, G. (1965). *J. Appl. Phys.,* **36,** 879
159. Frushour, R. H. and Vorres, K. S. (1965). Report *TID*-22207, Paper E.
160. Rabenau, A. and Eckerlin, P. (1958). *Acta Cryst.,* **11,** 304
161. Trzebiatowski, W. and Horyn, R. (1965). *Bull. Acad. Polon. Sci., Sér. Sci. Chim.,* **13,** 315
162. Han, Wen-lung and Kuo, Chu-Kun (1965). *Kuei Suan Yen Hsueh PaO,* **4,** 211
163. Warzee, M. H., Maurice, M., Halla, F. and Ruston, W. R. (1965). *J. Am. Ceram. Soc.,* **48,** 15
164. McDaniel, C. L. and Schneider, S. J. (1968). *J. Res. Nat. Bur. St.,* **72A,** 27

3
The Actinide Halides and their Complexes

D. BROWN
Harwell Atomic Energy Research Establishment

3.1 INTRODUCTION

The actinide halides and halogenocomplexes were exhaustively reviewed[1] as a group in 1968 whilst the actinide halide complexes with donor ligands were included in articles published[2] in 1967. More recent reviews which include information on the halide chemistry of certain actinide elements have appeared on the chemistry of protactinium[3], uranium(V)[4] and neptunium[5], and a new edition of the Chemistry of the Transuranium Elements has been published in the series edited by Pascal[6]. The present article covers mainly publications from 1968 to mid-1970 together with a few later ones in 1967; although this will inevitably lead to some overlap with the above reviews this is considered preferable to the possible omission of important work. The articles published in 1967[2] and 1968[1] will provide the reader with complete coverage of the earlier literature in this field. Thermodynamic data are not discussed in the present article since these are included elsewhere in this volume (Chapter 5).

In each of the following sections the compounds are discussed in order of decreasing valence state and increasing atomic number of both the actinide element and the halogen. Particular compounds are not mentioned unless new work has been reported.

3.2 HALIDES

The presently known halides are listed in Table 3.1. With the availability of increased amounts of some of the later actinide elements it is now possible to study their preparative chemistry (often on sub-milligramme amounts) and, for example, the compounds listed for berkelium (Bk), californium (Cf), and einsteinium (Es) have all been prepared only recently. It is in this area that the most significant contributions to the chemistry of the binary halides have been made. No halides have yet been characterised for the remaining actinide elements, fermium (Fm), mendelevium (Md), nobelium (No) and lawrencium (Lr).

3.2.1 Hexavalent

Steindler has presented a short review on the chemistry of plutonium hexa-fluoride[7] and a very brief review on neptunium hexafluoride has appeared[8].

The kinetics of the reactions between uranium or neptunium oxides and tetrafluorides and a variety of fluorinating agents such as bromine penta-fluoride[9-11], bromine trifluoride[11-13], chlorine monofluoride[14], chlorine trifluoride[15] and fluorine[11, 16, 17] are described in a series of papers. Earlier studies on uranium oxide-interhalogen reactions are summarised by Jarry and Steindler[9]. With the possible exception of the reaction between U_3O_8 and bromine trifluoride[12], uranium oxides are converted to the hexafluoride via UO_2F_2 whereas NpO_2 [11], in common with PuO_2 [18], is converted via the tetrafluoride. In addition, whilst the formation of intermediate fluorides is observed in the uranium tetrafluoride reactions, neptunium tetrafluoride

Table 3.1 The actinide halides

	Ac	Th	Pa	U	Np	Pu	Am	Cm	Bk	Cf	Es
Fluorides	AcF_3	ThF_4	— PaF_4 Pa_2F_9 PaF_5	UF_3 UF_4 U_4F_{17} U_2F_9 UF_5 UF_6	NpF_3 NpF_4 — NpF_5 NpF_6	PuF_3 PuF_4 Pu_4F_{17} — PuF_6	AmF_3 AmF_4	CmF_3 CmF_4	BkF_3 BkF_4	CfF_3 CfF_4	—
Chlorides	$AcCl_3$	$ThCl_4$	$PaCl_4$ $PaCl_5$	UCl_3 UCl_4 UCl_5 UCl_6	$NpCl_3$ $NpCl_4$	$PuCl_3$ $PuCl_4$*	$AmCl_3$	$CmCl_3$	$BkCl_3$	$CfCl_3$	$EsCl_3$
Bromides	$AcBr_3$	$ThBr_4$	$PaBr_4$ $PaBr_5$	UBr_3 UBr_4 UBr_5	$NpBr_3$ $NpBr_4$	$PuBr_3$	$AmBr_3$	$CmBr_3$	$BkBr_3$	$CfBr_3$	—
Iodides	AcI_3	ThI_2 ThI_3 ThI_4	PaI_3 PaI_4 PaI_5	UI_3 UI_4	NpI_3	PuI_3	AmI_3	CmI_3	BkI_3	CfI_3	—

*Exists in the gaseous state only in the presence of chlorine is converted directly to the hexafluoride[11]. The new U_3O_8/BrF_3 results[12] contradict those reported previously when a sharp drop in rate constant observed at c. 230 °C was attributed to the formation of UO_2F_2 [19]. However, the new results show no such rate drop and, since no UO_2F_2 has been detected as a reaction intermediate[12], they are probably the most reliable.

Derived activation energies are listed in Table 3.2 together with equations describing the dependence of the reaction rate on the partial pressure of the fluorinating agent and the temperature. Reaction rates for neptunium tetra-fluoride at 350 °C are in the order F_2 (100%) > BrF_3 (6–13 mole %) > BrF_5 (33–35 mole %); those for reactions involving fluorine are in the order UF_4 > NpF_4 > PuF_4 and for bromine pentafluoride UF_4 > NpF_4.

It is well known that fluorine oxidation of plutonium tetrafluoride pro-ceeds readily at room temperature under the influence of γ radiation[7] and it was recently demonstrated[20] that this reaction is also enhanced by ultra-violet light ($\lambda = 3125$ Å).

An investigation of the reactions between ionic chlorides of Groups I and II and molybdenum, tungsten and uranium hexafluorides has shown[21] that the order of reactivity is the reverse of that observed[22] for reactions of these hexafluorides with other fluorides. Thus, UF_6 does not react as extensively as MoF_6. Typical reactions are listed in Table 3.3 where it is seen that uranium hexafluoride is reduced to the tetrafluoride by Group II chlorides; chlorine was detected as a by-product of such reactions. Thermodynamic considerations indicate[21] that uranium hexachloride should be the preferred product and it is postulated, on the basis of previous results[22], that this is initially produced by a slow reaction and then reacts rapidly with the excess uranium hexafluoride:

$$2UF_6 + UCl_6 \rightarrow 3UF_4 + 3Cl_2$$

Reduction of gaseous uranium hexafluoride is also observed when it is mixed with gaseous thionyl chloride at room temperature[23]. The product is apparently β-UF_5 and not, as previously reported, UF_4.

Improved Raman data have been reported for gaseous uranium hexa-fluoride[24] and the Raman spectra of gaseous and liquid neptunium hexa-fluoride have been recorded for the first time[25]. The vapour phase data are listed in Table 3.4 together with up-to-date values for other gaseous hexa-fluorides. The three peaks for liquid NpF_6 occur at 651 (v_1), 524 (v_2) and 218 (v_5) cm^{-1}, respectively. Further calculation has led to minor amend-ments of the M—F bond distances derived for the octahedral hexafluoride molecules by electron diffraction studies[26]. The new values are 1.996 Å (U—F), 1.981 Å (Np—F) and 1.971 Å (Pu—F).

The density of liquid NpF_6 is reported[27] to be given by the equation ρ g cm^{-3} = $3.938 - 4.25 \times 10^{-3} T$ (56–77 °C), yielding $\rho = 3.640$ g cm^{-3} at 70 °C which can be compared with 3.595 g cm^{-3} for UF_6 at this temperature.

The phase relations of several binary uranium hexafluoride systems have been studied. No constant melting mixtures occur in the UF_6–PuF_6 system[28]; the resulting lens shaped diagram is characteristic of a continuous series of solid solutions between 51° and 64 °C. The liquidus points lie close to the ideal liquidus curve but the solidus points are all below the ideal solidus

Table 3.2 Activation energies for the fluorination of uranium and neptunium compounds

Reaction	Activation energy (kcal mol^{-1})	Temperature range (°C)	Partial pressure fluorinating agent (mmHg)	Equation for fluorinating agent (A), partial pressure and temperature dependence; (TK) log k(Torr) =	Ref
UO_2F_2/BrF_5	8.3	175–300	130–357	$0.38 \log P_A - 3690/T + 4.286$	9
UF_4/BrF_5	16.9	200–275	138–370	$0.71 \log P_A - 1810/T$	9
U_3O_8/BrF_5	9.2	225–350	89–369	$0.90 \log P_A - 2000/T - 0.22$	10
UO_2/BrF_5	7.5	225–375	74–369	$0.84 \log P_A - 1630/T - 0.27$	10
UO_3/BrF_5	7.7	225–300	90–277	$1.05 \log P_A - 1680/T - 0.767$	10
U_3O_8/BrF_3	0.7	100–400	18–44	$1.35 \log P_A - 152.3/T - 3.64$	12
UF_4/BrF_3	1.0	25–300	6–45	$0.80 \log P_A - 440/T - 5.98$	13
NpF_4/F_2	20	250–400	—	—	11
NpF_4/BrF_3	—	250–400	—	—	11
NpF_4/BrF_5	26	250–400	—	—	11

curve. However, in the absence of annealing to minimise effects of fractional crystallisation the authors state that there is no justification to treat these deviations in terms of thermodynamic activity coefficients.

Information on the UF_6–MoF_6 phase system[29, 30], the liquid-vapour phase equilibrium between UF_6 and WF_6 [30, 31], the liquid-solid UF_6–WF_6 [31]

Table 3.3   Products of reaction between ionic chlorides and hexafluorides[21]

Reactants	MoF_6	WF_6	UF_6
LiCl, NaCl(MCl)	$Mo_3Cl_9(MoF_6)_3$, MF	No reaction	MUF_5, MF
KCl, RbCl, CsCl	$Mo_3Cl_9(MoF_6)_3$, MF	No reaction	No reaction
$BeCl_2$	$Mo_3Cl_9(MoF_6)_3$, BeF_2	WCl_6, BeF_2	UF_4, BeF_2
$MgCl_2$	$Mo_3Cl_9(MoF_6)_3$, MgF_2	No reaction	UF_4, MgF_2
$CaCl_2$, $SrCl_2$, $BaCl_2$($M'Cl_2$)	$Mo_3Cl_9(MoF_6)_3$, $M'F_2$	No reaction	UF_4, $M'F_2$

Table 3.4   Fundamental frequencies of hexafluoride molecules in the vapour (cm^{-1})*†

Molecule	$v_1(a_{1_g})$	$v_2(e_g)$	$v_3(f_{1_u})$	$v_4(f_1^u)$	$v_5(f_{2_g})$	$v_6(f_{2_u})$
SF_6	773.5	641.7	939	614	525	(347)
SeF_6	706.9	658.7	780	437	405	(264)
TeF_6	697.1	670.3	752	325	314	(197)
MoF_6	741.5	651.6	741.1	264	318	(116)
TcF_6	712.9	(639)	748	265	(297)	(145)
RuF_6	(675)	(624)	735	275	(283)	(186)
RhF_6	(634)	(595)	724	283	(269)	(192)
WF_6	771	677.2	711	258	320	(127)
ReF_6	753.7	(671)	715	257	(295)	(147)
OsF_6	730.7	(668)	720	268	(276)	(205)
IrF_6	701.7	645	719	276	267	(206)
PtF_6	656.4	(601)	705	273	(242)	(211)
UF_6	667.1	532.5	624	186.2	202	(142)
NpF_6	654	535	624	198.6	208	(164)
PuF_6	(628)	(523)	616	206	(211)	(173)

*Values from ref. 24 which quotes references to original work
†Values in parentheses are calculated from combination bands or overtones

system and the UF_6–NbF_5 phase system[32] has also been published. In addition, the congruently melting compound (120 °C) $UF_6 \cdot XeF_2$ is reported to form in the UF_6–XeF_2 phase system[33]. It would be interesting to have information on the structure of this compound.

3.2.2   Pentavalent

The new, potentially useful preparation[23] of β-UF_5 has been mentioned above. However, the most important work on the pentafluorides is undoubtedly the first preparation of neptunium pentafluoride[34]. Preliminary results indicate that this is formed by reduction of NpF_6 by iodine in iodine pentafluoride at room temperature. Some reduction to the tetrafluoride is also observed but the resulting NpF_5, which is isostructural with β-UF_5, is

quite stable. Further work aimed at obtaining pure samples of the pentafluoride is in progress.

In addition to the UF_6 reactions discussed above O'Donnell and co-workers
have investigated the interaction of UF_5, U_2F_9 and U_4F_{17} with a variety
of covalent halides[35]. UF_5 reacts with BCl_3, $TiCl_4$ and PCl_3 to form a mixture
of UF_4 and UCl_6, possibly as a result of initial disproportionation, and is
reduced to the tetrafluoride by PF_3 and AsF_3, but does not react with SbF_3.
Reaction with $AsCl_3$, however, leads to the formation of UF_4, AsF_3 and
chlorine. U_2F_9 and U_4F_{17} also react with BCl_3 to yield a mixture of UF_4
and UCl_6.

An interesting, relatively simple preparation of uranium pentachloride
involving[36] the reduction of UO_3 by silicon tetrachloride at 400 °C has been

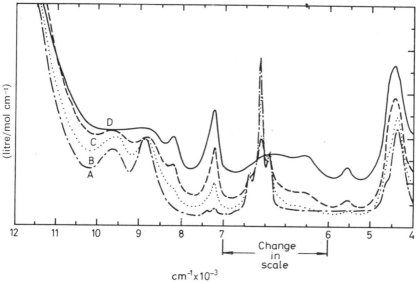

$$cm^{-1} \times 10^{-3}$$

Figure 3.1 Absorption spectra of gaseous uranium pentachloride at various temperatures:
A, 597 K; B, 747 K; C, 844 K; D, 1031 K. (From Gruen and McBeth[37], by courtesy of the American
Chemical Society.)

reported. The product, which is soluble in $SiCl_4$, is isolated by removal of
excess reagent *in vacuo*. This method offers several advantages over those
reported previously[1] and should be widely used.

In an interesting series of spectral investigations Gruen and McBeth[37]
have shown that uranium pentachloride vaporises as the dimer U_2Cl_{10}.
The spectrum of the gaseous compound, prepared *in situ* by reacting UCl_4
with a large excess of chlorine, undergoes drastic changes with temperature
as shown in Figure 3.1. No explanation for these changes has yet been advanced. The free energy change for the reaction,

$$2UCl_4(g) + Cl_2(g) = U_2Cl_{10}(g)$$

is calculated to be,

$$\Delta G = -15\ 132 - 15\ 38T\ cal\ mol^{-1}$$

between 450 and 650 K, whilst the pressure of U_2Cl_{10} as a function of temperature and chlorine pressure is given by the equation,

$$\log P_{U_2Cl_{10}} = -\frac{3307}{T} + 3\,361 + \log P_{Cl_2}$$

Although many other pentahalides are dimeric in the solid state uranium pentachloride is the first one known to exist as a dimer in the gas phase.

An improved method for the preparation of $PaBr_5$ has been described[38] and what is undoubtedly the most reliable information yet obtained on UBr_5 has recently been published[39]. The former is obtained by treatment of protactinium pentachloride, which is relatively easy to prepare, with refluxing boron tribromide; the product is completely free from oxybromide impurity. The use of this reagent for the preparation of actinide bromo-complexes is discussed on (page 113). Uranium pentabromide is conveniently obtained[39] by heating freshly cleaned uranium metal with an excess of bromine at c. 80 °C. It is a dark red solid which hydrolyses immediately on exposure to the atmosphere and is isostructural with α-$PaBr_5$. Full details of the structure of β-$PaBr_5$ have now been published[40]. It is isostructural with UCl_5, space group $C_{2h}^5 - P2_1/n$ with $a_0 = 8.385$, $b_0 = 11.205$, $c_0 = 8.950$ and $\beta = 91.1$ degrees. Each protactinium atom is octahedrally coordinated with Pa—Br bond lengths of 2.85–2.86 Å (bridging bromines) and 2.54–2.60 Å (terminal bromines). The main difference between this structure and that of UCl_5 is that in this case the bridging between the metal atoms is symmetrical. The calculated single-bond covalent radius for Pa(V) in β-$PaBr_5$ is 1.44 Å.

3.2.3 Tetravalent

Asprey and co-workers have prepared the first samples of both berkelium[41] and californium[42] tetrafluoride by heating microgramme amounts of the appropriate oxide (in sapphire crucibles) with fluorine at 400 °C. These are the first tetravalent halides known for their respective elements. A new set of unit cell parameters[43] for the isostructural tetrafluorides of Th to Bk inclusive have been shown to fit a volume v. atomic number plot appreciably better than the original values reported over a number of years by different groups. The new values are listed in Table 3.5 whilst an atomic volume v. atomic number plot is illustrated in Figure 3.2. On the basis of these results it is predicted that if the transberkelium tetravalent fluorides and fluoro-complexes maintain the same symmetry and coordination there will be little further decrease in over-all size.

Electron diffraction studies[44] on gaseous thorium and uranium tetrafluoride suggest C_{2v} symmetry for the molecules with M—F distances of 2.14 and 2.06 Å, respectively. Similar data on the tetrachlorides and tetrabromides are presented below. N.M.R. information reported for the uranium tetrafluoride hydrates does not contribute significantly to the structural knowledge on these compounds[45]. Raman and infrared vibrations have been recorded[46] for UF_4 at 614 (R), 420 (R;i.r.), 340 (R) and 180 (R;i.r.) cm^{-1} and assuming a U—F bond distance of 2.3 Å, the force constant of the valence

Table 3.5 Lattice constants for the actinide tetrafluorides (space group) $C_{2h}^6 - C2/c$:$Z = 12$)[43]

Compound	a_0, Å	b_0, Å	c_0, Å	β, °	Mol vol, Å³
ThF_4	12.90 ± 0.04	10.93 ± 0.03	8.58 ± 0.04	126.4 ± 0.2	81.1
PaF_4	12.83 ± 0.03	10.82 ± 0.02	8.45 + 0.03	126.4 ± 0.1	78.7
UF_4	12.73 ± 0.01	10.753 ± 0.007	8.404 ± 0.008	126.33 ± 0.05	77.2
NpF_4	12.64 ± 0.02	10.70 ± 0.02	8.36 ± 0.02	126.4 ± 0.1	75.8
PuF_4	12.59 ± 0.03	10.69 ± 0.02	8.29 ± 0.04	126.0 ± 0.2	75.2
AmF_4	12.56 ± 0.04	10.58 ± 0.03	8.25 ± 0.04	125.9 ± 0.2	74.0
CmF_4	12.51 ± 0.05	10.61 ± 0.03	8.20 ± 0.05	125.8 ± 0.2	73.6
BkF_4	12.47 ± 0.06	10.58 ± 0.04	8.17 ± 0.05	125.9 ± 0.2	72.8

force field for the U—F stretching vibration is calculated to be 2.419 mdyn $Å^{-1}$.

Little new preparative work has been reported for the actinide tetrachlorides apart from an investigation of the chlorination of thorium dioxide[47] and some results on the conversion of UC to UCl_4 [48]. Neither makes a significant contribution to the wealth of preparative methods already available.

Some excellent work on the structure of $ThCl_4$ has shown[49] the original chlorine parameters deduced from powder data to be incorrect. Each thorium atom is surrounded by two sets of four chlorine atoms at 2.718 and 2.903 Å, respectively, distances which are much more nearly equal than those reported

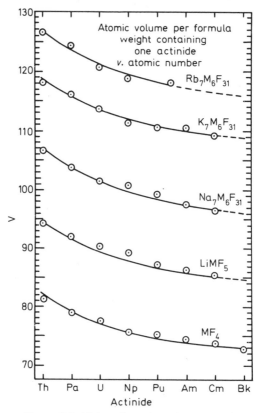

Figure 3.2 Volume/formula weight (containing 1 mol of actinide) ratio v. atomic number for tetravalent actinide fluorides. (From Keenan and Asprey[43], by courtesy of the American Chemical Society)

previously[50] (2.46 and 3.11 Å, respectively). Each chlorine atom has two thorium atoms and ten chlorine atoms as nearest neighbours; the coordination about one thorium atom is illustrated in Figure 3.3. The ratio d_A/d_B (2.903/2.718) = 1.06 and the angles θ_A and θ_B (Figure 3.3) made by the bond directions with the 4 axis, can be compared with the corresponding values for the 'most favourable' discrete polyhedron[51] (hard-sphere model),

$d_A/d_B = 1$, $\theta_A = 36.9°$ and $\theta_B = 69.5$ degrees. In PaBr$_4$, which is iso-structural[52, 53], the two sets of four bromine atoms are, respectively, 3.07 and 2.77 Å from the protactinium atom with $d_A/d_B = 1.11$, $\theta_A = 33.65$ and $\theta_B = 78.2°$. A value of 1.63 Å is deduced for the covalent single-bond radius of Pa(IV) in PaBr$_4$.

Electron diffraction studies on gaseous thorium and uranium tetrahalides indicate the possibility of C_{2v} symmetry with the bond distances Th—Cl, 2.58 Å, U—Cl, 2.53 Å, Th—Br, 2.72 Å and U—Br, 2.66 Å. The magnetic properties of uranium tetrachloride have been investigated[55] over an

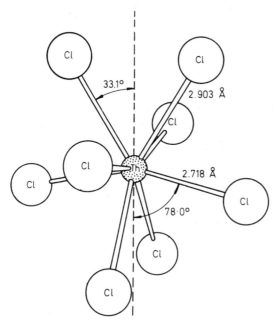

Figure 3.3 Coordination about Th in ThCl$_4$. The poly-hedron is a dodecahedron having $\bar{4}2m$ symmetry. (From Mucker, et al.[49], by courtesy of Munksgaard)

extended temperature range; between 76 and 550 K the Curie–Weiss law is followed with $\theta = -62°$ and $\mu_{eff} = 3.29$ BM.

Chlorine nuclear quadrupole resonance studies[56] on UCl$_4$ and PMR studies[57] on U^{4+} in CaF$_2$ have been reported and details of Mössbauer studies on NpCl$_4$ have been published[58]. Electronic spectra of UCl$_4$ and UBr$_4$ molecules isolated in nitrogen matrices cooled to liquid He temperature have been recorded from 400–50 000 cm^{-1} and the observed bands assigned to electronic transitions from the $^3H_4(T_5)$ ground state to excited states of the $5f^2$ configuration[59].

Neptunium tetrabromide is readily obtained[38] by direct union of the elements at 425 °C *in vacuo*, a method which offers distinct advantages over the only previous one employed which involved the interaction of neptunium dioxide and aluminium tribromide. In particular, the product can be sublimed without decomposition provided an excess of bromine is present. $NpBr_4$, a dark red, moisture-sensitive solid, is isostructural with uranium tetrabromide, possessing monoclinic symmetry with $a_0 = 10.89$, $b_0 = 8.74$, $c_0 = 7.05$ Å and $\beta = 94.19$ degrees.

In a somewhat similar experiment uranium tetraiodide has been prepared from the elements by heating the evacuated reaction vessel in a furnace graded from 500 to 180 °C. The resulting tetraiodide, which sublimes, condenses as finê, black needles; structural studies are currently in progress[60].

The only recent report concerning mixed tetrahalides deals with the UCl_4–UF_4 phase system in which the compounds UCl_2F_2, $UClF_3$ and UCl_3F have been observed; they melt incongruently at 460, 530 and 444 °C, respectively[61].

3.2.4 Trivalent

Details of an improved method for the preparation of UF_3 by the reaction between uranium tetrafluoride and the metal have been published[62] together with a critical survey of the older literature on this method. The most satisfactory results are achieved by repeated hydrogenation of freshly 'pickled' metal followed by decomposition to finely divided metal at 500 °C, the mixture with UF_4 then being heated at 900 °C in an evacuated stainless steel vessel which is continually rotated. Uranium tetrafluoride is also reduced

Table 3.6 Crystallographic data for actinide trihalides

Compound	Colour	Structure type	Lattice parameters (Å)			Reference
			a_0	b_0	c_0	
BkF_3	Yellow-green	YF_3	6.70	7.090	4.41	67
BkF_3		LaF_3	6.97	—	7.14	67
CfF_3	White	YF_3	6.651	7.040	4.393	66
$AmCl_3$	Pink	UCl_3	7.382	—	4.214	76
$BkCl_3$	Green	UCl_3	7.382	—	4.127	72
$CfCl_3$	Green	UCl_3	7.393	—	4.090	75
$CfCl_3$	Green	$PuBr_3$	11.750	3.869	8.561	75
$CfCl_3$	Green	UCl_3	7.393	—	4.090	74
$EsCl_3$*		UCl_3	7.47	—	4.10	73
$EsCl_3$†		UCl_3	7.40	—	4.07	73
$BkBr_3$	Yellow-green	$PuBr_3$	12.60	4.1	9.1	81
BkI_3	Yellow	BiI_3	7.50	—	20.4	81
CfI_3	Lemon-yellow	BiI_3	7.55	—	20.8	82

*Parameters at c. 425 °C
†Parameters calculated for 25 °C

to the trifluoride by zirconium[63]. The hydrofluorination of CmO_2, to yield CmF_3, proceeds rapidly to completion[64] at 435 °C.

The most interesting development in the chemistry of the trifluorides is the first preparation of both berkelium[65, 67] and californium[66] trifluoride by heating microgramme amounts of the appropriate oxide in a hydrogen/hydrogen fluoride mixture at 600 °C. Berkelium trifluoride is dimorphic, exhibiting both the orthorhombic YF_3-type (low temperature) and trigonal

LaF_3-type (high temperature) structures[67]; only the former type of structure has been recorded for CfF_3 (Table 3.6). Ionic radii calculated from the crystallographic data on the lanthanide and actinide trifluorides are listed in Table 3.7 from which it is apparent that the change in structural-type would be expected to occur at berkelium[67].

Table 3.7   Lattice parameters and ionic radii derived from the lanthanide and actinide trifluorides[67]

Trifluoride	Structure type	Lattice parameters (Å)			M^{3+} ionic radius* (Å) (CN = 6)
		a_0	b_0	c_0	
La	LaF_3	7.186	—	7.352	1.006
Ce	LaF_3	7.112	—	7.279	0.982
Pr	LaF_3	7.075	—	7.238	0.968
Nd	LaF_3	7.030	—	7.200	0.955
Pm	LaF_3	6.970	—	7.188	0.945
Sm	YF_3	6.669	7.059	4.405	0.921
Eu	YF_3	6.622	7.019	4.396	0.909
Gd	YF_3	6.570	6.984	4.393	0.900
Tb	YF_3	6.513	6.949	4.384	0.888
Dy	YF_3	6.460	6.906	4.376	0.877
Ho	YF_3	6.404	6.875	4.379	0.868
Er	YF_3	6.354	6.846	4.380	0.860
Tm	YF_3	6.283	6.811	4.408	0.855
Yb	YF_3	6.216	6.786	4.434	0.851
Lu	YF_3	6.151	6.758	4.467	0.848
Ac	LaF_3	7.41	—	7.55	1.076
U	LaF_3	7.181	—	7.348	1.005
Np	LaF_3	7.129	—	7.288	0.986
Pu	LaF_3	7.093	—	7.254	0.974
Am	LaF_3	7.044	—	7.225	0.962
Cm	LaF_3	6.999	—	7.179	0.946
Bk	LaF_3	6.97	—	7.14	0.935
Bk	YF_3	6.70	7.09	4.41	0.928
Cf	YF_3	6.653	7.041	4.395	0.915

*Calculated according to the following formula:

$$\text{Ionic radius} = (\text{weight} - \text{averaged M—F bond distance}) - 1.33 - 0.11$$

where the fluoride ionic radius is taken to be 1.33 Å and the nine-fold to six-fold coordination correction number is taken to be 0.11 Å

147812

Preliminary results[68] on the comparison of the ENDOR spectra of $U^{3+} - F^-$ and $Nd^{3+} - F^-$ in calcium fluoride show that the U^{3+} ion has the same C_{4v} symmetry as Nd^{3+} and that the superhyperfine tensor components are larger and more anisotropic for U^{3+}. Results of Mössbauer studies involving AmF_3 have been published[69]. The recent e.s.r. spectral studies on actinide ions in CaF_2 and similar matrices are discussed with the divalent halides for convenience.

Little new preparative work has been reported on the previously known actinide trichlorides although a fluidised-bed conversion of plutonium dioxide to $PuCl_3$ has been described[70] and americium trichloride has been prepared by heating the dioxide with aluminium trichloride[71]. This useful technique[71] has also been used for the preparation of $AmBr_3$ and AmI_3.

Interest has centred on the previously unknown trichlorides of berkelium, californium and einsteinium all of which have now been obtained in microgramme amounts by heating the appropriate oxide with anhydrous hydrogen chloride at c. 500 °C [65, 66, 72–74]. $BkCl_3$ and $EsCl_3$ have been observed to crystallise only with the 9-coordinate UCl_3-type structure (Table 3.6). The observations on $EsCl_3$, however, were made at 425 °C, as no powder photographs were obtained at room temperature and, in view of the recent observation[75] that $CfCl_3$ possesses a second crystal form, the 8-coordinate

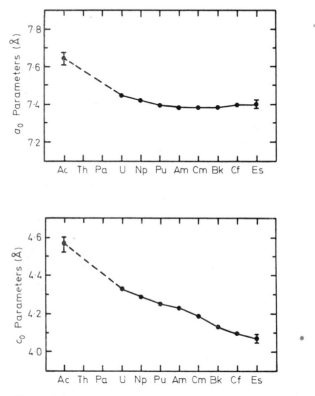

Figure 3.4   Lattice parameters of the UCl_3-type hexagonal actinide trichlorides. (From Fujita *et al.*[73], by courtesy of Pergamon Press)

$PuBr_3$-type structure (Table 3.6), it is likely that $EsCl_3$ will also exhibit this structure at room temperature.

Cunningham and co-workers[72, 73] have drawn attention to the anisotropic actinide contraction exhibited by the 9-coordinate trichlorides, the a_0 parameter passing through a minimum and then increasing with increasing atomic number whilst c_0 decreases continuously with increasing atomic number as shown in Figure 3.4. They point out that ionic radii values deduced from such a series of compounds should, therefore, be regarded with extreme caution since they are based on average interatomic distances and the radius concept begins to lose meaning when the distances averaged are considerably different, as are the M—3Cl and M—6Cl distances, and are diverging. In addition, of course, in the absence of single crystal studies for individual actinide trichlorides the positional parameters for UCl_3 have previously been assumed to hold throughout the series. The drawback of this assumption is illustrated by the recent structure determination[76] for $AmCl_3$, the resulting ionic radius 0.984 Å being appreciably smaller than that derived previously, 1.006 Å. This structural study, the first on the actinide trichlorides, shows that there are six chlorine atoms, forming a trigonal prism, 2.874 Å from the americium atom with three at 2.975 Å situated just outside the prism faces. The same authors report[77] structural details for the hexahydrate, $AmCl_3 \cdot 6H_2O$, which is isostructural with the lanthanide analogues, together with unit cell parameters for $BkCl_3 \cdot 6H_2O$ (Table 3.8). The bond distances in the $AmCl_2 (H_2O)_6$ unit are Am—Cl(2), 2.799 Å; Am—O(1), 2.471 Å; Am—O(2), 2.440 Å and Am—O(3), 2.474 Å. On the basis of ionic radii considerations Burns and Peterson[77] predict that both U and Np trichloride would be expected to form heptahydrates, the trichlorides of the heavier actinide elements forming only hexahydrates.

Low temperature absorption spectra have been recorded for the solid americium compounds[71] $AmCl_3$, $AmBr_3$ and AmI_3 and for thin films[78] of $PuCl_3$. Observed bathochromic shifts in the Am series are comparable to those in the lanthanide halides; theoretical analysis of the $AmCl_3$ and $PuCl_3$ results are presented. Values of the electrostatic, spin-orbit coupling and configuration-interaction parameters are given as, $PuCl_3$ [78]: E^1, 3634.5, E^2, 15.356, E^3 342.15, ζ 2272.2, α, 31.964, β, -675.0 and γ, 29.743 cm^{-1}; $AmCl_3$ [71]: E^1, 3582.8 E^2, 17.276, E^3, 334.30, ζ, 2593.3, α, 21.634, β, -158.48 and γ, 1240.4 cm^{-1}. Carnall has also presented an excellent appraisal of the current investigations and improved theoretical treatment in this interesting field[79].

Table 3.8 Lattice parameters for actinide trihalide hexahydrates*

Compound	Colour	Lattice parameters				Reference
		a_0 (Å)	b_0 (Å)	c_0 (Å)	β^0	
$AmCl_3 \cdot 6H_2O$	Yellow-rose	9.702	6.567	8.009	93.61	77
$BkCl_3 \cdot 6H_2O$	Yellow-green	9.66	6.54	7.97	93.77	77
$UBr_3 \cdot 6H_2O$	Red	10.061	6.833	8.288	92.99	80
$NpBr_3 \cdot 6H_2O$	Green	10.041	6.821	8.260	92.99	80
$PuBr_3 \cdot 6H_2O$	Blue	10.022	6.798	8.181	92.97	80
$AmBr_3 \cdot 6H_2O$	Light brown	9.955	6.783	8.166	92.75	80
$CfBr_3 \cdot 6H_2O$		9.992	6.716	8.146	93.50	66

*All possess monoclinic symmetry, space group $C_{2h}^1 - P2/n$

The anhydrous tribromides UBr_3, $NpBr_3$, $PuBr_3$ and $AmBr_3$ are obtained by controlled vacuum thermal decomposition of their hexahydrates, a method which is also useful for the preparation of the lanthanide tribromides[80]. The new compounds $BkBr_3$ and $CfBr_3$ are obtained[81, 82] by treatment of the appropriate oxide with anhydrous hydrogen bromide at c. 800 °C. The former possesses the 8-coordinate $PuBr_3$-type structure[81] in common with the tribromides of the earlier actinide elements, but the structure of

CfBr$_3$ is different. The x-ray powder data on this compound and on a second modification[66, 82] of BkBr$_3$ are similar to those reported for that crystal form of GdBr$_3$ for which structural information is not available. Tribromide hexahydrates have been prepared for several actinide elements and have been shown to be isostructural with the trichloride hexahydrates[66, 80]. Unit cell data are listed in Table 3.8.

An extremely thorough study of the polarised absorption and emission

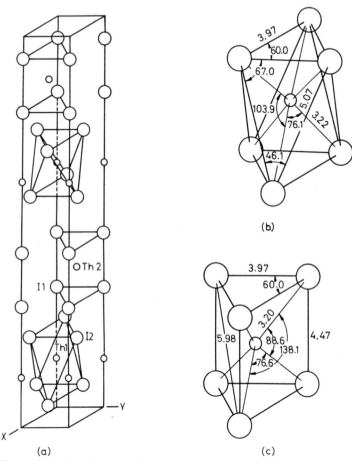

Figure 3.5 View of the crystal structure of ThI$_2$ showing (a) the contents of one unit cell; (b) coordination polyhedra in the trigonal-antiprismatic layers and (c) coordination polyhedra in the trigonal-prismatic layers. (From Guggenberger and Jacobsen[92], by courtesy of the American Chemical Society)

spectra and fluorescent excitation spectra (5000–25 000 cm^{-1}) of Np^{3+} in LaBr$_3$ at liquid nitrogen and helium temperatures has been reported; the results have been interpreted and the electronic energy levels assigned[83]. The values obtained for the electrostatic, spin-orbit coupling and configuration-interaction parameters are: E^0, 15 920, E^1, 3394.3, E^2, 14.19, E^3, 317.5, ζ, 1469.2, α, 35.0 and β, -802.4 cm^{-1}.

The new compounds BkI_3 and CfI_3 are obtained[81, 82] on the microgramme scale by heating the appropriate oxide in anhydrous hydrogen iodide at $c.\ 650\,°C$. They are both isostructural with other actinide tri-iodides; crystallographic data are given in Table 3.6.

3.2.5 Divalent

Studies of divalent actinide halides are obviously limited by the relative instability of this oxidation state for those actinide elements which are currently available in amounts sufficient for preparative studies. The only 'divalent' halide characterised to date is ThI_2 and, as discussed below, even this must be considered as $Th^{4+}(e^-)_2I_2$.

The search for divalent ions stabilised in CaF_2 and similar matrices has continued and several very reliable spectral studies have been reported[84–88]. Under those conditions where all the lanthanide elements exhibit divalency, such as reduction of M^{3+} by γ radiation, solid state electrolysis or exposure to alkaline-earth metal vapours, only americium[87] and einsteinium[88] have been observed as divalent ions. Cm^{3+}, Pu^{3+} and Np^{3+} in CaF_2 are, in fact, oxidised to the tetravalent state. Attempts to reduce $CfCl_3$ by hydrogen at elevated temperature have been unsuccessful[66] but the possibility of stabilising Cf^{2+} in CaF_2 is currently being investigated[89].

More conflicting evidence has been published on the existence of lower valence state thorium chlorides. Thus, Fuller[90], who presents an excellent survey of the earlier work, reports that he can find no evidence for the reduction of thorium tetrachloride either by metallic thorium or aluminium. These results are probably more reliable than those of Kudyakov et al.[91] who claim that $ThCl_2$ is formed in $ThCl_4$–KCl–Th melts at high temperature.

Structural work on ThI_2 has confirmed the view that this compound, although formally divalent, must be considered to contain the Th^{4+} ion and, therefore, is best represented as $Th^{4+}(e^-)_2I_2$ [92]. The compound possesses hexagonal symmetry, space group $D_{6h}^4 - P6_3/mmc$ with $a_0 = 3.97$ and $c_0 = 31.75\,Å$. The structure consists of four two-dimensionally infinite layers alternating between trigonal-antiprismatic and trigonal-prismatic layers. The thorium atoms are encompassed by trigonal-antiprismatic polyhedra of iodine atoms in the antiprismatic layers and by trigonal-prismatic polyhedra in the prismatic layers. Coordination polyhedra of both types are linked by edge-sharing. Th—I bond distances are 3.22 and 3.20 Å in the antiprismatic and prismatic layers, respectively. The structure is illustrated in Figure 3.5.

3.3 OXYHALIDES

Relatively little work has been reported on the actinide oxyhalides and most of this relates to the uranyl halides, new pentavalent oxyhalides, and the preparation of new trivalent oxyhalides of californium, berkelium and

Table 3.9 The actinide oxyhalides

	Ac	Th	Pa	U	Np	Pu	Am	Cm	Bk	Cf	Es
Fluorides	AcOF	ThOF $ThOF_2$	Pa_2OF_8 PaO_2F Pa_3O_7F	'UOF$_2$' U_2OF_8 UO_2F_2	$NpOF_3$ NpO_2F NpO_2F_2	PuOF PuO_2F_2	AmO_2F_2	—	—	CfOF	—
Chlorides	AcOCl	$ThOCl_2$	$PaOCl_2$ Pa_2OCl_8 $Pa_2O_3Cl_4$ PaO_2Cl	UOCl $UOCl_2$ $UOCl_3$ UO_2Cl UO_2Cl_2	$NpOCl_2$	PuOCl	AmOCl	CmOCl	BkOCl	CfOCl	EsOCl
Bromides	AcOBr	$ThOBr_2$	$PaOBr_2$ $PaOBr_3$ PaO_2Br	$UOBr_2$ $UOBr_3$ UO_2Br UO_2Br_2	$NpOBr_2$	PuOBr	—	CmOBr	BkOBr	CfOBr	—
Iodides	AcOI	$ThOI_2$	$PaOI_2$ $PaOI_3$ PaO_2I	$[UO_2I_2]$	—	PuOI	—	—	BkOI	CfOI	—

einsteinium on the microgramme scale. The presently known oxyhalides are shown in Table 3.9.

3.3.1   Hexavalent

Complete details of the preparation of AmO_2F_2 have now been published[93]. It is obtained by adding anhydrous hydrogen fluoride containing fluorine to $NaAmO_2(CH_3COO)_3$ at $-196\,°C$, allowing the mixture to warm to room temperature and then pumping off the excess fluorinating agent. The light brown product is isostructural with UO_2F_2, NpO_2F_2 and PuO_2F_2 (Table 3.10). Details of new preparative methods have also now been published[94] for NpO_2F_2.

In a novel preparation[95] uranyl chloride is formed in the reaction between

Table 3.10   Crystallographic data for actinide oxyhalides

Compound	Colour	Symmetry; space group	Lattice parameters (Å)			Reference
			a_0	b_0	c_0	
NpO_2F_2	Pink	$R; D_{3d}^5 - R\bar{3}m$	4.185	—	15.790	94
AmO_2F_2	Light brown	$R; D_{3d}^5 - R\bar{3}m$	4.136	—	15.850	93
PaO_2F	White	$O; \ —$	6.894	12.043	4.143	100
Pa_3O_7F	White	$O; C_{2v}^{11} - Cmm2$	6.947	12.030	4.203	100
$NpOF_3$	Green	$R; D_{3d}^5 - R\bar{3}m$	4.185	—	15.799	94
$'NpO_2F'$	Green	$T; \ —$	8.341	—	7.193	94
$ThOCl_2$	White	$O; D_{2h}^9 - Pbam$	15.494	18.095	4.078	113
$PaOCl_2$	Yellow-green	$O; D_{2h}^9 - Pbam$	15.332	17.903	4.012	113
$UOCl_2$	Green	$O; D_{2h}^9 - Pbam$	15.255	17.828	3.992	113
$NpOCl_2$	Light-brown	$O; D_{2h}^9 - Pbam$	15.209	17.670	3.948	113
$CfOF$		$FCC; O_h^5 - Fm3m$	5.561	—	—	118
$BkOCl$	Green	$T; D_{4h}^7 - P4/nmm$	3.966	—	6.710	121
$CfOCl$		$T; D_{4h}^7 - P4/nmm$	3.956	—	6.662	120
$EsOCl*$		$T; D_{4h}^7 - P4/nmm$	3.970	—	6.75	73
$EsOCl†$		$T; D_{4h}^7 - P4/nmm$	3.948	—	6.702	66
$CfOBr$	Green	$T; D_{4h}^7 - P4/nmm$	3.900	—	8.110	66
$CfOBr$	White	$T; D_{4h}^7 - P4/nmm$	3.90	—	8.12	82
$BkOBr$	White	$T; D_{4h}^7 - P4/nmm$	3.95	—	8.1	81
$CfOI$	Tan	$T; D_{4h}^7 - P4/nmm$	3.97	—	9.14	82
$CfOI$	Tan	$T; D_{4h}^7 - P4/nmm$	3.97	—	9.13	66
$BkOI$	White	$T; D_{4h}^7 - P4/nmm$	4.0	—	9.15	81

R, rhombohedral; O, orthorhombic; T, tetragonal; FCC, face centred cubic
*Parameters at 425 °C
†Room temperature parameters calculated from those at 425 °C

UO_3 and $MoOCl_4$ at 110–120 °C. The volatile molybdenum dioxydichloride is removed by heating *in vacuo*,

$$UO_3 + MoOCl_4 \rightarrow UO_2Cl_2 + MoO_2Cl_2$$

In view of the relative ease of preparing UO_2Cl_2[1] this reaction is unlikely to be widely used.

Infrared results have been reported for the uranyl halides[96, 97] and Bullock has reported Raman data[98] for UO_2Cl_2 and a wide range of other uranyl compounds; v_1 (sym. stretching vibration, R) occurs at $871\ cm^{-1}$ and v_3 (asymmetric stretching vibration, i.r.) at $960\ cm^{-1}$. The force constant is calculated to be 7.40 mdyn $Å^{-1}$ and the U=O bond distance[98] to be 1.72 Å. This value is possibly slightly high since similar calculations lead to a U=O distance of 1.74 Å for $Cs_2UO_2Br_4$ compared with the distance 1.69 Å obtained from single crystal x-ray results[99].

3.3.2 Pentavalent

Several methods for the preparation of new protactinium(V) oxyfluorides have been studied[100]. Many yield $PaF_5\cdot2H_2O$, $PaF_5\cdot H_2O$ or Pa_2OF_8 but controlled thermal decomposition of Pa_2OF_8 in the atmosphere results in the successive formation of $PaO_2F(250–290\ °C)$ and $Pa_3O_7F(500–600\ °C)$. These are both white, air stable solids, the latter being isostructural with high temperature ('trigonal') U_3O_8 which was recently shown[101] to possess orthorhombic symmetry. Lattice parameters are given in Table 3.10. An orthorhombic pseudo-cell for PaO_2F with $a_0 = 6.894$, $b_0 = 4.014$ and $c_0 = 4.143$ Å is virtually identical with that reported[102, 103] for β-Pa_2O_5 which was observed during attempts to fluorinate microgramme amounts of the pentoxide and identified only by x-ray methods [102]. Since numerous attempts to prepare β-Pa_2O_5 by other methods have been unsuccessful[104, 105] it appears likely that the original work actually yielded PaO_2F. This compound provides an attractive alternative to Pa_2O_5 for the preparation of the tetra-fluoride; above 650 °C Pa_3O_7F decomposes to the cubic form of Pa_2O_5 [100].

The initial product of the hydrofluorination of Np_2O_5 at 40 °C is the hydrated compound $NpOF_3\cdot1.5H_2O$; this is converted to anhydrous neptunium(V) oxytrifluoride, $NpOF_3$, in gaseous hydrogen fluoride at 150–200 °C [94]. At higher temperatures some reduction to NpF_4 is observed. $NpOF_3$, a green, air-stable solid is isostructural with NpO_2F_2 (Table 3.10); the Np=O stretching vibration is observed at $985\ cm^{-1}$ and bands at 350 and $300\ cm^{-1}$ are tentatively assigned to Np—F stretching vibrations. Controlled hydrogen reduction of NpO_2F_2 yields phases close to the composition NpO_2F[94]. The green products, $NpO_{1.96}F_{1.08} \rightarrow NpO_{1.8}F_{1.4}$, are not isostructural with PaO_2F but possess tetragonal symmetry (Table 3.10).

By heating together uranium tetrafluoride and either UO_3 or U_3O_8 in varying proportions at 400–500 °C in vacuo Kemmler-Sack[106] has prepared a range of uranium oxyfluoride complexes containing uranium in different valence states. Their compositions range from $UO_{2.5}F_{0.17}$ to UO_2F, the four major phases corresponding to the compositions or ranges of composition, I, $UO_{2.50}F_{0.5}$, II, $UO_{2.58}F_{0.17} - UO_{2.42}F_{0.33}$, III, $UO_{2.40}F_{0.50} - UO_{2.25}F_{0.50}$ and IV, $UO_{2.25}F_{0.67} - UO_{2.0}F_{1.0}$, respectively. The possible structures of these phases are discussed in relationship to those of uranium oxides on the basis of magnetic, spectral and x-ray powder data. It is interesting that the unit cell dimension listed for UO_2F, $a_0 = 4.11$ ($\times 2$), $b_0 = 6.81$, $c_0 = 4.01$ ($\times 8$)Å and $\beta = 90° 30'$, indicate that this compound is not isostructural with either PaO_2F or NpO_2F (Table 3.10).

An extremely attractive new method for the preparation of $UOCl_3$ involves the reaction between uranium trioxide and molybdenum pentachloride at 200–220 °C *in vacuo*[107]. The mechanism is as shown below,

$$UO_3 + MoCl_5 \rightarrow UO_2Cl_2 + MoOCl_3$$
$$UO_2Cl_2 + MoOCl_3 \rightarrow UOCl_3 + MoO_2Cl_2 \uparrow$$

Infrared vibrations have been observed for $UOCl_3$ at 965, 845, 750 and 615 cm^{-1}.

Uranium(V) dioxymonochloride, a violet-brown, water-stable solid is obtained[108] by the reaction between uranyl chloride and uranium dioxide at 590 °C in an argon atmosphere. It is oxidised by chlorine and oxygen at 400 °C to yield UO_2Cl_2 and U_3O_8, respectively. The magnetic moments for UO_2Cl and UO_2Br are[108], respectively, 1.86 BM ($\theta = -95°$) and 1.76 BM ($\theta = -200°$); UO_2Cl is not isostructural with PaO_2Cl.

3.3.3 Tetravalent

There have been two claims for the preparation of $UOF_2 \cdot H_2O$ from aqueous solution. In one instance[109] by photolysis of uranyl formate in aqueous hydrogen fluoride, to yield a green solid, and in the other[110] by precipitation of a black solid from molar perchlorate solution, $p_H = 1.7$, containing U and F in a 1:2 molar ratio. The latter product, identified by uranium analyses only, was reported to lose water at 100–280 °C, the resulting UOF_2 being stable to 700 °C in nitrogen. These results do not agree with the reported non-existence of UOF_2 in the UO_2–UF_4 phase system and those obtained during a study of the ThO_2–UF_4 system[111]. In addition, attempts to prepare UOF_2 by the reaction between UF_4 and antimony sesquioxide (cf. the preparation of $ThOF_2$ and other oxydihalides discussed below) have resulted[112] only in the formation of a mixture of UO_2 and UF_4 at temperatures up to c. 350 °C. Obviously further research is required to clarify the present confused state of affairs.

Actinide oxydihalides, MOX_2 (M = Th $\rightarrow$ Np incl.; X = variously, F, Cl, Br and I) are readily obtained[112, 113] by reaction of the appropriate actinide tetrahalide and antimony sesquioxide *in vacuo* at c. 400 °C. Infrared spectral results show that all the bands occur below 600 cm^{-1}, indicating the absence of discrete MO^{2+} groups. This is confirmed for the isostructural oxydichlorides by the structure reported[114] for $PaOCl_2$. This consists of infinite polymeric chains which extend along the short c axis and which are cross-linked in the *ab* plane by bridging chlorine atoms. The protactinium atoms are 7-, 8- and 9-coordinate and oxygen atoms are 3- or 4-coordinate. Protactinium-oxygen and protactinium-chlorine bond distances lie within the ranges 2.19–2.38 and 2.76–3.08 Å, respectively (Figure 3.6). Unit cell parameters for the oxydichlorides are listed in Table 3.10; no structural information is available for the presently known oxydibromides and oxydiiodides.

The red and white forms of $ThOI_2$ have been reinvestigated and it is concluded that the former, obtained by heating together ThO_2 and ThI_4 in the presence of (but not necessarily in contact with) thorium metal or ThI_2, is coloured because slight reduction occurs[115]. It is readily converted to white $ThOI_2$ by heating with a trace of iodine. Conversion of the white to

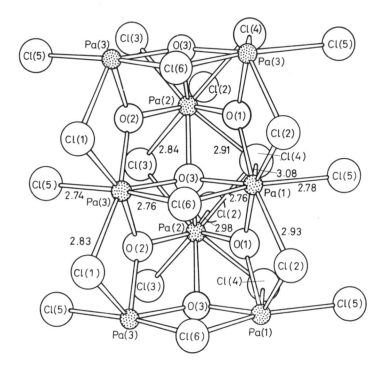

Selected bond lengths (Å)

Pa(1)—O(1)	2.191		Pa(2)—Cl(2')	2.982
Pa(1)—O(3)	2.376		Pa(2)—Cl(3)	2.843
Pa(1)—Cl(2)	2.925		Pa(2)—Cl(4)	2.913
Pa(1)—Cl(4)	3.082		Pa(3)—O(2)	2.226
Pa(1)—Cl(5)	2.778		Pa(3)—O(3)	2.341
Pa(1)—Cl(6)	2.762		Pa(3)—Cl(1)	2.826
Pa(2)—O(1)	2.261		Pa(3)—Cl(5)	2.743
Pa(2)—O(2)	2.268		Pa(3)—Cl(6)	2.760
Pa(2)—O(3)	2.337		Pa(3)—Cl(3)	3.462

Figure 3.6 The structure of PaOCl$_2$. (From Dodge et al.[114], by courtesy of Munksgaard)

the red form, although achieved in the presence of thorium metal, does not occur in the presence of hydrogen at 500–1000 °C or aluminium at 850 °C.

Dichlorodihydroxyuranium(IV), $U(OH)_2Cl_2$, is reported to form on evaporation of a solution of the tetrachloride in deoxygenated water at 100 °C followed by dissolution of the product in ethanol and removal of this solvent under reduced pressure at room temperature[116]. It is converted to the tetrachloride when treated with refluxing thionyl chloride.

3.3.4 Trivalent

Lucas and Rannon claim[117] to have obtained the first trivalent thorium oxyhalide, ThOF, by heating together stoicheiometric amounts of reagents for the reaction

$$Th + 2ThO_2 + ThF_4 \rightarrow 4ThOF$$

in a sealed nickel tube at 1200 °C. The product was FCC, as would be expected for ThOF, with $a_0 = 5.68$ Å. However, in view of the instability of trivalent thorium, more information is required on this reaction and on the properties of the product.

The structure of CfOF has been determined[114] by single crystal studies. In the FCC cell (fluorite structure) with $a_0 = 5.561$ Å the oxygen and fluorine atoms are randomly distributed over the anion sites 2.408 Å from the californium atom. A variety of trivalent oxyhalides have been prepared on the microgramme scale for curium[81], berkelium[65, 81, 121], californium[66, 82, 118–120], and einsteinium[66, 73] by heating the appropriate sesquioxide with a mixture of water vapour and the hydrogen halide. They are all isostructural with the earlier actinide (III) oxyhalides. Unit cell dimensions are listed in Table 3.10; the berkelium—oxygen and berkelium—chlorine distances in BkOCl (PbClF-type structure) are 2.32 Å (Bk—O), 3.07 Å (Bk—4Cl) and 3.05 Å (Bk—1Cl)[121].

3.4 HALOGENO-COMPLEXES

A wide range of preparative and structural studies have been reported during the past 3 years. Most interest, however, has centred on the tetravalent fluoro-complexes with relatively little work on trivalent complexes; in addition most of the work has been confined to the earlier actinide elements, Th–Pu inclusive.

3.4.1 Hexavalent

The existence of $UF_6 \cdot XeF_2$ is mentioned above[33]. The preparation of uranium(VI) fluoro-complexes with Group I and II metals has been described[122–124]. The compounds $Na_3U_2F_{15}$, and Na_3UF_9, and those of types M^IUF_7 (M^I = K, Rb and Cs) and $M_2^IUF_8$ (M^I = Na, K and Cs) have been prepared and their thermal stability investigated. Na_3UF_9 appears to form on decomposition of $Na_3U_2F_{15}$ [124] (158 °C) and of phases of composition $UF_6 \cdot 2.2 \to 2.5NaF$ [122] (170–260 °C); it loses UF_6 between 310 and 390 °C. The stability of the octafluoro-complexes, $M_2^I UF_8$, increases with the atomic weight of the alkali metal (Na to Cs); the lithium complex was not obtained[122]. The ultimate products of the thermal decomposition of the heptafluoro-uranates(VI), $M^I UF_7$ (M^I = K, Rb and Cs) are pentafluorouranates(IV)[124], $M^I UF_5$, the caesium salt decomposing initially to $CsUF_6$ which then loses fluorine above 585 °C.

Neptunium and plutonium hexafluoride undergo reduction on reaction with alkali-metal fluorides; these reactions are discussed below. The numerous separation procedures for actinide hexafluorides, based on adsorption-desorption processes, will not be discussed.

Sadikova and co-workers[125] report that $CsUF_7$ possesses cubic symmetry with $a_0 = 5.51$ Å at room temperature and tetragonal symmetry with $a_0 = 5.48$ Å, $c_0 = 5.33$ Å between 240 and 250 °C. Refinement of the structure of the former modification suggests, as might be expected, a statistical distribution of the seven fluorine atoms over the eight sites in a distorted cubic array.

3.4.2 Pentavalent

Uranium pentafluoride and xenon hexafluoride interact to form the yellow, air-sensitive solid $UF_5 \cdot XeF_6$ [126]. This complex, which can be recrystallised unchanged from anhydrous hydrofluoric acid, is also the ultimate product of the oxidation of uranium tetrafluoride by xenon hexafluoride; an unstable complex of approximate composition $UF_5 \cdot 1.75XeF_6$ is observed during this reaction. It will be interesting to have structural information on $UF_5 \cdot XeF_6$; the magnetic moment is reported to be 1.57 BM with $\theta = -55$ degrees.

Hydrazinium bishexafluorouranate(V), $N_2H_6(UF_6)_2$, is obtained when excess uranium hexafluoride is reacted with $N_2H_6F_2$ in anhydrous hydrogen fluoride whereas in the presence of excess hydrazinium fluoride the product is $N_2H_6UF_7$ [127,128]. The magnetic moment of the latter is 1.67 BM with $\theta = -111$ degrees. The ultimate product of thermal decomposition of these compounds is uranium tetrafluoride. U—F stretching vibrations have been assigned at 628 cm^{-1} (R) for $N_2H_6(UF_6)_2$, NH_3OHUF_6 and $N_2H_6UF_7$ and at 526 cm^{-1} (i.r.) for the first two and 435 cm^{-1} (i.r.) for the last[127].

The octafluoroneptunate(V), Na_3NpF_8, has been prepared by two methods. Thus, neptunium hexafluoride undergoes reduction in contact with sodium fluoride at 150–250 °C [129,130] and neptunium tetrafluoride–sodium fluoride mixtures are oxidised by fluorine at $c.$ 400 °C [131,132] (cf. the previous preparation of Rb_2NpF_7 and $CsNpF_6$ [1]). The former reaction reflects the relative stabilities of the actinide fluoro-complexes; with UF_6 stable hexavalent complexes are formed whilst with PuF_6 reduction to tetravalent complexes occurs. Some preliminary work[133], however, indicates that PuF_6 reacts with caesium fluoride to form a green solid which *may* be $CsPuF_6$; this decomposes in fluorine at temperatures up to 300 °C yielding hexagonal Cs_2PuF_6 which is reoxidised to $CsPuF_6$ above 300 °C in fluorine. It is interesting to note that an orthorhombic form of Cs_2PuF_6, prepared from aqueous solution[133], is stable on exposure to fluorine up to 500 °C. More information is obviously required on this potentially interesting system, particularly concerning the relative stabilities of $CsPuF_6$ in fluorine at temperatures below and above 300 °C.

Na_3NpF_8 is isostructural with Na_3PaF_8; single crystal studies have shown[132] that in the latter each protactinium atom is surrounded by eight equidistant-fluorine atoms (Pa—F = 2.21 Å) at the vertices of an almost perfect cube. A slight tetragonal distortion (F—F distances of 2.47 and 2.60 Å, respectively), as shown in Figure 3.7, results in D_{4h} symmetry. Metal—fluorine bond distances for Na_3UF_8 and Na_3NpF_8 are 2.21 and 2.19 Å, respectively.

The $KOsF_6$-type structure has been confirmed for $CsUF_6$ [134]. The uranium atom lies on a centre of symmetry with six equidistant fluorines (U—F,

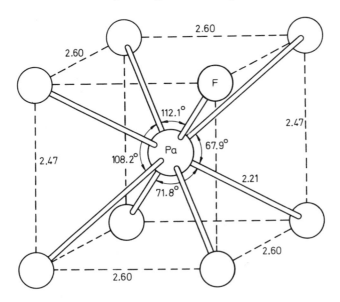

Figure 3.7 Coordination around the protactinium atom in Na_3PaF_8. (From Brown et al.[132], by courtesy of the Chemical Society)

2.057 Å) forming an octahedral array which is slightly flattened along the threefold axis; the F—U—F angles are 91.9 degrees and 88.1 degrees. The resulting axial distortion confirms deductions based on paramagnetic resonance studies[135]. The 'g' values obtained in the latter study were −0.768, −0.764 and −0.709, respectively, for $LiUF_6$, $NaUF_6$ and $CsUF_6$. Surprisingly, no signals were observed for the analogous potassium, ammonium, rubidium, silver and thallium complexes. Single crystal paramagnetic resonance results have also been reported[136] for certain hexafluorouranates(V).

Electronic spectral studies on $CsNpF_6$ have been reported[137] and the crystal-field splitting of the free-ion levels studied in detail[138]. A complete intermediate crystal-field coupling diagram for this f^2 configuration under O_h point-group symmetry has been constructed. Combined coulomb, spin-orbit and octahedral crystal-field spectroscopic parameters are[138], respectively, $E^1 = 3144$, $E^2 = 16.75$, $E^3 = 257.9$, $\zeta = 2230$, $B_0^4 = 1071$ and $B_0^6 = 551\ cm^{-1}$.

Montoloy and Plurien have obtained the first hydrated complexes of uranium(V)[139]. Compounds of the type $M^{II}U_2^V F_{12}\cdot 4H_2O$ (M^{II} = Co, Ni and Cu) precipitate on the addition of the appropriate difluoride to a solution of β-UF_5 in 48–50% HF; unfortunately the water content of the complexes was obtained by difference. $CoU_2F_{12}\cdot 4H_2O$ possesses triclinic symmetry with $a_0 = 11.52$, $b_0 = 10.11$, $c_0 = 5.22$, $\alpha = 91°\ 30'$, $\beta = 91°\ 30'$ and $\gamma = 90°$.

Selbin and co-workers have reported the preparation of $RbUCl_6$ and $(n\ C_3H_7)_4NUCl_6$ together with details of the visible and near infrared spectra of various U^V chloro-complexes in thionyl chloride solution[36]. Band assignments for $5f^1$ systems are discussed at length in this and other articles[4,140]; extensive tables of spectral data are given in the review article[4] together with an exhaustive list of references to the original publications. E.S.R. measurements[36] on a series of U^V chloro-complexes have indicated 'g' values of 1.1 (probable -ve) which compare with the value of -1.18 calculated previously for $UCl_5\cdot SOCl_2$.

Stumpp[141] has prepared a number of hexachlorouranates(V) of the type $M^I UCl_6$ (M^I = alkali metal, Tl, Ag and NH_4^+) and $Ba(UCl_6)_2$ by heating together UO_3 or U_3O_8, the appropriate metal chloride and thionyl chloride in a sealed tube at 180–200 °C. This method offers no advantages over the previous one involving iodine monochloride–thionyl chloride mixtures[142] or the one described by Selbin[36]. However, the preparation of hexachloro-uranates(V) by chlorine oxidation of UCl_6^{2-} complexes in nitromethane[143] provides an extremely attractive alternative to reactions in thionyl chloride. Ryan has prepared Ph_4AsUCl_6 and NEt_4UCl_6 by the chlorine oxidation technique but failed to isolate solid $CsUCl_6$ although this was obtained in propylene carbonate solution. The method is obviously widely applicable. Ph_4AsUCl_6 reacts slowly with 48% aqueous hydrofluoric acid to yield the corresponding hexafluoro-complex. Far infrared, assignments[140] have been made for NEt_4UCl_6 as $310\ cm^{-1}$ (v_3), and $122\ cm^{-1}$ (v_4) with other bands at 65 and $56\ cm^{-1}$. Attempts to prepare neptunium(V) chloro-complexes by this method have been unsuccessful[140,143].

Gruen and McBeth[37], who are investigating vapour phase complexes formed between various halides and aluminium trihalides, report the formation of $UCl_5\cdot AlCl_3$ in the vapour phase. The free energy change for the reaction,

$$UCl_{4(s)} + \tfrac{1}{2}Al_2Cl_{6(g)} + \tfrac{1}{2}Cl_{2(g)} = UCl_5\cdot AlCl_{3(g)}$$

is given by the equation,

$$\Delta G = 8914 - 10.74T\ cal\ mol^{-1}\ (440\text{–}630\ K)$$

and the partial pressure of the complex as a function of temperature, and $Al_2Cl_{6(g)}$ and $Cl_{2(g)}$ pressures is,

$$\log P_{UCl_5\cdot AlCl_3} = \frac{-1948}{T} + 2.347 + \tfrac{1}{2}\log P_{Al_2Cl_{6(g)}} + \tfrac{1}{2}\log P_{Cl_{2(g)}}$$

The complex is appreciably more volatile than uranium pentachloride itself under the same conditions; for example, the pressure of $UCl_5 \cdot AlCl_3$ (1 atm Cl_2; 1 atm Al_2Cl_6) is 34 mm at 500 K whilst for U_2Cl_{10} (1 atm Cl_2) it is only 0.44 mm. The vapour phase spectrum is remarkably similar to that of the U_2Cl_{10} dimer (Figure 3.1) and those of complexes of the type M^IUCl_6 suggesting the existence of species of the type $AlCl_4(UCl_6)$.

A new octachloroprotactinate(V), $(NO)_3PaCl_8$, forms when nitrosyl chloride is allowed to react with protactinium pentachloride[144]. Under the same conditions niobium(V) and tantalum(V) form only hexachloro-complexes. Raman spectral data are reported together with assignments for the species $PaCl_8^{3-}$, UCl_8^{3-} and UCl_6^-.

The preparation of hexabromouranates(V) has been achieved by bromine oxidation of uranium tetrabromide in the presence of suitable univalent bromides (e.g. NEt_4Br and Ph_4AsBr) in either nitromethane or methyl cyanide[140] and also by metathesis of hexachlorouranates(V) with liquid boron tribromide[38]. The complexes are dark brown-black, moisture sensitive solids which can be handled safely in oxygen-free, non-aqueous solvents. Their visible and near infrared spectra are very similar to those recorded for hexachlorouranates(V); the results are discussed in detail by Ryan[140]. Far infrared bands have been recorded for NEt_4UBr_6 at 214 cm^{-1} (ν_3), 87 cm^{-1} (ν_4), 68 and 62 cm^{-1} (ν_6)[140]. Analogous hexabromoprotactinates(V) can also be prepared by reactions in liquid boron tribromide[38].

Ryan[140] also reports evidence for the existence of UI_6^- complexes. Black solids are obtained by condensing anhydrous hydrogen iodide onto, for example, NEt_4UCl_6 or Ph_4AsUCl_6 at $-78\,°C$; these products are thermally unstable, decomposing irreversibly at $-30\,°C$ with the liberation of I_3^- species. Metatheses in solution have also produced intensely coloured, unstable products believed to be UI_6^- species.

3.4.3 Tetravalent

Several new tetravalent fluoro-complexes have been prepared during the past 3 years and some very interesting structural studies have been reported; in addition, Penneman[145] has published an important paper which high-lights the use of molar refractivity as a diagnostic tool for determining the composition of transition metal fluorides.

A range of not very interesting uranium(IV) fluoro-complexes of the type RUF_5 and R_2UF_6 (R = an organic base such as primary amines, substituted pyridines, etc.) have been prepared[146, 147] by photolysis of aqueous solutions of uranyl fluoride and the appropriate base; under similar conditions $UO_2F_2 \cdot (N_2H_4 \cdot 2HF) \cdot H_2O$ yields $UF_4 \cdot H_2O$.

Two new hydrazinium fluorouranates(IV), $(N_2H_5)_2UF_6$ and $(N_2H_5)_3UF_7$, have been identified during an investigation of the $N_2H_5F-UF_4-H_2O$ and $N_2H_5F-UF_4-N_2H_4$ systems[148]. Both decompose in vacuo with the inter-mediate formation of $N_2H_5UF_5$ at 110 °C. Infrared bands at 330 cm^{-1} have been assigned to U—F stretching vibrations for each compound.

Thalmayer and Cohen[149] have prepared two crystalline forms of K_2NpF_6. α-K_2NpF_6, which is prepared by oxidation of an unstable Np^{III} fluoro-complex obtained from potassium fluoride solution, is a green solid which possesses cubic symmetry, space group O_h^5-$Fm3m$ with $a_0 = 5.905$ Å, whilst β_1-K_2NpF_6, obtained by the addition of Np^{IV} in dilute DCl to a saturated KF solution, is hexagonal, space group $P\bar{6}2m$ with $a_0 = 6.56$ and $c_0 = 3.73$ Å. Each polymorph is isostructural with analogous Th^{IV} and U^{IV} hexafluoro-complexes[1].

The reduction of plutonium hexafluoride on contact with LiF [150-152] and NaF [153, 154] yields, respectively, $LiPuF_5$ (300 °C) and Li_4PuF_8 (400 °C), and Na_3PuF_7 (100 °C). Plutonium hexafluoride can be recovered from $LiPuF_5$ by heating this in fluorine at 300–400 °C,

$$4LiPuF_5 + 3F_2 \rightarrow Li_4PuF_8 + 3PuF_6$$

the resulting Li_4PuF_8 reacting with fluorine more slowly to release more hexafluoride. Na_3PuF_7 is converted to $Na_7Pu_6F_{31}$ when heated alone in fluorine at 150 °C but is stable in the presence of excess sodium fluoride. $Na_7Pu_6F_{31}$ reacts further with fluorine to yield β_3-Na_2PuF_6, which possesses hexagonal symmetry with $a_0 = 6.074$ and $c_0 = 7.15$ Å[154].

Cs_2PuF_6, prepared by treating Cs_2PuCl_6 with aqueous hydrofluoric acid, is reported to possess orthorhombic symmetry with $a_0 = 12.145$, $b_0 = 7.156$ and $c_0 = 4.056$ Å whereas the hexagonal form, $a_0 = 7.185$ and $c_0 = 4.11$ Å, is formed when $CsPuF_6$ is heated in fluorine at 300 °C[133]. The two forms are appreciably different in their reactivity towards fluorine, the former being stable to 500 °C whilst the latter is oxidised to $CsPuF_6$ at 400 °C. In the presence of fluorine at 500 °C caesium fluoride and an excess of plutonium tetrafluoride react to yield the pentafluoro-complex $CsPuF_5$. It would be interesting to have more information on the reactions of the various pentavalent and tetravalent plutonium fluoro-complexes.

$Na_7Cf_6F_{31}$, the first tetravalent fluoro-complex of californium, has been prepared[42] on the microgramme scale; it is isostructural with the analogous compounds formed by the earlier actinides[155].

Details of the preparation of complexes of the type $M^{II}M^{IV}F_6$ (M^{II} = Ca, Sr, Ba etc.; M^{IV} = variously Th–Pu incl.) have appeared[156]; the existence of $CaUF_6$ has been confirmed during an investigation of the CaF–UF_4 phase system[157]. Although the $M^{II}M^{IV}F_6$ complexes are all reported by Keller and Salzer[156] to possess the LaF_3-type structure, a single crystal structure study has indicated that $CaUF_6$ in fact possesses a smaller cell with the Na_3As-type structure[158]. The unit cell, space group D_{6h}^4-$C6/mmc$, has $a_0 = 3.997$ and $c_0 = 7.103$ Å, compared with that reported earlier[156], $a_0 = 6.928$, $c_0 = 7.171$ Å, space group D_{3d}^4-$P\bar{3}c1$. It is unfortunate that the authors[158] have used intensities measured from powder photographs and not single crystal reflections when deducing the atomic parameters and, in addition, base their arguments on the LaF_3 work of Oftedal[159] (space group D_{6h}^3-$P6_3/mmc$) and not on the more reliable work on LaF_3 by Zalkin and co-workers[160] who report the larger cell observed by Oftedal but with the space group D_{3d}^4-$P\bar{3}c1$ and different fluorine parameters. In view of this, and the 11-coordination reported for the U atom[158] (cf. the same coordination number reported for La by Oftedal[159]) these results are obviously unreliable.

Numerous single crystal studies have been reported for tetravalent thorium and uranium fluoro-complexes and for convenience the results are summarised in Table 3.11. A review by Penneman on actinide fluoro-complexes, including detailed discussion of the structural work, will appear shortly[161]. It is interesting to note the different structures exhibited by $(NH_4)_4ThF_8$[162] with 9-coordinate Th atoms (cf. $(NH_4)_3ThF_7$[163], CN = 9 and K_5ThF_9[164], CN = 8) and $(NH_4)_4UF_8$[165], which contains discrete units in which the U atoms are 8-coordinate. The structural unit in $(NH_4)_4ThF_8$, an infinite chain of composition $(ThF_7)_n^{3-}$, parallel to b_0, formed by the sharing of polyhedron edges, is similar to that found in K_2PaF_7. There is an ionic

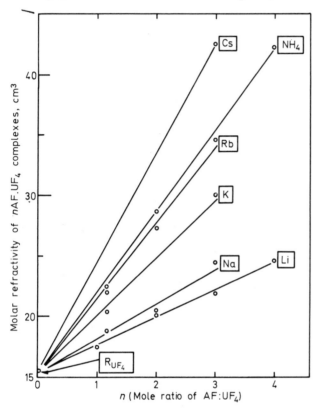

Figure 3.8 Molar refractivities of alkali fluoride-uranium tetra-fluoride complexes. (From Penneman[145], by courtesy of the American Chemical Society)

fluoride in both $(NH_4)_4ThF_8$ and K_5ThF_9, the nearest approaches to thorium atoms being 4.98 and 4.53 Å, respectively.

X-ray diffraction results have also been reported for $N_2H_5UF_5$[169] and NH_3OHUF_5[170], both of which possess orthorhombic symmetry.

Penneman[145] has drawn attention to the value of molar refractivity in determining the composition of transition metal fluorides. Essentially he demonstrates that the composition of alkali-metal (and ammonium) fluoro-complexes of tetravalent d- and f-transition elements can be determined by measurement of their molar refractivity, R, since $R_{nAF\cdot MF_4} = nR_{AF} + R_{MF_4}$.

Table 3.11 Structural data for actinide(IV) fluoro-complexes

Compound	Symmetry; Space group	Lattice parameters (Å)			$\rho g\ cm^{-3}$ calc.	Coordination number	Stereochemistry around M^{4+}	Bond lengths	Reference
		a_0	b_0	c_0					
$(NH_4)_3ThF_7$	Orthorhombic; D_{2h}^{16}–$Pnma$	13.944	7.928	7.041	3.57	9	Tricapped trigonal prismatic array of nine fluorines around each thorium. The chains run parallel to b_0 with Th atoms located on ac mirror planes.	Th—F unshared = 2.25 Å; Th—F shared = 2.59 Å	163
$(NH_4)_4ThF_8$	Triclinic; C_i^1–$P\bar{1}$	8.477 $\alpha = 88°\,23'$	8.364 $\gamma = 106°\,20'$	7.308 $\beta = 96°\,5'$	3.06	9	A distorted tricapped trigonal prismatic array with the three equatorial fluorines approximately normal to face centres. ThF_7^{3-} polyhedra joined into an infinite chain parallel to b_0.	Th—F unshared = 2.26–2.41 Å; Th—F shared = 2.42–2.45 Å; 1 non-bonded F is 4.98 Å from nearest Th	162
K_5ThF_9	Orthorhombic; C_{2v}^{12}–$Cmc2_1$	7.848	12.840	10.785	3.66	8	An irregular dodecahedral array with triangular faces.	Av. Th—F = 2.33 Å, 1 non-bonded F is 4.53 Å from nearest Th.	164
$(Na, Li)_7\ Th_6F_{31}$	Hexagonal; C_{3v}^3–$P3c1$	9.9056	—	13.2820	6.045	Th(1) 9 Th(2) 10	A tripyramidal array around Th(1) with an additional (fourth) fluorine in the centre girdle of this tripyramidal array for Th(2).	Th(1)—9F = 2.28–2.52 Å; Th(2)—10F = 2.29–2.57 Å	168
β_1–K_2UF_6	Hexagonal; D_{3h}^3–$P\bar{6}2m$	6.5528	—	3.749	5.123	9	A trigonal prismatic array with a pyramid on each of the prism faces. Chains run parallel to c_0.	U—6F shared = 2.38 Å; U—3F unshared = 2.22 Å	166

Table 3.11 Structural data for actinide(IV) fluoro-complexes

NaKThF$_6$	Hexagonal; C_{3i}^1–$P\bar{3}$	6.3073	—	7.8907	4.985	9	A trigonal prismatic array with a pyramid on each prism face. Chains run parallel to c_0.	Th—3F shared = 2.40 Å Th—3F shared = 2.42 Å Th—3F unshared = 2.28 Å	166
KU$_2$F$_9$	Orthorhombic; D_{2h}^{16}–$Pnma$	8.7021	11.4769	7.0350	6.485	9	A trigonal prismatic array with a pyramid on each of the three faces.	U—9F = 2.29–2.39 Å	167
(NH$_4$)$_4$UF$_8$	Monoclinic; C_{2h}^6–$C2/c$	13.126	6.692 $\beta = 121°\,19'$	13.717	2.982	8	A discrete distorted tetragonal antiprismatic array.	U—F = 2.25–2.33 Å	165

Thus, the absolute value of $R_{nAF\cdot MF_4}$ depends on R_{MF_4} and the mole ratio $AF:MF_4$, the increment per mole of AF being constant. This is illustrated in Figure 3.8 for uranium(IV) fluoro-complexes where it is clearly seen that different slopes are observed for different alkali-metal fluorides but that all plots have the same origin at R_{UF_4} (15.5 cm^3). It is also interesting that co-ordination number appears to have little or no effect on the value of $R_{nAF\cdot MF_4}$. This additivity also applies to penta- and trivalent fluoro-complexes.

Hydrated uranium(IV) fluoro-complexes of the type $M^{II}U_2F_{10}.8H_2O$ (M^{II} = Co, Ni and Cu) and the compounds $MnUF_6\cdot 3H_2O$ and $ZnUF_6\cdot 5H_2O$ have been prepared[171, 172]; unfortunately the water content of the complexes has not been confirmed analytically. The last two possess monoclinic and orthorhombic symmetry, respectively[173].

Uranium tetrachloride, like the pentachloride, forms a stable complex with aluminium trichloride in the vapour phase, which is appreciably more volatile than the binary halide itself[37, 174]. The free energy change for the reaction,

$$UCl_{4(s)} + Al_2Cl_{6(g)} \rightarrow UCl_4\cdot Al_2Cl_{6(g)}$$

is given by the equation,

$$\Delta G = 15\,780 - 15.30T \ (600\text{--}800 \text{ K})$$

The partial pressure of the complex can be represented as

$$\log P_{COMP} = -\frac{3450}{T} + 3.34 + \log P_{Al_2Cl_6}$$

typical volatility ratios, $V_r(P_{COMP}/P_{UCl4})$ are $\sim 10^7$ (500 K), $\sim 10^4$ (600 K) and $\sim 10^3$ (700 K).

Hexachlorouranates(IV) of the type $M_2^I UCl_6$ are obtained from reactions involving uranium tetra-acetate and the appropriate alkali metal acetate (M = K, Rb and Cs) in acetyl chloride[175]. $(NH_4)_2UCl_6$ and $(NMe_4)_2 UCl_6$ can be prepared in a similar manner and reactions in acetyl bromide have yielded $(NMe_4)_2UBr_6$ and $(NEt_4)_2UBr_6$. The previously reported[176] cubic symmetry for the tetramethylammonium hexahalogeno-complexes has been confirmed[175]. Hexachlorometallates(IV) of the type $(RH_2)_2MCl_6$ where R = 1,10 phenanthroline and 2,2' bipyridyl have also been prepared for thorium[177] and uranium[178] and a similar series of hexabromouranates(IV) has been reported[178].

The berkelium(IV) chloro-complex Cs_2BkCl_6, which precipitates from aqueous hydrochloric acid, is not isostructural with the analogous complexes formed by the elements Th-Pu inclusive[1]. It possesses hexagonal symmetry, space group C_{6v}^4-$P6_3mc$, with a_0 = 7.45 and c_0 = 12.095 Å. In the *dhcp* structure each berkelium atom is surrounded by six equidistant chlorine atoms, Bk—Cl = 2.55 Å [179].

Details of phase studies have been reported for the systems, $PbCl_2/ThCl_4$ [180], $PbCl_2/UCl_4$ [181] and $MgCl_2/UCl_4$ [181]. The last is a simple eutectic system whilst in the $PbCl_2/UCl_4$ system evidence is observed for the existence of $UCl_4\cdot 4PbCl_2$.

Magnetic moments (no quoted temperature range) of 2.05 and 2.44 BM, respectively, have been reported for $BipyH_2UCl_6$ and $PhenH_2UCl_6$ [178] and

Trzebiatowski and Mulak[55] have recorded magnetic susceptibilities for a range of complexes of the types $MUCl_5$ and M_2UCl_6 (M = an alkali metal) and for CsU_2Cl_9 and Cs_3UCl_7 over the temperature range 84–1200 K. Temperature independent paramagnetism is observed for the hexachloro-uranates(IV) below 350 K as reported previously for Cs_2UCl_6 [182] and $(Ph_3PH)_2UCl_6$ etc.[183], for example, but temperature dependence is found at higher temperatures due to thermal population of higher levels of the $^3H_4$ term.

A few recent publications deal with infrared and Raman data for both hexachloro- and hexabromo-uranates(IV) [184–186]. The most interesting results are those of Woodward and Ware[184] who studied Raman and infrared spectra of the complexes dissolved in methyl cyanide solution. Observed vibrations were assigned for UCl_6^{2-} species as follows, 262 (v_3, M—Cl, i.r.); 121 (v_5, δCl—M—Cl, R); 237 (v_2, M—Cl, R); 299 (v_1, M—Cl, R). Using the 121 cm^{-1} value (v_5), the i.r. and R forbidden transition v_6 was calculated to occur at 86 cm^{-1}.

Chlorine oxidation of plutonium metal suspended in propane-1,2-diol carbonate (PDC) yields solutions containing both Pu^{III}, 40%, and Pu^{IV}, 60%[187]. On the addition of benzene the mixed valence state compound $[Pu^{III}\cdot7PDC]_2\cdot[Pu^{IV}Cl_6]_3$ precipitates. In the presence of a suitable cation (e.g. tetraethylammonium) and chloride ion oxidation to Pu^{IV} is complete.

3.4.4 Trivalent

Relatively little work has been reported on trivalent halogeno-complexes but, nonetheless, the first examples of complexes of uranium(III), americium(III) and berkelium(III) have been characterised. The most interesting report deals with the preparation of trivalent uranium complexes of the type $M^IUCl_4\cdot5H_2O$ (M^I = K, NH_4 and Rb) by the addition of oxygen-free, concentrated hydrochloric acid to trivalent suphate complexes[188]. The sodium and caesium salts are too unstable to permit characterisation. It is likely, in view of these results, that trivalent neptunium halogeno-complexes could be prepared from aqueous solution and, it would be worth while re-investigating the preparation of plutonium(III) complexes from aqueous solution. Ryan[189] has prepared hexachloro- and hexabromo-complexes of Pu^{III} and Am^{III} from non-aqueous solvents saturated with the appropriate halogen acid but failed to obtain the analogous U^{III} compounds owing to rapid oxidation even in the absence of air. The $PuCl_6^{3-}$ and $PuBr_6^{3-}$ species are oxidised to plutonium(IV) species on exposure to air. Other trivalent americium complexes have been prepared[190, 191]; $CsAmCl_4\cdot xH_2O$ (x = c. 4) precipitates from 11 M hydrochloric acid solution saturated with hydrogen chloride; Cs_3AmCl_6 is obtained by the addition of ethanol to hydrochloric acid solutions of Am^{III} containing caesium chloride whilst in the presence of both sodium and caesium chlorides $Cs_2NaAmCl_6$ crystallises[191]. This complex, and the corresponding berkelium(III) compound, $Cs_2NaBkCl_6$, which has recently been prepared[179], both possess face-centred cubic symmetry with q_0 = 10.86 and 10.805 Å, respectively.

$CsAmCl_4 \cdot xH_2O$ can be dehydrated in hydrogen chloride at about 320 °C. The anhydrous product readily regains close to four molecules of water in an atmosphere containing a low partial pressure of water vapour.

Cs_3AmCl_6 also forms on evaporation of a 3 M hydrochloric acid solution containing stoicheiometric amounts (1:3) of Am^{III} and caesium chloride[190]. The complexes $(NEt_4)_3AmCl_6$ and $(NEt_4)_2LiAmCl_6$ are obtained from ethanol–acetone mixture; both solids, which are yellow, are non-hygroscopic[190]. Solution and solid state spectra have been recorded for these compounds and for others of less well-defined composition.

The magnetic moments of the above tetrachlorouranate(III) hydrates range from 3.32 to 3.36 BM with θ values between 80 and 68 °. Metal chlorine stretching vibrations are observed at 230 cm^{-1} (UCl$_4^-$ complexes), 218 cm^{-1} ($CsAmCl_4$), 235 and 197 cm^{-1} ($CsAmCl_4 \cdot xH_2O$), 242 cm^{-1} ($Cs_2NaAmCl_6$) and 214 cm^{-1} (Cs_3AmCl_6).

3.5   OXYHALOGENO–COMPLEXES

Although the majority of the recent publications deal with uranyl complexes, particularly spectral properties, the most interesting developments are concerned with the first preparation of an americium(VI) oxychloro-complex and a variety of protactinium(V), uranium(V) and neptunium(V) oxyhalogeno-complexes.

3.5.1   Hexavalent

A variety of uranyl fluoro-complexes have been identified[192–195] in the $M^IF–UO_2F_2–H_2O$ systems (M^I = K and Rb); the various complex potassium, rubidium and caesium salts isolated from such systems have been summarised[192]. Hydrated complexes of the type $M^{II}UO_2F_4 \cdot 4H_2O$ (M^{II} = Zn, Cd, Cu, Mn, Co and Ni) crystallise from acqueous solution[196] and several anhydrous and hydrated uranyl complexes with organic bases have been described[197]. Ozone oxidation of Np^{IV} or Np^V in hot solutions of saturated potassium fluoride results[149] in the formation of the bright green complex $K_3NpO_2F_5$, which is isostructural with its uranium (VI) analogue, possessing tetragonal symmetry (Table 3.12).

$Cs_3UO_2F_5$ is reported to be cubic (Table 3.12) whilst $(NH_4)_3UO_2F_5$ [199, 200], previously thought to be isostructural with $K_3UO_2F_5$ [201], does, in fact, possess monoclinic symmetry with $a_0 = 29.22$, $b_0 = 9.48$, $c_0 = 13.51$ Å and $\beta = 136° \, 7'$. The O and F atoms (U—O = 1.9 and U—F = 2.2 Å) are arranged as a slightly distorted pentagonal bipyramid.

Vorobei et al. report the formation of complexes of the types $M_2^IUO_2Cl_4$ (M^I = Na and K), $M^ICl \cdot 2UO_2Cl_2$ (M^I = Na and K) and $M^ICl \cdot 3UO_2Cl_2$ in the phase systems $NaCl–UO_2Cl_2$ and $KCl–UO_2Cl_2$ [202] and a series of non-stoicheiometric complexes $M_x^IUO_3Cl_x$ (M^I = K, Rb and Cs; x = ~0.8) have been prepared[203] by heating UO_3 with the appropriate alkali-metal chloride in oxygen above 500 °C. The analogous bromo-complex $K_xUO_3Br_x$ forms under similar conditions. When heated in oxygen and water vapour above 300 °C, the chloro-complexes decompose to compounds of the type $M_4^I U_5 O_{16} Cl_2$. Unit cell dimensions are listed in Table 3.12 together with those reported recently for $(NHEt_3)_2 UO_2 Cl_4$ [204]. The structure of $K_x UO_3 Cl_x$ has been described previously[205].

By treating the americium(V) complex $Cs_3 AmO_2 Cl_4$ with concentrated hydrochloric acid a dark red, crystalline solid, $Cs_2 AmO_2 Cl_4$ is obtained[191]. This reaction appears to involve oxidation rather than disproportionation

Table 3.12   X-ray diffraction data for some oxyhalogeno-complexes

Compound	Symmetry; Space group	Lattice parameters				Reference
		a_0 (Å)	b_0 (Å)	c_0 (Å)	β^0	
$Cs_3 UO_2 F_5$	C; —	9.869	—	—	—	198
$(NH_4)_3 UO_2 F_5$	M; C_s^3-Cm	29.22	9.48	13.51	136.11	200
$(Ph_4 As)_2 UOCl_5$	O; —	12.16	20.51	9.02	—	140
$K_x UO_3 Cl_x$	M; C_{2h}^3-$P2_1/m$	8.55	4.10	7.00	104.22	203
$Rb_x UO_3 Cl_x$	M; C_{2h}^3-$P2_1/m$	8.56	4.11	7.35	104.56	203
$Cs_x UO_3 Cl_x$	M; C_{2h}^3-$P2_1/m$	8.74	4.11	7.71	105.34	203
$K_x UO_3 Br_x$	M; C_{2h}^3-$P2_1/m$	9.57	4.14	6.89	111.17	203
$K_4 U_5 O_{16} Cl_2$	M; C_{2h}^5-$P2_1/c$	9.96	6.99	19.60	134.97	203
$Rb_4 U_5 O_{16} Cl_2$	M; C_{2h}^5-$P2_1/c$	10.21	7.01	19.64	134.22	203
$(NHEt_3)_2 UO_2 Cl_4$	T; C_{4h}^6-$I4_1/a$	13.50	—	24.32	—	204
$(NEt_4)_2 PaOCl_5$	M; C_{2h}^6-Cc	14.131	14.218	13.235	91.04	223
$K_3 NpO_2 F_5$	T; C_{4h}^6-$I4_1/a$	9.12	—	18.12	—	149
$KNpO_2 F_2$	R; D_{3d}^5-$R\bar{3}m$	6.80; $\alpha = 36.32°$	—	—	—	149
$Cs_2 AmO_2 Cl_4$	BCC; —	15.10	—	—	—	191

T: tetragonal; C: cubic; M: monoclinic; R: rhombohedral; BCC: body centred cubic; O: orthorhombic

since the supernatant contains less than one third of the original americium, the quantity required for the disproportionation,

$$3Am^V \rightarrow 2Am^{VI} + Am^{III}$$

$Cs_2 AmO_2 Cl_4$ is slightly soluble in concentrated hydrochloric acid, in which it is rapidly reduced to Am^{III}. It is dimorphic, one form possessing cubic symmetry (Table 3.12); the AmO_2^{2+} stretching vibration occurs at 902 cm^{-1} and americium-chlorine vibrations are observed at 313 and 244 cm^{-1} (cubic modification) and 303 and 230 cm^{-1} (monoclinic modification).

The majority of the papers on hexavalent oxyhalogeno-complexes are concerned with infrared and Raman results[97, 98, 192, 206–220]. They deal, for example, with the effects of cation size on the position of the asymmetric uranium-oxygen stretching vibration (v_3) in complexes of the types $M_2^I UO_2 Cl_4$ [206], and $M_3^I UO_2 F_5$ [218] and calculation of uranium-oxygen bond lengths[97, 98, 208] and force constants[97, 98, 208–210]. In complexes of the type $M_2^I UO_2 X_4$ (X = Cl or Br) the symmetric, Raman active uranium—oxygen stretching vibration (v_1) occurs between 831 and 840 cm^{-1} [98, 215], the bending mode (i.r. active) is observed at 256–265 cm^{-1} [194, 195] with i.r. active U—Cl [215] and U—Br [207, 215] stretching vibrations at c. 240 and 170 cm^{-1}, respectively, and the two Raman active vibrations[215] at c. 264 and 130 cm^{-1} (U—Cl) and 198 and 89 cm^{-1} (U—Br). The assignment[217] of an infrared vibration at 265 cm^{-1} to the U—Cl stretching vibration in $Cs_2 UO_2 Cl_4$ is probably a consequence of lack of resolution coupled with lack of Raman data. Uranium—oxygen bond lengths calculated for $Cs_2UO_2Cl_4$ are $1.72\,Å$ [97] and $1.74\,Å$ [98] whilst that for $Cs_2UO_2Br_4$ is $1.74\,Å$ [98] (cf. $1.69\,Å$ from x-ray studies[99]). The asymmetric uranium-oxygen stretching vibration in complexes of the type $M_3^IUO_2F_5$ occurs between 847 and 887 cm^{-1} [210, 212, 213, 218] (i.r.) and the O—U—O bending mode at 266–289 cm^{-1} [216, 218] (i.r.). Uranium–fluorine stretching vibrations are observed in their infrared spectra between 350 and 430 cm^{-1} [209, 210, 216, 218] with O—U—F and F—U—F vibrations in the ranges 211–221 cm^{-1} and 188–200 cm^{-1}, respectively[218]. Terminal U—F stretching vibrations occur between 350 and 430 cm^{-1} in complexes containing the $UO_2F_3^-$ ion with bridging fluorines around 250 cm^{-1} and the O—U—O bending mode at c. 250 cm^{-1} [216].

Electronic[221] and e.s.r.[222] spectral results have been reported for NpO_2^{2+} in a $Cs_2UO_2Cl_4$ matrix.

3.5.2 Pentavalent

The first examples of protactinium(V)[223] and uranium(V)[140] oxypenta-halogeno-complexes, $M_2^IM^VOX_5$ (M^I = univalent cation; M^V = Pa and U; X = variously F, Cl and Br) and the first americium(V) oxyhalogeno-complex, $Cs_3AmO_2Cl_4$ [191], have recently been characterised. Some penta-valent bisethoxytetrahalogenoprotactinates(V), $NEt_4Pa(OEt)_2X_4$ have also been prepared[223].

$NEt_4Pa(OEt)_2Cl_4$ and $NEt_4Pa(OEt)_2Br_4$ crystallise when the appropriate pentahalide and tetraethylammonium halide are mixed in alcohol[223]. Boiling with an excess of tetraethylammonium halide does not result in the formation of oxypentahalogenoprotactinates(V), behaviour which is quite different from that of niobium(V) [224]. However, treatment of $NEt_4Pa(OEt)_2X_4$-NEt_4X mixtures with commercial methyl cyanide (water content c. 0.5%) yields the bright yellow complexes $(NEt_4)_2PaOCl_5$ and $(NEt_4)_2PaOBr_5$, respectively. Oxypentahalogenouranates(V), $M_2^IUOX_5$ (M^I = Ph_4As and NEt_4; X = F, Cl and Br) and the complexes $(NEt_4)_2UOF_5\cdot2H_2O$, $(C_5H_5NH)_2UOCl_5\cdot2.5C_5H_5NHCl$ and $(NEt_4)_2UOBr_5\cdot2.5NEt_4Br$ are also obtained from the appropriate hexahalogenouranate(V) by reactions involving undried solvents (e.g. nitromethane or acetone) and undried reagents (NEt_4Cl, Ph_4AsF, NEt_4Br, etc.) at $-10\,°C$. Electronic and infrared spectra of these uranium(V) oxypentahalogeno-complexes are discussed by Ryan[140]. It is interesting to note that whereas two metal-oxygen stretching vibrations are observed for the uranium(V) complexes, only a single peak is found in the spectra of the protactinium(V) complexes (Table 3.13). This protactinium–oxygen band is some 100 cm^{-1} lower than expected for discrete $Pa = O^{3+}$ groups. However, conductivity and solution infrared data indicate the presence of this group (cf. complexes of the type $PaOX_3\cdot$ 2TPPO discussed later). $(NEt_4)_2PaOCl_5$ possesses monoclinic symmetry, space group C_{2h}^6-Cc with $a_0 = 14.131$, $b_0 = 14.218$, $c_0 = 13.235\,Å$ and $\beta = 91.04$ degrees. Full structural information will shortly be available[225] for this complex; in view of the infrared differences, it would be extremely interesting to have similar information on the U^V analogue.

The preparation of analogous oxyfluoroprotactinates(V) is currently being investigated[226].

$KNpO_2F_2$, which is isostructural with $KAmO_2F_2$, is obtained when Np^V in dilute acid solution is added to a saturated, aqueous potassium fluoride solution. A second phase obtained from similar reactions has not

Table 3.13 Infrared stretching vibrations for pentavalent oxypentahalogeno-complexes

Compound	Colour	$\nu_{M=O}(cm^{-1})$	$\nu_{M-X}(cm^{-1})$	Reference
$(NEt_4)_2UOF_5$	Violet-blue	853, 760	—	140
$(NEt_4)_2PaOCl_5$	Yellow	830	251s, 289w	223
$(NEt_4)_2UOCl_5$	Blue	913, 813	296sh, 253s, 197sh, 120m	140
$(NEt_4)_2PaOBr_5$	Yellow	840	—	223
$(NEt_4)_2UOBr_5 \cdot$ 2.5NEt_4Br	Green	919, 817	190s, 80m	140

X = F, Cl or Br; M = U or Pa
s, strong; m, medium; sh, shoulder

yet been characterised[149]. Complexes of the types $M_2^INpOF_5$ and $M_2^INpOBr_5$ still appear to be unknown.

Uranium(V) oxyfluoro-complexes of the type $M^IM^{II}U_2O_6F$ (M^I = Na, K, Rb and Tl; M^{II} = Ba, Sr and Pb) are obtained[227] by the following types of reaction carried out at 500–900 °C *in vacuo*,

$$\left.\begin{array}{l} 2M^IF + 2M^{II}UO_4 + 2UO_2 \\ M^{II}F_2 + M_2^IUO_4 + M^{II}UO_4 + 2UO_2 \\ M_2^IUO_4 + 2M^{II}O + UO_3 + 1.5UO_2 + 0.5UF_4 \end{array}\right\} \rightarrow 2M^IM^{II}U_2O_6F$$

No lithium or caesium salts can be obtained. The complexes all possess cubic symmetry, space group O_h^7-$Fd3m$ with a_0 values in the range 11.1–11.4 Å. Magnetic and spectral data[227, 228] indicate that uranium–oxygen octahedra exist in $KSrU_2O_6F$, $RbSrU_2O_6F$, $KBaU_2O_6F$ and $RbBaU_2O_6F$ but that this arrangement is distorted in the remaining compounds. The influence of ionic radius ratios on the existence of compounds of this type is discussed by Kemmler-Sack[227]. Other cubic oxyfluoride phases of the type $M_{0.33}^IPbU_2O_{5.67}F$ (M^I = K, Rb and Tl) have also been prepared by high temperature reactions[229]. $Pb_2U_2O_7$ forms only solid solutions with $NaUO_3F$ up to a 1:1 mole ratio. Magnetic and spectral data have been recorded for the above compounds which are stated to contain UO_2^+ groups.

$Cs_3AmO_2Cl_4$, the first pentavalent americiumoxychloro-complex to be prepared, is obtained[191] by addition of ethanol to a molar hydrochloric acid solution of Am^V containing caesium chloride. Americium – oxygen and americium—chlorine stretching vibrations occur at 800 and 290 cm^{-1}, respectively.

3.5.3 Tetravalent

No work has been reported on tetravalent oxyhalogeno-complexes but $U(OH)_2Cl_2$ is reported to react with triphenylphosphine and hydrogen chloride in aqueous solvents to yield $(Ph_3PH)_2U(OH)_2Cl_4$, a green, air-stable solid[116] which behaves as a 2:1 electrolyte in nitrobenzene.

3.6   HALIDE AND OXYHALIDE COMPLEXES WITH DONOR LIGANDS

Although a large number of new complexes have been reported together with details of infrared spectra in certain instances, detailed structural information is still not available for any actinide halide or oxyhalide complexes with donor ligands. Since many of the complexes have potentially high coordination numbers this type of information would be very valuable.

3.6.1   Hexavalent

Several new uranyl halide complexes with oxygen and nitrogen donor ligands have been reported during the past few years. Thus, uranyl chloride forms 1:1 complexes with $CNCl^{230}$, o-toluidine[231], m and p-phenylenediamine[231], hydrazine[232] and pyridine[233] (the last two being hydrated), 2:1 complexes with diphenylsulphoxide[234, 235], triphenylarsine oxide[214], quinolinol-N-oxide[236], a variety of aromatic amines[231] (e.g. diphenylamine, o-phenylenediamine, dibenzylamine, tribenzylamine etc.), and o-, m-, and p-aminobenzoic acid, o-, m-, and p-aminophenol, α, β and γ-picoline, 2-amino-pyridine and p-aminoacetanilide etc.[237], and a 1:4 complex with

Table 3.14   Infrared results for some uranyl halide complexes (cm^{-1})

Compound	Colour	$\nu_{UO^{2+}}$	$\nu_{L=O}$*	$\Delta\nu_{L=O}$	ν_{U-Cl}	Reference
$UO_2Cl_2 \cdot 2Ph_2SO$	Yellow	940	990	52	—	234
$UO_2Cl_2 \cdot 2Ph_3PO$	Yellow	921	1070	—	243	214, 239, 240
$UO_2Br_2 \cdot 2Ph_3PO$	Yellow	931	1060	—	238	214, 239, 240
$UO_2I_2 \cdot 5CO(NH_2)_2$	Orange-red	915	—	—	—	238
$UO_2I_2 \cdot 5CO(NH_2) \cdot H_2O$	Yellow	935	—	—	—	238
$UO_2Cl_2 \cdot NCCl$	Yellow	973	—	—	—	230
$UO_2Cl_2 \cdot N_2H_4 \cdot 6H_2O$	Yellow	925	—	—	—	232

*L = O; S = O, P = O or C = O as appropriate

pyridine[233]. Uranyl bromide forms 1:2 complexes with triphenylphosphine oxide and triphenylarsine oxide[214], and a 1:3 complex with diphenylsulphoxide[235] whilst uranyl iodide forms both an anhydrous and a monohydrated 1:5 complex with urea[238]. Apart from the uranyl iodide complexes, none of the above represent particularly original work or are of very much interest. $UO_2I_2 \cdot 5CO(NH_2)_2 \cdot H_2O$, an orange-red solid, which crystallises from aqueous solution, loses the water of hydration at room temperature when stored over sulphuric acid to yield the yellow compound $UO_2I_2 \cdot 5CO(NH_2)_2$. The latter forms the monohydrate on exposure to the atmosphere or on crystallisation from water. These complexes are probably the most stable uranyl iodide compounds presently known.

Infrared results for a few of the above complexes are listed in Table 3.14. Hart and Newberry[214], who have compared the spectra of the uranyl chloride and uranyl bromide complexes $UO_2X_2 \cdot 2Ph_3PO$ and $UO_2X_2 \cdot 2Ph_3AsO$ (X = Cl or Br), point out that the previous assignment[239] of bands around $274 \, cm^{-1}$ to U—Cl stretching vibrations is incorrect. They state that bands between 250 and $265 \, cm^{-1}$ are associated with O—U—O bending modes and that the U—Cl stretching vibrations occur between 238 and $243 \, cm^{-1}$ (Table 3.14). These assignments are more in line with those discussed above for U^{VI} oxytetrahalogeno-complexes. These authors, however, assign a band around $417 \, cm^{-1}$ to the U—OPPh$_3$ vibration in line with previous work by Day and Venanzi[239]. As discussed below for the tetrahalide complexes this must be viewed with some reservation.

3.6.2   Pentavalent

Protactinium pentachloride and pentabromide form both 1:1 and 1:2 complexes with triphenylphosphine oxide[223], the latter crystallising from non-aqueous solvents such as methylene dichloride and methyl cyanide; uranium pentachloride also forms a 1:2 complex, $UCl_5 \cdot 2Ph_3PO$, in addition to the previously known[241] 1:1 complexes with phosphine oxides. The first complexes of uranium pentabromide, $UBr_5 \cdot Ph_3PO$ and $UBr_5 \cdot [N(CH_3)_2]_3$ PO, have recently been prepared[38, 143] by oxidation of uranium tetrabromide

Table 3.15   Infrared data for some pentavalent halide and oxyhalide complexes
          (cm^{-1})

Compound	Colour	$v_{L=O}$*	$\Delta v_{L=O}$	v_{M-X}	Reference
$PaCl_5 \cdot Ph_3PO$	Yellow	975	217	—	223
$PaCl_5 \cdot 2Ph_3PO$	Bright yellow	1038	154	—	223
$UCl_5 \cdot Ph_3PO$	Yellow	970	222	285	241
$UCl_5 \cdot 2Ph_3PO$	Yellow	1055	137	—	223
$PaBr_5 \cdot Ph_3PO$	Orange	960	232	—	223
$PaBr_5 \cdot 2Ph_3PO$	Orange	985	217	—	223
$UBr_5 \cdot Ph_3PO$	Dark red	960	232	214	38, 143
$UBr_5 \cdot [N(CH_3)_2]_3PO$	Dark red	925	276	215	38, 143
$PaBr_5 \cdot 3CH_3CON(CH_3)_2$	Orange	1608	39	—	246
$UCl_5 \cdot TCAC$		1560, 1520	200, 240	320	36
$PaOCl_3 \cdot 2Ph_3PO$	Bright yellow	1075	117	254, 261	223
$PaOBr_3 \cdot 2Ph_3PO$	Bright yellow	1065	127	—	223
$Pa(OEt)_2Cl_3 \cdot Ph_3PO$	White	1068	124	—	223
$Pa(OEt)_2Br_3 \cdot Ph_3PO$	Pale yellow	1065	127	2?4?	223

*L = O refers to C = O or P = O as appropriate

by liquid bromine in methyl cyanide containing the appropriate ligand or by the reaction between liquid boron tribromide and the appropriate pentachloride complex. The dark red, crystalline products are extremely moisture-sensitive. Infrared spectral results for these phosphine oxide complexes are listed in Table 3.15. Preliminary work has indicated that the protactinium pentahalides also form complexes with ligands such as Ph_3PS and Ph_3PSe[242], analogous to those formed by niobium and tantalum pentahalides[243].

The complex formed between uranium pentachloride and trichloro-acrylyl chloride (TCAC = $Cl_2 \cdot C \cdot CClCOCl$), has been shown[36] to be $UCl_5 \cdot TCAC$ and not, as previously reported[244], $5UCl_5 \cdot TCAC$. This complex exhibits a remarkably large C=O shift in the infrared spectrum just as the pentahalide phosphine oxide complexes exhibit a large P = O shift (Table 3.15). Spectral studies have been reported for $UCl_5 \cdot TCAC$ and $UCl_5 \cdot SOCl_2$ in a variety of solvents; the magnetic moment for the former is 1.5 BM and the 'g' value obtained from e.s.r. studies is c. 1.1 for both complexes and for $UCl_5 \cdot PCl_5$[36, 140].

Selbin and co-workers have reported[245] the first examples of uranium pentachloride complexes with ligands containing N, P, As, S, Se and Te as the donor atom. Thus, 1:1 complexes are formed with phenazine, 2,2′-bipyridyl, ethylene bis(diphenylphosphine), triphenylphosphine, diar-sine, triphenylbismuthine, Ph_2Se_2 and Ph_2Te_2 whilst 1:2 complexes are formed with pyridine, o-phenanthroline, 2-mercaptopyridine, 8-hydroxy-quinoline, pyrazine and phthalazine. All the new complexes are obtained by addition of the appropriate ligand to a benzene solution of $UCl_5 \cdot TCAC$; magnetic moments (RT values only) range from 1.095 to 3.77 BM. On the basis of optical spectra and e.s.r. spectra (g values all close to 2.0 with very narrow line widths) it is suggested that the single f-electron is heavily delocalised onto the ligands. The complexes $2UCl_5 \cdot Ph_3As$, possibly best represented as $[U_2Cl_9(Ph_3As)]Cl$, and $UCl_5 \cdot (oxine H)_4$ have also been characterised[245].

Protactinium ethoxyhalogeno-complexes of the type $Pa(OEt)_2X_3 \cdot Ph_3PO$ crystallise from ethanol solution at room temperature[233]. When heated in commercial methyl cyanide (water content c. 0.5%) with an excess of tri-phenylphosphine oxide they are converted to oxytrihalogeno-complexes of the type $PaOX_3 \cdot 2TPPO$ (X = Cl and Br). These compounds, the first examples of actinide(V) oxytrihalogeno-complexes, exhibit a smaller P= O shift than the pentachloride complexes (Table 3.15), as might be expected from the presence of the Pa=O group and the extra ligands. The Pa=O stretching frequency occurs around 830 cm^{-1} as observed for the oxypenta-halogeno-complexes discussed above.

The only other new pentavalent complex is $PaBr_3 \cdot 3[CH_3CON(CH_3)_2]$, an air-sensitive solid which is obtained from the components in methylene dichloride solution[246].

Attempts to prepare neptunium pentachloride or pentabromide tri-phenylphosphine oxide complexes by chlorine or bromine oxidation of the appropriate tetrahalide in non-aqueous solvents containing the ligand have been unsuccessful[143]. No complexes with donor ligands are known for these thermodynamically unstable pentahalides or for protactinium pentaiodide and uranium pentaiodide (the last also thermodynamically unstable).

3.6.3 Tetravalent

Numerous new complexes with oxygen and nitrogen donor ligands have been reported for both thorium and uranium tetrahalides but relatively few studies have been extended to include protactinium, neptunium and plutonium. In those instances where this has been done, however, interesting differences in behaviour have been observed.

Protactinium tetrachloride[247] forms the complexes $PaCl_4 \cdot 3DMA$ and $PaCl_4 \cdot 2.5DMA$ (DMA = dimethylacetamide), behaviour intermediate between that of thorium tetrachloride ($ThCl_4 \cdot 4DMA$) and uranium, neptunium and plutonium tetrachloride ($MCl_4 \cdot 2.5DMA$). Depending on the preparative conditions and solvent employed a range of actinide tetrabromide-DMA complexes are obtained; these are listed in Table 3.16 together with the tetrachloride complexes. In view of the changes discussed below for the tetrachloride-DMSO (DMSO = dimethylsulphoxide) complexes it would be interesting to have information on the neptunium and plutonium tetrabromide-DMA complexes. During this more recent study[247] it was shown that complexes originally reported as $ThBr_4 \cdot 4DMA$ and $UBr_4 \cdot 4DMA$ do,

Table 3.16   Actinide tetrahalide N,N-dimethylacetamide complexes*

Thorium	Protactinium	Uranium	Neptunium	Plutonium
$ThCl_4 \cdot 4DMA$	—	—	—	—
$ThCl_4 \cdot 3DMA$†	$PaCl_4 \cdot 3DMA$	—	—	—
—	$\gamma PaCl_4 \cdot 2.5DMA$	α, β and γ-$UCl_4 \cdot 2.5DMA$	α and $\gamma NpCl_4 \cdot 2.5DMA$	$\alpha PuCl_4 \cdot 2.5DMA$
α and β-$ThBr_4 \cdot 5DMA$	$\alpha PaBr_4 \cdot 5DMA$	$\alpha UBr_4 \cdot 5DMA$	—	—
$ThBr_4 \cdot 4DMA \cdot Ac$‡	—	$UBr_4 \cdot 4DMA \cdot Ac$‡	—	—
—	$PaBr_4 \cdot 2.5DMA$	$UBr_4 \cdot 2.5DMA$	—	—

*α, β and γ refer to different crystallographic modifications
†Observed only during vacuum thermal decomposition of $ThCl_4 \cdot 4DMA$ (160–204 °C)
‡Ac, acetone

in fact, crystallise with a molecule of acetone in the lattice; this is almost completely removed by prolonged pumping at room temperature. $\Delta\nu_{C=O}$ ranges from 39 to 45 cm^{-1} for the tetrachloride- and tetrabromide-DMA complexes[247], a surprising observation in view of the range of stoicheiometries observed, the differences noted for similar series of complexes with triphenylphosphine oxide (Table 3.19) and dimethylsulphoxide (Table 3.17) and, the probable dimeric nature of those complexes of the type $MX_4 \cdot 2.5DMA$.

Other new complexes with C=O ligands include the 1:3 complexes $UCl_4 \cdot 3L$ and $UBr_4 \cdot 3L$ (L = ethyl acetate)[248] and 1:8 complexes of the types $ThX_4 \cdot 8U$ [249-251] (U = urea, X = Cl, Br and I) and $UX_4 \cdot 8U$ [252] (X = Br and I only), and $UICl_3 \cdot 8U$. Numerous hydrated urea complexes containing fewer molecules of ligand have also been characterised[249-252]. Infrared results for the anhydrous complexes are listed in Table 3.17. $ThCl_4$ forms a 1:1 complex with 8-quinolinol[253] (ν_{Th-Cl} 240s, 250sh) and there is some evidence for the existence of $UCl_4 \cdot CH_3NO_2$, but this complex has not been fully characterised[248].

A reinvestigation of the thorium, uranium and plutonium tetrachloride-dimethylsulphoxide (DMSO) systems, together with the first results on the analogous protactinium and neptunium complexes has shown that three

Table 3.17 Infrared data for tetravalent complexes with oxygen donor ligands

Compound	Colour	$\nu_{L=O}$*	$\Delta\nu_{L=O}$	Reference
ThCl₄·8U	White	1640	50	251
ThBr₄·8U	White	1640	50	251
ThI₄·8U	Yellow	1645	45	251
UI₄·8U	Green	1641	49	252
UCl₄·3EA	Green	1631	111	248
UBr₄·3EA	Green	1617	124	248
ThCl₄·5DMSO	White	945	100	254
ThCl₄·3DMSO	White	960	85	254
PaCl₄·5DMSO	Yellow	940	105	254
PaCl₄·3DMSO	Yellow	950	95	254
UCl₄·7DMSO	Green	1047, 942	—, 103	254
UCl₄·5DMSO	Green	940	105	254
UCl₄·3DMSO	Green	960	85	254
NpCl₄·7DMSO	Brownish-pink	1047, 942	—, 103	254
NpCl₄·5DMSO	Pink	945	100	254
NpCl₄·3DMSO	Pink	962	83	254
PuCl₄·7DMSO	Red	1050, 945	—, 100	254
PuCl₄·3DMSO	Red	965	80	254

L = O refers to C = O or S = O as appropriate

Table 3.18 Actinide tetrachloride-dimethyl sulphoxide complexes and a comparison of their chemical behaviour

—	—	UCl₄·7DMSO	NpCl₄·7DMSO	PuCl₄·7DMSO
ThCl₄·5DMSO*	PaCl₄·5DMSO*	UCl₄·5DMSO	NpCl₄·5DMSO†	†
ThCl₄·3DMSO	PaCl₄·3DMSO	UCl₄·3DMSO*	NpCl₄·3DMSO*	PuCl₄·3DMSO*

Reaction conditions	Product, MCl₄·xDMSO, for				
	ThCl₄	PaCl₄	UCl₄	NpCl₄	PuCl₄
Tetrahalide‡ + excess of hot DMSO	§	**	7	7	7
MCl₄·7DMSO‖ + excess of CCl₄ (12 h)	5	5	5	3	3
Tetrahalide + excess of DMSO in acetone	5	5	3	3	—
MCl₄·3DMSO + excess of DMSO in acetone¶	5	5	3	3	3
MCl₄·7DMSO or MCl₄·5DMSO dissolved in hot CH₃CN or CH₃NO₂	5	**	3	3	3
MCl₄·7DMSO or MCl₄·5DMSO boiled with acetone	3	**	3	3	3

*The most stable complex
†Very unstable, readily converted into MCl₄·3DMSO
‡Cs₂PuCl₆ or PuCl₄·3DMSO in the case of plutonium
§Products of variable composition obtained
‖ThCl₄·8 → 10DMSO used
¶Initial solid complex not dissolved
**Oxidised to an uncharacterised white protactinium(V) compound

series of complexes are formed[254]. These are $MCl_4 \cdot 3DMSO$ (M = Th–Pu inclusive), $MCl_4 \cdot 5DMSO$ (M = Th–Np inclusive) and $MCl_4 \cdot 7DMSO$ (M = U–Pu inclusive). The compounds are listed in Table 3.18 together with some pertinent reactions.

The decrease in stability of the 1:7 complexes and increase in stability of the 1:3 complexes with increasing atomic number is probably a consequence of the decrease in ionic radius of the tetravalent elements (cf. the similar change in the DMA complexes discussed above). Infrared results (Table 3.17) indicate that in the 1:7 complexes two of the DMSO molecules are very weakly held in the lattice. The unambiguous assignment of the S=O stretching vibration in the spectra of the complexes was achieved by studies involving the deuterated ligand. It would be interesting to have structural information on $ThCl_4 \cdot 5DMSO$ which is monomeric non-electrolyte in nitromethane. $ThCl_4 \cdot 8DMSO$ is reported to exist[255] although it was not obtained pure in the above investigation. Thorium tetrachloride and tetrabromide 1:4 complexes with diphenylsulphoxide are also known[177]. The effect of ligand size on the nature of the complexes formed between the actinide tetrachlorides and sulphoxide ligands, e.g. diethyl-, diphenyl-, di-t-butyl- and di-α-naphthyl-sulphoxide is currently being investigated[256].

Protactinium, neptunium and plutonium tetrabromide and the tetrachlorides of the first two elements form 1:2 complexes with triphenylphosphine oxide and hexamethylphosphoramide[143, 257] analogous to those previously known for the uranium tetrahalides. Thorium tetrabromide, however, forms both 1:2 and 1:3 complexes with these ligands[177, 257], presumably as a consequence of the larger ionic radius of tetravalent thorium relative to the later actinides. The plutonium compounds, which are the first examples of complexes of plutonium tetrabromide, itself thermodynamically unstable, are obtained[143, 257] by bromine oxidation of plutonium tribromide in anhydrous methyl cyanide containing the appropriate ligand. The remaining complexes are obtained by direct reaction between ligand and tetrahalide in solvents such as acetone, methyl cyanide or ethyl acetate[177, 257]. Uranium tetrachloride-tributylphosphine oxide complexes (1:2, 1:3.5 and 1:4) have been isolated[258] from reactions in aqueous hydrochloric acid or benzene; oily liquids of composition $UCl_4 \cdot 5TBPO$ and $UCl_4 \cdot 8TBPO$ have also been recorded but these may not have been pure phases. The tribromoethoxy complexes $ThBr_3(OEt) \cdot 3Ph_3PO$ and $ThBr_3$ (OET)$\cdot 2Ph_3PO$ have been isolated from boiling ethanol solution[177].

Pertinent infrared data are listed in Table 3.19. The previous assignment[239] of a band at 417 cm^{-1} to the U—O stretching vibration of $UCl_4 \cdot 2Ph_3PO$ has been queried[257] in view of other changes which occur in this region of the spectrum on coordination of triphenylphosphine oxide and, in particular, since similar bands are found in the infrared spectra of triphenylphosphine oxide complexes of, for example, various pentahalides, tin tetrahalides, indium trihalides and uranyl halides. On the basis of $\Delta\nu_{P=O}$ values for these various complexes (ranging from 60 to 220 cm^{-1}) and in view of the large differences in the molecular weights of the binary halides, one would certainly anticipate a large variation in the positions of the metal-oxygen stretching vibrations which have, for example, been assigned to bands in the region 311–346 cm^{-1} for the tin tetrahalide complexes. It would be valuable to have structural information coupled with a full co-ordinate analysis of infrared and Raman data on one of the actinide tetra-halide complexes to resolve this question.

Actinide tetrachloride complexes (Th–Pu inclusive) formed with trimethyl-phosphine oxide and octamethylpyrophosphoramide are currently being studied in order to obtain information on the effect of ligand size and the effect of the actinide contraction on the nature of the complexes formed[259].

Several new complexes with nitrogen donors have been reported for thorium and uranium tetrahalides but the only new results on protactinium, neptunium and plutonium are concerned with the preparation of 1:4

Table 3.19　Infrared results for actinide tetrahalide-phosphine oxide complexes[257, 258]

Compound	Colour	$\nu_{P=O}$	$\Delta\nu_{P=O}$	ν_{M-X}
$ThCl_4 \cdot 2TPPO$	White	1070	121	261s, 256sh
$PaCl_4 \cdot 2TPPO$	Yellow	1070	121	265s, 256sh
$UCl_4 \cdot 2TPPO$	Pale green	1068	124	265s, 258sh
$NpCl_4 \cdot 2TPPO$	Pale green	1068	124	258s, br*
$ThBr_4 \cdot 3TPPO$	White	1055, 1070	147, 122	—
$ThBr_4 \cdot 2TPPO$	White	1040	152	—
$PaBr_4 \cdot 2TPPO$	Yellowish-green	1038	154	—
$UBr_4 \cdot 2TPPO$	Pale green	1038	154	187s
$NpBr_4 \cdot 2TPPO$	Pink	1040	152	—
$PuBr_4 \cdot 2TPPO$	Red	1038	154	—
$ThCl_4 \cdot 2HMPA$	White	1042	159	258s, br*
$PaCl_4 \cdot 2HMPA$	Yellowish-green	1042	159	260s, br*
$UCl_4 \cdot 2HMPA$	Pale green	1039	163	255s, br*
$NpCl_4 \cdot 2HMPA$	Pale green	1037	165	257s, br*
$ThBr_4 \cdot 3HMPA$	White	1071	130	—
$ThBr_4 \cdot 2HMPA$	White	1027	174	—
$PaBr_4 \cdot 2HMPA$	Yellowish-green	1023	178	—
$UBr_4 \cdot 2HMPA$	Pale green	1020	181	184s
$NpBr_4 \cdot 2HMPA$	Pink	1023	178	—
$PuBr_4 \cdot 2HMPA$	Red	1025	176	—
$UCl_4 \cdot 2TBPO$	Green	1045	115	—
$UCl_4 \cdot 3.5TBPO$	Green	1050	110	—
$UCl_4 \cdot 4TBPO$	Green	1060	100	—

*The centre of the broad M—X bands is quoted; it is possible, of course, that these bands are composed of more than one independent vibration.

TPPO = triphenylphosphine oxide; HMPA = hexamethylphosphoramide; TBPO = tributylphosphine oxide

tetrabromide-methyl cyanide complexes[38, 143] analogous to those known[2] for other actinide tetrachlorides and tetrabromides. Uranium tetrafluoride does not react with either liquid or gaseous ammonia at temperatures between 300 and 600 °C but ammoniates of UCl_4[260] and UI_4[261] have been investigated. In an interesting study involving reactions with CO_2 Berthold and Knecht[260] have shown that although ammoniates of composition $UCl_4 \cdot 7.3$–$7.5NH_3$ and $UCl_4 \cdot 9$–$10NH_3$ are formed under certain conditions only four molecules of ammonia are in fact coordinated to the uranium atom. $UCl_4 \cdot 4NH_3$ is formed when the above ammoniates are heated at 45 °C in nitrogen; at higher temperatures this decomposes to the 1:1 complex which is stable to 300 °C. UI_4 reacts[261] with ammonia at 50–100 °C to form a 1:10 adduct and at 150–200 °C to form complexes of composition $UI_4 \cdot 4$–$5NH_3$. The number of coordinated ammonia molecules in the tetra-iodide complexes has not been determined.

$ThCl_4$ forms 1:2 complexes with phenanthroline[177] and coordination via the azomethine nitrogen has been demonstrated[262] for a series of complexes of the type $ThCl_4 \cdot 2[R \cdot C_6H_4CH:NC_6H_3R'(OH)$-n] where $R' = H$ with $R = H$, p-NO_2, p-Br, p-NMe_2 and p-OMe, $R' = m$-NO_2 with $R = H$, p-Br and p-NMe_2 and, $R' = p$-NO_2 with $R = p$-NO_2 and p-Br. Similar complexes are known with the heterocyclic Schiff's bases $C_6H_5CH:NC_6H_5$ and $CH_3 \cdot C_6H_4CH:NC_6H_5$ [263].

Selbin and co-workers[245] report that $UCl_5 \cdot TCAC$ reacts with diarsine to yield a light yellow, very unstable solid which may be $UCl_4 \cdot diars$. This observation should stimulate a search for further actinide tetrahalide complexes with arsenic ligands.

3.6.4 Trivalent

Actinide trihalide complexes with donor ligands is still a very much neglected field and the only known examples are a few uranium compounds formed with the nitrogen donors ammonia and methyl cyanide.

Uranium trichloride[264, 265] and tribromide[265] form ammoniates of approximate composition $UCl_3 \cdot 7NH_3$ (6.6–7.4) and $UBr_3 \cdot 8NH_3$, respectively, at room temperature. Reactions between the uranium trichloride ammoniates and carbon dioxide have shown[264] that only three molecules of ammonia are coordinated and $UCl_3 \cdot 3NH_3$ is, in fact, formed when the ammoniates are heated in nitrogen at 45 °C. At higher temperatures this decomposes to the 1:1 complex, $UCl_3 \cdot NH_3$. Presumably $UBr_3 \cdot 8NH_3$ also contains loosely bound ammonia.

Uranium trichloride reacts with oxygen-free methyl cyanide only slowly at 80 °C to form $UCl_3 \cdot CH_3CN$ [266]. The $C \equiv N$ stretching vibrations are observed at 2270 and 2280 cm^{-1} for this complex, which is isostructural with $CeCl_3 \cdot CH_3CN$ obtained by the thermal decomposition of $CeCl_3 \cdot 2CH_3CN$.

References

1. Brown, D. (1968). *Halides of the Lanthanides and Actinides* (London: Wiley & Sons)
2. Bagnall, K. W. (1967). *Co-ord. Chem. Rev.*, **2**, 145; *Halogen Chemistry* (Ed. V. Gutmann), Vol. 3, 303 (London: Academic Press) 1969
3. Brown, D. (1969). Chapter 1 in *Advances in Inorganic and Radiochemistry*, Vol. 12, 1
4. Selbin, J. and Ortego, J. D. (1969). *Chem. Rev.*, **69**, 657
5. Keller, C. (1969). *Fortsch. Chem. Forsch.*, **13**, 1
6. 'Nouveau Traite de Chimie Minerale', (Ed. P. Pascal), (1970). Vol. XV (Paris: Masson)

7. Steindler, M. J. (1968). Conf. -680610, *Proc. Rocky Flats Fluoride Volatility Conference*
8. Macheret, G. and Bourgeois, M. (1969). *French report CEA-BIB-174*
9. Jarry, R. L. and Steindler, M. J. (1967). *J. Inorg. Nucl. Chem.*, **29**, 1591
10. Jarry, R. L. and Steindler, M. J. (1968). *J. Inorg. Nucl. Chem.*, **30**, 127
11. Trevorrow, L. E., Gerding, T. J. and Steindler, M. J. (1968). *J. Inorg. Nucl. Chem.*, **30**, 2671
12. Jarry, R. L. and Steindler, M. J. (1969). *J. Inorg. Nucl. Chem.*, **31**, 1847
13. Sakurai, T. and Iwasaki, M. (1968). *J. Phys. Chem.*, **72**, 1491
14. Luce, M., Benoit, R. and Hartsmann, O. (1968). *French report CEA-R-3558*
15. Luce, M. and Hartmanshenn, O. (1967). *J. Inorg. Nucl. Chem.*, **29**, 2823
16. Iwasaki, M. (1968). *J. Nucl. Mater.*, **25**, 216
17. Macheret, G. (1969). *Radiochem. Radioanalyt. Lett.*, **2**, 249
18. Steindler, M. J., Steidl, D. V. and Steunenberg, R. K. (1959). *Nucl. Sci. Eng.*, **6**, 333
19. Iwasaki, M. and Sakurai, T. (1965). *J. Nucl. Sci. Technol. Tokyo*, **2**, 225
20. Trevorrow, L. E., Gerding, T. J. and Steindler, M. J. (1969). *Inorg. Nucl. Chem. Lett.*, **5**, 837
21. O'Donnell, T. A. and Wilson, P. W. (1968). *Aust. J. Chem.*, **21**, 1415
22. O'Donnell, T. A., Stewart, D. F. and Wilson, P. W. (1966). *Inorg. Chem.*, **5**, 1438
23. Moncela, B. and Kikindai, J. (1968). *Compt. Rend.*, **267C**, 1485
24. Claasen, H. H., Goodman, G. L., Holloway, J. H. and Selig, H. (1970). *J. Chem. Phys.*, **53**, 341
25. Gasner, E. L. and Frlec, B. (1968). *J. Chem. Phys.*, **49**, 5135
26. Kimura, M., Schomaker, V., Smith, D. W. and Weinstock, B. (1968). *J. Chem. Phys.*, **48**, 4007
27. Frlec, B. (1967). *J. Inorg. Nucl. Chem.*, **29**, 1804
28. Trevorrow, L. E., Steindler, M. J., Steidl, D. V. and Savage, J. T. (1967). *Inorg. Chem.*, **6**, 1060
29. Trevorrow, L. E., Steindler, M. J. and Steidl, D. V. (1967). *Adv. Chem. Ser. 71*; Lanthanide/Actinide Chemistry, 308, Washington, *Amer. Chem. Soc.*
30. Wedge, W. D. (1968). *USAEC Report K-1697*
31. Prusakov, V. N. and Ezhov, V. K. (1968). *Atomneya Energ.*, **25**, 35 and 64
32. Prusakov, V. N. and Ezhov, V. K. (1970). *Atomneya Energ.*, **28**, 496
33. Ezhov, V. K., Prusakov, V. N. and Chaivanova, B. B. (1970). *Atomneya Energ.*, **28**, 497
34. Fried, S. and Holloway, J. (1970). Personal communication
35. O'Donnell, T. A. and Wilson, P. W. (1969). *Aust. J. Chem.*, **22**, 1877
36. Selbin, J., Ortego, J. D. and Gritzner, N. (1968). *Inorg. Chem.*, **7**, 976
37. Gruen, D. M. and McBeth, R. L. (1969). *Inorg. Chem.*, **8**, 2625
38. Brown, D., Hill, J. and Rickard, C. E. F. (1970). *J. Chem. Soc. (A)*, 476
39. Lux, F., Wirth, G. and Bagnall, K. W. (1970). *Chem. Ber.*, **103**, 2807
40. Brown, D., Petcher, T. J. and Smith, A. J. (1969). *Acta Crystallogr.*, **25B**, 178
41. Asprey, L. B. and Keenan, T. (1968). *Inorg. Nucl. Chem. Lett.*, **4**, 537
42. Asprey, L. B. (1970). Personal communication
43. Keenan, T. and Asprey, L. B. (1969). *Inorg. Chem.*, **8**, 235
44. Ezhov., U. S., Akishin, P. A. and Rambidi, N. S. (1969). *Zh. strukt. Khim.*, **10**, 571
45. Gabuda, S. P., Matsutin, A. A. and Zadneprovskii, G. M. (1969). *Zh. strukt. Khim.*, **10**, 1115
46. Krasser, W. and Nuernberg, H. W. (1970). *Spectrochim. Acta.*, **26**, 1059
47. Ketov, A. N., Malitseva, N. A. and Shilgevskii, A. S. (1970). *Izv. Vyssh. Ucheb. Zaved Tsvet. Met.*, **13**, 94
48. Mitamura, T. (1969). *Denki Kagaku*, **37**, 235 (acc. to *NSA*, **24**, 13866)
49. Mucker, K., Smith, G. S., Johnson, Q. and Elson, R. E. (1969). *Acta Crystallogr.*, **25B**, 2362
50. Mooney, R. L. (1949). *Acta Crystallogr.*, **2**, 189
51. Hoard, J. L. and Silverton, J. V. (1963). *Inorg. Chem.*, **2**, 235
52. Brown, D., Petcher, T. J. and Smith, A. J. (1969). *Nature (London)*, **217**, 738
53. Brown, D., Petcher, T. J. and Smith, A. J. (1971). *J. Chem. Soc. (A)*, in press
54. Ezhov, U. S., Akishin, P. A. and Rimbidi, N. S. (1969). *Zh. strukt. Khim.*, **10**, 763
55. Trzebiatowski, W. and Mulak, J. (1970). *Bull. Acad. Pol. Sci. Ser. Sci. Chim.*, **18**, 121
56. Carlson, E. H. (1969). *Phys. Lett.*, **29A**, 696
57. Meyer, H. C., McDonald, P. F. and Settler, J. D. (1967). *Phys. Lett.*, **24A**, 569
58. Stone, J. A. and Pillinger, W. L. (1968). *U.S. Report Conf.* 660633-13

59. Clifton, J. R., Gruen, D. M. and Ron, A. (1969). *J. Chem. Phys.*, **51**, 224
60. Brown, D. and Moseley, P. T. (1970). Unpublished results
61. Khripin, L. A., Gagarinskii, Yu. V., Zadneprovskii, G. M. and Luk'yanova, L. A. (1968). *Sov. At. Energy*, **19**, 1398
62. Friedman, H., Weaver, C. F. and Grimes, W. R. (1970). *J. Inorg. Nucl. Chem.*, **32**, 3131
63. Khripin, L. A., Poduzova, S. A. and Zadneprovskii, G. M. (1968). *Zh. Neorg. Khim.*, **13**, 1439
64. Moseley, J. D. and Robinson, H. N. (1968). *J. Inorg. Nucl. Chem.*, **30**, 2277
65. Peterson, J. R. (1967). *U.S. Report UCRL*-17875
66. Fujita, D. K. (1969). *U.S. Report UCRL*-19507
67. Peterson, J. R. and Cunningham, B. B. (1968). *J. Inorg. Nucl. Chem.*, **30**, 1775
68. Secemski, E., Kiro, D., Low, W. and Schipper, D. J. (1970). *Phys. Lett.*, **31A**, 45
69. Kalvius, G. M., Ruby, S. L., Dunlap, B. D., Shenoy, G. K., Cohen, D. and Brodsky, M. B. (1969). *Phys. Lett.*, **29B** 489
70. Rasmussen, M. J., Stiffler, G. L. and Hopkins, H. H., Jr. (1968). *U.S. Report HW*-69738
71. Pappalardo, R. G., Carnell, W. T. and Fields, P. R. (1969). *J. Chem. Phys.*, **51**, 1182
72. Peterson, J. R. and Cunningham, B. B. (1968). *J. Inorg. Nucl. Chem.*, **30**, 823
73. Fujita, D. K., Cunningham, B. B. and Parsons, T. C. (1969). *Inorg. Nucl. Chem. Lett.*, **5**, 307
74. Green, J. L. and Cunningham, B. B. (1967). *Inorg. Nucl. Chem. Lett.*, **3**, 343
75. Burns, J. H. (1970). Personal communication
76. Burns, J. H. and Peterson, L. R. (1969). *U.S. Report ORNL*-4437, p. 33, (1970). *Acta Crystallogr.*, **26B**, 1885
77. Burns, J. H. and Peterson, L. R. (1969). *U.S. Report ORNL*-4437, p. 32. (1971). *Inorg. Chem.*, **10**, 147
78. Carnell, W. T., Fields, P. R. and Pappalardo, R. G. (1971). *J. Chem. Phys.*, in press
79. Carnell, W. T. (1970). Paper presented at the *Int. Conf. Coord. Chem.* (Poland: Krakow-Zakopane)
80. Brown, D., Fletcher, S. and Holah, D. G. (1968). *J. Chem. Soc. (A)*, 1889
81. Cohen, D., Fried, S., Siegel, S. and Tani, B. (1968). *Inorg. Nucl. Chem. Lett.*, **4**, 257
82. Fried, S., Cohen, D., Siegel, S. and Tani, B. (1968). *Inorg. Nucl. Chem. Lett.*, **4**, 495
83. Krupke, W. F. and Gruber, J. B. (1967). *J. Chem. Phys.*, **46**, 542
84. McLaughlin, R., White, R., Edelstein, N. and Conway, J. G. (1968). *J. Chem. Phys.*, **48**, 967
85. Edelstein, N. and Easley, W. C. (1968). *J. Chem. Phys.*, **48**, 2110
86. Edelstein, N., Mollet, H. F., Easley, W. C. and Mehlhorn, R. J. (1969). *J. Chem. Phys.*, **51**, 3281
87. Edelstein, N., Easley, W. C. and McLaughlin, R. (1966). *J. Chem. Phys.*, **44**, 3130
88. Edelstein, N., Conway, J. G., Fujita, D., Kolbe, W. and McLaughlin, R. (1970). *J. Chem. Phys.*, **52**, 6425
89. Edelstein, N. (1970). Personal communication
90. Fuller, J. E. (1968). *Report IS-T*-301
91. Kudyakov, Y. Ya., Smirnov, M. V., Chukreev, N. Ya. and Posokhin, Yu. V. (1968). *Sov. At. Energy*, **24**, 551
92. Guggenberger, L. J. and Jacobsen, R. A. (1968). *Inorg. Chem.*, **7**, 2257
93. Keenan, T. K. (1968). *Inorg. Nucl. Chem. Lett.*, **4**, 381
94. Bagnall, K. W., Brown, D. and Easey, J. F. (1968). *J. Chem. Soc. (A)*, 2223
95. Glukhov, J. A., Eliseev, S. S. and Vozhdaeva, E. E., (1968). *Russ. J. Inorg. Chem.*, **13**, 1488
96. Bullock, J. I. (1967). *J. Inorg. Nucl. Chem.*, **29**, 2257
97. Ohwada, K. (1968). *Spectrochim. Acta*, **24A**, 595
98. Bullock, J. I. (1969). *J. Chem. Soc. (A)*, 781
99. Mikhailov, Yu. N., Kuznetsov, V. G. and Kovaleva, E. S. (1965). *J. Struct. Chem. (USSR)*, **6**, 752
100. Brown, D. and Easey, J. F. (1970). *J. Chem. Soc. (A)*, 3378
101. Herak, R. (1969). *Acta Crystallogr.*, **25B**, 2505
102. Sellers, P. A., Fried, S., Elson, R. E. and Zachariasen, W. H. (1954). *J. Amer. Chem. Soc.*, **76**, 5935
103. Roberts, L. E. J. (1961). *Quart. Rev. Chem. Soc.*, **15**, 442
104. Roberts, L. E. J. and Walter, A. (1966). In 'Physico-Chimie du Protactinium', Publ. No. 154, *C.N.R.S., Paris*, 51

105. Stchouzkoy, T., Pezerat, H. and Muxart, R. (1966). In 'Physico-Chimie du Protactinium', Publ. No. 154, *C.N.R.S., Paris,* 61
106. Kemmler-Sack, S. (1969). *Z. Anorg. Allgem. Chem.,* **364,** 88
107. Glukhov, I. A., Eliseev, S. S. and Vozhdaeva, E. E. (1968). *Russ. J. Inorg. Chem.,* **13,** 483
108. Levet, J. C. (1969). *Compt. Rend.,* **268C,** 703
109. Satpathy, K. C. (1967). *Ind. J. Chem.,* **5,** 278
110. Vdovenko, V. M., Romanov, G. A. and Solutseva, L. V. (1967). *Radiokhimiya,* **9,** 727
111. Fonteneau, G. and Lucas, J. (1969). *Compt. Rend.,* **269C,** 760
112. Brown, D. and Easey, J. F., unpublished observations
113. Bagnall, K. W., Brown, D. and Easey, J. F. (1968). *J. Chem. Soc. (A),* 288
114. Dodge, R. P., Smith, G. S., Johnson, Q. and Elson, R. E. (1968). *Acta Crystallogr.,* **24B,** 304
115. Corbett, J. D., Guidotti, R. A. and Adolphson, D. (1969). *Inorg. Chem.,* **8,** 163
116. Hayton, B. and Smith, B. C. (1970). *J. Inorg. Nucl. Chem.,* **32,** 1219
117. Lucas, J. and Rannon, J. P. (1968). *Compt. Rend.,* **266C,** 1056
118. Peterson, J. R. and Burns, J. H. (1968). *J. Inorg. Nucl. Chem.,* **30,** 2955
119. Copeland, J. C. (1967). *U.S. Report UCRL*-17718
120. Copeland, J. C. and Cunningham, B. B. (1969). *J. Inorg. Nucl. Chem.,* **31,** 733
121. Peterson, J. R. and Cunningham, B. B. (1967). *Inorg. Nucl. Chem. Lett.,* **3,** 579
122. Peka, I. and Vachuska, J. (1967). *Coll. Czech. Chem. Comm.,* **32,** 426
123. Korinak, K. and Peka, I. (1968). *Coll. Czech. Chem. Comm.,* **33,** 3113
124. Nikolaev, N. S. and Sadikova, A. T. (1968). *Sov. At. Energy,* **25,** 1232
125. Sadikova, A. T., Sadikov, G. G. and Nikolaev, N. S. (1969). *Sov. At. Energy,* **26,** 313
126. Slivnik, J. and Frlec, B. (1970). *J. Inorg. Nucl. Chem.,* **32,** 1397
127. Frlec, B. B. and Hyman, H. (1967). *Inorg. Chem.,* **6,** 2233
128. Frlec, B. B., Brcic, B. and Slivnik, J. (1967). *Nukl. Inst.* Joseph Stefan, NIJS Porocilo *(acc. to Chem. Abstr.,* **69,** 64237q)
129. Trevorrow, L. E., Gerding, T. J. and Steindler, M. J. (1968). *Inorg. Chem.,* **7,** 2226
130. Katz, S. and Cathers, G. I. (1968). *Nucl. Appl.,* **5,** 206
131. Easey, J. F. (1968). *Ph.D. Thesis,* Leicester Univ.
132. Brown, D., Easey, J. F. and Rickard, C. E. F. (1969). *J. Chem. Soc. (A),* 1161
133. Riha, J. (1967). *U.S. Report ANL-7375,* 54
134. Rosenzweig, A. and Cromer, D. T. (1967). *Acta Crystallogr.,* **23,** 865
135. Rigny, P. and Plurien, P. (1967). *J. Phys. Chem. Solids,* **28,** 2589
136. Drifford, M., Rigny, P. and Plurien, P. (1968). *Phys. Rev. Lett.,* **27A,** 620
137. Varga, L. P., Asprey, L. B., Keenan, T. K. and Penneman, R. A. (1970). *J. Chem. Phys.,* **52,** 1664
138. Varga, L. P., Brown, J. D., Reisfeld, M. J. and Swan, R. D. (1970). *J. Chem. Phys.,* **52,** 4233
139. Montoloy, F. and Plurien, P. (1968). *Compt. Rend.,* **267C,** 1036
140. Ryan, J. (1971). *J. Inorg. Nucl. Chem.,* in press
141. Stumpp, E. (1969). *Naturwiss.,* **5,** 370
142. Bagnall, K. W., Brown, D. and du Preez, J. G. H. (1964). *J. Chem. Soc.,* 2603
143. Brown, D., Holah, D. G. and Rickard, C. E. F. (1968). *Chem. Comm.,* 651
144. MacCordick, J., Kaufman, G. and Rohmer, R. (1969). *J. Inorg. Nucl. Chem.,* **31,** 3059
145. Penneman, R. A. (1969). *Inorg. Chem.,* **8,** 1379
146. Satapathy, K. C. and Sahoo, B. (1970). *J. Inorg. Nucl. Chem.,* **32,** 549
147. Sahoo, B. and Satapathy, K. C. (1968). *Ind. J. Chem.,* **6,** 460
148. Glavnič, P. and Slivnik, J. (1970). *J. Inorg. Nucl. Chem.,* **32,** 2939
149. Thalmayer, C. E. and Cohen, D. (1967). *Adv. Chem. Ser.,* 71, Lanthanide/Actinide Chemistry, 256. Washington: Amer. Chem. Soc.
150. Steindler, M. J. (1967). *U.S. Report ANL-7438* and (1968) *ANL-7595*
151. Riha, J. and Trevorrow, L. E. (1968). *U.S. Report ANL-7575*
152. Katz, S. and Cathers, G. I. (1968). *Nucl. Appl.,* **5,** 5
153. *U.S. Report ANL-7279,* 1966, 68
154. Riha, J. and Trevorrow, L. E. (1968). *U.S. Report ANL-7425,* 52
155. Keenan, T. K. (1966). *Inorg. Nucl. Chem. Lett.,* **2,** 153, 211
156. Keller, C. and Salzer, M. (1967). *J. Inorg. Nucl. Chem.,* **29,** 2925
157. Nekrasova, N. P., Oblomeev, E. N., Golovanova, V. N. and Beznosikova, A, V. (1967). *Sov. At. Energy,* **22,** 367

158. Chebotarev, N. T. and Beznosikova, A. V. (1968). *Sov. At. Energy*, **25**, 1119
159. Oftdel, I. (1929). *Z. Phys. Chem.*, **B5**, 272; 1931, **B13**, 190
160. Zalkin, A., Templeton, D. H. and Hopkins, T. E. (1966). *Inorg. Chem.*, **5**, 1467
161. Penneman, R. A., to be published.
162. Ryan, J. J., Penneman, R. A. and Rosenzweig, A. (1968). *Chem. Comm.*, 990; (1969) *Acta Crystallogr.*, **25B**, 1958
163. Penneman, R. A., Ryan, J. J. and Kressin, I. K. (1970). *Acta Crystallogr.*, in press
164. Ryan, J. J. and Penneman, R. A. (1970). *Acta Crystallogr.*, in press
165. Rosenzweig, A. and Cromer, D. T. (1970). *Acta Crystallogr.*, **26B**, 38
166. Brunton, G. (1969). *Acta Crystallogr.*, **25B**, 2163; (1970). **26B**, 1185
167. Brunton, G. (1969). *Acta Crystallogr.*, **25B**, 1819
168. Brunton, G. and Sears, D. R. (1969), *Acta Crystallogr.*, **25B**, 2519
169. Ratho, T. and Patel, T. (1968). *Ind. J. Phys.*, **42**, 240
170. Ratho, T., Patel, T. and Sahu, B. (1969). *Ind. J. Phys.*, **43**, 164
171. Montoloy, F., Maraval, S. and Capestan, M. (1968). *Compt. Rend.*, **266C**, 787
172. Montoloy, F. and Maraval, S. (1968). *Compt. Rend.*, **267C**, 1309
173. Charpin, P., Montoloy, F. and Nierlich, M. (1969). *Compt. Rend.*, **268C**, 156
174. Gruen, D. M. and McBeth, R. L. (1968). *Inorg. Nucl. Chem. Lett.*, **4**, 299
175. Hardt, D. H. and Hefer, E. (1969). *Naturwissenschaften*, **56**, 88
176. Brown, D. (1966). *J. Chem. Soc. (A)*, 766
177. Smith, B. C. and Wassef, M. A. (1968). *J. Chem. Soc. (A)*, 1817
178. Saha, K. H. (1970). *J. Ind. Chem. Soc.*, **47**, 88
179. Morss, L. R. and Fuger, J. (1969). *Inorg. Chem.*, **8**, 1433
180. Desyatnik, V. N., Melnikov, V. T. and Raspopin, S. A. (1969). *Izv. Vyssck, Ucheb. Zaved. Tsvet. Met.*, **12**, (3), 99
181. Sterlin, Ya. M. and Artamonov, V. V. (1967). *Sov. At. Energy*, **22**, 589
182. Bagnall, K. W., Brown, D. and Colton, R. (1964). *J. Chem. Soc.*, 2527
183. Day, P. J. and Venanzi, L. M. (1966). *J. Chem. Soc. (A)*, 197
184. Woodward, L. A. and Ware, M. J. (1968). *Spectrochim. Acta*, **24A**, 921
185. Staffsudd, O. M. (1967). *U.S. Report UCLA*-34-P-103-3
186. Pandey, A. N., Singh, H. S. and Samyal, K. N. (1969). *Curr. Sci.*, **38**, 108
187. Cleveland, J. M., Bryan, G. H. and Eggerman, W. G. (1970). *Inorg. Chem.*, **9**, 964
188. Barnard, R., Bullock, J. I. and Larkworthy, L. F. (1968). *Chem. Comm.*, 960
189. Ryan, J. L. (1967). *Adv. in Chem. Ser.*, 71, Lanthanide/Actinide Chemistry (Washington: *Amer. Chem. Soc.)* 331
190. Marcus, Y. and Shiloh, M. (1969). *Isr. J. Chem.*, **7**, 37
191. Bagnall, K. W., Laidler, J. B. and Stewart, M. A. A. (1968). *J. Chem. Soc. (A)*, 133
192. Davidovich, R. L., Sergienko, V. I. and Kalacheva, T. A. (1968). *Bull. Acad. Sci. USSR*, 1588
193. Opalovskii, A. A. and Nestevenko, M. N. (1968). *Izv. Sib. Otd. Akad. Nauk. SSSR Ser. Khim. Nauk.*, **(4)**, 13
194. Opalovskii, A. A., Batsanova, S. S., Kuznetsova, Z. M. and Nesterenko, M. N. (1968). *Izv. Sib. Otd. Acad. Nauk. SSSR Ser. Khim. Nauk.*, **(1)**, 15
195. Davidovich, R. L. and Kalacheva, T. A. (1968). *Izv. Sib. Otd. Acad. Nauk. SSSR Ser. Khim. Nauk.*, **(5)**, 44
196. Davidovich, R. L., Buslaev, Yu. A. and Murzakhanova, L. M. (1968). *Bull. Acad. Sci. USSR*, **(3)**, 675
197. Chakravorti, M. C. and Bandyopadhyay, N. (1969). *J. Ind. Chem. Soc.*, **46**, 961
198. Rebenko, A. N., Brusentsev, F. A. and Opalovskii, A. A. (1968). *Izv. Sib. Otd. Akad. Nauk. SSSR, Ser, Khim. Nauk.*, **(1)**, 136
199. Nguyen-Quy-Dao (1968). *Bull. Soc. Chim. Fr.*, 3542
200. Brusset, H., Gillier-Pandraud, H. and Nguyen-Quy-Dao. (1969). *Acta Crystallogr.*, **25B**, 67
201. Baker, A. E. and Haendler, H. M. (1962). *Inorg. Chem.*, **1**, 127
202. Vorobei, M. P., Skiba, O. V., Kapshnkov, I. I., Desyatnik, V. M. and Yakolov, G. N. (1969). *Sov. At. Energy*, **27**, 827
203. Allpress, J. G., Anderson, J. S. and Hambly, A. N. (1968). *J. Inorg. Nucl. Chem.*, **30**, 1195
204. Brusset, H., Nguyen-Quy-Dao and Hoffner, F. (1970). *Bull. Soc. Chim. Fr.*, 1759
205. Allpress, J. G. and Wadsley, A. D. (1964). *Acta Crystallogr.*, **17**, 41
206. Vdovenko, V. M., Mashirev, L. G., Skoblo, A. I. and Suglobov, D. N. (1967). *Russ. J. Inorg. Chem.*, **12**, 1542

207. Belyaev, Yu. I., Vdovenko, V. M., Ladygin, I. N. and Suglobov, D. N. (1967). *Russ. J. Inorg. Chem.*, **12**, 1705
208. Kharitonov, Yu. Ya., Kuyazeva, N. A., Tsapkin, V. V. and Ellert, G. V. (1967). *Sov. Radiochem.*, **9**, 316
209. Brusset, H., Gillier-Pandraud, H. and Nguyen-Quy-Dao (1967). *Compt. Rend.*, **265C**, 1209
210. Nguyen-Quy-Dao (1968). *Bull. Soc. Chim. Fr.*, 3976
211. Sergienko, V. I. and Davidovitch, R. L. (1968). *Izv. Sib. Otd. Acad. Nauk. SSSR, Ser. Khim. Nauk.*, **(4)**, 71
212. Vdovenko, V. M., Skoblo, A. I., Suglobov, D. N., Shcherbakova, L. L. and Shcherbakova, V. A. (1968). *Sov. Radiochem.*, **10**, 518
213. Kharitonov, U. Y., Khyazeva, N. A. and Buslaev, V. A. (1969). *Russ. J. Inorg. Chem.*, **14**, 539
214. Newberry, J. E. and Hart, F. A. (1968). *J. Inorg. Nucl. Chem.*, **30**, 318
215. Newberry, J. E. (1969). *Spectrochim. Acta.* **25A**, 1699
216. Vdovenko, V. M., Ladygin, I. N. and Suglobov, D. N. (1970). *Zh. Neorg. Khim.*, **15**, 265
217. Kaufmann, G., Leroy, M. J. F. and Rohmer, R. (1967). *Bull. Soc. Chim. Fr.*, 2880
218. Sergienko, V. I. and Davidovich, R. L. (1970). *Spectrosc. Lett.*, **3**, 27
219. Kharitonov, Yu. Ya. and Knyazeva, N. A. (1970). *Russ. J. Phys. Chem.*, **44**, 323
220. Sergienko, V. I. and Davidovich, R. L. (1970). *Spectrosc. Lett.*, **3**, 35
221. Staffsudd, O. M., Leung, A. F. and Wong, E. Y. (1969). *Phys. Rev.*, **180**, 339
222. Leung, A. F. and Wong, E. Y. (1969). *Phys. Rev.*, **180**, 380
223. Brown, D. and Rickard, C. E. F. (1971). *J. Chem. Soc. (A)*, 81
224. Furlani, C. and Zinato, E. (1967). *Z. Anorg. Allgem. Chem.*, **351**, 210
225. Brown, D. Moseley, P. T. and Reynolds, C. (1970). Unpublished observations
226. Brown, D. (1970). Unpublished observations
227. Kemmler-Sack, S. (1968). *Z. Anorg. Allgem. Chem.*, **363**, 282
228. Kemmler-Sack, S. (1969). *Z. Anorg. Allgem. Chem.*, **363**, 295
229. Kemmler-Sack, S. (1969). *Z. Anorg. Allgem. Chem.*, **364**, 135
230. MacCordick, J., Kaufmann, G. and Rohmer, R. (1968). *Rev. Chim. Miner*, **5**, 629
231. Prasad, S. and Pandey, P. L. (1969). *Ladbev*, **7A**, 164
232. Athavak, V. T. and Iyer, C. S. P. (1967). *J. Inorg. Nucl. Chem.*, **29**, 1003
233. Camelot, M. (1967). *Compt. Rend.*, **265C**, 812
234. Savant, V. V. and Patel, C. C. (1969). *J. Inorg. Nucl. Chem.*, **31**, 2319
235. Majumdar, A. K. and Battacharyya, R. G. (1970). *Chem & Ind.*, 95
236. Majumdar, A. K. and Battacharyya, R. G. (1969). *Science and Culture*, **35**, 271
237. Prasad, S. and Pandey, P. L. (1969). *Ladbev*, **7A**, 82
238. Tsapkina, I. V. and Ellert, G. V. (1968). *Russ. J. Inorg. Chem.*, **13**, 730
239. Day, J. P. and Venanzi, L. M. (1966). *J. Chem. Soc. (A)*, 1363
240. Fitzsimmons, B. W., Gans, P., Hayton, B. and Smith, B. C. (1966). *J. Inorg. Nucl. Chem.*, **28**, 915
241. Bagnall, K. W., Brown, D. and du Preez, J. G. H. (1965). *J. Chem. Soc.*, 5217
242. Brown, D. and Rickard, C. E. F., unpublished observations
243. Brown, D., Hill, J. and Rickard, C. E. F. (1970). *J. Less-Common Metals*, **20**, 57
244. Panzer, R. E. and Suttle, J. F. (1960). *J. Inorg. Nucl. Chem.*, **13**, 244
245. Selbin, J., Ahmad, N. and Pribble, M. J. (1969). *Chem. Comm.*, 759; (1970). *J. Inorg. Nucl. Chem.*, **32**, 3249
246. Bagnall, K. W., Brown, D. and Lux, F., unpublished observations
247. Bagnall, K. W., Brown, D., Lux, F. and Wirth, G. (1969). *Naturforsch.* **24B**, 216
248. Vdovenko, V. M., Volkov, V. A., Suglobova, I. G. and Suglobova, D. N. (1969). *Radiokhimiya*, **11**, 26
249. Molodkin, A. K., Ivanova, O. M., Kuchumova, A. N. and Kozina, L. E. (1967). *Russ. J. Inorg. Chem.*, **12**, 963
250. Molodkin, A. K., Ivanova, O. M. and Kozina, L. E. (1968). *Russ. J. Inorg. Chem.*, **13**, 1192
251. Petrov, K. I., Molodkin, A. K., Ivanova, O. D. and Saralidze, O. D. (1969). *Russ. J. Inorg. Chem.*, **14**, 215
252. Golovnya, V. A. and Bolotova, G. T. (1966). *Russ. J. Inorg. Chem.*, **11**, 1419
253. Frazer, M. J. and Rimmer, B. (1968). *J. Chem. Soc. (A)*, 2273
254. Bagnall, K. W., Brown, D., Holah, D. G. and Lux, F. (1968). *J. Chem. Soc. (A)*, 465

255. Ivanova, O. M., Petrov, K. I., Molodkin, A. K., Saralidze, O. D. and Kozina, L. E. (1968). *Russ. J. Inorg. Chem.,* **13**, 693

256. Alvey, P. J., Bagnall, K. W. and Brown, D., unpublished observations

257. Brown, D., Hill, J. and Rickard, C. E. F. (1970). *J. Chem. Soc. (A),* 497

258. Ellert, G. V., Bolotova, G. T. and Krasovskaya, T. I. (1968). *Russ. J. Inorg. Chem.,* **13** 726

259. Bagnall, K. W., Brown, D. and Al-Kazzaz, Z. M. S., unpublished observations

260. Berthold, H. J. and Knecht, H. (1969). *Z. Anorg. Allgem. Chem.,* **366**, 249

261. Burk, W. (1969). *Z. Chem.,* **233B**, 9

262. Ovlova, L. V., Garnovskii, A. D., Osipov, O. A. and Kukushkina, F. I. (1968). *Zh. Obshch. Khim.,* **38**, 1850 (according to *Chem. Abstr.,* **69**, 113055q)

263. Garnovskii, A. D., Minkin, V. I., Osipov, O. A., Panyushkin, V. T., Isaeva, L. K. and Knyazhanskii, M. I. (1967). *Russ. J. Inorg. Chem.,* **12**, 1288

264. Berthold, H. J. and Knecht, H. (1968). *Z. Anorg. Allgem. Chem.,* **356**, 151

265. Burk, W. (1967). *Z. Anorg Allgem. Chem.,* **350**, 62

266. MacCordick, J. and Brun, C. (1970). *Compt. Rend.,* **270C**, 620

4
Complexes of the Actinide Cyanides and Thiocyanates

K. W. BAGNALL
University of Manchester

4.1 INTRODUCTION

There has been very little work published on the actinide pseudohalides or their complexes, and these compounds represent one of the almost completely neglected areas of actinide chemistry research, in marked contrast to the actinide halides, on which the amount of published information is very large indeed. Even now, the majority of published papers refer only to thorium and uranium thiocyanates, leaving the transuranium element thiocyanates virtually unknown. This is certainly surprising in the case of the actinide cyanides, for these compounds, and cyano-complexes derived from them, should be particularly interesting for the investigation of ligand field effects on actinide metal ions.

No simple thiocyanates are known for thorium, uranium or neptunium, but neutral complexes with oxygen or nitrogen donor ligands, and anionic thiocyanato-complexes are all easily prepared. In some cases these last include neutral ligands or other anions in the complex anion.

The metal ions in these compounds generally exhibit higher coordination numbers than their chloride analogues, so providing a potentially interesting field for the structural chemist, a field which is, as yet, almost completely unexplored.

4.2 CYANIDES

There is an early[1], and unconfirmed, report of the precipitation of a uranyl(VI) cyano-complex, $K_2[UO_2(CN)_4]$, from aqueous solutions of uranyl(VI) acetate on the addition of a large excess of potassium cyanide. The reported stoicheiometry is analogous to that of the halide complexes of uranium(VI), but the composition appears unlikely in view of the higher coordination number exhibited by uranium(VI) in the thiocyanates (Section 3.10); the product is most probably a hydrolysed species. An inconclusive attempt to prepare a uranium(IV) cyanide complex with N,N-dimethylacetamide (DMA) by reaction of the uranium(IV) chloride or thiocyanate–DMA complex with iodine monocyanide or mercuric cyanide has also been reported[2]. However, marked changes in the visible spectrum of uranium(IV) perchlorate in methyl cyanide solution occur on the addition of hydrocyanic acid, a possible indication of the formation of a chlorocyanocomplex in solution[3].

Neither uranium tetracyanide, nor complexes derived from it, are known as yet, but a uranium(IV) chlorocyanide, $[UCl_3(CN)\cdot4NH_3]$, and a chlorocyanouranate(IV), $[(C_2H_5)_4N]_2[UCl_4(CN)_2]$ have recently been reported[4]. The light green ammine complex is precipitated, together with alkali chloride, when sodium or potassium cyanide is added to uranium tetrachloride in anhydrous liquid ammonia. Further replacement of chloride in the complex by cyanide does not occur at high temperatures and pressures. The compound is insoluble in acetone, dichloromethane, ethyl acetate, liquid ammonia, methyl cyanide and tetrahydrofuran, but dissolves, with decomposition, in DMA and in nitromethane. The complex is very sensitive to atmospheric oxidation and to moisture, which suggests a relatively weak π-bonding contribution from the metal to the carbon atom. In the infrared spectrum the C—N stretching mode appears at 2120 cm^{-1}, while the u.v.-visible reflectance spectrum (maxima at 22 500; 19 800; 17 400 and 14 750 cm^{-1}) is very similar to that[2] of $[U(NCS)_4\cdot4DMA]$.

A second uranium(IV) chlorocyano-complex remains in the supernatant from the preparation of $[UCl_3(CN)\cdot4NH_3]$; in this the C—N stretching mode appears at 2180 cm^{-1}. A product with the same C—N stretching frequency is also obtained when $[UCl_3(CN)\cdot4NH_3]$ is heated to 70 °C under vacuum, but the compositions are unknown.

Uranium tetrabromide and tetraiodide, and the corresponding thorium halides, appear to react with potassium cyanide in anhydrous liquid ammonia in the same way as uranium tetrachloride, but the reaction products, in which the C—N stretching mode appears at 2120 cm^{-1}, have not been identified. No reaction occurs between uranium tetrachloride and alkali metal cyanides in anhydrous liquid hydrocyanic acid. The only isolable product is an HCN adduct of uranium tetrachloride[4].

Trichlorocyanotetrammine uranium(IV) does not react with tetraethyl-ammonium cyanide in liquid ammonia, in contrast to uranium tetrachloride, which appears to form a tetrachlorodicyano-complex in liquid hydrocyanic acid; the C—N stretching mode in this complex appears at 2150 cm^{-1} after dissolution in methyl cyanide and evaporation of the solvent. This complex has not been obtained pure[4].

4.3 THIOCYANATES

The greater part of the recent literature on the actinide thiocyanates has been concerned with the complexes of thorium and uranium; only a few neptunium thiocyanate complexes have been reported and studies of the thiocyanates of the later actinides appear to be restricted to observations of complexing in aqueous solution. For example, studies of the extraction of trivalent plutonium, americium, curium and californium from aqueous thiocyanate media with di(2-ethylhexyl)orthophosphoric acid indicate that the complexes $[M(SCN)]^{2+}$ and $[M(SCN)_2]^+$ exist in the aqueous phase[5] and earlier cation exchange studies[6] have also demonstrated the existence of $[M(SCN)]^{2+}$ ions (M = Am,Cm) in relatively dilute solutions of thiocyanate, whereas at higher (4–5 molar) concentrations of thiocyanate the anionic complexes $[Am(SCN)_4]^-$ and $[Cm(SCN)_4]^-$ appear to be formed[6].

Solution studies of the complexation of the uranyl, UO_2^{2+}, ion in per-chloric acid–thiocyanate media[7] have likewise provided evidence for the stepwise formation of the species $[UO_2(SCN)]^+$, $[UO_2(SCN)_2]$ and $[UO_2(SCN)_3]^-$ in aqueous solution, while the observation[8] that the precipi-tation of plutonium(IV) phenylarsonate is inhibited in the presence of thiocyanate ion has been interpreted as evidence for the complexation of plutonium(IV) by that ion.

Apart from this brief, and very incomplete, mention of aqueous solution work, this section will cover only the preparation and properties of actinide thiocyanate complexes which have been isolated as solid compounds, usually from non-aqueous solvents. In this connection it should be mentioned that fused salts have not been used for such preparative work owing to thermal decomposition of the actinide thiocyanates; as an example[9], the use of fused potassium thiocyanate (at 190 °C) as a solvent for the uranyl(VI) ion led to the precipitation of UO_2S.

The available crystallographic and infrared spectral evidence indicates that in all of the known actinide thiocyanate complexes which have been isolated and characterised in this manner the thiocyanate ion is nitrogen bonded, and their formulae are so written. The ionic and neutral species mentioned earlier in this section also probably involve nitrogen bonded thiocyanate but, in the absence of evidence for this, their formulae have been written as recorded in the original papers.

4.3.1 Hydrates of thorium(IV) and uranyl(VI) thiocyanates

The white thorium(IV) compound $[Th(NCS)_4(H_2O)_4]$, is prepared by melting hydrated thorium tetranitrate with the stoicheiometric quantity of sodium thiocyanate and extracting the thorium compound into ethanol, in which sodium nitrate is insoluble. It can also be prepared from aqueous solution, by treating thorium(IV) sulphate with an excess of barium thiocyanate in the presence of sufficient dilute sulphuric acid to ensure precipitation of all the barium; the free thiocyanic acid so formed appears to prevent hydrolysis of the thorium compound and to improve its crystallisation from solution[10]. The compound is very soluble in water, in which it is extensively dissociated, and is also readily soluble in acetone, ethanol and methanol, but is insoluble in benzene, carbon tetrachloride and toluene. It is very hygroscopic, deliquescing in moist air; no water is lost when the complex is allowed to stand over concentrated sulphuric acid in a desiccator, but the compound dehydrates, with some decomposition, at 100 °C and the compound also decomposes in bright sunlight[10]. From the available evidence it seems probable that the water molecules are coordinated to the central thorium ion, making the coordination number at least 8.

The uranium(IV) analogue has not been recorded, although there is evidence for the existence of species such as $[U(SCN)]^{3+}$ and $[U(SCN)_2]^{2+}$ in aqueous solution[11,12], which suggests that the hydrated tetrathiocyanate should be preparable.

Hydrated uranyl(VI) thiocyanate $[UO_2(NCS)_2(H_2O)_3]$, has been known for over 50 years. It was originally obtained by mixing equimolar amounts of uranyl(VI) sulphate and barium thiocyanate in water and evaporating the filtrate to a red, tarry mass which solidified on standing[13]. The solidification process is accelerated by grinding the tarry mass with benzene, a procedure which yields the trihydrate as an orange powder. Recrystallisation from ether likewise yields a tarry product, but this is also susceptible to treatment with benzene[14]. An alternative preparation is by reaction of hydrated uranyl nitrate with the stoicheiometric amount of potassium thiocyanate in methanol[15].

The trihydrate dissociates in aqueous solution, behaving as a three ion electrolyte. When heated, the solid compound loses two molecules of water at 85–100 °C and the last molecule of water is lost at 115–125 °C. The resulting anhydrous product is dark green, almost black, and is insoluble in water[14]. In the infrared spectra of the monohydrate and trihydrate, the asymmetric O—U—O stretching mode appears at 944 and 937 cm^{-1} respectively, while the C—N stretching mode appears at 2103 (also a weak band at 1910) and 2057 cm^{-1} respectively. The C—N mode at 2103 cm^{-1} indicates the presence of bridging thiocyanate groups in the monohydrate, whereas that at 2057 cm^{-1} is typical of terminal thiocyanate groups[16]. The uranium atom in the trihydrate is presumably in a 7-coordinate environment, probably pentagonal bipyramidal.

4.3.2 Amide complexes

A hydrated acetamide complex of uranyl(VI) thiocyanate, $[UO_2(NCS)_2(H_2O) (CH_3CONH_2)_2]$, is obtained as a yellow, crystalline solid when acetamide is added to an aqueous solution of uranyl(VI) thiocyanate trihydrate and the solution is evaporated; it can be recrystallised from water without decomposition[15]. The compound melts at 105 °C, loses water at 150–160 °C and decomposes at 240 °C. The asymmetric O—U—O stretching mode appears at 939 cm^{-1} in the infrared spectrum[17] and the C—N stretching modes appear at 2072 and 2095 cm^{-1}. The carbonyl frequency shift is only 20 cm^{-1}, the C—O stretching mode appearing at 1660 cm^{-1}. The metal is probably 7-coordinate in this compound.

N,N-dimethylacetamide (DMA) complexes of thorium(IV) and uranium(IV) thiocyanates $[M(NCS)_4(DMA)_4]$, are prepared by reaction of the corresponding chloride complexes, $[ThCl_4(DMA)_4]$ and $[(UCl_4)_2$ $(DMA)_5]$, with potassium thiocyanate in, respectively, acetone and a hot mixture of acetone, DMA and nitromethane. In the case of the uranium compound chloroform is added to precipitate any excess potassium thiocyanate; after filtration, addition of isopentane precipitates the complex, a method used for the direct isolation of the thorium complex from the acetone supernatant.

The white thorium compound[18] melts at 162–3 °C and is isostructural with the grey-green uranium compound, which melts at 158 °C. Neither compound is hygroscopic, and both are very soluble in methyl cyanide, nitromethane and in acetone, although the uranium complex is appreciably less soluble than the thorium one in the last. However, the uranium compound is appreciably the more soluble in methanol and is also very soluble in chloroform and in ethanol. The two complexes are insoluble in benzene, carbon tetrachloride, diethyl ether, ethyl or methyl acetate, and methylisobutylketone.

The C—N stretching modes appear at 2033 cm^{-1} (Th) and 2047 cm^{-1} (U), and the carbonyl stretching frequencies are 1605 and 1606 cm^{-1} respectively, a shift of 41–42 cm^{-1} from that observed for the free ligand[2, 18].

The magnetic susceptibility of the uranium compound is temperature dependent, but the large Weiss constant (-126 K) means that the effective magnetic moment, 3.48 BM, has little significance[2]. The visible spectrum of the uranium compound in nitromethane solution has been reported[2], the principal bands being at 450, 500, 565, 655 and 685 nm (molar extinction coefficients 28.2, 46.3, 23.3, 57 and 125 respectively).

4.3.3 Urea complexes of uranyl(VI) thiocyanate

At present only the uranyl thiocyanate complexes, $[UO_2(NCS)_2(H_2O)$ $(CO(NH_2)_2)]$ and $[UO_2(NCS)_2(CO(NH_2)_2)_3]$ have been recorded[15]. The first of these separates as yellow plates when a substoicheiometric quantity of urea is added to a methanol solution of $[UO_2(NCS)_2(H_2O)_3]$ and the solution is evaporated. The tris-urea complex, a dark orange solid, is formed in the same way when two or more moles of urea are added for each mole of the uranium(VI) thiocyanate present. The tris-urea complex is described as readily soluble in methanol and in water, but in the preparative procedure the products are said to have been washed with the latter. The bis-urea complex apparently melts, with decomposition, at 90–125 °C, while the tris-urea complex melts at 125–140 °C and decomposes at 200–240 °C. Their infrared spectra have also been recorded[17]; in the bis- and tris-urea complexes respectively the asymmetric O—U—O stretching modes appear at 927 and 910 cm^{-1}, the C—N stretching modes at 2070, 2094 and 2068, 2091 cm^{-1}, and the urea carbonyl stretching modes at 1635, 1650 and 1632, 1648 cm^{-1}. There is no evidence for the presence of bridging thiocyanate groups in either compound and the coordination number of the uranium atom is probably 7 in both compounds.

Anionic thiocyanato-complexes in which urea is also coordinated to the central uranium atom are described in Section 3.10.

4.3.4 Antipyrine complex of uranyl(VI) thiocyanate

Antipyrine (1,5-dimethyl-2-phenyl–3-pyrazolone) behaves like an amide ligand with uranium, coordinating by way of the carbonyl oxygen atom. The tris-antipyrine complex of uranyl thiocyanate[15], $[UO_2(NCS)_2(C_{11}H_{12}N_2O)_3]$, precipitates as a pale yellow powder when the ligand is added to a cold, dilute aqueous solution of $[UO_2(NCS)_2(H_2O)_3]$; from warm, more concentrated, aqueous solutions the product first separates as a sticky mass which solidifies when heated and then cooled. The complex is almost insoluble in water, but is soluble in acetone, methanol and nitro-benzene. It decomposes above 270 °C. In its infrared spectrum[17], the O—U—O asymmetric stretching mode appears at 916 cm^{-1}, the ligand carbonyl stretching mode is at 1604 cm^{-1} (free ligand 1667 cm^{-1}) and the C—N stretching mode is a single band at 2058 cm^{-1}.

4.3.5 Tetrahydrofuran (THF) and furfural complexes

The yellow complex $[UO_2(NCS)_2(THF)_3]$ separates when a solution of $[UO_2(NCS)_2(H_2O)_3]$ in THF is evaporated. The compound melts, with decomposition, at 175 °C and is hygroscopic; it is readily soluble in water and in methanol[15]. The O—U—O asymmetric stretching mode appears at 935 cm^{-1} and the C—N stretching modes are at 2031, 2060 cm^{-1} in its infrared spectrum[17]. Furfural (2-furaldehyde) extracts thorium(IV) and uranyl(VI) thiocyanates from aqueous solution, apparently forming the bis-furfural complexes[22] $[Th(SCN)_4 \cdot 2L]$ and $[UO_2(SCN)_2 \cdot 2L]$ but these do not appear to have been isolated in the solid state.

4.3.6 Quinoline N–oxide (QNO) complex of uranyl(VI) thiocyanate

The complex $[UO_2(NCS)_2(QNO)_3]$, which melts at 196 °C, is prepared[19] by boiling ethanol solutions of the ligand and uranyl thiocyanate (presumably $[UO_2(NCS)_2(H_2O)_3]$). The appearance of the C—N stretching mode at 2072, 2100 cm^{-1} and of the C—S stretching mode at 839(w), 806(s) cm^{-1} is consistent with N-bonded thiocyanate.

Table 4.1 Phosphine oxide complexes of the actinide thiocyanates

Number	*Ligand	Complex	Colour	m.p., °C	$\nu_{P=O}$ cm^{-1}	$\Delta\nu_{P=O}$ cm^{-1}	$\nu_{C≡N}$ cm^{-1}	Reference
1	$(CH_3)_3PO$	$[Th(NCS)_4(TMPO)_4]$	white	250–253	1098	62	2038, 2056, 2078	20
2	(TMPO)	$[U(NCS)_4(TMPO)_4]$	grey-green	—	1090	70	2025, 2060, 2080	20
3		$[Np(NCS)_4(TMPO)_4]$	yellow-brown	238–242(d)	1094	66	2035, 2060, 2085	20
4	$[(CH_3)_2N]_3PO$	$[Th(NCS)_4(HMPA)_4]$	white	254(d)	1098	103	2060, 2085	20
5	(HMPA)	$[U(NCS)_4(HMPA)_4]$	grey-green	195–8(d)	1090	111	2060, 2070	20
6		$[Np(NCS)_4(HMPA)_4]$	yellow-brown	228	1098	103	2073	20
7	$(C_6H_5)_3PO$	$[Th(NCS)_4(TPPO)_4]$	white		1068	123	2060, 2080	20
8	(TPPO)	$[U(NCS)_4(TPPO)_4]$	yellow-green	145–7(d)	1065	126	2020, 2050, 2070	20
9		$[Np(NCS)_4(TPPO)_4]$	yellow	231–3	1078	113	2029, 2064	20
10		$[UO_2(NCS)_2(TPPO)_2]$	yellow		1138, 1085	53, 106	2052	37
11	$[((CH_3)_2N)_2PO]_2O$	$[Th(NCS)_4(OMPA)_2]$	white	145(d)	1150	83	2005, 2035, 2060	20
12	(OMPA)	$[U(NCS)_4(OMPA)_2]$	yellow-green		—	—	—	20
13		$[Np(NCS)_4(OMPA)_2]$	pale-yellow	155–160(d)	1163	70	2053	20
14	$(nC_4H_9)_3PO$ (TBPO)	$[UO_2(NCS)_2(TBPO)_3]$	—	(d) at 325	—	—	—	21
15	$(C_6H_5)_3AsO$ (TPAsO)	$[UO_2(NCS)_2(TPAsO)_2]$	yellow	235–240	833†	45†	2019	37

*The numbers are to facilitate references to these complexes in the text

†As=O

(d) = with decomposition

4.3.7 Phosphine and arsine oxide complexes

The known phosphine oxide complexes of thorium(IV), uranium(IV), neptunium(IV) and uranium(VI) and an arsine oxide complex of the last, are listed in Table 4.1, which includes infrared spectral data and melting points, where known. The actinide(IV) complexes (1–9, 11–13, Table 4.1) are of the form $[M(NCS)_4(R_3PO)_4]$ or, where the ligand (OMPA) is bidentate $[M(NCS)_4(OMPA)_2]$, and the metal is presumably 8-coordinate in these compounds.

The complexes are prepared[20] by reaction of the corresponding complex of the metal tetrachloride with the stoicheiometric quantity of potassium thiocyanate, in the presence of an excess of ligand, in a non-aqueous solvent, either acetone (complexes 3,4,6,7,12,13; Table 4.1), acetone/methanol (9,11), methyl cyanide (2,8) or nitromethane (1,5). After removal of the precipitated potassium chloride, the supernatant is evaporated to small volume and the

Table 4.2 Solvent data for actinide thiocyanate—phosphine oxide complexes

Complex No.*	Chloroform	Dichloromethane	Nitromethane	Acetone	Methyl cyanide	Ethyl acetate
1	insol.	insol.	v. sol.	insol.	sol.	—
2	—	—	sol.	sl. sol.	sol.	insol.
3	—	—	—	sol.	sol.	—
4	—	—	sol. (hot)	sol.	sol. (hot)	insol.
5	—	—	—	sol.	sol.	sol.
6	—	—	sol.	v. sol.	sol.	sl. sol.
7	v. sol.	v. sol.	sol. (hot)	sol.	sol.	insol.
8	—	sol.	sol.	sol.	—	sl. sol.
9	—	sol.	—	sl. sol.	sol.	—
10†	—	—	—	—	—	—
11	—	v. sol.	v. sol.	v. sol.	v. sol.	insol.
12	—	—	—	sol.	sol.	insol.
13	—	sol.	sol.	insol.	sol.	insol.
14	sol.	—	—	—	—	—
15†	—	—	—	—	—	—

*See Table 4.1
†10 and 15 are insoluble in non-polar organic solvents, methanol and ethanol
sol. = soluble; sl. = slightly; v. = very; insol. = insoluble

complexes crystallise out on cooling. The products so obtained are recrystallised from acetone (complexes 4,7,9,11; Table 4.1), methyl cyanide (3,7,8), ethyl acetate (5,6), nitromethane (2), a mixture of dichloromethane (13) or acetone (12) and isopentane, or acetone containing a little nitromethane (1). Some solubility data for these complexes are given in Table 4.2

The shifts in the P—O stretching band in the infrared spectra of these complexes (Table 4.1) are quite substantial, but are much smaller than those observed in the spectra of the complexes of the corresponding tetrahalides, possibly a result of the higher coordination number exhibited by the metals in the tetra-N-thiocyanate compounds as compared with the tetrahalide analogues.

Complex 10 is precipitated when a solution of TPPO in ethanol is added to the filtrate from the reaction between uranyl nitrate hexahydrate and an excess of potassium thiocyanate in ethanol. It is a non-electrolyte in benzene[37]. The analogous arsine oxide complex (15) is prepared in the same way[37].

Complex 14 (Table 4.1) separates at the phase boundary when the stoicheiometric quantity of TBPO, dissolved in heptane, is added to $[UO_2(NCS)_2(H_2O)_3]$. A small volume of water is added to permit separation of the product, which is insoluble in water or saturated hydrocarbons[21]. No x-ray crystallographic data have been reported for any of the complexes 1–15.

Tri-n-butyl phosphate (TBP), commonly used for the separation of the actinide elements by solvent extraction from nitrate solution, has donor properties resembling those of the phosphine oxides; a uranyl(VI) N-thiocyanate complex $[UO_2(NCS)_2(TBP)_2]$, is obtained when a benzene solution of TBP (1 molar) is saturated with hydrated uranyl(VI) thiocyanate[16]. Two C—N stretching frequencies are observed in its infrared spectrum, at 2036 and 2110 cm^{-1}, the latter being assigned to a bridging thiocyanate group, for this band disappears when the ratio TBP:U^{VI} in the solution is increased to 3:1, indicating the formation of the complex $[UO_2(NCS)_2(TBP)_3]$, in the spectrum of which there is only a single C—N stretching frequency, observed at 2043 cm^{-1}. The asymmetric O—U—O stretching modes are at 944 and 934 cm^{-1} respectively in the bis- and tris-TBP complexes, while the corresponding shifts in the P—O stretching mode, as compared with the free ligand, are 86 and 73 cm^{-1} respectively[16].

Anionic N-thiocyanatouranyl(VI) complexes in which TBPO and TBP are coordinated to the metal are also known; these are described in Section 3.10.

4.3.8   Dimethylsulphoxide (DMSO) complex of uranyl(VI) thiocyanate

The complex $[UO_2(NCS)_2(DMSO)_3]$ is obtained on addition of DMSO to a solution of $[UO_2(NCS)_2(H_2O)_3]$ in water or methanol; the product, which can be recrystallised from methanol, melts at 105–120 °C and decomposes[15] at 270 °C. The infrared spectrum has been reported[17]; the C—N stretching frequencies appear at 2042 and 2062 cm^{-1}, and the asymmetric O—U—O stretching mode at 914 cm^{-1}. The shift in the S—O stretching mode on complexation is 66 cm^{-1}.

4.3.9   Complexes with nitrogen donor ligands

The α,α dipyridyl (dipy) complex $[UO_2(NCS)(dipy)_2]NCS$, is a yellow solid precipitated when methanol solutions of the ligand and $[UO_2(NCS)_2(H_2O)_3]$ are mixed[15]. It decomposes at 300 °C.

The analogous o-phenanthroline (4,5 diazaphenanthrene,phen) complex $[UO_2(NCS)(phen)_2]NCS$, is obtained as a pale yellow precipitate in the same way as the dipy complex; it decomposes at 325 °C. Both complexes are insoluble in water and only sparingly soluble in methanol[15]. The metal is presumably 7-coordinate in both cases.

4.3.10   Anionic thiocyanato–complexes

Thorium(IV) and uranium(IV) anionic complexes of the form $[M^{IV}(NCS)_8]^{4-}$ are quite well known and a number of alkali metal and other salts of this type of anion have been isolated from aqueous and non-aqueous solvents. Penta- and hepta-N-thiocyanatothorates(IV) are also known, and in these the thorium atom is also 8-coordinate, the remaining coordination positions being occupied by water molecules. The known salts are listed in Table 4.3; although neptunium(IV) and plutonium(IV) analogues have not yet been recorded, there is no reason to suppose that they cannot be prepared.

Table 4.3   Anionic N-thiocyanato-thorates(IV) and uranates(IV)

Compound	Colour	Reference
$Rb[Th(NCS)_5(H_2O)_3]$	colourless	10
$(NH_4)_3[Th(NCS)_7(H_2O)]\cdot4H_2O$	colourless	10
$K_4[Th(NCS)_8]\cdot3.5H_2O$	colourless	10
$Rb_4[Th(NCS)_8]\cdot2H_2O$	colourless	10, 24
$Cs_4[Th(NCS)_8]\cdot2H_2O$	colourless	24
$(NH_4)_4[Th(NCS)_8]\cdot2H_2O$	colourless	10, 24
$K_4[U(NCS)_8]$	grey-green	2
$Rb_4[U(NCS)_8]\cdot H_2O$	green	23
$Cs_4[U(NCS)_8]\cdot H_2O$	green	23
$(NH_4)_4[U(NCS)_8]$	grey-green	25
$(dipy\ H)_4[U(NCS)_8]*$	green	23
$[(C_2H_5)_4N]_4[U(NCS)_8]$	dark-green	26

*dipy = α, α dipyridyl

The octa-N-thiocyanato-complex salts listed in Table 4.3 are usually prepared by reaction of thorium(IV) or uranium(IV) sulphate with the stoicheiometric quantities of alkali metal or ammonium sulphate and barium thiocyanate in sulphuric acid, according to the equation:

$$M^{IV}(SO_4)_2 + 2M_2^I SO_4 + 4Ba(SCN)_2 \rightarrow M_4^I[M^{IV}(NCS)_8] + 4BaSO_4$$

The solutions are either heated to boiling (Th) or maintained at 40–50 °C (U) prior to filtration, and the filtrate is finally evaporated on a water bath[10, 23] or at room temperature[24, 25] until crystals of the product separate. An alternative[10] for the preparation of $(NH_4)_4[Th(NCS)_8]\cdot2H_2O$ consists in adding the stoicheiometric quantity of ammonium thiocyanate to aqueous $[Th(NCS)_4(H_2O)_4]$ and evaporating as in the previous method. The potassium and tetraethylammonium salts of the $[U(NCS)_8]^{4-}$ ion are prepared in non-aqueous solvents; the potassium compound[2] is obtained by treating a solution of uranium tetrachloride in acetone with an excess of potassium thiocyanate, also in acetone, followed by addition of chloroform to the filtrate to precipitate the excess of potassium thiocyanate, after which the complex is precipitated with isopentane. The tetraethylammonium salt[26] is prepared in methyl cyanide solution by reaction of the hexachloro-uranate(IV) with potassium thiocyanate, followed by addition of tetraethyl-ammonium bromide to eliminate potassium ion:

$$[(C_2H_5)_4N]_2[UCl_6] + 8KCNS \rightarrow [(C_2H_5)_4N]_2K_2[U(NCS)_8] + 6KCl$$

and

$$[(C_2H_5)_4N]_2K_2[U(NCS)_8] + 2(C_2H_5)_4NBr \rightarrow [(C_2H_5)_4N]_4[U(NCS)_8] + 2KBr$$

Table 4.4 Infrared spectra of octa-N-thiocyanato complexes

Compound	ν_{C-N}, acetone solution, cm^{-1}	ν_{C-N}, solid, cm^{-1}	ν_{C-S}, cm^{-1}	δ_{NCS}, cm^{-1}	Reference
Rb^4[Th(NCS)$_8$]·2H$_2$O	2088m, 2055vs	2097sh, 2086s, 2073s, 2060s	797w	478m	25
Cs^4[Th(NCS)$_8$]·2H$_2$O	2087m, 2048vs	2092sh, 2083s, 2062s, 2055s	794w	480m	25
K$_4$[U(NCS)$_8$]	—	—	815w	474m	2
Rb$_4$[U(NCS)$_8$]·H$_2$O	2088wm, 2050vs	2097sh, 2087s, 2073w, 2061s	796w	478m	25
Cs$_4$[U(NCS)$_8$]·H$_2$O	2087m, 2052vs	2098s, 2086sh, 2064s, 2012s	797w, 834w	482m	25
(NH$_4$)$_4$[U(NCS)$_8$]	2087m, 2049vs	2097sh, 2072s	796w	480m	25
(Et$_4$N)$_4$[U(NCS)$_8$]	—	2095sh, 2060s	—	—	26

w = weak: m = medium; s = strong: vs = very strong; sh. = shoulder

The penta- and hepta-N-thiocyanatothorates(IV) are prepared from aqueous solutions by the same methods as were used for the corresponding octa-N-thiocyanato-complexes, but using the stoicheiometric amounts of reagents appropriate to the desired final product[10]. These two compounds, and the ammonium and potassium[10] octa-N-thiocyanatothorates(IV), are very hygroscopic, whereas the rubidium[10] and caesium[24] salts, and the salts

Table 4.5   X-ray crystallographic data for anionic thiocyanato complexes

Compound	Crystal symmetry	Lattice parameters, Å			Z	Reference
		a_0	b_0	c_0		
$Cs_4[Th(NCS)_8]\cdot2H_2O$	Monoclinic, $P2_1/n$	13.520	13.696	16.226	4	24, 28
$Cs_4[U(NCS)_8]\cdot2H_2O$	Orthorhombic, $Pnab$	13.15	13.20	15.75	4	28
$(Et_4N)_4[U(NCS)_8]$	Tetragonal, $I4/mmm$	11.64	—	23.01	2	27
$Cs_3[UO_2(NCS)_5]$	Orthorhombic, $Pnma$-D_{2h}^{16}	13.629	13.249	11.556	4	28

of the corresponding[24] uranates(IV) appear to be stable in air, apart from $K_4[U(NCS)_8]$ which is rapidly oxidised[2] to a uranate(VI).

All of these compounds are readily soluble in water and in methanol, but insoluble in benzene, carbon tetrachloride or diethyl ether. Methanol is also a useful solvent for the thorium complexes and for $Cs_4[U(NCS)_8]\cdot H_2O$; the other salts are also probably soluble in this solvent. Conductance measurements in methanol indicate that the salts are 5 ion electrolytes[23, 24]. The complex anions are slowly dissociated in aqueous solution due to replacement of coordinated thiocyanate ion by water, but even so $Cs_4[Th(NCS)_8]\cdot2H_2O$ can be recrystallised from water[24]. $K_4[U(NCS)_8]$ is also soluble in chloroform, ethyl acetate and nitromethane, and slightly soluble in methyl cyanide[2], in which the tetraethylammonium salt also appears to be appreciably soluble. The other recorded salts will probably also be found to dissolve in these solvents.

The hydrated salts (Table 4.3) appear to lose water slowly at 60–100 °C or in a desiccator over concentrated sulphuric acid, the anhydrous products rehydrating in air; none of the compounds is particularly stable to heat, decomposition occurring between 100 and 300 °C.

The infrared spectra (2200–265 cm^{-1}) of the octa-N-thiocyanato-complexes are given in Table 4.4. From the band splittings of the C—N vibrations, recorded for both acetone solution and the solids, it has been concluded[25] that in the rubidium and caesium compounds investigated, the $[M(NCS)_8]^{4-}$ anion adopts an Archimedean antiprismatic arrangement (D_{4d}) in acetone solution but is of lower symmetry (D_{2d}, dodecahedral) in the solid. However, the infrared spectra of the ammonium and tetraethylammonium salts of the $[U(NCS)_8]^{4-}$ ion appear to be much simpler than those of the rubidium and caesium salts. The visible spectrum of the tetraethylammonium salt indicated[26] that the ion might be centrosymmetric, a view confirmed by an x-ray structure determination[27], which showed that the arrangement of NCS groups about the uranium atom in the $[U(NCS)_8]^{4-}$ ion is an almost ideal cube. This arrangement is rare and it is very probable that f-orbitals are involved in the bonding, possibly hybrid metal orbitals between fd^3sp^3 (f_{xyz} specifically) and f^4d^3s. Both of these would provide eight equivalent orbitals directed to the corners of a cube[27]. X-ray crystallographic data for this, and the two other actinide(IV) complexes which have been investigated, are summarised in Table 4.5.

Magnetic susceptibility data (286–95 K) have been reported[2] for $K_4[U(NCS)_8]$; the susceptibility is temperature dependent, but the large Weiss constant (-119 K) makes the value of the effective magnetic moment (3.26 BM) rather meaningless.

Evidence for the formation of mixed N-thiocyanato-chloro-complexes of the type $K_m[Th(NCS)_4Cl_m]$ has been obtained from conductivity and spectrophotometric ($\lambda = 325$ nm) studies[29] of the reaction of thorium tetrachloride with potassium thiocyanate in acetone, N, N-dimethylformamide and methanol; the complexes concerned do not appear to have been isolated. Mixed thiocyanato-formatothorates(IV) have been mentioned occasionally in the early literature (e.g. reference 30), but these species are all somewhat complex analytically and are probably rather doubtful entities.

The only recorded anionic actinide(VI) N-thiocyanato-complexes are those formed by the uranyl(VI) ion; the compounds which have been reported, with infrared spectral data where known, are given in Table 4.6.

The tri-N-thiocyanato-complexes, $M^I[UO_2(NCS)_3(H_2O)_2]$, which behave as 1 : 1 electrolytes in methanol[31], are made by reaction of the stoicheiometric quantities of the appropriate alkali metal or ammonium uranyl sulphate (e.g. $K_2[(UO_2)_2(SO_4)_3(H_2O)_5]$), or of uranyl sulphate and the alkali metal sulphate, with barium thiocyanate in water, the filtrate being evaporated at 50 °C and then left to crystallise in a desiccator. An alternative method is to mix the stoicheiometric amounts of the alkali metal or ammonium thiocyanate with $[UO_2(NCS)_2(H_2O)_3]$ in ethanol and evaporate the solution. The ammonium salt is dark yellow, while the potassium and rubidium salts are orange red[31]. The three salts are very soluble in water, acetone, ether, ethanol and methanol, but are insoluble in benzene and toluene; the anion is appreciably dissociated in aqueous solution. The compounds dehydrate in two stages when heated; the first molecule of water is lost at 87 °C (K salt) or 100 °C (NH$_4$,Rb), and the second at 125 °C (K) or 140 °C (NH$_4$,Rb). The anhydrous products are black, so that at least partial decomposition occurs on dehydration. There is no proof that the complex anion is as shown in Table 4.6, but it seems probable that this is correct in view of the existence of analogous urea complex anion species.

$(NH_4)_2[UO_2(NCS)_4(H_2O)_2]$ is said to be obtained from $(NH_4)_2[UO_2(SO_4)_2(H_2O)_2]$ by treatment with the stoicheiometric amount of barium thiocyanate in water; its aqueous solution is apparently acid and it behaves as a 3 ion electrolyte in methanol[31]; this compound requires further investigation, preferably crystallographic, before it can be regarded as substantiated.

The orange alkali metal and ammonium penta-N-thiocyanato-complexes are prepared in the same way as the tri-N-thiocyanatodioxouranates(VI); they are readily soluble in water, acetone, methanol and ethanol but, in contrast to the tri-N-thiocyanato-complexes, are insoluble in diethyl ether. They are insoluble in other non-polar organic solvents. The dihydrates lose water on prolonged storage over concentrated sulphuric acid and on heating,

when the first molecule of water is lost at 55–65 °C (NH_4 salt) or 90–110 °C (K), and the second at 80–90 °C and at 140–145 °C respectively[32]. The anhydrous compounds decompose at 190 °C (NH_4), 235 °C (K) and 280 °C (Cs).

The yellow dipyridylium, acridinium and triethylammonium salts[14], which melt at 158 °C, 235–245 °C (with decomposition) and 235–250 °C respectively, are precipitated when an aqueous solution of the amine hydrochloride, or a solution of the amine in hydrochloric acid, is added to an aqueous mixture of $[UO_2(NCS)_2(H_2O)_3]$ and ammonium thiocyanate. The monohydrates lose water at 50 °C (acridinium) and 80–90 °C (triethylammonium salt). The three compounds are more soluble in ethanol than in water, and are insoluble in diethyl ether; the dipyridylium salt behaves as a 4 ion electrolyte in methanol[14].

The pyridinium, piperidinium and quinolinium salts, $(AmH)_3[UO_2(NCS)_5]H_2O$ have been prepared in much the same way[39], by dropwise addition of the amine to an aqueous mixture of uranyl acetate and ammonium thiocyanate acidified with concentrated hydrochloric acid. These compounds crystallise as yellow needles (the quinolinium salt separates initially as an oil which crystallises after 15 min); they are more soluble in ethanol than in water, like the salts previously described, and are insoluble in nonpolar solvents such as benzene, carbon tetrachloride or hexane.

The infrared spectra of these complexes, and of the triethylammonium salt, are given in Table 4.7. The polarographic reduction of the triethylamine complex in aqueous solution shows a step at -0.185 mV (probably a misprint for -0.185 V) corresponding to reduction to uranium(V) and a second step at -1.190 V corresponding to reduction of uranium(V) to uranium(IV). The pyridinium salt shows similar reduction behaviour[39].

The $[UO_2(NCS)_5]^{3-}$ anion in the caesium salt is a pentagonal bipyramid, with the five NCS groups in the equatorial plane[28]; crystallographic data are given in Table 4.5. It is probable that this configuration is adopted by the other 7-coordinate anionic species.

The rather unusual caesium ammonium compound, $Cs_3(NH_4)_3[UO_2(NCS)_5]_2$, prepared in the same way as the ammonium salt[32], appears to be genuine and not a mixture; when the aqueous solution is evaporated in the final stages of the preparation, some $(NH_4)_3[UO_2(NCS)_5]\cdot2H_2O$ crystallises with the caesium ammonium salt but is easily removed from the latter by washing the crystals with ethanol, in which the unwanted ammonium salt is much the more soluble[32]. This compound decomposes at 240 °C, an appreciably higher temperature than that at which the ammonium salt decomposes.

The tetradecylammonium salts make an interesting study in themselves; the compound $R_4N[UO_2(NCS)_3]$ is formed[33] when a benzene solution of the ammonium thiocyanate is mixed with an excess of $[UO_2(NCS)_2(H_2O)_3]$. The latter slowly goes into the solution, which becomes bright orange, and on standing the complex salt separates; a band in the infrared spectrum at 2100 cm^{-1} is assigned[16] to a bridging NCS group (or groups) and coordinated water is presumably absent. On prolonged contact with the benzene solution of the ammonium thiocyanate, a second product, a dark orange oil, is obtained and this analyses[33] as $(R_4N)_2[(UO_2)_3(NCS)_8]$. The band[16] at

Table 4.7 Infrared spectra of uranyl N-thiocyanatocomplexes[39]

Complex	$\nu_{as, O-U-O}$ cm^{-1}	ν_{C-N} cm^{-1}	ν_{C-S} cm^{-1}	δ_{NCS} cm^{-1}
*(AmH)$_3$[UO$_2$(NCS)$_5$]H$_2$O	925	2095, 2070	815, 805	482–490
†(Am'H)$_3$[UO$_2$(NCS)$_5$]H$_2$O	920	2100, 2075	808	—
‡(Am''H)$_3$[UO$_2$(NCS)$_5$]H$_2$O	925	2085, 2065, 2045	813	—
§(Am'''H)$_3$[UO$_2$(NCS)$_5$]	928	2095, 2080, 2065	803–808	493, 482

*AmH = C$_5$H$_5$N, pyridine
†Am'H = C$_5$H$_{11}$N, piperidine
‡Am''H = C$_9$H$_7$H, quinoline
§Am'''H = triethylamine

2100 cm^{-1} is more intense in the spectrum of this compound than in the tri-N-thiocyanato-complex, presumably indicating more extensive NCS bridging. This band is no longer observed[16] if an excess of the alkylammonium thiocyanate is used, when the product is the penta-N-thiocyanato-complex[33].

Extraction of uranium(VI) from aqueous solutions of uranyl(VI) thiocyanate into a carbon tetrachloride solution of tri-n-octylammonium (TOAH) N-thiocyanate, followed by vacuum evaporation of the solvent, yields an intense orange, viscous oil[38]; the product is $(TOAH)_2(UO_2(NCS)_4]$ when an excess of uranyl(VI) thiocyanate is used, otherwise the product is $(TOAH)_3[UO_2(NCS)_5]$. From the infrared spectra, bridging NCS (band at 2122 cm^{-1}) and terminal NCS (bands at 2076, 2048 cm^{-1}) groups are present in the former, but only terminal NCS groups (single band at 2055 cm^{-1}) are present in the latter[38].

The yellow caesium and guanidinium salts[15] of the bis (urea) tri-N-thiocyanatodioxouranate(VI) anion are prepared by mixing aqueous solutions of $[UO_2(NCS)_2(H_2O)_3]$, urea and the appropriate thiocyanate; both are soluble in water. From their infrared spectra (Table 4.6) it seems evident that bridging NCS groups are not present in these compounds. The analogous TBP and TBPO complexes are also known[16, 33].

Complex anions in which thiocyanate replaces acetato-groups have been reported; the compound $Cs[UO_2(NCS)(CH_3COO)_2]\cdot 3H_2O$, a yellow-orange solid, separates on cooling a hot, aqueous mixture of uranyl acetate and caesium thiocyanate (mole ratio 1 : 3). With less caesium thiocyanate present, the yellow-orange $Cs_3[(UO_2)_3(NCS)_2(CH_3COO)_7]$ results; the analogous rubidium and ammonium salts, which are monohydrates, are obtained in the same way from an equimolar mixture of uranyl acetate and the appropriate thiocyanate[35]. Infrared data are given in Table 4.6; all NCS groups in these compounds appear to be terminal. Oxalato-N-thiocyanato-complexes, in which the anion is of the form $[UO_2(NCS)(C_2O_4)_2]^{3-}$, result from the replacement of water in the oxalatoanion, $[UO_2(C_2O_4)_2(H_2O)]^{2-}$ by thiocyanate groups[36]. Little is known about these species.

A trichloro-N-thiocyanato-complex,$(R_4N)_2[UO_2Cl_3(NCS)](R=C_{10}H_{21})$, is obtained from an equimolar mixture of the ammonium trichlorouranate(VI) and thiocyanate[16]; there is also evidence for the existence of nitrato-N-thiocyanato-complex ions of the type $[UO_2(NO_3)(NCS)_3]^{2-}$, formed[33] in mixtures of tetradecylammonium nitrate and the tri-N-thiocyanatodioxouranate(VI).

The cobalt(III) tris ethylenediamine cation stabilises anionic thiocyanato species in which the uranium atom appears to be 8-coordinate, as in $[Co(en)_3]_2[UO_2(NCS)_6](NO_3)_2\cdot 5H_2O$ and $[Co(en)_3(NH_4)[UO_2(NCS)_5Cl]$. These are obtained from $[(Co(en)_3]Cl_3$ and a mixture of ammonium thiocyanate with, respectively, uranyl nitrate and $[UO_2(NCS)_2 (H_2O)_3]$ in aqueous solution; both are yellow solids. The mother liquor from the preparation of the first compound also yields the cherry-red complex $[Co(en)_3](NH_4)[UO_2(NCS)_5NO_3]$ on standing and a guanidinium analogue of this is also known[34]. The uranium atom in these last two complexes may be 9-coordinate.

References

1. Aloy, J. (1901). *Ann. Chim. Phys.*, **24**, 417
2. Bagnall, K. W., Brown, D. and Colton, R. (1964). *J. Chem. Soc.*, 2527
3. Day, J. P. (1965). *Thesis*, D. Phil., Oxford, 51
4. Bagnall, K. W. and Baptista, J. L. (1970). *J. Inorg. Nucl. Chem.*, **32**, 2283
5. Choppin, G. R. and Ketels, J. (1965). *J. Inorg. Nucl. Chem.*, **27**, 1335
6. Lebedev, I. A. and Yakovlev, G. N. (1962). *Radiokhimiya*, **4**, 304
7. Ahrland, S. (1949). *Acta Chem. Scand.*, **3**, 1067
8. Harvey, B. G., Heal, H. G., Maddock, A. G. and Rowley, E. L. (1947). *J. Chem. Soc.*, 1010
9. Yanagi, T. (1969). *Bunseki Kagaku*, **18**, 195
10. Molodkin, A. K. and Skotnikova, G. A. (1964). *Zhur. Neorgan. Khim.*, **9**, 60; *Russ. J. Inorg. Chem.*, **9**, 32
11. Ahrland, S. and Larsson, R. (1954). *Acta Chem. Scand.*, **8**, 137
12. Day, R. A., Jr., Wilhite, R. N. and Hamilton, F. D. (1955). *J. Am. Chem. Soc.*, **77**, 3180
13. Pascal, P. (1914). *Compt. Rend.*, **158**, 1672
14. Markov, V. P. and Traggeim, E. N. (1960). *Zh. Neorgan. Khim.*, **5**, 1467; *Russ. J. Inorg. Chem.*, **5**, 712
15. Shchelokov, R. N., Shul'gina, I. M. and Chernyaev, I. I. (1967). *Zh. Neorgan. Khim.*, **12**, 1246; *Russ. J. Inorg. Chem.*, **12**, 660
16. Vdovenko, V. M., Skoblo, A. I. and Suglobov, D. N. (1968). *Zh. Neorgan. Khim.*, **13**, 3059; *Russ. J. Inorg. Chem.*, **13**, 1577
17. Shchelokov, R. N., Shul'gina, I. M. and Chernyaev, I. I. (1966). *Dokl. Akad. Nauk S.S.S.R.*, **168**, 1338; *Dokl. Chem. Proc. Acad. Sci. USSR (English Transl.)*, **166-8**, 640
18. Bagnall, K. W., Brown D., Jones, P. J. and Robinson, P. S. (1964). *J. Chem. Soc.*, 2531
19. Majumdar, A. K. and Bhattacharya, R. G. (1969). *Sci. Cult. (Calcutta)*, **35**, 271
20. Al-Kazzaz, Z. M. S., Bagnall, K. W., Brown D. and Day, J. P. (1971) (to be published)
21. Sinitsyna, S. M. and Sinitsyn, N. M. (1966). *Dokl. Akad. Nauk S.S.S.R.*, **168**, 110; *Dokl. Chem. Proc. Acad. Sci. USSR (English Transl.)*, **166-8**, 467
22. Spaccamela Marchetti, E. (1968). *Ann. Chim. (Rome)*, **58**, 801
23. Markov, V. P. and Traggeim, E. N. (1961). *Zh. Neorgan. Khim.*, **6**, 2316; *Russ. J. Inorg. Chem.*, **6**, 1175
24. Molodkin, A. K. and Skotnikova, G. A. (1962). *Zh. Neorgan. Khim.*, **7**, 1548; *Russ. J. Inorg. Chem.*, **7**, 800
25. Grey, I. E. and Smith, P. W. (1969). *Australian J. Chem.*, **22**, 311
26. Gans, P. and Marriage, J. W. (1970) (personal communication)
27. Countryman, R. and McDonald, W. S. (1971). *J. Inorg. Nucl. Chem.* (in press)
28. Arutyunyan, E. G. and Porai-Koshits, M. A. (1963). *Zh. Strukt. Khim.*, **4**, 110; *J. Struct. Chem.*, **4**, 96
29. Golub, A. M. and Kalibabchuk, V. A. (1966). *Zh. Neorgan. Khim.*, **11**, 590; *Russ. J. Inorg. Chem.*, **11**, 320
30. Reihlem, H. and Debus, M. (1929). *Z. Anorg. Chem.*, **178**, 157
31. Markov, V. P. and Traggeim, E. N. (1960). *Zh. Neorgan. Khim.*, **5**, 1493; *Russ. J. Inorg. Chem.*, **5**, 724
32. Markov, V. P., Traggeim, E. N. and Shul'gina, I. M. (1964). *Zh. Neorgan. Khim.*, **9**, 550; *Russ. J. Inorg. Chem.*, **9**, 305
33. Vdovenko, V. M., Skoblo, A. I. and Suglobov (1967). *Radiokhimiya*, **9**, 119
34. Markov, V. P. and Traggeim, E. N. (1961). *Zh. Neorgan. Khim.*, **6**, 1244; *Russ. J. Inorg. Chem.*, **6**, 636
35. Shchelokov, R. N. and Orlova, I. M. (1969). *Zh. Neorgan. Khim.*, **14**, 1436; *Russ. J. Inorg. Chem.*, **14**, 753
36. Shchelokov, R. N., Shul'gina, I. M. and Chernyaev, I. I. (1966). *Zh. Neorgan. Khim.*, **11**, 2652; *Russ. J. Inorg. Chem.*, **11**, 1424
37. Hart, F. A. and Newbery, J. E. (1966). *J. Inorg. Nucl. Chem.*, **28**, 1334
38. Lipovskii, A. A. and Kuzina, M. G. (1968). *Zh. Neorgan. Khim.*, **13**, 222; *Russ. J. Inorg. Chem.*, **13**, 116
39. Andrä, K. and Böhland, H. (1965). *Z. Chem.*, **5**, 145

5
Thermodynamic Properties of Simple Actinide Compounds

J. FUGER
Laboratory of Nuclear Chemistry, University of Liège

5.1   INTRODUCTION

Our knowledge of the thermodynamic properties of the actinides is still at present quite fragmentary and there are essentially no thermodynamic data available for the elements above curium. Even for some of the lighter elements of the series, experimental data are extremely scarce, as in the case of actinium or protactinium compounds, for which no experimental heats of formation are so far available. The intense radioactivity of the available isotopes and their scarcity are, of course, two factors leading to this situation. Even with isotopes which have relatively long half-lives, radiation damage has been shown to be the origin of discrepancies observed in the low temperature thermodynamic properties of some compounds and radiation heat itself can be in these cases the cause of experimental difficulties. The chemical and physical integrity of the compounds under study can be seriously altered even at room temperature by radiolytic effects, leading to inaccuracies in the measurements. In addition, there is the problem of the accumulation of the daughter isotopes; one month after purification a ^{244}Cm compound will contain more than 0.3% ^{240}Pu in a not necessarily known valency state.

In recent years a considerable amount of experimental data has been reported, mainly for thorium, uranium and plutonium compounds and, to a much lesser extent, for protactinium, neptunium, americium and curium compounds, either replacing estimates based on analogies with similar compounds or modifying previously obtained experimental values. Some of the recent measurements have helped to resolve inconsistencies in the existing data. The best example of this is probably the case of the uranyl(VI) ion for which, as noted by Rand and Kubaschewski[1], the accepted enthalpy of formation, based on $\Delta H_f^\circ(UCl_4(s))$, leads to a value for the enthalpy of formation of uranium trioxide which is approximately 6 kcal mol^{-1} more negative than that deduced from other measurements. Indeed, recent data[2] indicate for $\Delta H_f^\circ(UCl_4(s), 298)$ a value of -243.6 ± 0.6 kcal mol^{-1}, instead of the previously accepted value of -251.3 ± 1 kcal mol^{-1}.

The thermodynamic data on uranium compounds were critically assessed by Rand and Kubaschewski[1] in 1963 and the thermodynamic properties of plutonium compounds have been reviewed more recently by Rand[3] and by Oetting[4]. Other reviews have also appeared during the past 5 years dealing with thermodynamic properties of particular classes of compounds such as chalcogenides and pnictides[5], oxides[6, 7], carbides[8, 9] and there is a section on the available thermodynamic data in a recent book[10] on the halides of the lanthanides·and actinides. These previous reviews are taken as a background for the present study and are referred to when more detailed information is required on a particular topic. In the present review, the author has not considered as such the thermodynamic properties of the gaseous species obtained upon vaporisation of refractory materials, but has only referred to these processes when drawing conclusions as to the thermodynamics of the condensed phase. In view of the large amount of literature still using the calorie instead of the SI recommended joule (J), all the data are reported in terms of calories (1 cal = 4.184 J) or its multiples. Unless otherwise indicated, the degree kelvin (K) has been used. In logarithmic expressions, only decimal logarithms have been used. To simplify notations, especially in formulae, 298 K has often been written instead of 298.15 K. All the thermodynamic symbols used have their usual significance.

5.2   IONS IN SOLUTION

5.2.1   Thorium

Twenty years ago Eyring and Westrum[11] reported the heat of formation of tetrapositive thorium ions from the heat of solution of the metal in 6 MHCl–0.005 MNa_2SiF_6. The value obtained, -181.4 ± 0.1 kcal mol^{-1} (after correction for the presence of the SiF_6^{2-} ions), together with the presently accepted[12] value of ΔH_f° ($ThCl_4(s)$, 298) $= -283.8 \pm 0.4$ kcal mol^{-1} and its heat of solution in infinitely dilute acid, yields ΔH_f° (Th^{4+} (aq), 298) $= -184.0 \pm 1$ kcal mol^{-1}. There is no experimental value for the entropy of Th^{4+} (aq), S° (298), which has been estimated to be -75 cal mol^{-1} K^{-1} by Latimer[13].

5.2.2   Protactinium

There are no experimental data on the enthalpy of formation of protactinium ions in solution. This situation is due in part to the extreme difficulty of dissolving the metal[14], even in HCl–HF media, and to the marked hydrolytic behaviour of the more stable pentavalent valency state.

5.2.3   Uranium

From their critical analysis of the literature data prior to 1962 for the various uranium ions, Rand and Kubaschewski[1] give the values listed in Table 5.1.

Table 5.1   Enthalpies of formation and entropies of uranium ions (ref. 1)

	U^{3+}(aq)	U^{4+}(aq)	UO_2^{2+}(aq)
$-\Delta H_f^\circ(298)$/kcal mol^{-1}	122.6 ± 3	146.3 ± 3	250.0 ± 2
$S^\circ(298)$/cal $mol^{-1}K^{-1}$	-35 ± 6	-81 ± 7	-20 ± 5

There has been no recent determination of data for the trivalent uranium ion.

From their value of the enthalpy of solution of U (s, α) in 6 MHCl–0.005 MNa_2SiF_6, -146.88 ± 0.33 kcal mol^{-1}, and the heat of solution of UCl_4 in various HCl and $HClO_4$ solutions (down to 0.01 $MHClO_4$) Argue et al.[15] have obtained, for the enthalpy of formation of tetravalent uranium in aqueous solution, $\Delta H_f^\circ(U^{4+}$(aq),298), after correction of their data for hydrolysis phenomena, a value of $-157.1 \pm$ kcal mol^{-1}, which is more than 10 kcal mol^{-1} more negative than the above listed value. It should be pointed out, however, that the validity of such a treatment of the data rests entirely on the accuracy with which the hydrolysis constants of uranium are known, together with the heat effect associated with that hydrolysis[16]. Along the same lines, Argue et al. report $\Delta H_f^\circ(\text{U(OH)}^{3+}(\text{aq}),298) = -213.7 \pm 1.0$ kcal mol^{-1}.

Very recently, Fitzgibbon et al.[2] have measured the heat of solution of U(s,α) in 4 M and 6 MHCl, obtaining -137.5 ± 0.5 and -136.2 ± 1.0 kcal mol^{-1}, respectively. Taking $\Delta H_f^\circ(\text{UCl}_4(\text{s}),298)$ to be -243.6 ± 0.6 kcal mol^{-1} from these authors and, for the heat of solution of UCl$_4$ in infinitely dilute acid, a value of 55.0 ± 1.5 kcal mol^{-1}, obtained by extrapolation of its enthalpy of solution in 1 MHCl[17], 4 MHCl and 6 MHCl[2], one obtains $\Delta H_f^\circ(\text{U}^{4+}(\text{aq}),298) = -138.8 \pm 2$ kcal mol^{-1}, a value which is 7.5 kcal mol^{-1} less negative than that previously assessed[1]. The large difference between these two results is ascribed to the fact that in the dissolution of uranium metal a reaction between U^{3+} ions and dissolved oxygen occurs unless the solution has been de-aerated with the utmost care. The validity of these new results is confirmed by the fact that the enthalpy of formation of UCl$_4$(s) obtained from the dissolution of the metal and of UCl$_4$ in hydrochloric acid solutions agrees extremely well with the value obtained, also by Fitzgibbon et al.[2], from the solution of UO$_2$ and UCl$_4$ in sulphuric acid solutions and based on the enthalpy of formation of UO$_2$(s) from combustion data[18]. Of course, as the values for the other aquated uranium ions are partly based on the enthalpy of formation of UCl$_4$(s), their values should be modified accordingly.

Similarly, Cordfunke[19] has reported the heat of solution of γ-UO$_3$ in 6 MHNO$_3$ to be -17.04 ± 0.1 kcal mol^{-1} and this value, with the enthalpy of formation of γ-UO$_3(-293.5 \pm 1$ kcal mol^{-1} from dissociation pressure measurements), gives $\Delta H_f^\circ(\text{UO}_2^{2+},6 \text{ MHNO}_3,298) = -242.2 \pm 1.0$ kcal mol^{-1}. Although only strictly applicable in 6 MHNO$_3$, this value also appears low as compared to the assessed value of $\Delta H_f^\circ(\text{UO}_2^{2+}(\text{aq}),298)$, -250 ± 2 kcal mol^{-1}. There is still no experimental value in the literature for the UO$_2^+$ ion.

5.2.4   Neptunium

The enthalpy of formation of the Np^{3+} ion has been obtained by Argue et al.[15] from the heat of solution of the metal (α-phase) in 1 MHCl–0.005 M NaHSO$_2$· CH$_2$O·2H$_2$O, the sodium formaldehyde sulphoxylate being present to prevent oxidation of the tervalent neptunium by traces of dissolved oxygen. The value obtained for $\Delta H_f^\circ(\text{Np}^{3+},1 \text{ MHCl},298)$, -125.6 ± 0.3 kcal mol^{-1}, assuming a negligible effect from the reducing agent, is in fair agreement with the figure obtained more recently by Fuger et al.[20] for the solution of the metal in oxygen free 1 MHCl, -126.2 ± 0.2 kcal mol^{-1}.

The enthalpy of formation of the tetrapositive neptunium ion in 1 MHCl has also been obtained by Fuger et al.[20] from the heat of solution of the metal in 1 MHCl–0.005 MNa$_2$SiF$_6$. After correction for the presence of the fluoride ion, these authors obtain a value for ΔH_f°(Np^{4+}, 1 MHCl,298), -132.3 ± 0.3 kcal mol^{-1}, in agreement with the previously reported values of -132.4[21] and -131.9[15] kcal mol^{-1} in the same medium, also corrected for the presence of the fluoride ion. The difference between the experimentally determined heats of formation of the tetra- and tervalent neptunium ions in 1 MHCl is also consistent with the heat of oxidation of tervalent neptunium, -5.7 ± 0.2 kcal mol^{-1}, deduced from the temperature coefficient of the Np^{3+}–Np^{4+} cell[22].

There are still no experimental values for the entropy of the Np^{3+}(aq) and Np^{4+}(aq) ions; these have been estimated[13] to be -31 and -78 cal mol^{-1} K^{-1}, respectively.

The only values available for the enthalpy of formation of the NpO$_2^+$(aq) and NpO$_2^{2+}$(aq) ions are those deduced from the data of Cohen and Hindman[23] on the temperature coefficients of the Np^{4+}–NpO$_2^+$ and NpO$_2^+$–NpO$_2^{2+}$ cells in 1 MHClO$_4$, namely ΔH_f°(NpO$_2^+$, 1 MHClO$_4$, 298) = -233.6 kcal mol^{-1} and ΔH_f°(NpO$_2^{2+}$,1 MHClO$_4$,298) = -205.5 kcal mol^{-1}.

Estimated values[24] in 1 MHCl have also been reported for NpO$_2^+$ and NpO$_2^{2+}$ ions, -230.8 and -208.5 kcal mol^{-1}, respectively. From the measurement of the solubility and of the heat of the solution of NpO$_2$-(NO$_3$)$_2\cdot$6H$_2$O, and using for the neptunium compound the activity coefficient values of uranyl nitrate solutions of the same molarity, Brand and Cobble[25] obtained a value for S°(NpO$_2^{2+}$(aq),298) of -20 cal mol^{-1} K^{-1}, as compared with -17, previously estimated by Latimer[13]. From the enthalpy of the reduction of the hexavalent neptunyl ion to the pentavalent state by hydrogen peroxide, the enthalpy for the reaction

$$\text{NpO}_2^+(\text{aq}) + \text{H}^+(\text{aq}) = \text{NpO}_2^{2+}(\text{aq}) + \tfrac{1}{2}\text{H}_2(\text{g})$$
$$\Delta H^\circ(298) = 28.07\pm0.14 \text{ kcal mol}^{-1}$$

was obtained by the same authors. This value, together with the standard potential of the NpO$_2^+$–NpO$_2^{2+}$ couple, $E^\circ = -1.236\pm0.010$ abs. volt, yields for the entropy of NpO$_2^+$(aq) the value -6.2 ± 2 cal mol^{-1} K^{-1} as compared with the estimated[13] value of 12 cal mol^{-1} K^{-1}. It is worth noting here that the measured enthalpy of reduction of NpO$_2^{2+}$(aq) to NpO$_2^+$(aq) is remarkably consistent with the values of Cohen and Hindman[23] for the enthalpy of formation of these two ions.

5.2.5 Plutonium

The heat of solution of plutonium metal, according to the reaction

$$\text{Pu(s, }\alpha) + 3\text{H}^+(\text{in 6 MHCl}) = \text{Pu}^{3+}(\text{in 6 MHCl}) + \tfrac{3}{2}\text{H}_2(\text{g}), \Delta H(298)$$

has been reported by several authors (Table 5.2).

From this table it can be seen that the early results of Westrum and Robinson are in remarkable agreement with those of Akhachinskii et al., while those of Fuger and Cunningham and of Hinchey and Cobble are about 2.5 kcal mol^{-1} less negative. Initially, Fuger and Cunningham had proposed that the difference between these results and those of Westrum and Robinson could be due to a failure to characterise the phase of the starting plutonium metal in the early measurements of Westrum and Robinson. This point seems to be excluded, however, in view of the results of Akhachinskii. Hinchey and Cobble have noted that more negative values were obtained when oxygen is not strictly excluded from the solutions, possibly due to oxidation of the trivalent hydroxide formed in the acid depleted zone near

Table 5.2   Heat of solution of Pu(s, α) in 6 MHCl

$-\Delta H(298)$ kcal mol^{-1}	Authors	Year	Reference
141.64 ± 0.2	Westrum and Robinson	1949	26
141.02 ± 0.19	Akhachinskii and Kopitin	1960	27
141.14 ± 0.14	Akhachinskii et al.	1962	28
138.9 ± 0.9	Fuger and Cunningham	1963	29
141.5 ± 0.4	Akhachinskii	1965	30
139.4 ± 0.8	Hinchey and Cobble	1970	31

the dissolving metal sample. In any case, in view of the fact that the less negative results were obtained with small samples (less than 1 mg) and that the error limits reported for these experiments are larger, it is suggested that the most recent value of Akhachinskii, -141.5 ± 0.4 kcal mol^{-1}, be used in thermodynamic calculations involving Pu^{3+} ions. Hinchey and Cobble[31] have determined the heat of solution of PuCl$_3 \cdot 6$H$_2$O in water at infinite dilution:

$$PuCl_3 \cdot 6H_2O(1) = Pu^{3+}(aq) + 3Cl^-(aq) + 6H_2O(1)$$

$$\Delta H^\circ(298) = -8.260 \pm 0.058 \text{ kcal mol}^{-1}$$

The value of the free energy change, $\Delta G^\circ(298)$, is -7.207 kcal mol^{-1} for the above reaction; this was obtained from solubility determinations, using activity coefficients extrapolated from osmotic coefficient data [32] for the isomorphous salt SmCl$_3 \cdot 6$H$_2$O. From the lattice entropy of SmCl$_3 \cdot 6$H$_2$O[33] and after correcting for the difference in mass between plutonium and samarium and for the magnetic entropy of Pu^{3+}(^{6}H$_{\frac{5}{2}}$), Hinchey and Cobble obtain $S^\circ(PuCl_3 \cdot 6H_2O(s),298) = 99.6$ cal mol^{-1} K^{-1}. From this value and the entropy change upon dissolution of PuCl$_3 \cdot 6$H$_2$O, the value of the entropy, $S^\circ(Pu^{3+}(aq),298)$ is found to be -44.6 cal mol^{-1} K^{-1} as compared with the value of -39 cal mol^{-1} K^{-1} given by Latimer[13], based on $S^\circ(Gd^{3+}(aq),298) = -43$ cal mol^{-1} K^{-1}[34]. Hinchey and Cobble[33] have also recently determined, from the heats and free energies of solution of several of the hydrated lanthanide trichlorides, the entropies of the aqueous lanthanide ions. These values are 10–15 cal mol^{-1} K^{-1} more negative than the previous estimates based on the old experimental value for Gd^{3+}. In particular the recently determined value for gadolinium is $S^\circ(Gd^{3+}(aq),298) = -53.39 \pm 0.8$ cal mol^{-1} K^{-1}.

Recalling that the entropies of all the plutonium ions are based on that of Pu^{3+} and accepting the new determination of Hinchey and Cobble, the previously accepted values[3] should be amended as in Table 5.3. Since there has been no recent determination of the enthalpy of formation of other plutonium ions, the values given here are those assessed by Rand[3].

Table 5.3   Enthalpies of formation and entropies of plutonium ions

	Pu^{3+}(aq)	Pu^{4+}(aq)	PuO_2^+(aq)	PuO_2^{2+}(aq)
$-\Delta H_f^{\circ}(298)$/kcal mol^{-1}	141.7 ± 1.1	128.2 ± 1.2	219.9 ± 2	197.8 ± 1.9
$S^{\circ}(298)$/cal mol^{-1}K^{-1}	-44.6 ± 2	$-91\ \pm3$	$-13\ \pm6$	$-26\ \pm6$

The entropy values, especially those of the plutonyl ions, should be treated with extreme caution since, for instance, the value obtained for $S^{\circ}(PuO_2^+(aq),$ $298) = -13$ cal mol^{-1} K^{-1} appears surprisingly negative when compared to the value of $S^{\circ}(NpO_2^+(aq),298)$, -6.2 ± 2 cal mol^{-1} K^{-1}, recently obtained by Brand and Cobble[25].

From their determination of the entropy of Pu^{3+}(aq), Hinchey and Cobble[31] have estimated the entropies of the various trivalent actinide ions, assuming that the entropy due to the size and charge of the ions has the same dependence as observed for the lanthanides[33] and that the magnetic degeneracy of the aqueous ion is that of the free ion. The values obtained are summarised in Table 5.4.

Table 5.4   Estimated entropies of the trivalent actinide ions, according to Hinchley and Cobble (ref. 31)

Me^{3+}(aq)	$S^{\circ}(298)$, cal mol^{-1}K^{-1}	Me^{3+}(aq)	$S^{\circ}(298)$, cal mol^{-1}K^{-1}
Ac	-43.3	Np	-43.3
Th	-40.9	Pu	-44.6
Pa	-41.3	Am	-48.7
U	-42.1	Cm	(-45)

This series of values can probably be considered as more accurate and consistent than the early estimates of Latimer based on the Gd^{3+} ion.

5.2.6   Americium

The only experimental result available for the enthalpy of formation of the trivalent americium ion is that obtained by Lohr and Cunningham[35] from the solution of the metal, of unspecified structure, in 1.5 MHCl, namely $\Delta H_f(Am^{3+},1.5$ MHCl,298$) = -162.3\pm2.7$ kcal mol^{-1}. From this value and from the heat of solution of $AmCl_3$ [29] in 10^{-3}MHCl (-33.36 ± 0.25 kcal mol^{-1}) and 1.5 MHCl (-30.60 ± 0.20 kcal mol^{-1}), one obtains $\Delta H_f^{\circ}(Am^{3+}$ (aq),298$) = -162.4\pm3$ kcal mol^{-1}. From the heat of solution of AmO_2 in HNO_3 solutions[36], the enthalpy of formation of the tetravalent americium ion has been estimated as -116 ± 6 kcal mol^{-1} and, from the enthalpy of reduction[37] of AmO_2^+ and AmO_2^{2+} by ferrous ion in 1 MHClO$_4$, the enthalpy of formation of these ions has been estimated as -207.7 ± 2.9 and -170.8 ± 2.7 kcal mol^{-1}, respectively.

Now that americium metal of high purity has become available[38], a new determination of its heat of solution is urgently needed.

5.2.7 Curium

The heat of solution of curium (s, α)[39] in hydrochloric acid has been measured, according to the reaction:

$$Cm(s, \alpha) + 3H^+ (\text{in } 1 \text{ MHCl}) = Cm^{3+}(\text{in } 1 \text{ MHCl}) + \tfrac{3}{2}H_2(g),$$

$$\Delta H(298) = -140.2 \pm 1 \text{ kcal mol}^{-1}.$$

Using this value, and the extrapolation of existing data on the heat of solution of $PuCl_3$ [40] and $AmCl_3$ [29], one obtains

$$\Delta H_f^\circ (Cm^{3+}(aq), 298) = -140.9 \pm 1.3 \text{ kcal mol}^{-1}.$$

It is interesting to note that, although americium and curium metal have similar structures and similar lattice parameters, the enthalpy of formation of $Cm^{3+}(aq)$ is about 20 kcal mol^{-1} less negative than that of $Am^{3+}(aq)$. The large difference between the enthalpies of vaporisation of these two elements[41] is obviously a major contributing factor.

5.3 HALIDES, OXYHALIDES AND RELATED COMPOUNDS

5.3.1 Trivalent compounds

5.3.1.1 Solids

Table 5.5 summarises the available information concerning the enthalpies of formation and entropies of these compounds. Estimated data are indicated in parentheses. Only the limited number of data which have appeared in the last few years and which have not been reviewed elsewhere[1, 3, 4, 10, 42] will be discussed here.

Khanaev and Khripin[43] have measured the heat of solution of UF_3 in concentrated hydrochloric acid (HCl, $3.91H_2O$) containing 10% $FeCl_3$ and 1% H_3BO_3 at 323K and for the reaction

$$UF_3(s) + [(\tfrac{3}{4}H_3BO_3 + 3FeCl_3) \text{ in } (HCl, 3.91H_2O)]$$
$$= [(UO_2Cl_2 + 3FeCl_2 + HCl + \tfrac{1}{4}H_2O + \tfrac{3}{4}HBF_4)$$
$$\text{in } (HCl, 3.91H_2O)] + \tfrac{1}{2}H_2(g)$$
$$\Delta H(323) \text{ is } -34.43 \pm 0.06 \text{ kcal mol}^{-1}.$$

From the measured heats of solution of UO_2Cl_2 (-8.79 ± 0.03 kcal mol^{-1}) and UCl_4(-18.44 ± 0.02 kcal mol^{-1}) in the same medium, the heats of formation of these two compounds (taken as -301.8 and -251.2 kcal mol^{-1} respectively), and using auxiliary thermodynamic data, Khanaev and Khripin obtain, as a weighted average, a value of -356.8 ± 5 kcal mol^{-1} for

$\Delta H^\circ_f(\text{UF}_3,\text{s},298)$.

However, as we shall see in Section 3.2.1, the value used by these authors for the heat of formation of $\text{UCl}_4(\text{s})$ is about 7 kcal mol^{-1} more negative than the most recently reported value.

The preceding result is in fair agreement with the value of -351 ± 5 kcal mol^{-1} derived from studies[41] of the equilibrium pressures of HF in the reaction: $\text{UF}_4(\text{s}) + \frac{1}{2}\text{H}_2(\text{g}) = \text{UF}_3(\text{solid soln. in } \text{UF}_4) + \text{HF(g)}, 840–1060$ K. After correction for the formation of the solid solution, the free energy,

Table 5.5 Enthalpies of formation and entropies of the trivalent halides and oxyhalides

Compound	$\dfrac{-\Delta H^\circ_f(298)}{\text{kcal mol}^{-1}}$	Reference	$\dfrac{S^\circ(298)}{\text{cal mol}^{-1}\text{K}^{-1}}$	Reference
AcF$_3$	(420 ± 10)	42	—	—
UF$_3$	$\begin{cases} 356.8 \pm 5 \\ 351 \ \pm 5 \end{cases}$	$\begin{cases} 43 \\ 44 \end{cases}$	(28 ± 2)	1
NpF$_3$	(360 ± 2)	42	—	—
PuF$_3$	371 ± 3	3	(30.7 ± 3)	3
AmF$_3$	(394)	42	—	—
AcCl$_3$	(260 ± 10)	42	—	—
UCl$_3$	213.5 ± 2	1	38.0 ± 0.2	1
NpCl$_3$	(216 ± 1)	42	—	—
PuCl$_3$	229.4 ± 0.8	12	(39.2 ± 3)	3
AmCl$_3$	249.2 ± 3	29	—	—
CmCl$_3$	226.4 ± 1.2	39	—	—
AcBr$_3$	(220 ± 10)	42	—	—
UBr$_3$	172.3 ± 2	1	(45.0 ± 2)	1
NpBr$_3$	(174 ± 1)	42	—	—
PuBr$_3$	187.7 ± 1	3	(45.7 ± 5)	3
AmBr$_3$	(202)	65	—	—
AcI$_3$	(169 ± 10)	42	—	—
UI$_3$	114.2 ± 2	1	(57.0 ± 3)	1
NpI$_3$	(120 ± 1)	42	—	—
PuI$_3$	130 ± 3	3	(56 ± 5)	3
AmI$_3$	(146)	65	—	—
AcOF	(420 ± 10)	42	—	—
PuOCl	222.7 ± 1	3	(25.2 ± 3)	3
AmOCl	241.5 ± 3	29	—	—
PuOBr	206.4 ± 3	3	(27 ± 3)	3
PuOI	(183)	42	—	—
Cs$_2$NaPuCl$_6$	548	45	(107)	45

$\Delta G(\text{UF}_3,\text{s})$, was found to be $-350 + 50 \times 10^{-3} T$ kcal mol^{-1}, which yields, after correction to 298 K,

and
$$\Delta H^\circ_f(\text{UF}_3(\text{s}),298) = -351 \pm 5 \text{ kcal mol}^{-1}$$
$$\Delta S^\circ_f(\text{UF}_3(\text{s}),298) = -54.2 \pm 0.7 \text{ cal mol}^{-1}\text{K}^{-1}$$

The enthalpy of formation, $\Delta H^\circ_f(\text{AmCl}_3(\text{s}),298)$, is -249.2 ± 3 kcal mol^{-1}, obtained from the heat of solution of the metal[35] (-162.3 ± 2.7 kcal mol^{-1}) in 1.5 MHCl and the heat of solution of AmCl$_3$ in the same medium[29] (-30.60 ± 0.20 kcal mol^{-1}), while for $\Delta H^\circ_f(\text{CmCl}_3(\text{s}),298)$ the value is

-226.4 ± 1.2 kcal mol^{-1}, which rests on the heat of solution of CmCl$_3$ in 1 MHCl (-140.2 ± 1 kcal mol^{-1}) and on an estimated value (-32.6 ± 0.5 kcal mol^{-1}) for the solution of CmCl$_3$ using known heat of solution data for AmCl$_3$[29] and PuCl$_3$[40].

ΔH_f°(AmOCl(s),298) is -241.5 ± 3 kcal mol^{-1} (erroneously reported[29] as -227.6 ± 3 kcal mol^{-1}) obtained from the enthalpy of formation of AmCl$_3$(s) reported above and the data of Koch and Cunningham[46] on the temperature dependence of the reaction:

$$AmCl_3(s) + H_2O(g) = AmOCl(s) + 2HCl(g)(680-800\ K)$$

for which these authors give $\Delta H(298) = 21.38$ kcal mol^{-1}.

Morss[47] has found the heat of solution of Cs$_2$NaPuCl$_6$ in 10^{-3} MHCl to be -13.4 ± 0.1 kcal mol^{-1}. From this value, and using the enthalpy of formation of PuCl$_3$(s)[29] and of CsCl(s) and NaCl(s) and their heats of solution at infinite dilution, Morss has obtained the enthalpy of formation of this compound.

The free energy of formation of UI$_3$(s) has been obtained[48] from e.m.f. measurements of the solid galvanic cell:

$$Pt\,|\,(U,UI_3)(s)\,|\,NaI(s)\,|\,AgI(s)\,|\,Ag(s)\,|\,Pt$$

over the range 493–643 K, the cell reaction being U(s) + 3AgI(s) = UI$_3$(s) + 3Ag(s). Combining the experimental results with the standard free energy of formation of AgI(s) yields a value for the free energy, ΔG_f°(UI$_3$(s),643), -98.8 kcal mol^{-1}. This value is about 7 kcal mol^{-1} more positive than previous estimates[49] based on the calorimetric heat of formation[50] of UI$_3$(s), which itself depends on the enthalpy of formation of UCl$_3$(s). Similar measurements of the free energy of formation [51] of UCl$_3$, ΔG_f°(UCl$_3$(s),673) $= -168$ kcal mol^{-1}, had revealed an analogous discrepancy.

5.3.1.2   Gaseous species

Zmbov[52] has recently estimated the enthalpies of formation of ThF$_3$(g) and ThF$_2$(g) using a combination of mass spectrometric and Knudsen effusion techniques. The heats corresponding to the following experimental reactions:

$$Ca(g) + ThF_4(g) = CaF(g) + ThF_3(g)$$
$$\Delta H(298) = 32.3\ kcal\ mol^{-1}$$
$$2Ca(g) + ThF_4(g) = 2CaF(g) + ThF_2(g)$$
$$\Delta H(298) = 45.6\ kcal\ mol^{-1}$$

were obtained at 298 K, using various published and estimated free energy functions. The reported error limits are ±5 kcal mol^{-1} for the former reaction and unspecified, but larger, for the latter. From these data, combined with the dissociation energy[53, 54] of CaF(g,298) (127.3 kcal mol^{-1}), the enthalpy of sublimation[55] of ThF$_4$(298) (82.6 kcal mol^{-1}), the enthalpy of formation[56] of F(g,298) (18.86 kcal (g at.)$^{-1}$), the recently published calorimetric value[57] $\Delta H_f^\circ(\text{ThF}_4(\text{s}),298) = -504.5$ kcal mol^{-1} and the enthalpy of sublimation[58] of Th (137.5 kcal mol^{-1}), one obtains:

$\text{Th(s)} + \frac{3}{2}\text{F}_2(\text{g}) = \text{ThF}_3(\text{g})$	$\Delta H_f^\circ(\text{ThF}_3(\text{g}),298) = -281.2$ kcal mol^{-1}
$\text{ThF}_3(\text{g}) = \text{Th(g)} + 3\text{F(g)}$	$\Delta H_{\text{atom}}^\circ(\text{ThF}_3(\text{g}),298) = 475.1$ kcal mol^{-1}
$\text{Th(s)} + \text{F}_2(\text{g}) = \text{ThF}_2(\text{g})$	$\Delta H_f^\circ(\text{ThF}_2(\text{g}),298) = -159.4$ kcal mol^{-1}
$\text{ThF}_2(\text{g}) = \text{Th(g)} + 2\text{F(g)}$	$\Delta H_{\text{atom}}^\circ(\text{ThF}_2(\text{g}),298) = 334.6$ kcal mol^{-1}

Kent[59] has reported a mass spectrometric investigation of the sublimation of PuF$_3$ in the temperature range 1243–1475 K. The vapour pressure was given by the equation:

$$\log P \text{ (atm)} = \left(\frac{-20734+118}{T}\right) + (9.288 \pm 0.087)$$

and from the experimental data, the enthalpy and entropy of sublimation at 298 K, assuming $\Delta C_p = -6$ cal mol^{-1} K^{-1} for the sublimation process, were calculated to be 101 ± 3 kcal mol^{-1} and 51.6 cal mol^{-1}K^{-1}. These new values are in good agreement with those deduced by Rand[3] from earlier results obtained by Carniglia and Cunningham[60] and by Phipps et al.[61]. From the enthalpy of formation[3] of PuF$_3$(s) (-371 ± 3 kcal mol^{-1}) and the enthalpy of sublimation[62] of Pu(s) (83.0 ± 0.5 kcal mol^{-1}), one obtains:

$$\text{Pu(s)} + \tfrac{3}{2}\text{F}_2(\text{g}) = \text{PuF}_3(\text{g})$$
$$\Delta H_f^\circ(\text{PuF}_3(\text{g}),298) = -270 \pm 5 \text{ kcal mol}^{-1}$$
and
$$\text{PuF}_3(\text{g}) = \text{Pu(g)} + 3\text{F(g)}$$
$$\Delta H_{\text{atom}}^\circ(\text{PuF}_3(\text{g}),298) = 410 \pm 5 \text{ kcal mol}^{-1}$$

A study of the species above the system PuF$_3$ – Pu as a function of temperature led to the following enthalpy changes for the following reactions:

$$3\,\text{PuF(g)} = \text{PuF}_3(\text{g}) + 2\text{Pu(g)}$$
$$\Delta H^\circ(298) = -29.6 \pm 5 \text{ kcal mol}^{-1}$$
$$3\text{PuF}_2(\text{g}) = 2\text{PuF}_3(\text{g}) + \text{Pu(g)}$$
$$\Delta H^\circ(298) = -32.6 \pm 3 \text{ kcal mol}^{-1}$$

these were obtained at 298 K using estimated C_p values for the fluoride species and published data for plutonium[62]. From these data one obtains immediately:

$$\text{PuF}_2(\text{g}) = \text{Pu(g)} + 2\text{F(g)}$$
$$\Delta H_{\text{atom}}^\circ(\text{PuF}_2(\text{g}),298) = 263 \pm 10 \text{ kcal mol}^{-1}$$
$$\text{PuF(g)} = \text{Pu(g)} + \text{F(g)}$$
$$\Delta H_{\text{atom}}^\circ(\text{PuF(g)},298) = 128 \pm 7 \text{ kcal mol}^{-1}$$

A comparison of these results with those of Zmbov[52] shows that the atomisation energies are much higher for the thorium fluorides, the average bond energy in $ThF_3(g)$ being about 22 kcal mol^{-1} greater than that in $PuF_3(g)$.

5.3.2   Tetravalent compounds

5.3.2.1   Solids

Table 5.6 summarises the available data on the enthalpies of formation and entropies of these compounds. In recent years, many important thermodynamic data have been obtained by calorimetric techniques.

The determination, by fluorine bomb calorimetry, of the enthalpy of formation of $ThF_4(s)$, was recently reported by Van Deventer et al.[57]:

$$Th(s) + 2F_2(g) = ThF_4(s,monocl.)$$
$$\Delta H_f^{\circ}(ThF_4(s),monocl.,298) = -504.5 \pm 1.2 \text{ kcal mol}^{-1}.$$

The standard enthalpy of formation of $ThF_4(s)$ had previously been obtained from a study of high temperature equilibria and these data have been recalculated by Van Deventer et al., using the most recently available auxiliary thermodynamic data. The measurements of Darnell[70] from the study of the equilibrium:

$$ThF_4(s) + SiO_2(s) \rightleftharpoons ThO_2(s) + SiF_4(g)$$

yield a value of -500.0 ± 0.6 kcal mol^{-1} for the enthalpy of formation of $ThF_4(s)$, while a similar treatment of the data of Heus and Egan[71], based on e.m.f. measurements with cells of the type:

$$(Mg,MgF_2)(s) \,|\, CaF_2(s) \,|\, (ThF_4,Th)(s)$$

or analogous cells, gives enthalpy of formation values ranging between -495.5 and -501.1 kcal mol^{-1}. It is obvious, however, that owing to the usual difficulties inherent in equilibrium or e.m.f. data, the direct calorimetric result should, at present, be preferred.

Laser and Merz[64] have estimated the enthalpies of formation, entropies and heat capacities at 298 K of PaO_2 and PaF_4 by interpolation of the existing data for ThO_2 and UO_2, and for ThF_4 and UF_4, respectively. From a study of the fluorination of protactinium (^{233}Pa) in various uranium and thorium compounds such as fluorides, oxides and mixed oxides, using various fluorinating agents above 1000 K, these authors have estimated the enthalpies of formation and entropies of a series of protactinium fluorides, oxyfluorides and oxides. It is obvious, however, that while such calculations are useful for the prediction of the behaviour of protactinium at high temperature under various fluorinating conditions, they only lead to very approximate thermodynamic data. These values have not, therefore, been reproduced here, except that for PaF_4, for which the given estimate is probably as good as the value given by Maslov and Maslov[65]. The values given for the other tetrahalides of protactinium are also estimates[65].

Fitzgibbon *et al.*[2] have determined the enthalpy of formation of $UCl_4(s)$ from the heat of solution of the metal and of UCl_4 in oxygen free 4 M and 6 MHCl containing 0.005 MNa_2SiF_6. They obtain (as a weighted mean of two sets of measurements) a value of -243.4 ± 0.7 kcal mol^{-1}. The enthalpy of formation of $UCl_4(s)$ was also determined by these authors using another path which included the heat of solution of UO_2 and UCl_4 in sulphuric acid solutions and depended on the enthalpy of formation of $UO_2(s)(-259.3 \pm 0.2$ kcal $mol^{-1})$ obtained by combustion calorimetry[18]. These measurements lead to a value of -244.0 ± 1.5 kcal mol^{-1}. The weighted average of the two sets of determinations, which are remarkably consistent, leads to a

Table 5.6　Enthalpies of formation and entropies of the tetravalent halides and oxyhalides

Compound	$-\Delta H_f^\circ(298)$ kcal mol^{-1}	Reference	$S^\circ(298)$ cal $mol^{-1}K^{-1}$	Reference
ThF_4	504.5 ± 1.2	57	33.95	63
PaF_4	$\begin{cases}(465.3)\\(460)\end{cases}$	$\begin{cases}64\\65\end{cases}$	(35.8)	64
UF_4	450 ± 5	1	36.3 ± 0.1	1
$UF_{4.25}$	461.5 ± 6	1	(37.7 ± 1)	1
$UF_{4.50}$	472.5 ± 6	1	(39.4 ± 2)	1
NpF_4	(428)	42	—	—
PuF_4	425 ± 8	3	(38.7 ± 0.5)	3
AmF_4	(400)	42	—	—
$ThCl_4$	283.8 ± 0.4	12	45.5	63
$PaCl_4$	(268)	65	—	—
UCl_4	243.6 ± 0.6	2	47.4 ± 0.3	1
$NpCl_4$	235.7 ± 0.3	20	—	—
$ThBr_4$	230.2 ± 0.5	66	54.5	63
$PaBr_4$	(212)	65	—	—
UBr_4	191.1 ± 1.1	66	56.0 ± 2	1
$NpBr_4$	184.5 ± 0.5	66	—	—
ThI_4	160.3	63	63.5	63
PaI_4	(144)	65	—	—
UI_4	121.6 ± 1	66	67 ± 3	1
$ThOF_2$	401 ± 2	63	24.2 ± 2	63
UOF_2	352.0 ± 5	67	—	—
$UOF(OH)$	$333.3 \pm 5*$	68	—	—
$ThOCl_2$	296.1 ± 1	63	27.7	63
$UOCl_2$	$252 \pm 2*$	1	33.1 ± 0.1	1
$NpOCl_2$	245.6 ± 0.3	20	—	—
$ThOBr_2$	270 ± 3	63	33.0	63
$UOBr_2$	$233 \pm 2*$	1	37.7 ± 0.1	1
$ThOI_2$	237.6	63	40	63
Cs_2ThCl_6	508.7 ± 0.5	69	—	—
Cs_2UCl_6	476.7 ± 0.7	69	—	—
Cs_2NpCl_6	468.5 ± 0.4	69	—	—
Cs_2PuCl_6	466.9 ± 1	69	—	—
K_2UCl_6	$461.8 \pm 0.7*$	70	—	—
$KUCl_5$	$353.3 \pm 0.7*$	70	—	—
Na_2UCl_6	$441.6 \pm 0.7*$	70	—	—
$KNaUCl_6$	$452.0 \pm 0.7*$	70	—	—

*Recalculated from indicated reference using listed value for $\Delta H_f^\circ(UCl_4, s)$

value of -243.6 ± 0.6 kcal mol^{-1} for $\Delta H_f^{\circ}(UCl_4(s),298)$ which differs greatly from all the previously reported values, which were -251.3 [73], -251.1 [50], -251.2 [15] and quite recently, -251.5 kcal mol^{-1} [74]. Such good agreement between the earlier data could appear convincing. However, the fact that the value of Fitzgibbon et al. is obtained from two independent reaction schemes and that only one of them depends on the heat of solution of the metal in acid solution (for which, as pointed out in Section 5.2.3 difficulties may arise unless oxygen is excluded), indicates that their value is at present the best available figure. In addition, the fact that this result substantiates Rand and Kubaschewski's opinion[1] that the previously accepted values for the enthalpy of formation of uranium ions were approximately 6 kcal mol^{-1} too negative, can be considered as a supplementary argument.

Fuger and Brown[17, 20, 66, 69] have measured the heats of solution in 1 M and 6 MHCl of various tetravalent compounds. Their results are summarised in Table 5.7. The enthalpy of formation of the thorium compounds were obtained using the heat of formation[11] of Th^{4+} in 6 MHCl(-181.4 kcal mol^{-1}), the enthalpy of formation of ThCl$_4$(s) (-283.8 ± 0.4 kcal mol^{-1}) [12], and the heats of solution of ThCl$_4$ [75] in 1 M and 6 MHCl(-57.42 ± 0.1 and -44.27 ± 0.1 kcal mol^{-1}, respectively). For the protactinium compounds, unfortunately, no heat of formation can be calculated since the heat of formation of the Pa^{IV} ion has not yet been measured. In fact, these data were the first heat of solution measurements reported for protactinium compounds. The heats of formation of the uranium compounds are based on the data of Fitzgibbon et al.[2] for the heat of formation of UCl$_4$(s). For the neptunium compounds the value -132.3 kcal mol^{-1} was used for the

Table 5.7 Heats of solution and heats of formation of some actinide(IV) halides and related compounds (ref. 17, 20, 66, 69)

Compound	$-\Delta H_{soln}(298)$ kcal mol^{-1}		$-\Delta H_f^{\circ}(298)$ kcal mol^{-1}
	1 M HCl	6 M HCl	
PaCl$_4$	55.59 ± 0.06	49.29 ± 0.06	—
UCl$_4$	54.46 ± 0.23	—	—
NpCl$_4$	53.87 ± 0.13	—	235.7 ± 0.3
ThBr$_4$	67.76 ± 0.15	55.91 ± 0.10	230.2 ± 0.5
PaBr$_4$	65.44 ± 0.13	51.70 ± 0.24	—
UBr$_4$	63.24 ± 0.09	49.17 ± 0.20	191.1 ± 1.1
NpBr$_4$	61.86 ± 0.08	47.58 ± 0.08	184.5 ± 0.5
PaI$_4$	75.94 ± 0.17	—	—
UI$_4$	69.20 ± 0.13	57.27 ± 0.09	121.6 ± 1
Cs$_2$ThCl$_6$	31.37 ± 0.12	20.97 ± 0.06	508.7 ± 0.5
Cs$_2$PaCl$_6$	23.94 ± 0.31	12.74 ± 0.24	—
Cs$_2$UCl$_6$	19.91 ± 0.04	8.11 ± 0.05	476.7 ± 0.7
Cs$_2$NpCl$_6$	20.16 ± 0.07	7.48 ± 0.10	468.5 ± 0.4
Cs$_2$PuCl$_6$	13.67 ± 0.19	0.18 ± 0.03	—
NpOCl$_2$	33.64 ± 0.12	—	245.6 ± 0.3
	1 M HClO$_4$		
Cs$_2$PuCl$_6$	17.74 ± 0.05	—	466.9 ± 1

heat of formation[20] of Np^{IV} in 1 MHCl. Since $PuCl_4$ does not exist as a condensed phase[76, 77] the thermodynamic cycle used for the enthalpy of formation of $Cs_2PuCl_6(s)$ is based on its heat of solution in 1 $MHClO_4$, on the heat of oxidation of Pu^{3+} to Pu^{4+} in that medium[78] ($+13.5 \pm 0.5$ kcal mol^{-1}), on the heat of formation[12] of $PuCl_3(s)$ (-229.4 ± 0.8 kcal mol^{-1}), and its heat of solution[40] in 1 MHCl (-30.3 ± 0.3 kcal mol^{-1}), assuming it to be the same as in 1 $MHClO_4$.

The heat of solution of the various tetrahalides decreases with the decreasing ionic radii of the cations. This trend may be due, at least partly, to the increased complexing of the cation by the chloride ions when the ionic radius decreases. In the case of the bromides and iodides, however, it must be remembered that the various compounds are not isomorphous[10] before drawing any conclusions. For the hexachloro-complexes the trend is similar but more important and reflects the decreasing hygroscopicity of the salts with increasing atomic number.

Martynova et al.[70] have also measured the heats of solution (298 K) of several double chlorides of uranium(IV) with potassium and sodium chlorides in 2%HCl containing 0.5%$FeCl_3$, and have obtained the enthalpies of formation of these compounds. Comparison of the results obtained for the various uranium(IV) alkali metal chloro-complexes shows the dramatic increase of their stability towards decomposition into the simple chlorides with the increasing size of the alkali metal ion:

$2NaCl(s)+UCl_4(s) = Na_2UCl_6(s) \qquad \Delta H^{\circ}(298) = -1.5 \pm 0.9$ kcal mol^{-1}
$NaCl(s)+KCl(s)+UCl_4(s) = NaKUCl_6(s)$
$\qquad\qquad\qquad\qquad\qquad \Delta H^{\circ}(298) = -6.0 \pm 0.9$ kcal mol^{-1}
$2KCl(s)+UCl_4(s) = K_2UCl_6(s) \qquad \Delta H^{\circ}(298) = -9.9 \pm 0.9$ kcal mol^{-1}
$2CsCl(s)+UCl_4(s) = Cs_2UCl_6(s) \qquad \Delta H^{\circ}(298) = -26.1 \pm 0.9$ kcal mol^{-1}
$KCl(s)+UCl_4(s) = KUCl_5(s) \qquad \Delta H^{\circ}(298) = -5.5 \pm 0.9$ kcal mol^{-1}

Vdovenko et al.[67] have determined the enthalpy of formation of $UOF_2 \cdot H_2O(s)$ from its heat of solution in dilute HCl and auxiliary thermodynamic data. Recalculating their figures and using the results of Fitzgibbon et al. for $UCl_4(s)$, yields $\Delta H_f^{\circ}(UOF_2 \cdot H_2O(s),298) = -420.9 \pm 5$ kcal mol^{-1}. Vdovenko et al. also measured the temperature dependence of the dissociation pressure of $UOF_2 \cdot H_2O$ in the temperature range 283–363 K and obtained:

$$UOF_2 \cdot H_2O(s) = UOF_2(s) + H_2O(g), \begin{cases} \Delta G(T) = 2.82 \text{ kcal mol}^{-1} \\ \Delta H(T) = 11.1 \text{ kcal mol}^{-1} \\ \Delta S(T) = 27.8 \text{ cal mol}^{-1} \text{ K}^{-1} \end{cases}$$

which yields $\Delta H_f^{\circ}(UOF_2(s),298) = -352.0 \pm 5$ kcal mol^{-1}. Using the value of Huber and Holley[18] for $\Delta H_f^{\circ}(UO_2(s),298)$ (-259.3 ± 0.2 kcal mol^{-1}) and the accepted value[1] of $\Delta H_f^{\circ}(UF_4(s),298)$($-450 \pm 5$ kcal mol^{-1}), it can be seen immediately that UOF_2 should be marginally stable towards the decomposition:

$$2UOF_2(s) = UO_2(s) + UF_4(s)$$
$$\Delta H^{\circ}(298) = -5 \pm 7 \text{ kcal mol}^{-1}$$

The same authors[68] have used an analogous method to measure the enthalpy of formation of $UOF(OH)\cdot 0.5H_2O$, the value of which is, upon recalculation, -367.5 ± 5 kcal mol^{-1}. They have also measured its dehydration pressure between 283 and 363 K:

$$UOF(OH)\cdot 0.5H_2O(s) = \tfrac{1}{2}H_2O(g) + UOF(OH)(s),$$
$$\begin{cases} \Delta G(T) = 1.46 \text{ kcal mol}^{-1} \\ \Delta H(T) = 5.19 \text{ kcal mol}^{-1} \\ \Delta S(T) = 12.48 \text{ cal mol}^{-1}\,K^{-1} \end{cases}$$

which gives, on combination with the above data:

$$\Delta H_f^{\circ}(UOF(OH)(s),298) = -333.4 \pm 5 \text{ kcal mol}^{-1}$$

A simple calculation also shows, however, that this compound should also be very marginally stable towards the decomposition:

$$4UOF(OH)(s) = 3UO_2(s) + UF_4(s) + 2H_2O(l),$$
$$\Delta H^{\circ}(298) = -31 \pm 20 \text{ kcal mol}^{-1}.$$

Gagarinskii and Khanaev[79] have measured the heat of the dissolution (at 298 and 323 K), in HCl(33.06%) containing 1.03% H_3BO_3. of UF_4 (s, monocl.) and of three metastable forms of this compound obtained by dehydration in vacuo of various hydrates. The following results were obtained:

$$UF_4(s,\text{amorphous,metastable}) = UF_4(s,\text{monocl.}),$$
$$\Delta H(323) = -2.17 \pm 0.04 \text{ kcal mol}^{-1}$$
$$UF_4(s,\text{cub.,metast.}) = UF_4(s,\text{monocl.}),$$
$$\Delta H(323) = -2.56 \pm 0.04 \text{ kcal mol}^{-1}$$
$$UF_4(s,\text{isotropic,metast.}) = UF_4(s,\text{monocl.}),$$
$$\Delta H(323) = -0.51 \pm 0.03 \text{ kcal mol}^{-1}.$$

Using a similar technique these authors also measured the enthalpy for the hydration of the various forms of UF_4 to the hydrates with 1.33, 1.5 and 2.5 H_2O molecules per uranium atom.

5.3.2.2   Gaseous species

Gruen and DeKock[77] deduced, from spectroscopic measurements on $PuCl_4(g)$, thermodynamic data for the reaction:

$$PuCl_3(l) + \tfrac{1}{2}Cl_2(g) = PuCl_4(g),$$
$$\Delta G(T) = 23\,000 - 14.1\,T \text{ cal mol}^{-1}$$

in the temperature range 1052–1187 K. Combining this with the data obtained earlier by Benz[80] for the reaction

$$PuCl_3(g) + \tfrac{1}{2}Cl_2(g) = PuCl_4(g)$$
$$\Delta G(T) = 37\,700 - 28.3\,T \text{ cal mol}^{-1}(670\text{–}1025 \text{ K}),$$

Gruen and DeKock obtained the enthalpy change on melting[81] $PuCl_3$

(1040 K), $+14.7$ kcal mol^{-1}, in agreement with the value $+15.2\pm0.7$ kcal mol^{-1} from vapour pressure measurements[61].

Chudinov and Choporov[82] have recently reported measurements of the vapour pressure of PuF_4 by effusion techniques:

$$\log P(\text{mm}) = -\frac{14577}{T} + 12.1203 \qquad (779\text{–}1125 \text{ K})$$

The observed pressures are much higher than those previously observed by others[83, 84], who probably encountered difficulties due to the reaction of PuF_4 with the cell material. These new measurements are in acceptable agreement with the earlier data of Chudinov and Choporov[85] on which Rand's assessment[3] for PuF_4 was based. These authors also confirm the decomposition, already observed by others[86, 87], of PuF_4 at higher temperatures under their experimental conditions and for the pressure of PuF_3 report:

$$\log P(\text{mm}) = \frac{-21323}{T} + 12.22 \qquad (1402\text{–}1523 \text{ K})$$

The sublimation of AmF_4 has also been studied by Chudinov and Choporov[88]. They reported the vapour pressure equation:

$$\log P(\text{mm}) = \frac{-11911.5}{T} + 9.336 \qquad (729\text{–}908 \text{ K})$$

and the values of the enthalpy and entropy changes in the temperature range studied are 54.5 kcal mol^{-1} and 29.5 cal mol^{-1} K^{-1}, respectively. Assuming ΔC_p to be -6 cal mol^{-1} K^{-1}, these authors obtain, for sublimation in the temperature range 298–908 K:

$$\Delta G(T) = 59\,340 + 13.8\,T\log T - 75.7\,T \text{ cal mol}^{-1}.$$

Above 908 K AmF_4 is thermodynamically unstable and decomposes to the trifluoride.

5.3.3 Pentavalent and hexavalent compounds

Table 5.8 summarises the available data on the enthalpy of formation and entropy of these compounds. Data for the gaseous hexafluorides have also been included. Few thermodynamic data have been reported recently for these compounds, but some of them are very important, as will be seen.

The vapour pressure of UF_5 was studied by Wolf et al.[91] by measuring its transpiration rate over the temperature range 555–685 K, using UF_6 as carrier gas under sufficient pressure to prevent the disproportionation of UF_5; for the solid:

$$\log P(\text{mm}) = \frac{-(8001\pm664)}{T} + (13.994\pm1.119)$$

and for the liquid:

$$\log P(\text{mm}) = \frac{-(5388 \pm 664)}{T} + (9.819 \pm 1.236)$$

These equations yield a value for the enthalpy of sublimation of $+36.6 \pm 3$ kcal mol^{-1} and, for vaporisation, $+24.7 \pm 3.8$ kcal mol^{-1}, in the temperature range studied. The melting point of this compound was found to be 621 K. The behaviour of UF$_5$ differs markedly from that of PaF$_5$ which is much less volatile[92], subliming only above 773 K in a vacuum.

Table 5.8 Enthalpies of formation and entropies of pentavalent halides and oxyhalides

Compound	$-\Delta H_f^\circ(298)$ kcal mol^{-1}	Reference	$S^\circ(298)$ cal mol^{-1}K^{-1}	Reference
PaF$_5$(s)	(540)	64	(80)	64
UF$_5$(s)	491.5 $\pm$ 6	1	45.0 $\pm$ 3	1
UCl$_5$(s)	261.5 $\pm$ 2	1	(58.0 $\pm$ 1.5)	1
UOCl$_3$(s)	284.2 $\pm$ 3	1	(42 $\pm$ 3)	1
UOBr$_3$(s)	236.0 $\pm$ 2.5	1	(49 $\pm$ 3)	1
UF$_6$(s)	522.6$_4$ $\pm$ 0.4$_3$	89	54.4 $\pm$ 0.5	3
UF$_6$(g)	510.7$_7$ $\pm$ 0.4$_5$	89	90.4 $\pm$ 0.2	1
NpF$_6$(s)	—	—	54.76 $\pm$ 0.11	90
NpF$_6$(g)	(463 $\pm$ 3)	42	89.99 $\pm$ 0.12	90
PuF$_6$(s)	430 $\pm$ 8	3	53.0 $\pm$ 0.5	3
PuF$_6$(g)	418 $\pm$ 8	3	88.4 $\pm$ 0.2	3
UO$_2$F$_2$(s)	399 $\pm$ 4	1	32.4 $\pm$ 0.2	1
UO$_2$Cl$_2$(s)	302.9 $\pm$ 3	1	36.0 $\pm$ 0.1	1
UO$_2$Br$_2$(s)	276.6 $\pm$ 3	1	40.25 $\pm$ 2.5	1

Weigel et al.[93] have measured the vapour pressure of PaCl$_5$ between 500 and 600 K, obtaining the relations:

$$\log P(\text{mm}) = -\frac{4843}{T} + 10.35$$

for the solid, and

$$\log P(\text{mm}) = -\frac{3204}{T} + 7.500$$

for the liquid.

Extrapolation of the latter equation to a pressure of 760 mmHg yields 693 K for the boiling point of the compound. These equations give also, for the sublimation process:

$$\Delta H_s(T) = 22.17 \text{ kcal mol}^{-1}, \qquad \Delta S_s(T) = 34.2 \text{ cal mol}^{-1}\text{ K}^{-1}$$

and for vaporisation:

$$\Delta H_v(T) = 14.65 \text{ kcal mol}^{-1}\text{ K}^{-1}, \qquad \Delta S_v(T) = 21.1 \text{ cal mol}^{-1}\text{ K}^{-1}$$

The enthalpy of formation of UF$_6$ was determined by Settle et al.[89] from direct combination of the elements in a fluorine bomb calorimeter. The values obtained, $\Delta H_f^\circ(UF_6(s),298 = -522.64\pm0.43$ kcal mol^{-1} and ΔH_f° $(UF_6(g),298) = -510.77\pm0.45$ kcal mol^{-1}, are in remarkable agreement with the value given by Gross et al.[94] (-522.08 ± 0.43 kcal mol^{-1}) from fluorine flow calorimeter measurements and also with the figure (-523 ± 6 kcal mol^{-1}) assessed by Rand and Kubaschewski[1] from heat of solution data. The fact that this value is known with such precision is extremely important since it allows accurate determination of the heat of formation of a number of uranium compounds by fluorine calorimetry.

The heat capacity of NpF$_6$ has been measured by Osborne et al.[90] by adiabatic calorimetry in the temperature range 7–350 K. The measurements revealed no anomaly in the heat capacity curve. The thermodynamic

Table 5.9   Thermodynamic properties[90] of NpF$_6^*$

Solid		Liquid	
$C_p^\circ(298)$	40.02 ± 0.08	Triple point: 327.91 ± 0.02	
$S^\circ(298)$	54.76 ± 0.11	$\Delta H_m(327.91) = 4188.4\pm4.0$	
$[H^\circ(298)-H^\circ(0)]$	7436 ± 15	$\Delta S_m(327.91) = 12.77$	
$-[G^\circ(298)-H^\circ(0)]/T$	29.82 ± 0.06	$\Delta S^\circ(340) = 73.07\pm0.15$	
	Gas		
	$\Delta H_v(340) = 6829.6$		
	$\Delta S_v(340) = 20.27\pm0.34$		
	$S^\circ(340) = 94.09\pm0.38$		
	$S^\circ(298) = 89.99\pm0.12$		

*Units: calorie, mole, K

functions at 298.15 K, together with important thermodynamic data, are summarised in Table 5.9. The experimental entropy for the gas at 340 K agrees remarkably well with the value (94.13 ± 0.12) calculated for the same temperature from the fundamental vibration frequencies and symmetry information on the NpF$_6$ molecule. Although the low temperature thermo-dynamic properties of UF$_6$ were obtained long ago[95, 96], such accurate data for PuF$_6$ will probably have to be obtained with an isotope longer lived than ^{239}Pu to avoid important radiolytic (^{239}PuF$_6$ decomposes at the rate of 1.5 % per day) and self-heating effects.

5.4   OXIDES

5.4.1   Heat capacities and entropies

5.4.1.1   Low temperature data

Low temperature heat capacity measurements have been carried out on a number of actinide oxides and detailed information on these matters can be found in reviews devoted exclusively to the properties of actinide oxide[6, 7]. Figure 5.1 shows the shapes of the heat capacity curves of the various actinide dioxides. The absence of a λ type anomaly in PuO_2 appears surprising, for in this compound the plutonium atom has four 5f electrons and this anomaly, present in UO_2 and NpO_2, has been attributed[7] to an antiferromagnetic ordering of the electron spins. It has been suggested[6] that either radiation damage in the sample under study prevented long range order or that self-heating is the origin of a substantial error in the temperature measurement.

More recently, the heat capacity of $UO_{2.250}$ (U_4O_9) has been determined calorimetrically from 1.6 to 24 K[101]. The results extend the temperature

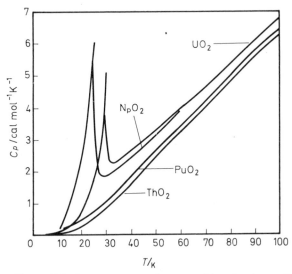

Figure 5.1 Low temperature heat capacities of various actinide dioxides (after ref. 6, by courtesy of International Atomic Energy Agency – Vienna)
ThO_2: Osborne and Westrum[97]
UO_2: Jones, Gordon and Long[98]
NpO_2: Westrum, Hatcher and Osborne[99]
PuO_2: Sandenaw[100]

range over previous data[102], and, due to improved experimental technique, have a greater accuracy. No λ type anomaly was observed in this compound at the temperature at which (6.4 K) other authors found a peak in the magnetic susceptibility[103]. The same situation exists for β-$UO_{2.333}$ (β-U_3O_7) for which a peak in the magnetic susceptibility was observed[103] at 6.4 K, and for α-$UO_{2.667}$(α-U_3O_8), for which the magnetic susceptibility peak is observed at 4.2 K. In $UO_{2.667}$, Westrum and Grønvold[104], who measured the heat capacity from 5 to 350 K, observed a λ type anomaly at 25.3 K ($\Delta H_t = 12$ cal mol^{-1}, $\Delta S_t = 0.56$ cal mol^{-1} K^{-1}) which could not be interpreted from magnetic susceptibility data. The heat capacity of $UO_{2.25}$ at low temperature (< 10 K) is several times larger than that of UO_2 or ThO_2, the difference being attributed[101] to a large magnetic contribution. Approximate resolution of the entropies of $UO_{2.250}$ and $UO_{2.333}$ into lattice and magnetic contributions shows that at 200 K the magnetic entropy of each of these compounds is about 70% of that of UO_2.

Standard entropy values of several actinide oxides are given in Table 5.10.

Table 5.10   Entropies of various actinide oxides

Compound	$\dfrac{S°(298)}{\text{cal mol}^{-1}\text{K}^{-1}}$	Reference	Remarks
ThO_2	15.59 ± 0.02	97	Experimental value
UO_2	18.63 ± 0.1	98	Experimental value
$UO_{2.25}$	19.96 ± 0.04	101	Experimental value
$UO_{2.667}(\alpha)$	22.51 ± 0.04	105	Experimental value
$UO_3(\gamma)$	23.6 ± 0.1	98	Experimental value
NpO_2	19.19 ± 0.1	99	Experimental value
$PuO_{1.5}$(hex.)	18.1	6	Assessed value
$PuO_{1.52}$(b.c. cub.)	18.1	6	Assessed value
$PuO_{1.61}$(b.c. cub.)	18.9	6	Assessed value
PuO_2	19.7	100	Assessed by (6) from exp. data of (100)

5.4.1.2   High temperature data

(a) *Uranium*—High temperature enthalpy determinations, especially on uranium oxides, have been quite numerous, but the results are sometimes controversial. Frederickson and Chasanov[106] have recently obtained new data on the enthalpy of $UO_{2.007}$ from 500 to 1500 K by drop calorimetry. Using the assessed values[7] for $C_p°(UO_2,298)$ (15.20 cal mol^{-1}K^{-1}) and $S°(UO_2,298)$ (18.41 cal mol^{-1}K^{-1}), they obtained:

$$H°(T)-H°(298) = 18.4272\,T + 1.010\,27 \times 10^{-3}T^2$$
$$+ 3.404\,29 \times 10^5 T^{-1} - 6725.69 \text{ cal mol}^{-1}$$

(standard deviation: 29 cal mol^{-1}) for the temperature range 298–1434 K.

These results are in substantial agreement with the early data of Moore and Kelley[107], the small difference being attributed to the fact that these authors were using slightly hypostoicheiometric oxide samples (O/U = 1.98). Agreement is also satisfactory with the heat capacity data of Engel[108], who used a thermal analysis technique. The very recent data of Grønvold et al.[109] on $UO_{2.017}$ from 300 to 1000 K also substantiate remarkably the data of Frederickson and Chasanov. On the other hand, the agreement with the data of Ogard and Leary[110] is less satisfactory, while the results of Conway et al.[111] and especially the fairly recent results of Affortit[112] (thermal analysis) yield substantially different C_p versus temperature curves. A summary of the various results obtained for the range up to 2000 K is presented in Figure 5.2. Grønvold et al.[109] have attempted to resolve the various contributions to the heat capacity of UO_2, showing that the electronic contribution is very large. No satisfactory explanation is offered, however, for the enhanced heat capacity above 2000 K.

Several workers have extended the temperature range studied up to the melting point of UO_2. The results of Hein *et al.*[115] yield for the range 1175–3115 K the relation:

$$H°(T) - H°(298) = -11\ 688 + 31.937\ T - 9.753 \times 10^{-3}T^2$$
$$+ 2.577 \times 10^{-6}T^3\ \text{cal mol}^{-1}$$

(standard deviation: 396 cal mol^{-1}), the observed melting point being $(3115 \pm {}^{25}_{15})$ K with $\Delta H_m = 18.2 \pm 0.5$ kcal mol^{-1}. The data of Leibowitz *et al.*[116]

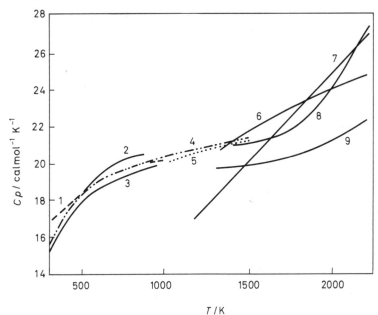

Figure 5.2   High temperature heat capacity of UO_2 (after Grønvold *et al.*[109], by courtesy of Academic Press, London.)
  1. Engel[108]
  2. Popov, Galichenko and Senin[113]
  3. Grønvold, Kveseth, Sveen and Tichy[109]
  4. Moore and Kelley[107]
  5. Frederickson and Chasanov[106]
  6. Ogard and Leary[110]
  7. Conway and Hein[111]
  8. Hein, Sjodahl and Szwarc[114]
  9. Affortit[112]

for the temperature range 2500–3100 K agree remarkably with those of Hein *et al.* On the other hand, Grossmann and Kaznoff[117] give an enthalpy of fusion of 25 ± 3 kcal mol^{-1} from thermal analysis data.

The heat capacity of UO_{2+x} in the region $O \leqslant x \leqslant 0.1$ from 1600 K to the melting point has been studied by Affortit and Marcon[118] using adiabatic heating by the Joule effect. Their results can be expressed by the analytical expression:

$$C_p = a + bT + \frac{c}{T^2} \exp\left(\frac{-Q}{RT}\right), \text{ in cal mol}^{-1}\text{K}^{-1},$$

where the constants a, b, c and Q are as under,

	a	b	c	Q
$UO_{2.00}$	17.40	2.13×10^{-3}	6.48×10^{10}	38 400
$UO_{2.03}$	17.69	2.49×10^{-3}	6.27×10^{10}	37 000
$UO_{2.04}$	17.93	2.56×10^{-3}	3.46×10^{10}	32 800
$UO_{2.10}$	18.18	3.10×10^{-3}	1.83×10^{10}	29 200

The C_p versus temperature curves for the various compositions are roughly parallel, the heat capacity increasing steadily with the value of x. However, the results for the stoicheiometric composition do not agree with the data of other authors[115, 116], the values of Affortit and Marcon being about 5% lower at 2000 K and 15% lower at 3000 K.

New data for the heat capacity of $UO_{2.254}$ have also been obtained by Grønvold et al.[109] in the temperature range 300–1000 K. The λ type transition around 350 K (348 K,$\Delta S_t = 0.52 \pm 0.05$ cal mol^{-1}K^{-1}, $\Delta H_t = 165 \pm 17$ cal mol^{-1}) was observed as in previous studies by Westrum et al.[119] (348 K, $\Delta S_t = 0.45 \pm 0.01$ cal mol^{-1}K^{-1},$\Delta H_t = 150 \pm 2$ cal mol^{-1}) and Gotoo and Naito[120] (330 K, $\Delta S_t = 0.50 \pm 0.01$ cal mol^{-1}K^{-1}, $\Delta H_t = 170 \pm 5$ cal mol^{-1}). The shape of the transition differs appreciably in the three investigations, the broadness of the transition in the case of the observations of Westrum being tentatively attributed by Grønvold et al. to radiation damage and in the case of the observations of Gotoo and Naito to the experimental technique. The heat capacity data show a large electronic contribution and are tentatively analysed assuming the presence of U(IV) and U(V) in equal amounts, corresponding to the formula $2UO_2 \cdot U_2O_5$.

Heat capacity measurements on $\alpha - UO_{2.667}$ have been made by various workers and in their assessment Rand and Kubaschewski[1] took $C_p^\circ(UO_{2.667}, 298)$ to be 18.97 cal mol^{-1}K^{-1} and, for the temperature range 298–900 K,

$$C_p(UO_{2.667}) = 22.5 + 2.94 \times 10^{-3}T - 11.94 \times 10^5 T^{-2} \text{ cal mol}^{-1}\text{K}^{-1}$$

However, disagreement exists in the literature as to the presence of anomalies in the heat capacity curve either between 590–670 K[113] or at 1046 and 1213 K[121]. The temperature range 300–555 K has been investigated by Girdhar and Westrum[105] using adiabatic calorimetry. A λ type anomaly of uncertain origin was reported at 482.7 K. Apart from this region of anomaly the recent data conform well with the earlier assessment[1]. At 298.15 K, Girdhar and Westrum give $C_p^\circ(UO_{2.667}) = 18.99$ cal mol^{-1}K^{-1}. Very recently, Maglic and Herak[122] have reported enthalpy measurements on $UO_{2.667}$ between 273 and 1000 K using a Bunsen ice drop calorimeter. They observed a discontinuity in the curve in the vicinity of 480 K thus confirming the previous results. The smoothed data are well represented by

$$H^\circ(T) - H^\circ(273.15) = -8.91533 \times 10^3 + 26.41616T - 26.93690 \times 10^{-4}T^2$$
$$+ 5.19993 \times 10^5 T^{-1} \text{ cal mol}^{-1} \text{ (273–480 K)}$$

and

$$H^\circ(T) - H^\circ(273.15) = -4.60079 \times 10^3 + 18.97866T + 17.32147 \times 10^{-4}T^2$$
$$- 31.68186 \times 10^4 T^{-1} \text{ cal mol}^{-1} \text{ (480–1000 K)}$$

As to the origin of this transition, no definitive interpretation can be given until a better knowledge of the structural parameters in $UO_{2.667}$ at high temperature has been acquired.

(b) *Plutonium*—The heat capacity of PuO_2 is distinctly higher than that of UO_2. Engel[108] has shown the difference to amount to 8–10% in the temperature range 500–1100 K. The results of Ogard[123] on the heat content of PuO_2 between 1500 and 2715 K (melting point) can be represented by:

$$H°(T)-H°(298) = -6360+18.0T+1.95 \times 10^{-3}T^2$$
$$+2.46 \times 10^5 T^{-1} \text{ cal mol}^{-1}$$

For a lower temperature range (192–1400 K) Kruger and Savage[124] give:

$$H°(T)-H°(298) = 8468+22.18T+1.040 \times 10^{-4}T^2$$
$$+4.935 \times 10^5 T^{-1} \text{ cal mol}^{-1}$$

(standard deviation: 136 cal mol^{-1}). These results are in particularly good agreement with the low temperature data of Sandenaw[100] for the temperature range covered by both groups.

High temperature heat capacity data for mixed oxides (mainly UO_2–PuO_2) have been given by several authors, i.e.[110, 118]. These systems are not considered in the present study.

5.4.2 Thermodynamics of formation

5.4.2.1 Thorium

No recent value for the heat of formation of $ThO_2(s)$ has been obtained since the combustion calorimetry work of Huber et al.[125] ($\Delta H_f°(ThO_2(s),298)$ $= -293.2 \pm 0.4$ kcal mol$^{=1}$) which was in good agreement with the much earlier result of Rothe and Becker[126] (-292.1 ± 1.4 kcal mol^{-1}).

5.4.2.2 Uranium

In the case of the uranium oxides numerous data have appeared recently. Huber and Holley[18] have re-determined the enthalpy of formation of $U_3O_8(s)$ from combustion calorimetry and their results yield:

$$3U(s,\alpha)+4O_2(g) = U_3O_8(s)$$
$$\Delta H_f°(U_3O_8(s),298) = -854.4 \pm 0.6 \text{ kcal mol}^{-1}$$

This value is clearly more accurate than a previous result by the same team[125] (-853.5 ± 1.6 kcal mol^{-1}) and than the data of Popov and Ivanov[127] (-856.5 ± 3.1 kcal mol^{-1}), also obtained by combustion calorimetry. From a series of combustions of slightly hyperstoicheiometric samples of UO_x ($2.005 \leqslant x \leqslant 2.041$), Huber and Holley[18] have shown that the enthalpy change

$$3UO_x+\left(\frac{8-3x}{2}\right)O_2(g) = U_3O_8(s)$$

fits the equation $\Delta H° = -[74.49 - 90.96\,(x - 2.0216)]$ kcal mol^{-1} very well. This yields, for the standard heat of combustion of the stoicheiometric dioxide, the value -25.48 ± 0.13 kcal mol^{-1} and upon combination with the enthalpy of formation of U_3O_8(s) there results $\Delta H_f°(UO_2$(s),298) $= -259.3 \pm 0.2$ kcal mol^{-1}, thus confirming the previously reported[125] value -259.2 ± 0.6 kcal mol^{-1}.

The various oxides of intermediate composition between UO_2 and UO_3, as well as UO_3 and its hydrates, have been the subject of several studies. All these data have been collected in Table 5.11 together with the previously assessed data[1]. Where applicable, the most recent data of Huber and Holley[18] have been used in the calculations. Fitzgibbon et al.[128], by combining the

Table 5.11 Enthalpies of formation of various uranium oxides and hydrates, $-\Delta H$ f (298) kcal mol^{-1}

	UO_2	U_4O_9	β-U_3O_7	U_3O_8
Huber and Holley[18]	259.3 ± 0.2	—	—	854.4 ± 0.6
Fitzgibbon et al.[128]	—	1078.8 ± 1.4	818.9 ± 1	—
Mukaibo et al.[129]	—	—	815.7 ± 2.4	—
			821.1 ± 2.1	—
Rand and Kubaschewski[1]	259.0 ± 0.6	1078.0 ± 4	—	854.1 ± 2

	Amorph.-UO_3	α-UO_3	β-UO_3	γ-UO_3	δ-UO_3
Cordfunke et al.[19, 131]	289.6 ± 1	291.8 ± 1	292.6 ± 1	293.5 ± 1	290
Vidavskii et al.[130]	—	—	—	295.8 ± 0.45	—
Fitzgibbon et al.[128]	—	—	—	292.4 ± 0.5	—
Rand and Kubaschewski[1]	290	292	292.5	293.0 ± 2	292.5

	ε-UO_3	$UO_3 \cdot H_2O$	$UO_{2.92}$	α-$UO_3 \cdot 0.85H_2O$	$UO_3 \cdot 2H_2O$
Cordfunke[19]	291 ± 1	$\begin{cases}\beta\text{-}367.4 \pm 1 \\ \varepsilon\text{-}366.8 \pm 1\end{cases}$	292.0 ± 1	367.4 ± 1	437.3 ± 1
Rand and Kubaschewski[1]	292.0	367.5 ± 5	—	—	437 ± 5

known enthalpies of formation of UO_2(s) and U_3O_8(s) with the data obtained from the solution of samples of U_4O_9, β-U_3O_7 and U_3O_8, and of mixtures of $UO_{2.028}$ and $UO_{2.993}$ of corresponding overall composition, have obtained the enthalpies of formation of U_4O_9, β-U_3O_7 and γ-UO_3.

Mukaibo et al.[129] have reported the measurement of the heat of oxidation of UO_2 to β-U_3O_7 and of β-U_3O_7 to U_3O_8 according to the reactions:

$$UO_2(s) + \tfrac{1}{6}O_2(g) = \tfrac{1}{3}U_3O_7(s)$$
$$\Delta H(503) = -13.5 \pm 1.8 \text{ kcal mol}^{-1}$$
$$\tfrac{1}{3}U_3O_7(s) + \tfrac{1}{6}O_2(g) = \tfrac{1}{3}U_3O_8(s)$$
$$\Delta H(653) = -10.8 \pm 1.3 \text{ kcal mol}^{-1}$$

by a differential thermal analysis technique. Using the enthalpies of formation of UO_2 and U_3O_8 and correcting to 298 K, the standard enthalpy of formation of U_3O_7 is obtained. Within the experimental errors reported, however, the two results are not entirely consistent although their average fits Huber and Holley's calorimetric result satisfactorily.

The value obtained by Vidavskii et al.[130] for the enthalpy of formation of γ-UO_3 does not depend on the enthalpy of formation of other uranium oxides, since it was obtained by dissolution of γ-UO_3 in aqueous hydrofluoric acid and uses the enthalpy of formation and solution of UF_6 in the calculations; the reason for the discrepancy between this value and the others is not clear. The enthalpy of formation of γ-UO_3 was also obtained by Cordfunke and Ailing[131] from its dissociation pressure between 773 and 873 K, according to the reaction: $3UO_3(s) = U_3O_8(s) + \frac{1}{2}O_2(g)$.

As noted by Fitzgibbon et al., the small difference between their data for γ-UO_3 and the result of Cordfunke and Ailing could be due to the fact that U_3O_8 exhibits a trigonal modification above 673 K. Cordfunke[19] has

Table 5.12 Partial molar thermodynamic properties of oxygen[135] in UO_{2-x}

O/U	$\dfrac{\log P(O_2)}{atm}$	$\dfrac{\Delta \bar{H}(O_2, 2175)}{kcal\ mol^{-1}}$	$\dfrac{\Delta \bar{S}(O_2, 2175)}{cal\ mol^{-1}K^{-1}}$
1.900	$20.68 - 81\ 260\ T^{-1}$	372.0	94.6
1.925	$20.18 - 78\ 980\ T^{-1}$	361.0	92.3
1.950	$19.70 - 76\ 260\ T^{-1}$	349.0	90.2
1.975	$19.49 - 72\ 860\ T^{-1}$	333.3	89.3
1.989	$18.04 - 67\ 720\ T^{-1}$	309.4	82.5
1.997	$13.50 - 55\ 260\ T^{-1}$	253.0	61.8
2.000	$11.17 - 48\ 660\ T^{-1}$	222.5	51.2

also measured the heat of solution of various forms of UO_3 and its hydrates in HNO_3 and calculated their heats of formation, using his value for the heats of formation of γ-UO_3. These results are also summarised in Table 5.11. The figure reported for δ-UO_3 was corrected for the presence of about 30% α-UO_3, since δ-UO_3 could not be obtained as a pure phase.

Partial molar thermodynamic quantities of oxygen in the nonstoicheiometric oxides have been studied for many years and have been frequently reviewed[6, 7, 132, 133]; these are not discussed in this paper. However, subsequent to these reviews Gerdanian and Dodé[134] have obtained the direct calorimetric measurement at 1373 K of the partial molar enthalpy of mixing of oxygen in UO_{2+x} near the stoicheiometric composition. The value for $-\Delta \bar{H}(O_2)$, which is about 200 kcal mol^{-1} with the starting material $UO_{2.0003}$, decreases rapidly to reach a pronounced minimum of 7 kcal mol^{-1} at O/U = 2.018. From O/U = 2.01, where a value of 50 kcal mol^{-1} is observed, the variation is linear and much slower with the increasing oxygen to uranium ratio. Such drastic variation in the partial molar thermodynamic quantities near the stoicheiometry had been predicted[133]. Pattoret et al.[135] have obtained the partial thermodynamic properties of oxygen in the system UO_{2-x} in the temperature range 1950–2450 K, from mass spectrometric data on the various gaseous species above the UO_{2-x} system, as shown in Table 5.12.

5.4.2.3 Neptunium

The enthalpy of formation of $NpO_2(s)$, $(Np(s,\alpha) + O_2(g) = NpO_2(s)$; $\Delta H_f^\circ(NpO_2(s),298) = -256.7 \pm 0.6$ kcal mol^{-1}) has been obtained by Huber and Holley[136] from oxygen bomb calorimetry measurements. This value can only be compared with an early estimate[42] of -246 ± 10 kcal mol^{-1}.

Fuger et al.[20] have measured the heat of dissolution of $NpO_3 \cdot H_2O(s)$ in 1 MHCl and 1 MHClO$_4$, obtaining -12.70 ± 0.10 and -12.85 ± 0.10 kcal mol^{-1}, respectively. Using the assessed[24] value of -208.5 kcal mol^{-1} for the heat of formation of NpO_2^{2+} in 1 MHCl they obtained:

$$\Delta H_f^\circ(NpO_3 \cdot H_2O(s),298) = -332.4 \text{ kcal mol}^{-1}.$$

5.4.2.4 Plutonium

A new value for the enthalpy of formation[137] of $PuO_2(s)$ $(Pu(s,\alpha) + O_2(g) = PuO_2(s)$; $\Delta H_f^\circ(PuO_2(s),298) = -252.35 \pm 0.17$ kcal mol^{-1}) has been obtained by combustion calorimetry. Previous results, using the same technique, had yielded values of -251.6 ± 1.6 [127], -253.1 ± 1.6 [127] and -252.7 ± 0.38 [138] kcal mol^{-1}.

The enthalpies of formation of the lower oxides of plutonium have not been measured directly but it appears that fairly accurate estimates[6] have been obtained from measurements[139] of the oxygen potentials over the single-phase oxides between $PuO_{1.61}$ and PuO_2 above 950 K, using an e.m.f. method with ThO_2–Y_2O_3 solid electrolytes. The data yielded fairly precise $\Delta \bar{H}(O_2)$ and $\Delta \bar{S}(O_2)$ values which, upon integration and correction to 298 K, gave the values summarised in Table 5.13, which are based on the above reported data for the enthalpy of formation and entropy of $PuO_2(s)$.

Table 5.13   Enthalpies of formation and entropies of plutonium sesquioxide and related phases (ref. 6, 139)

Compound	$-\Delta H_f^\circ(298)$ kcal mol^{-1}	$S^\circ(298)$ cal mol^{-1}K^{-1}
Pu_2O_3(s, hex.)	408.7 ± 3	34.0 ± 5
Pu_2O_3(s, b.c. cub)	412.6 ± 5	34.4 ± 3
$PuO_{1.61}$(s, b.c. cub.)	213.9 ± 1.5	19.2 ± 2.5

It is probably useful to recall here that the hexagonal and body centered cubic 'Pu_2O_3' are not allotropes, the cubic phase being better represented by $PuO_{1.52}$. Furthermore, the $PuO_{1.61}$ phase is only stable above 620 K and is very closely related to $PuO_{1.52}$. Very recently Dean et al.[140] have measured calorimetrically the partial molar heats of solution over the

$PuO_{1.5}$–PuO_2 system. Integration of the data obtained, and correction to 298 K, gave

$$\Delta H_f^{\circ}(Pu_2O_3(s),\text{hex},298) = -402.7 \pm 5 \text{ kcal mol}^{-1}$$

and

$$\Delta H_f^{\circ}(PuO_{1.61}(s),\text{cub},298) = -211.4 \pm 5 \text{ kcal mol}^{-1}$$

in fair agreement with previous results, although the measured $\Delta\bar{H}(O_2)$ values were up to 15% higher than those deduced from e.m.f. data.

5.4.2.5 Americium

The only value available for the heat of formation of $AmO_2(s)$ is still that $(-239.9 \pm 3 \text{ kcal mol}^{-1})$ obtained more than 18 years ago by Eyring et al.[36] from its solution in 6 MHNO$_3$–0.1 MHBF$_4$ using the enthalpy of formation of $Am^{3+}(aq)$ obtained by the same group of authors[35] and correcting for the presence of BF_4^- ion.

Partial molar thermodynamic quantities of oxygen in substoicheiometric americium oxides have been obtained by Chikalla and Eyring[141] by measuring the oxygen dissociation pressure by means of a thermogravimetric isopiestic technique over the temperature range 1039–1445 K, the oxygen pressure ranging between 1 and 10^{-6} atm. The partial molar free energy of solution of oxygen was found to be quite linear over the temperature range studied. Consequently the $-\Delta\bar{S}(O_2)$ values were independent of temperature but varied almost linearly with the composition, from a value of 49.5 ± 1.5 kcal mol^{-1} K^{-1} at O/Am = 1.84 to 71.5 ± 1.5 cal mol^{-1} K^{-1} at O/Am = 1.96. Above this composition, $-\Delta\bar{S}(O_2)$ varied sharply and became very large. The value of $-\Delta\bar{H}(O_2)$ followed a smooth curve as a function of the composition from 98 ± 1.5 kcal mol^{-1} at O/Am = 1.84 to 106 ± 1.5 kcal mol^{-1} at O/Am = 1.90, and reached 110 ± 1.5 kcal mol^{-1} at O/Am = 1.96. Above this composition $-\Delta\bar{H}(O_2)$ also increased sharply. The evolution of the partial thermodynamic quantities in the AmO_{2-x} system is fairly analogous to that observed for PuO_{2-x}[133, 139].

5.5 SULPHIDES

5.5.1 Heat capacities and entropies

5.5.1.1 Thorium

Experimental data on the heat capacity of the thorium sulphides are completely lacking except for one set of measurements[142] on ThS$_2$ in the temperature range 53–279 K. The reported heat capacity and entropy at 298 K are 16.80 and 23.0 ± 0.2 cal mol^{-1} K^{-1}, respectively. Westrum and Grønvold[143] have estimated the entropies of other thorium sulphides at 298 K and give $S^{\circ}(ThS(s),298) = 18.0$, $S^{\circ}(ThS_{1.5}(s),298) = 20.0$, $S^{\circ}(ThS_{1.7}(s),298) = 21.0$ and $S^{\circ}(ThS_{2.5}(s),298) = 23.0$ cal mol^{-1} K^{-1}.

More recently, Aronson and Ingraham[144] have estimated the entropy of ThS from its characteristic temperature determined from x-ray diffraction data. The characteristic temperature was calculated by using in the formula either an average value for the masses of the two types of atoms in the lattice or only the mass of the heavier atom. These two treatments of the data yielded, respectively $S°(ThS(s),298) = 19.3$ and 16.9 cal mol^{-1} K^{-1}. Aronson and Ingraham note, however, that the heavy mass model applied to other similar compounds yields values closer to the experimental results and indicate that the value of 16.9 cal mol^{-1} K^{-1} should therefore be preferred.

5.5.1.2 Uranium

The heat capacity of US(cub.,NaCl type)[145] in the temperature range 1.5–350 K, of US$_{1.9}$(tetrag.)[146], US$_2$(orth.) and US$_3$(monocl.)[147] in the temperature range 5–350 K have been measured by Westrum et al. Table 5.14 summarises for each of these compounds the data on the observed transition characteristics and the thermodynamic functions at 298.15 K. For these, the reported errors are about 0.2 % for US and less than 0.1 % for the other compounds.

Table 5.14 Thermodynamic properties of uranium sulphides* (ref. 145, 146, 147)

	US			US$_{1.9}$	
Transition temp.	180.1			≈ 25	
ΔH_t	231 ± 20	}	ferromag.	—	Schottky
ΔS_t	1.62 ± 0.2			—	
$C_p°(298)$	12.08			17.68	
$[S°(298) - S°(0)]$	18.64			25.91	
$[H°(298) - H°(0)]$	2665			3605	
$-[G°(298) - H°(0)]/T$	9.70			13.82	

	US$_2$			US$_3$
Transition temp.	≈ 25			
ΔH_t	—	}	Schottky	
ΔS_t	0.5			
$C°_p(298)$	17.86			22.85
$[S°(298) - S°(0)]$	26.42			33.09
$[H°(298) - H°(0)]$	3698			4663
$-[G°(298) - H°(0)]/T$	14.01			17.45

*Units: calorie, mole, K

In the case of US, the heat capacity below 9 K has been resolved into conduction electronic, magnetic and lattice contributions ($C_p = 5.588 \times 10^{-3}T + 2.627 \times 10^{-4}T^{\frac{3}{2}} + 6.752 \times 10^{-5}T^3$). Entropies are reported as $S°(298) - S°(0)$, because, as noted by Westrum et al.[146, 147], there is uncertainty concerning the degree of order of the spins at the lowest experimental temperature (5 K) in US$_{1.9}$, US$_2$ and US$_3$. In the case of US$_{1.9}$ Westrum et al. estimate a zero point entropy of 0.28 cal mol^{-1} K^{-1} due to the fact that in this compound approximately one fifth of the atoms in the tetragonal cell are reported to be in a fourfold position[148]. The shape of the heat capacity curve of $US_{1.9}$ falls slightly below that of US_2 above 30 K. In the temperature range 5–30 K the molar heat capacity of $US_{1.9}$ is slightly above that of US_2, this excess being tentatively attributed to Schottky type excitations.

MacLeod and Hopkins[149] have measured the enthalpy of US from 417 to 1415 K by adiabatic drop calorimetry. Their result, using the value of $C_p^\circ(US(s),298)$ of Westrum et al. (12.08 cal mol^{-1} K^{-1}) is best represented by:

$$H^\circ(T) - H^\circ(298) = -4139 + 12.633\ T +$$
$$7.7865 \times 10^{-4} T^2 + 9.0413 \times 10^4 T^{-1}\ \text{cal mol}^{-1}\ \text{(standard deviation: } 0.55\%\text{)}.$$

Moser and Kruger[150] have obtained heat capacity data for US from room temperature to 873 K by a transient technique using a laser as a heat pulse source. They reported:

$$C_p = 12.12 + 7.107 \times 10^{-4} T - 1.781 \times 10^3 T^{-2}\ \text{cal mol}^{-1}\ \text{K}^{-1}.$$

Their value at 298 K (12.1 cal mol^{-1} K^{-1}) is remarkably close to that of Westrum et al. but the observed slope of C_p versus temperature is rather low and the difference between these results and those obtained by MacLeod and Hopkins amounts to 10% at 1000 K.

5.5.1.3 Plutonium

Estimated values for the entropy of PuS and $PuS_{1.5}$ have been given by Oetting[4] by comparison with the analogous sulphides of the lanthanides and actinides. Suggested values are $S^\circ(PuS(s),298) = 18.7 \pm 1.0$ and $S^\circ(Pus_{1.5}(s),298) = 23.0 \pm 2.0$ cal mol^{-1} K^{-1}.

Moser and Kruger have measured the heat capacity of PuS by the technique used for US. The observed variation is linear in the temperature range studied, from 14.6 cal mol^{-1} K^{-1} at 298 K to 15.8 cal mol^{-1} K^{-1} at 923 K. The 298 K value is somewhat higher than that estimated by Oetting[4], 13 ± 1 cal mol^{-1} K^{-1}.

5.5.2 Thermodynamics of formation

5.5.2.1 Thorium

The enthalpy of formation of $Th_2S_3(s)$, according to the reaction $2Th(s) + 3S(s,\text{rhomb.}) = Th_2S_3(s)$, was determined more than 20 years ago by Eyring and Westrum[152] from its heat of solution in 6 MHCl and found to be -258.6 ± 0.5 kcal mol^{-1}. Somewhat more recently, Bear and McTaggart[153] have reported for the same reaction -238 kcal mol^{-1} from heat of combustion data.

The enthalpies of formation of the other sulphides have been estimated by Eastman *et al.*[154] on the basis of the Westrum and Eyring experimental value for Th_2S_3, in accordance with the high temperature chemical and physical properties of these compounds. Calculated from $S_2(g)$, they were given as:

$$\Delta H_f(ThS(s),298) = -120 \pm 5 \text{ kcal mol}^{-1}$$
$$\Delta H_f(Th_7S_{12}(s),298) = -665 \pm 35 \text{ kcal mol}^{-1}$$
$$\Delta H_f(ThS_2(s),298) = -170 \pm 20 \text{ kcal mol}^{-1}$$

Free energies of formation were obtained by Aronson[155] from e.m.f. measurements of solid state cells of the type $(Th,ThF_4)(s) \mid CaF_2(s) \mid (ThF_4,Th_2S_3, ThS)(s)$, or analogous cells containing the various sulphides, in the temperature range $998 \leqslant T \leqslant 1098$ K. The thermodynamic parameters were deduced for the following reactions:

$$Th(s) + Th_2S_3(s) = 3ThS(s) \begin{cases} \Delta G(1173) = -19.8 \pm 2 \text{ kcal(g.at.Th)}^{-1} \\ \Delta H(T) = -26.7 \pm 4 \text{ kcal(g.at.Th)}^{-1} \\ \Delta S(T) = -5.9 \pm 2 \text{ cal(g.at.Th)}^{-1} \text{K}^{-1} \end{cases}$$

$$Th(s) + Th_7S_{12}(s) = 4Th_2S_3(s) \begin{cases} \Delta G(1173) = -39.7 \pm 2 \text{ kcal(g.at.Th)}^{-1} \\ \Delta H(T) = -47.1 \pm 4 \text{ kcal(g.at.Th)}^{-1} \\ \Delta S(T) = -6.3 \pm 3 \text{ cal(g.at.Th)}^{-1} \text{K}^{-1} \end{cases}$$

$$Th(s) + 6ThS_2(s) = Th_7S_{12}(s)$$
$$\begin{cases} \Delta G(1173) = -95.2 \pm 4 \text{ kcal(g.at.Th)}^{-1} \\ \Delta H(T) = (-102) \text{ kcal(g.at.Th}^{-1}\text{(estimated))} \end{cases}$$

From the standard enthalpy of formation of Th_2S_3 given by Westrum and Eyring and taking into account the vaporisation of sulphur to $S_2(g)$ ($+30.9$ kcal(mol.S_2)$^{-1}$) and a value for ΔC_p of 1.6 cal mol^{-1} K^{-1} Aronson obtains:

$$2Th(s) + 3S_2(g) = Th_2S_3(s), \quad \Delta H_f(Th_2S_3(s),1173) = -300.5 \text{ kcal mol}^{-1}$$

which leads, using the cell data, to the following values:

$$Th(s) + \tfrac{1}{2}S_2(g) = ThS(s)$$
$$\Delta H_f(ThS(s),1173) = -109 \text{ kcal mol}^{-1}$$
$$7Th(s) + 6S_2(g) = Th_7S_{12}(s)$$
$$\Delta H_f(Th_7S_{12}(s),1173) = -1155 \text{ kcal mol}^{-1}$$
$$Th(s) + S_2(g) = ThS_2(s)$$
$$\Delta H_f(ThS_2(s),1173) = -175 \text{ kcal mol}^{-1}$$

These values, when compared with those of Eastman *et al.*, make the inconsistencies existing in the data on the thorium sulphides even more obvious.

5.5.2.2 *Uranium*

O'Hare *et al.*[156] have obtained, from fluorine flow calorimetry, the heat associated with the reaction:

$$US_{1.011}(s) + 6.033F_2(g) = UF_6(g) + 1.011 SF_6(g)$$

From the literature data for the standard enthalpy of formation of $UF_6(g)$[89] $(-510.77\pm0.45$ kcal mol$^{-1})$ and of $SF_6(g)$[157] $(-291.77\pm0.24$ kcal mol$^{-1})$ they then calculated the enthalpy of the reaction:

$$U(s,\alpha)+1.011S(s,rhomb) = US_{1.011}(s)$$
$$\Delta H_f^\circ(US_{1.011}(s),298) = -73.2\pm3.5\,\text{kcal mol}^{-1}.$$

Fluorine bomb calorimetry was used by the same authors to check the above value, the results being -74.8 and -74.5 kcal mol^{-1}. These data are in marked disagreement with the estimate given by Rand and Kubaschewski[1] $(-93\pm5$ kcal mol$^{-1})$.

From a mass spectrometric study of the sublimation of US in the temperature range 1825–2400 K, Cater et al.[158] obtained:

$$US(s) = U(g)+S(g)\begin{cases}\Delta H(2020) = 270.9\pm0.16\,\text{kcal mol}^{-1}\\ \Delta S(2020) = 65.0\pm0\,7\,\text{cal mol}^{-1}\,\text{K}^{-1}\end{cases}$$

which, taking ΔC_p to be -2 cal mol^{-1} K^{-1}, gives:

$$\Delta H(298) = 274.3\pm2\,\text{kcal mol}^{-1}$$

and

$$\Delta S(298) = 68.8\pm1\,\text{cal mol}^{-1}\,\text{K}^{-1}.$$

From the enthalpies of sublimation of uranium[159] $(126\pm3$ kcal mol$^{-1})$ and of sulphur to $S(g)$[160]$(66.64$ kcal mol$^{-1})$ one then obtains

$$U(s)+S(s,rhomb.) = US(s)$$
$$\Delta H_f^\circ(US(s),298) = -81.7\pm5\,\text{kcal mol}^{-1}$$

a value distinctly more negative than the calorimetric one. The enthalpy of formation of orthorhombic $US_{2.000\leqslant0.002}$ has also been determined by fluorine bomb calorimetry by O'Hare and Settle[161];

$$U(s,\alpha)+2S(s,rhomb) = US_2(s)$$
$$\Delta H_f^\circ(US_2(s),298) = -124.2\pm2.1\,\text{kcal mol}^{-1}$$

Kolar et al.[162] have reported the heat of oxidation of US and US_2 (identified as the tetragonal α phase) using a differential thermal analysis technique. The following data were reported:

$$US(s)+\tfrac{7}{3}O_2(g) = \tfrac{1}{3}U_3O_8(s)+SO_2(g)$$
$$\Delta H(643) = -276\pm12\,\text{kcal mol}^{-1}$$

and

$$US_2(s,\alpha)+\tfrac{10}{3}O_2(g) = \tfrac{1}{3}U_3O_8(s)+2SO_2(g)$$
$$\Delta H(629) = -345\pm24\,\text{kcal mol}^{-1}$$

Using known values for the thermodynamic functions of U_3O_8 and SO_2, and estimated data for the sulphides, one obtains for $\Delta H_f^\circ(US(s),298)$, -78 ± 14 kcal mol^{-1} and for $\Delta H_f^\circ(US_2(s),\alpha,298)$, -82 ± 26 kcal mol^{-1}. The value for US_2 is distinctly smaller than the calorimetric one. However, in view of the large experimental errors, these values can only be considered as approximate.

5.6 SELENIDES

5.6.1 Heat capacities and entropies

5.6.1.1 Thorium

The entropies of the various thorium selenides have been estimated by Westrum and Grønvold[143] to be $S°(ThSe(s),298) = 22.0$, $S°(ThSe_{1.5}(s), 298) = 25.5$, $S°(ThSe_{1.7}(s),298) = 27.0$, $S°(ThSe_2(s),298) = 30.0$, and $S°(ThSe_{2.5}(s),298) = 35.0$ cal mol^{-1} K^{-1}. There are no experimental thermodynamic data for these compounds.

5.6.1.2 Uranium

The low temperature heat capacities of USe (cub.,NaCl type)[163] and USe_2 (tetrag.)[146] have been measured by Westrum et al. from 5 to 350 K by adiabatic calorimetry. USe exhibits a λ type anomaly near 160.5 K due to a transition from the ferromagnetic to the paramagnetic state. In USe_2, a λ type anomaly occurs at 13.1 K and is tentatively attributed to an antiferromagnetic transition. Table 5.15 summarises these data together with the values for the thermodynamic functions of these compounds at 298 K. The experimental errors are reported to be less than 0.2% in USe and 0.1% in USe_2.

Table 5.15 Thermodynamic properties of uranium selenides* (ref. 146, 163)

	USe		USe$_2$	
Transition temperature	60.5 ⎫		13.1 ⎫	
ΔH_t	154 ⎬ ferromag.		— ⎬ antiferromag.	
ΔS_t	1.05 ⎭		0.19 ⎭	
$C_p°(298)$	13.10		18.92	
$[S°(298) - S°(0)]$	23.07		31.98	
$[H°(298) - H°(0)]$	3097		4209	
$-[G°(298) - H°(0)]/T$	12.68		17.86	

*Units: calorie, mole, K

In the case of USe_2 the use of $S°(T) - S°(0)$ to describe entropies is justified by the same considerations applied to the isomorphous compound $US_{1.9}$. The lowest temperature heat capacity values, in the case of USe may be represented by the expression $C_v = 2.075 \times 10^{-2}T + 8.04 \times 10^{-5}T^3$ cal

$mol^{-1} K^{-1}$, which can be used to extrapolate data below 5 K and in which the linear term is the electronic contribution and the T^3 term the lattice contribution.

5.7 NITRIDES

5.7.1 Heat capacities and entropies

5.7.1.1 Thorium

The existing experimental data are quite fragmentary. The entropy of ThN was estimated to be 12.0 cal $mol^{-1} K^{-1}$ by Aronson and Ingraham[144] from the characteristic temperature obtained from x-ray measurements. As far as Th_3N_4 is concerned, we have the relation for the specific heat given by Satoh[164] 40 years ago,

$$C_p = 0.04895 + 4.436 \times 10^{-5}T - 1.834 \times 10^{-8}T^2 \text{ (in cal g}^{-1}\,°C^{-1}),$$

valid between 0 and 500 °C, from which $C_p(298)$ is 37.3 cal $mol^{-1} K^{-1}$.

5.7.1.2 Uranium

More information is available for this element. Counsell et al.[165] have measured the low temperature heat capacity of UN(cub.,NaCl type), $UN_{1.59}$ (b.c.cub.) and $UN_{1.73}$(b.c.cub.) from 11 to 370 K. Westrum and Barber[166] have reported similar measurements on UN in the range 5–350 K. The two sets of results for UN, together with the data for $UN_{1.59}$ and $UN_{1.73}$,

Table 5.16 Thermodynamic properties of uranium nitrides*

	UN		$UN_{1.59}$	$UN_{1.73}$
	Counsell et al.[165]	Westrum et al.[166]	Counsell et al.[165]	
Transition temperature (antiferromag.)	52	52	94	33
ΔH_t	7.1	—	10.1	1.7
ΔS_t	0.15	0.17	0.12	0.05
$C_p°(298)$	11.31	11.43	12.95	13.77
$S°(298)$	14.87	14.97	15.54	15.74
$[H°(298) - H°(0)]$	2167	2179	2354	2410
$-[G°(298) - H°(0)]/T$	7600	7664	7650	7660

*Units: calorie, mole, K

are shown in Table 5.16. The transitions are due to antiferromagnetic ordering. The experimental value for the entropy of UN at 298.15 K is somewhat higher than the estimate (13.0 ± 1.5) of Rand and Kubaschewski[1]. Errors in the thermodynamic functions are reported to be about 0.2%.

Measurement of the enthalpy of UN by a drop calorimetry technique have been reported[167]. The derived heat capacity equation,

$$C_p = 13.32 + 1.19 \times 10^{-3}T - 2.10 \times 10^5 T^{-2} \text{ cal mol}^{-1} K^{-1} \text{ (273-1423 K)},$$

gives for $C_p^\circ(298)$ a value of 11.31 cal mol^{-1} K^{-1}, which agrees with the low temperature results. Above 1000 K, however, this equation does not fit the more recent data of Affortit[168] on the specific heat up to 2000 K (using adiabatic heating by the Joule effect) which can be expressed by the relation

$$C_p(T) = 10.2 + 4.2 \times 10^{-3}T \text{ cal mol}^{-1} \text{ K}^{-1} \quad (1000\text{--}2000 \text{ K}),$$

the values of Affortit being about 8% higher. On the other hand, enthalpy measurements up to 2600 K made by Conway and Flagella[169], using drop calorimetry, yield the relation:

$$H^\circ(T) - H^\circ(298) = 8915.6 + 24.45T - 7.70 \times 10^{-3}T^2 +$$
$$2.09 \times 10^{-6}T^3 \text{ cal mol}^{-1} \quad (1100\text{--}2600 \text{ K})$$

which gives heat capacity values in excellent agreement with the results of Affortit, the difference between the two sets of data being a fraction of 1% at 2000 K.

5.7.1.3   Plutonium

There are no experimental data for plutonium nitride. Several authors, however, have estimated thermodynamic data for PuN. Oetting[4] gives for $S^\circ(\text{PuN(s)},298)$, 14.2 ± 1, and for $C_p^\circ(\text{PuN(s)},298)$, 12.75 cal mol^{-1} K^{-1}, while Spear and Leitnaker[170] estimate $S^\circ(\text{PuN(s)},298)$ to be 15.2 ± 1 and $C_p^\circ(\text{PuN(s)},298)$, 10.6 cal mol^{-1} K^{-1}. The various thermodynamic functions for PuN have been calculated up to 3000 K by these authors.

5.7.2   Thermodynamics of formation

5.7.2.1   Thorium

Calorimetric measurements of the enthalpy of formation of thorium nitrides were reported almost 40 years ago[171, 172]. By combustion of thorium in oxygen at 1240 K with combustion products in the range $0.7 \leqslant \text{N/Th} \leqslant 0.9$ a heat of -154.2 ± 1 kcal(mol N$_2$)$^{-1}$ was reported and from the combustion of an analysed sample of ThN$_{1.32}$ a value of -156.2 ± 1 kcal (mol N$_2$)$^{-1}$ was reported for the enthalpy of formation of this compound at 292 K. Later estimates [173] for thorium nitrides are based on these experimental values.

  Thermodynamic data on the formation of ThN(s) have been obtained more recently by Olson and Mulford[174] from the study of the decomposition pressure of this compound according to the equation ThN(s) = Th(l) + $\frac{1}{2}$N$_2$(g). In this study the apparent melting point of the system was measured with varying nitrogen pressure in the temperature range 2689–3063 K. The decomposition pressure is described by:

$$\log P(\text{atm}) = 8.586 - 33224T^{-1} + 0.958 \times 10^{-17}T^5.$$

From their experimental point for the lowest temperature studied, at which less nitrogen is dissolved in the metal and hence the activity of the metal is closer to unity, Olson and Mulford calculate a limit for the standard free energy of formation of ThN. Using an estimated entropy of formation of -20 cal mol^{-1} K^{-1} and 1.5 cal mol^{-1} K^{-1} for ΔC_p, they conclude that the enthalpy of formation at 298 K should be more negative than -73 kcal mol^{-1}.

Aronson and Auskern[175] have studied the N_2 pressure above the two-phase system ThN–Th$_3$N$_4$ in the temperature range 1723–2073 K. From the slope and intercept of the log P versus $1/T$ plot, they obtained:

$$6\text{ThN(s)} + N_2(g) = 2\text{Th}_3N_4(s) \begin{cases} \Delta H(T) = -72.7 \pm 2 \text{ kcal(mol } N_2)^{-1} \\ \Delta S(T) = -32.7 \pm 2 \text{ cal(mol } N_2)^{-1} \text{ K}^{-1} \end{cases}$$

Correcting to 298 K, and taking ΔC_p to be 3 cal(mol N_2)$^{-1}$ K^{-1}, yields $\Delta H(298) = -77.5$ kcal(mol N_2)$^{-1}$ and $\Delta S(298) = -38.2$ cal(mol N_2)$^{-1}$ K^{-1} for the above reaction. Taking[173] -155.2 kcal(mol N_2)$^{-\frac{1}{4}}$ and -44.8 cal (mol N_2)$^{-1}$ K^{-1}, respectively, for the enthalpy and entropy of formation of Th$_3$N$_4$ they obtain:

$$\text{Th(s)} + \tfrac{1}{2}N_2(g) = \text{ThN(s)} \begin{cases} \Delta H_f^\circ(\text{ThN(s)},298) = -90.6 \text{ kcal mol}^{-1} \\ \Delta S_f^\circ(\text{ThN(s)}, 298) = -23.5 \text{ cal mol}^{-1} \text{ K}^{-1} \end{cases}$$

From this result, it appears, as will be seen later, that the enthalpy of formation of ThN is approximately 20 kcal mol^{-1} more negative than that of the corresponding uranium, neptunium and plutonium compounds, thus making it more like a Group IV transition metal nitride.

5.7.2.2 Uranium

The standard enthalpies of formation of UN$_{0.965}$, UN$_{1.510}$ and UN$_{1.690}$ were obtained by O'Hare et al.[156] from fluorine bomb calorimetry measurements. From the experimentally observed reactions:

$$\text{UN}_x(s) + 3F_2(g) = UF_6(s) + \tfrac{x}{2}N_2(g)$$

and using the enthalpy of formation of UF$_6$(s) [89], they obtained:

$$U(s,\alpha) + 0.483N_2(g) = \text{UN}_{0.965}(s)$$
$$\Delta H_f^\circ(\text{UN}_{0.965}(s),298) = -71.5 \pm 1.1 \text{ kcal mol}^{-1}$$
$$U(s,\alpha) + 0.755N_2(g) = \text{UN}_{1.510}(s)$$
$$\Delta H_f^\circ(\text{UN}_{1.510}(s),298) = -90.1 \pm 1.7 \text{ kcal mol}^{-1}$$
$$U(s,\alpha) + 0.845N_2(g) = \text{UN}_{1.69}(s)$$
$$\Delta H_f^\circ(\text{UN}_{1.69}(s),298) = -94.1 \pm 1.7 \text{ kcal mol}^{-1}$$

From the value reported for UN$_{0.965}$, O'Hare et al., extrapolating to stoicheiometric UN, obtained a value of -72.8 ± 1.2 kcal mol^{-1}, and, taking into account the value of -72.0 ± 1.2 kcal mol^{-1} obtained by Frederickson[176] from oxygen combustion calorimetry, they suggest an average value for $\Delta H_f^\circ(\text{UN(s)},298)$ of -72.4 ± 0.9 kcal mol^{-1}. This value is in moderate agreement with the earlier result of Gross et al.[177] (-69.6 ± 0.8 kcal mol^{-1}), who used a hot zone calorimeter in which N_2(g) combined directly with an excess of uranium powder, and with the value given by Hubbard[178] ($-70.95\pm$ 0.67 kcal mol^{-1}) from combustion of UN(s) in an oxygen bomb calorimeter.

Vapour pressure measurements of N_2 above UN have also given information on the enthalpy of formation of this compound. Inouye and Leitnaker[178] obtain a value of -68.2 kcal mol^{-1} for $\Delta H_f^\circ(\text{UN(s)},298)$ from the slope of the nitrogen pressure versus $1/T$ in the temperature range 1573–1773 K, and a third law calculation on their experimental data gives -69.2 kcal mol^{-1}. More recently, Gingerich[179] has used mass spectrometry to study the vapour pressure of U and N_2 over the two-phase system $UN_{0.4}$–$UN_{0.9}$ in the temperature range 1910–2230 K; a third law treatment of his data yields $\Delta H_f^\circ(\text{UN(s)},298) = -68.9\pm0.4$ kcal mol^{-1}. It is obvious that the lowering of the activity of liquid uranium due to saturation by nitrogen at the temperature of the experiments can explain the generally lower values for the heat of formation of UN, as compared to calorimetric data.

The value obtained by O'Hare et al.[156] for the enthalpy of formation of $UN_{1.51}$ and $UN_{1.00}$ gives, for the reaction:

$$UN_{1.00}(s)+0.255N_2(g) = UN_{1.51}(s)$$
$$\Delta H^\circ(298) = -17.3\pm2 \text{ kcal mol}^{-1}$$

This value is in poor agreement with that reported by Gross[177] for the same reaction (-14.57 ± 0.25 kcal mol^{-1}) and with the value of Lapat and Holden[180] (-14.25 kcal mol^{-1}) obtained from equilibrium dissociation pressure measurements over the UN–$UN_{1.5}$ system.

From their calorimetric study of the various uranium nitrides O'Hare et al. propose for the heat of formation of hypothetical, stoicheiometric UN_2: $\Delta H_f^\circ(\text{UN}_2(s),298) = -107\pm3$ kcal mol^{-1}. Naoumidis and Stöcker[181] have reported the partial molar thermodynamic properties of nitrogen from equilibrium pressure measurements of nitrogen over UN_x ($1.5\leqslant x\leqslant1.75$) between 773 and 1273 K. They have shown that $\Delta\bar{H}(N_2)$ varies almost linearly from the value 68.91 kcal $(\text{mol N}_2)^{-1}$ at $x = 1.53$ to 22.66 kcal $(\text{mol N}_2)^{-1}$ at $x = 1.65$. In the same composition range, $\Delta\bar{S}(N_2)$ varies from 53.4 to 23.9 cal $(\text{mol N}_2)^{-1}$ K^{-1}. It is worth noting that the pressure observed by these authors for a given temperature and composition are distinctly higher than those observed by others, in particular Lapat and Holden[180]. Moreover, the slopes of the pressure versus $1/T$ curves are also different. The reason for these discrepancies is, however, not clear.

5.7.2.3 Neptunium

The only experimental result on NpN is the study by Olson and Mulford[182] of its decomposition pressure according to the reaction: NpN(s) = Np(l)+ N_2(g), in the temperature range 2483–3103 K. The relation:

$$\log P(\text{atm}) = 8.193 - 29.54\times10^3 T^{-1} + 7.87\times10^{-18}T^5$$

describes the observed data. Using the same procedure as for ThN, Olson and Mulford conclude that the standard enthalpy of formation of NpN(s) at 298 K should be more negative than -61 kcal mol^{-1}.

5.7.2.4 Plutonium

The enthalpy of formation of PuN(s) at 298 K, has been obtained by two groups of workers from the reaction $PuN(s)+O_2(g) = PuO_2(s)+\frac{1}{2}N_2(g)$ in an oxygen bomb calorimeter. Lapage and Bunce[183] have reported ΔH_f° (PuN(s),298) to be -70.2 ± 1.5 kcal mol^{-1}, while more recently Johnson et al.[137] obtained a value of -71.51 ± 0.62 kcal mol^{-1}. The two results are only in moderate agreement, especially if one considers that the use of the recent value of Johnson et al.[137] for ΔH_f°(PuO$_2$(s),298) (-252.35 ± 0.17 kcal mol^{-1}) in the calculations, makes Lapage and Bunce's result 0.45 kcal mol^{-1} less negative. In view of the fact that the stoicheiometry of the nitride used by Johnson et al. is better established and also that the reproducibility of the results of these authors is better, it is considered that, at present, their value should be preferred.

Using a chronopotentiometric technique, Campbell[184] measured the e.m.f. of the cell $Pu(s,l)\,|\,(PuCl_3,\ LiCl-KCl)(s)\,|\,N_2(g)$, PuN(s) over the temperature range 714–1032 K. Above the melting point of plutonium, the standard free energy change, $\Delta G_f(T)$, corresponding to the reaction:

$$Pu(l)+\tfrac{1}{2}N_2(g) = PuN(s), \text{ is} -73.8+0.0225T \text{ kcal mol}^{-1}$$

From literature thermodynamic data for plutonium and values estimated for PuN by analogy with UN, they obtain -72.5 kcal mol^{-1} for the enthalpy of formation of PuN at 298 K, in good agreement with the calorimetric results. Previous data from Campbell and Leary[185], using static e.m.f. measurements, had yielded a more negative value, -76 kcal mol^{-1}. Campbell[184] had later recognised, however, that in this technique involving electrode equilibria which are slow to establish, uncontrolled side reactions could occur which, together with the build up of impurities at the electrolyte–electrolyte interface, could affect the validity of the results obtained.

Recent vapour pressure measurements also give results in agreement with the calorimetric values. For the reaction, $PuN(s) = Pu(g)+\frac{1}{2}N_2(g)$, studied by a mass spectrometry-Knudsen effusion technique, Kent and Leary[186] give:

$$\Delta G(T) = 150.713 - 44.24 \times 10^{-3}T \text{ kcal mol}^{-1} \quad (1600-2000 \text{ K})$$

This ultimately yields ΔH_f°(PuN(s),298) $= -71.0\pm0.6$ kcal mol^{-1}.

For the same reaction studied by the Knudsen effusion technique, Marcon and Poitreau[187] report:

$$\Delta G(T) = 157.500-50.75 \times 10^{-3}T \text{ kcal mol}^{-1} \quad (1650-2000 \text{ K})$$

from which they calculate ΔH_f°(PuN)(s),298) to be -72.84 ± 2 kcal mol^{-1}.

5.8 PHOSPHIDES

5.8.1 Heat capacities and entropies

5.8.1.1 Thorium

Aronson and Ingraham[144] have obtained characteristic temperatures for $ThP_{0.95}$ and $ThP_{1.33}$ (Th_3P_4) from x-ray measurements and, from these data, suggest that the values of $S°(ThP_{0.95}(s),298)$ and $S°(ThP_{1.33}(s),298)$ are 12.1 and 16.6 cal mol^{-1} K^{-1} respectively.

5.8.1.2 Uranium

Low temperature heat capacity measurements (11–320 K) on UP (cub., NaCl type) have been reported by Counsell et al.[188], and on U_3P_4(b.c.cub.) by the same authors and by Stalinski et al.[189] Counsell et al. have shown that the heat capacity of UP undergoes a very sharp anomaly at 22.15 K, tentatively attributed to a transition in the electronic ground state, and a

Table 5.17   Thermodynamic properties of uranium phosphides*

	$UP_{1.007}$ Counsell et al.[188]		$UP_{1.35}$ Counsell et al.[188]		$UP_{1.333}$ Stalinski et al.[189]	
Transition temperature	22.5		136.5		138.3	
ΔH_t	10.2	electronic	64.1	ferromag.	114.7	ferromag.
ΔS_t	0.45		0.51		1.03	
Transition temperature	121					
ΔH_t	87.3	anti-				
ΔS_t	0.74	ferromag.				
$C_p°(298)$	11.86		13.95		14.24	
$S°(298)$	18.71		20.61		20.75	
$[H°(298)-H°(0)]$	2579		2959		2967	
$-[G°(298)-H°(0)]/T$	10.06		10.69		10.75	

*Units: calorie, mole, K

broader anomaly at 121 K related to antiferromagnetic ordering. In the case of U_3P_4, only one transition is observed, corresponding to ferromagnetic ordering of the unpaired electrons at the Curie temperature. Table 5.17 summarises these data together with thermodynamic functions for these compounds at 298.15 K. Reported errors in the thermodynamic functions are about 0.2%.

Brugger[190] has reported for UP, in the temperature range 1073–1473 K, a mean heat capacity of 13.8 cal mol^{-1} K^{-1}. However, using a laser as a heat pulse source, Moser and Kruger[150] obtain for this compound an essentially constant heat capacity, 12.1 cal mol^{-1} K^{-1}, in the temperature range 298–973 K.

5.8.1.3 Plutonium

Moser and Kruger[151] have shown that the heat capacity of PuP increases from 14.80 cal mol^{-1} K^{-1} at 298 K to 15.09 at 923 K, using the same technique as for UP.

5.8.2 Thermodynamics of formation

5.8.2.1 Thorium

There has been no calorimetric determination of the enthalpy of formation of these compounds. Gingerich and Aronson[191] have obtained thermodynamic information on the Th–P system from e.m.f. measurements with solid type electrochemical cells. From the results for a cell of the type (Th, ThF$_4$)(s) | CaF$_2$(s) | (ThF$_4$,ThP,Th$_3$P$_4$)(s) the thermodynamic data for the reaction:

$$Th(s)+Th_3P_4(s) = 4ThP(s) \begin{cases} \Delta G(1173) = -53.7 \pm 2.8 \text{ kcal (g at.Th)}^{-1} \\ \Delta H(1173) = -64.6 \pm 7.1 \text{ kcal (g at.Th)}^{-1} \\ \Delta S(1173) = -9.3 \pm 3.7 \text{ cal (g at.Th)}^{-1} \text{K}^{-1} \end{cases}$$

were obtained. Previous investigations by Gingerich[192, 193] on the partial pressure of P$_2$(g) above the system ThP–Th$_3$P$_4$ had yielded information on the reaction:

$$\begin{aligned} Th_3P_4(s) &= 3ThP(s) \\ &+0.5P_2(g) \end{aligned} \begin{cases} \Delta G(1173) = 17.8 \pm 1.5 \text{ kcal (g at.Th)}^{-1} \\ \Delta H(1173) = 41.5 \pm 3 \text{ kcal (g at.Th)}^{-1} \\ \Delta S(1173) = 20.2 \pm 5 \text{ cal (g at.Th)}^{-1} \text{K}^{-1} \end{cases}$$

and allowed the immediate derivation of the function for the formation of ThP(s) at 1173 K:

$$Th(s)+0.5P_2(g) = ThP(s) \begin{cases} \Delta G_f^\circ(1173) = -71.5 \pm 4.8 \text{ kcal mol}^{-1} \\ \Delta H_f^\circ(1173) = -106 \pm 10 \text{ kcal mol}^{-1} \\ \Delta S_f^\circ(1173) = -29.5 \pm 10 \text{ cal mol}^{-1} \text{K}^{-1} \end{cases}$$

and of Th$_3$P$_4$(s)

$$3Th(s)+2P_2(g) = Th_3P_4(s) \begin{cases} \Delta G_f^\circ(1173) = -232 \pm 15 \text{ kcal mol}^{-1} \\ \Delta H_f^\circ(1173) = -359 \pm 45 \text{ kcal mol}^{-1} \\ \Delta S^\circ(1173) = -109 \pm 40 \text{ cal mol}^{-1} \text{K}^{-1} \end{cases}$$

5.8.2.2 Uranium

O'Hare et al.[156] have obtained by fluorine bomb calorimetry the heat associated with the reaction:

$$UP_{0.992}(s)+5.48F_2(g) = UF_6(s)+0.992PF_5(g)$$

and from the enthalpy of formation of UF$_6$(s)[89] (-522.64 ± 0.43 kcal mol^{-1}) and of PF$_5$(g)[194] (-380.8 ± 0.3 kcal mol^{-1}) they obtained:

$$U(s,\alpha)+0.992P(s,white) = UP_{0.992}(s)$$
$$\Delta H_f^\circ(UP_{0.992}(s),298) = -75.2 \pm 1.3 \text{ kcal mol}^{-1}$$

and by extrapolation $\Delta H_f^\circ(UP(s),298 = -75.5 \pm 1.4 \, kcal \, mol^{-1}$.

Following other observations by Gingerich et al.[195, 196], Reishus and Gundersen[197] have studied the vaporisation of UP and have shown that phosphorus vaporises preferentially, leaving behind the mutually saturated diphasic system U-UP. From the partial pressure equations for the species $U(g)$, $P(g)$ and $P_2(g)$ in the temperature range 2073–2423 K, they derived:

$$U(l) + \tfrac{1}{2}P_2(g) = UP(s)$$
$$\Delta H_f(UP(s),2250) = -101.4 \, kcal \, mol^{-1}$$

from which they obtained:

$$U(s,\alpha) + P(s,white) = UP(s)$$
$$\Delta H_f^\circ(UP(s),298) = -76 \pm 4 \, kcal \, mol^{-1}$$

in agreement with the calorimetric value.

The enthalpies of formation of $U_3P_4(s)$ and $UP_2(s)$ have been determined by Dogu et al.[198], using differential thermal analysis which involved the heating at variable rates of uranium and phosphorus mixtures in variable proportions at temperatures up to 800 K. The proportions of each uranium phosphide phase formed were determined by quantitative x-ray analysis. The values obtained for the formation enthalpies of U_3P_4 and UP_2 are based on the calorimetric heat of formation of O'Hare[156] for UP. Using published values for the heat capacity of UP[188] and estimated ones for U_3P_4 and UP_2, Dogu et al. obtained:

$$3U(s) + 4P(s,white) = U_3P_4(s)$$
$$\Delta H_f^\circ(U_3P_4(s),298) = -266 \pm 25 \, kcal \, mol^{-1}$$

and

$$U(s) + 2P(s,white) = UP_2(s)$$
$$\Delta H_f^\circ(UP_2(s),298) = -100 \pm 12 \, kcal \, mol^{-1}$$

It is probably worth noting that possible deviations from stoicheiometry for the various compounds were not taken into account.

5.9 CARBIDES

5.9.1 Heat capacities and entropies

5.9.1.1 Thorium

For ThC (cub., NaCl type) there is no other measurement of the heat capacity than that of Harness et al.[199] in the narrow temperature range of 1.8–4.2 K and for which a low electronic specific heat of $(0.70 \pm 0.05) \times 10^{-3}$ cal $mol^{-1}K^{-1}$ was found. For the ThC_2 phase (monocl.) results are more complete. Westrum and Takahashi[200] have reported the heat capacity of an analysed $ThC_{1.93}$ sample from 5 to 350 K and calculated its thermodynamic functions in that temperature range. At 298.15 K, these are $C_p^\circ = 13.55$ cal

$mol^{-1}K^{-1}$, $S° = 16.37$ cal $mol^{-1}K^{-1}$, $[H° - H°(0)] = 2447$ cal mol^{-1} and $-[G° - H°(0)]/T = 8.168$ cal $mol^{-1}K^{-1}$. Earlier measurements by Takahashi and Westrum[201] on a less characterised $ThC_{1.98}$ sample yielded essentially the same results. The heat capacity of ThC_2 is very close to that of UC_2 and the slight difference can be attributed to the fact that these compounds are not isostructural. From their experimental data, Takahashi and Westrum calculate for the high temperature heat capacity of ThC_2, $C_p = 15.17 + 2.89 \times 10^{-3}T - 2.21 \times 10^5 T^{-2}$ cal $mol^{-1}K^{-1}$. Thermodynamic functions for ThC_2 up to 2500 K have also been calculated by Holley and Storms[8] in their review on actinide carbides.

5.9.1.2 Uranium

The low temperature heat capacity of UC(cub.,NaCl type), U_2C_3(b.c.cub.) and UC_2 (tetrag.) up to 350 K have been obtained by two groups of workers and the data agree quite well[202, 204].

Table 5.18 summarises the results obtained by Andon et al.[202] for the thermodynamic functions at 298.15 K. The accuracy of these data is believed to be about 0.3%.

At 298.15 K, Westrum et al[203] reported $S°(UC_{1.039}(s),298)$ to be 14.28 and $S°(UC_{1.90}(s),298)$, 16.30 cal $mol^{-1}K^{-1}$, while Farr et al.[204] give $S°(UC_{1.50}(s),298) = 16.46$ and $S°(UC_{1.94}(s),298) = 16.33$ cal $mol^{-1}K^{-1}$.

The measurements of de Combarieu[205] (1.5–4.2 K and 20–85 K) on UC yield C_p values distinctly higher than those reported above, while the results of Harness et al.[199] (1.8–4.2 K) are only slightly higher.

The high temperature heat content of UC has been measured by several groups of authors. The results of Levinson[206] and of Harrington and Vozella[207]

Table 5.18 Thermodynamic properties of uranium carbides* (ref. 202)

	UC	$UC_{1.50}$	$UC_{1.91}$
$C_p°(298)$	11.84	12.34	14.50
$S°(298)$	14.03	16.45	16.31
$[H°(298) - H°(0)]$	2159	2417	2522
$-[G°(298) - G°(0)]/T$	6.786	8.345	7.848

*Units: calorie, mole, K

agree well, and have been critically analysed by Leitnaker and Godfrey[208], who give:

$$H°(T) - H°(298 = 4.83 \times 10^3 + 14.430T - 1.074 \times 10^{-3}T^2 + 1.890 \times 10^5 T^{-1} + 3.473 \times 10^{-5}T^{\frac{5}{2}} \text{ cal mol}^{-1}$$

(average error: 0.84%), for the temperature range 298–2481 K. A relation in accordance with the preceding one is given by Storms[9]. Recent measurements by Affortit[168] of the heat capacity of UC by adiabatic heating, using the Joule effect, yield C_p values in agreement with the above data in the

temperature range 700–1100 K. From 1700 K to 2800 K Affortit observed
a much more rapid increase in the heat capacity as a function of temperature;

$$C_p = 11.2 + 3.45 \times 10^{-3} T$$
$$+ 4.6 \times 10^{-10} T^{-2} \exp(-6 \times 10^4/RT) \text{ cal mol}^{-1} \text{K}^{-1}$$
$$(800\text{–}2800 \text{ K})$$

Earlier data by other authors, obtained over a lower temperature range,
were either much less accurate[209] or very fragmentary[210]. There is no high
temperature heat content measurement on the U_2C_3 phase. The high
temperature heat content of UC_2 phases (tetragonal to cubic transition near
2060 K) has been the subject of controversy[211] following the measurements
of Levinson[206, 212], which indicated a high value and an unusual rise in the
C_p versus temperature curve. This behaviour has been explained in terms
of a composition change of the phase boundary as a function of temperature
by Leitnaker and Godfrey[208] and by Storms[9]. On the basis of Levinson's
results (obtained on the system 0.055 $UC + 0.945$ $UC_{1.91} + 0.07C$, overall
composition $UC_{1.93}$), Leitnaker and Godfrey give:

$$H°(T) - H°(298) = -3.013 \times 10^3 + 4.076\,T + 2.631 \times 10^{-2} T^2 - 2.332$$
$$\times 10^{-5} T^3 + 1.025 \times 10^{-8} T^4 - 1.573 \times 10^{-12}\,T^5 \text{ cal mol}^{-1}$$

(average error: 0.25%) in the temperature range 298–2060 K and $H°(T)$
$- H°(298) = 1.806 \times 10^4 - 2.512\,T + 6.894 \times 10^{-3} T^2$ cal mol^{-1} (average
error: 0.30%) in the temperature range 2060–2581 K. Holley and Storms[8]
also give the various thermodynamic functions for the uranium carbides.

5.9.1.3 Neptunium

Information on the low temperature heat capacity of $NpC_{0.89}$ has recently
been reported from adiabatic calorimetry measurements by Sandenaw and
Gibney[213], who observed a broad transition with a maximum around 225 K.
At low temperature the heat capacity is very close to that of UC. In the
temperature range 250–300 K (assumed antiferromagnetic region) the results
were highly irreproducible.

5.9.1.4 Plutonium

The only available low temperature heat capacity data were obtained by
Sandenaw and Gibney[213] on a sample of composition $PuC_{0.81}$. However,
difficulties in obtaining reproducible results were attributed to the fact
that $PuC_{0.81}$ is very close to the lower phase boundary and that during runs
a crossing of the boundary can occur, with production of a mixture of the
PuC and Pu_3C_2 phases, and, eventually, traces of plutonium metal. Self-
radiation damage is thought to influence this boundary crossing at low
temperature. Thermal cycling of the sample gave more reproducible results,
which were thought to be representative of single-phase $PuC_{0.81}$. The heat
capacity for this compound is distinctly higher than that of UC (more than

25% at 50 K), this high value being attributed to important electronic and magnetic contributions. At 298.15 K, Sandenaw and Gibney reported $C_p^{\circ}(PuC_{0.81}(s),298)$ to be 11.04 and $S^{\circ}(PuC_{0.81}(s),298)$ to be 16.84 cal mol^{-1} K^{-1}. The value for the entropy is in good agreement with the estimate of Storms[9] ($S^{\circ}(PuC(s),298) = 17.0$ cal mol^{-1}K^{-1}), allowing for differences in stoicheiometry. A value as high as 18.2 ± 1 cal mol^{-1}K^{-1} for $S^{\circ}(PuC_{0.87}(s),298)$ has been proposed in an assessment of data for plutonium carbides[214].

Kruger and Savage[215] have measured enthalpy increments, using a drop calorimeter, for $PuC_{0.869}$ in the temperature range 400–1300 K. They give $H^{\circ}(T) - H^{\circ}(298) = -5035 + 13.08T + 5.718 \times 10^{-4}T^2 + 3.232 \times 10^5 T^{-1}$ cal mol^{-1}. The slope of the heat capacity curve derived from these data is in reasonable agreement with the slope above 300 K in Sandenaw and Gibney's low temperature heat capacity data. Storms[9] has also given thermodynamic functions for $\widetilde{PuC}_{0.87}$ up to 1900 K.

5.9.2   Thermodynamics of formation

5.9.2.1   Thorium

Huber et al.[216] have obtained the enthalpies of formation of a series of analysed thorium carbides, using oxygen bomb calorimetry. They obtained:

$$\Delta H_f^{\circ}(ThC_{0.75}(s),298) = -16.6 \pm 1.6$$
$$\Delta H_f^{\circ}(ThC_{0.81}(s),298) = -23.5 \pm 0.9$$
$$\Delta H_f^{\circ}(ThC_{0.91}(s),298) = -28.4 \pm 1.9$$
$$\Delta H_f^{\circ}(ThC_{1.00}(s),298) = -29.6 \pm 1.1 \text{ kcal mol}^{-1}$$

and for the dicarbide phase

$$\Delta H_f^{\circ}(ThC_{1.91}(s),298) = -29.9 \pm 1.3 \text{ kcal mol}^{-1}$$

(weighted average of two different samples).

Free energies of formation of the di- and monocarbides have been obtained from e.m.f. data on solid state cells of the type:

$$(Th,ThF_4)(s) \,|\, CaF_2(s) \,|\, (ThF_4,ThC_2,C)(s)$$

and

$$(Th,ThF_4)(s) \,|\, CaF_2(s) \,|\, (ThF_4,ThC_2,ThC)(s)$$

From the slope of the e.m.f. data as a function of temperature the other thermodynamic functions were also obtained as shown in Table 5.19; the data of the various experimenters are in satisfactory agreement. Holley and Storms[8], however, have shown that the e.m.f. values pertaining to the formation of ThC_2 are low compared to the ΔG_f values obtained from calorimetric data using reasonable estimates for the thermal properties of ThC_2. Estimates obtained from high temperature equilibria [220, 221] involving the Th–U–C system yielded a value for $\Delta G_f(ThC(s),1273)$ of about -28 kcal mol^{-1} and from the systems Th–W–C, Th–Mo–C, $-32 < \Delta G_f(ThC(s),1773) < -23$ kcal mol^{-1}.

E.M.F. measurements on solid state cells of the type

$$(Th,ThF_4)(s) \mid CaF_2(s) \mid (ThF_4,ThC_x)(s)$$

with $0.13 \leqslant x \leqslant 1.60$ have also been carried out and have yielded information on the partial molar thermodynamic quantities of thorium. However, the data obtained by two groups[222, 223] of authors are in considerable disagreement,

Table 5.19 Thermodynamics of formation of ThC and ThC$_2$ from e.m.f. data

$\dfrac{T}{K}$	$\dfrac{-\Delta G_f^\circ(T)_1}{\text{kcal mol}^{-1}}$	$\dfrac{-\Delta H_f^\circ(T)}{\text{kcal mol}^{-1}}$	$\dfrac{-\Delta S_f^\circ(T)}{\text{cal mol}^{-1}\text{K}^{-}}$	Reference
Th(s) + 2C(s) = ThC$_2$(s)				
1073	29.4	37.1	7.2	Egan[217]
1100	28.8	35.2	5.8	Satow[218]
1173	27.7±2	35±4	6±3	Aronson[219]
Th(s) + C(s) = ThC(s)				
1100	25.9	32.9	6.4	Satow[218]
1173	23.8±2	29±4	4±2	Aronson[219]

free energy values differing by a factor up to four, for reportedly analogous experimental conditions. It is beyond the scope of this study to discuss these results in detail.

5.9.2.2 Uranium

Enthalpies of formation of the various uranium carbides have been determined by Holley and co-workers by oxygen bomb calorimetry and are summarised in Table 5.20. From these data, $\Delta H_f^\circ(UC_{1.00}(s),298) = -23.2 \pm 0.7$ kcal mol^{-1}. The currently reported value for U_2C_3 supersedes another value (-49 ± 1 kcal mol^{-1}) which was only considered to be tentative[228].

Table 5.20 Enthalpies of formation of uranium carbides from combustion calorimetry

Compound	$\dfrac{-\Delta H_f^\circ(298)}{\text{kcal mol}^{-1}}$	Reference
UC$_{0.96}$	21.0±1.0	224
UC$_{0.996}$	23.3±0.9	225
UC$_{1.032}$	23.0±1	225
U$_2$C$_3$	43.4±1.8	226
UC$_{1.90}$	21.1±1.4	227

Values for the enthalpy of formation of UC and UC$_{1.86}$, -23.2 and -20.5 kcal mol^{-1}, respectively, have also been derived at 298 K by Leitnaker and Godfrey[229] from their analysis of the results of other authors[230] on equilibrium data in the systems UO$_2$–UC$_2$–C–CO and UO$_2$–UC$_2$–UC–CO.

Free energies of formation have been determined from e.m.f. measurements.

The following data have been obtained from the cell[231] $U(s) | (UCl_3,LiCl-KCl)(l) | (UO_xC_{1-x},C)(s)$, extrapolating to zero oxygen content:

$\Delta G_f^\circ(UC,s) = -16\,915 - 1.9\,T$ cal mol^{-1} (730–1030 K); from the cell[232], $(U,UF_3)(s) | CaF_2(s) | (UF_3,U_2C_3,C)(s)$,
$\Delta G_f^\circ(U_2C_3,s) = -43\,860 - 7\,T$ cal mol^{-1} (953–1073 K) and from the cell[232], $(U,UF_3)(s) | CaF_2(s) | (UF_3,UC_2,C)(s)$,
$\Delta G_f^\circ(UC_2,s) = -15\,820 - 8.2\,T$ cal mol^{-1} (973–1045 K) and
$\Delta G_f^\circ(UC_2,s) = -18\,980 - 5.2\,T$ cal mol^{-1} (1045–1193 K).

Holley and Storms[8] have shown that the above e.m.f. data for UC are low compared to the values obtained by calorimetry using the published data for the thermal properties of UC. E.M.F. data for UC_2 appear to be much more satisfactory.

Storms[9] and Leitnaker and Godfrey[208] have analysed the existing data on the vaporisation of the uranium carbon system. Storms has obtained the enthalpy and free energy of formation for carbides with various carbon content. In particular, he gives for $UC_{1.00}$:

$\Delta G_f^\circ(2100) = -25.15$ kcal mol^{-1}, $\Delta H_f^\circ(298) = -23.0$ kcal mol^{-1}
$\Delta G_f^\circ(2300) = -25.39$ kcal mol^{-1}, $\Delta H_f^\circ(298) = -23.3$ kcal mol^{-1}

and for $UC_{1.88}$:

$\Delta G_f^\circ(2100) = -27.13$ kcal mol^{-1}, $\Delta H_f^\circ(298) = -19.0$ kcal mol^{-1}
$\Delta G_f^\circ(2300) = -28.05$ kcal mol^{-1}, $\Delta H_f^\circ(298) = -18.7$ kcal mol^{-1}

the enthalpies of formation were obtained by third law calculations. Allowing for error limits, these data appear to agree with the calorimetric values. Subsequent to Storms' study, Vozzella et al.[233] have deduced $\Delta H_f^\circ(UC_{1.1}(s)$, 298) to be -21.2 ± 4.1 kcal mol^{-1} from a thermogravimetric study of the vaporisation of $UC_{1.1}$ in the temperature range 2250–2525 K.

5.9.2.3   Plutonium

Enthalpies of formation of plutonium carbides of various compositions have been obtained by combustion calorimetry. The results are summarised in Table 5.21.

Table 5.21   Enthalpies of formation of plutonium carbides by combustion calorimetry

Compound	$-\Delta H_f^\circ(298)$ kcal mol^{-1}	Reference
$PuC_{0.7}$	4	Marcon et al.[234]
$PuC_{0.87}$	12.4	Marcon et al.[234]
$PuC_{0.878 \pm 0.002}$	11.41 ± 0.68	Johnson et al.[235]
$PuC_{1.5}$	24.4	Marcon et al.[234]

As can be seen, no error limits were given by Marcon et al.[234], which prevents an accurate evaluation of their data as compared with those of Johnson et al.[235]. Preliminary values[228] for the enthalpies of combustion of $PuC_{0.77}$ and $PuC_{1.5}$ had yielded values far more positive, but their authors had used. E.M.F. studies on plutonium carbides, using chronopotentiometric and E.M.F. studies on plutonium carbides, using chronopotentiometric and voltammetric techniques, have recently been reported by Campbell et al.[236], with the cells:

$$Pu(l) \,|\, (PuCl_3,LiCl-KCl)(l) \,|\, (PuC_{1.5},C)(s)$$

and

$$Pu(l) \,|\, (PuCl_3,LiCl-KCl)(l) \,|\, (PuC_{1.5},PuC_{0.9})(s)$$

in the temperature range 920–1060 K, yielding:

$$\Delta G_f^\circ(PuC_{1.5},s) = -(16\,650 \pm 90) + (0.352 \pm 0.09)\,T \text{ cal mol}^{-1}$$

and

$$\Delta G_f^\circ(PuC_{0.9},s) = -(14\,059 \pm 200) + (1.60 \pm 0.20)\,T \text{ cal mol}^{-1}.$$

Previous e.m.f. data obtained by Campbell et al.[237], using a static method of measurement, had yielded substantially different results due[236] to competing reactions at the electrodes and impurities adsorbed during the long equilibration times.

Vapour pressure measurements over plutonium carbides of various compositions have been carried out by several authors. The very recent data of Campbell et al.[236], who covered the whole composition range from Pu to $PuC_2 + C$ are used here specifically. Their results are in general agreement with the previous data of Olson and Mulford[238] ($PuC-Pu_2C_3$ and Pu_2C_3-C systems) and of Harris[239] ($PuC-Pu_2C_3$, Pu_2C_3-C and PuC_2-C systems). The agreement with the results of Battles et al.[240] (Pu_2C_3-C and PuC_2-C systems) is less satisfactory, the vapour pressures observed by these authors being lower by a factor of 2 or 3. Using the most recently available free energy of vaporisation[241] of Pu(l) in their calculations:

$$Pu(l) = Pu(g) \quad \Delta G(T) = 8\,0154 - 22.64\,T \text{ cal mol}^{-1} \quad (1426-1658 \text{ K})$$

Campbell et al. obtained:

$$Pu(l) + 1.5C(s) = PuC_{1.5}(s)$$
$$\Delta G(T) = -14\,098 - 2.20\,T \text{ cal mol}^{-1} \ (1668-1933 \text{ K})$$
$$Pu(l) + 0.9C(s) = PuC_{0.9}(s)$$
$$\Delta G(T) = -11\,487 - 0.74\,T \text{ cal mol}^{-1} \ (1450-1848 \text{ K})$$

and for the formation of PuC_2:

$$\Delta G(T) = -7117 - 5.84\,T \text{ cal mol}^{-1}$$
$$Pu(l) + 2C(s) = PuC_2(s), \quad \text{or} \quad \quad \quad (1933-2170 \text{ K})$$
$$\Delta G(T) = -7574 - 5.58\,T \text{ cal mol}^{-1}$$
$$(1933-2140 \text{ K})$$

depending on the vaporisation process observed.

Campbell *et al.*[236] have critically compared their e.m.f. and vapour pressure data with the vapour pressure data of the other authors and have calculated in each case the thermodynamic functions for the formation of the plutonium carbides at 298 K; from this analysis, they recommend the following data as 'best values', which, in fact, represent a fairly consistent set:

$$Pu(s)+0.9C(s) = PuC_{0.9}(s) \quad \begin{cases} \Delta H_f^\circ(PuC_{0.9}(s),298 = -10.9 \pm 1.0 \text{ kcal mol}^{-1} \\ \Delta S_f^\circ(PuC_{0.9}(s),298) = +2.9 \pm 0.5 \text{ cal mol}^{-1}\text{K}^{-1} \\ S^\circ(PuC_{0.9}(s),298) = +17.3 \pm 0.5 \text{ kcal mol}^{-1}\text{K}^{-1} \end{cases}$$

$$Pu(s)+1.5C(s) = PuC_{1.5}(s) \quad \begin{cases} \Delta H_f^\circ(PuC_{1.5}(s),298) = -13.2 \pm 2.5 \text{ kcal mol}^{-1} \\ \Delta S_f^\circ(PuC_{1.5}(s),298 = +5.5 \pm 1 \text{ cal mol}^{-1}\text{K}^{-1} \\ S^\circ(PuC_{1.5}(s),298) = +20.7 \pm 1 \text{ cal mol}^{-1}\text{K}^{-1} \end{cases}$$

$$Pu(s)+2C(s) = PuC_2(s) \quad \begin{cases} \Delta H_f^\circ(PuC_2(s),298) = -6.8 \pm 2.5 \text{ kcal mol}^{-1} \\ \Delta S_f^\circ(PuC_2(s),298) = +8.2 \pm 1 \text{ cal mol}^{-1}\text{K}^{-1} \\ S^\circ(PuC_2(s),298) = +24.1 \pm 2.5 \text{ cal mol}^{-1}\text{K}^{-1} \end{cases}$$

The enthalpy of formation value so obtained for $PuC_{0.9}$ is in fair agreement with the calorimetric value. In the case of $PuC_{1.5}$, however, the only calorimetric value[234] reported (-24.4 kcal mol^{-1}) is 11 kcal mol^{-1} more negative than the value recommended by Campbell *et al.* and is, in fact, surprisingly negative in view of the general trend in the properties of the various carbides from thorium to plutonium.

References

1. Rand, M. H. and Kubaschewski, O. (1963). *The Thermochemical Properties of Uranium Compounds* (Edinburgh and London: Oliver and Boyd)
2. Fitzgibbon, G. C. Pavone, D. and Holley, C. E., Jr. (1970). *Rep.* LA-DC-1121; To appear in *J. Chem. Thermodyn.*
3. Rand, M. H. (1966). *At. Energy Rev.*, **4**, Special Issue No. 1, 7
4. Oetting, F. L. (1967). *Chemical Reviews*, **67**, 261
5. Westrum, E. F., Jr. and Lyon, W. G. (1967). *Symp. Thermodynamics of Nuclear Materials*, 239 (Vienna: IAEA)
6. International Atomic Energy Agency (1967). *The Plutonium–Oxygen and Uranium–Plutonium–Oxygen Systems: A Thermochemical Assessment—Technical Reports Series No. 79* (Vienna: IAEA)
7. International Atomic Energy Agency (1965). *Thermodynamic and Transport Properties of Uranium Dioxide and Related Phases* (Vienna: IAEA)
8. Holley, C. E., Jr. and Storms, E. K. (1967). *Symposium on Thermodynamics of Nuclear Materials*, 397 (Vienna: IAEA)
9. Storms, E. K. (1967). *The Refractory Carbides* (New York: Academic Press)
10. Brown, D. (1968). *Halides of the Lanthanides and Actinides*, 237 (London: Wiley–Interscience)
11. Eyring, L. and Westrum, E. F., Jr. (1950). *J. Am. Chem. Soc.*, **72**, 5555
12. Rand, M. H. (1970). *Private Communication*; Assessed data to be published
13. Latimer, W. H. (1959). *Oxidation Potentials* (Englewood Cliffs, N.J.: Prentice Hall)
14. Cunningham, B. B. (1966). *Colloque International sur la Physicochimie du Protactinium*, 45 (Paris: CNRS)
15. Argue, G. R., Mercer, E. E. and Cobble, J. W. (1961). *J. Phys. Chem.*, **65**, 2041
16. Kraus, K. A. and Nelson, F. R. (1950), *J. Am. Chem. Soc.*, **72**, 3901
17. Fuger, J. and Brown, D. (1970). *J. Chem. Soc. (A)*, 763
18. Huber, E. J., Jr. and Holley, C. E., Jr. (1969). *J. Chem. Thermodyn.*, **1**, 267
19. Cordfunke, E. H. P. (1964). *J. Phys. Chem.*, **68**, 3353
20. Fuger, J., Brown, D. and Easey, J. F. (1969). *J. Chem. Soc. (A)*, 2995
21. Westrum, E. F., Jr. and Eyring, L. (1952). *J. Am. Chem. Soc.*, **74**, 2045
22. Cohen, D. and Hindman, J. C. (1952). *J. Am. Chem. Soc.*, **74**, 4682

23. Cohen, D. and Hindman, J. C. (1952). *J. Am. Chem. Soc.*, **74**, 4679
24. Rossini, F. D., Wagman, D. D., Evans, W. H., Levine, S. and Jaffe, I. (1952). *Selected Values of Chemical Thermodynamic Properties*, NBS Circular 500 (Washington: NBS)
25. Brand, J. R. and Cobble, J. W. (1970). *Inorg. Chem.*, **9**, 912
26. Westrum, E. F., Jr. and Robinson, H. P. (1949). *Paper 6.53—The Transuranium Elements*, 914 (New York: McGraw-Hill)
27. Akhachinskii, V. V. and Kopitin, L. M. (1960). *At. Energ. (USSR)*, **9**, 504
28. Akhachinskii, V. V., Kopitin, L. M., Ivanov, M. I. and Podolskaya, N. C. (1962). *Symp. Thermodynamics of Nuclear Materials*, 309 (Vienna: IAEA)
29. Fuger, J. and Cunningham, B. B. (1963). *J. Inorg. Nucl. Chem.*, **25**, 1423
30. Akhachinskii, V. V. (1965). *Symp. Thermodynamics with Emphasis on Nuclear Materials*, 561 (Vienna: IAEA)
31. Hinchey, R. J. and Cobble, J. W. (1970). *Inorg. Chem.*, **9**, 922
32. Saegar, V. W. and Spedding, F. H. (1960). *Rep*. IS-338
33. Hinchey, R. J. and Cobble, J. W. (1970). *Inorg. Chem.*, **9**, 917
34. Latimer, W. M. and Coulter, L. V. (1940). *J. Am. Chem. Soc.*, **62**, 2557
35. Lohr, H. R. and Cunningham, B. B. (1951). *J. Am. Chem. Soc.*, **73**, 2025
36. Eyring, L., Lohr, H. R. and Cunningham, B. B. (1952). *J. Am. Chem. Soc.*, **74**, 1186
37. Gunn, S. R. and Cunningham, B. B. (1957). *J. Am. Chem. Soc.*, **79**, 1563
38. Wade, W. Z. and Wolf, T. T. (1967). *J. Inorg. Nucl. Chem.*, **29**, 2577
39. Wallmann, J. C. Fuger, J., Haug, H., Marei, S. A. and Bansal, B. M. (1967), *J. Inorg. Nucl. Chem.*, **29**, 2097
40. Westrum, E. F., Jr. and Robinson, H. P. (1949). *Paper 6.54—The Transuranium Elements*, 922 (New York: McGraw-Hill)
41. Smith, P. K., Hale, W. H. and Thompson, M. C. (1969). *J. Chem. Phys.*, **50**, 5066
42. Katz, J. J. and Seaborg, G. T. (1957). *The Chemistry of the Actinide Elements*, 426 (London: Methuen)
43. Khanaev, E. I. and Khripin, L. A. (1970). *Radiokhimiya*, **12**, 178
44. Long, G. and Blankenship, F. F. (1969). *Rep*. ORNL-TM-2065
45. Morss, L. R. (1969). *Rep*. UCRL-18951
46. Koch, C. W. and Cunningham, B. B. (1954). *J. Am. Chem. Soc.*, **76**, 1470
47. Morss, L. R. (1969). *Rep*. UCRL-18951
48. Tveekrem, J. O. and Chandrasekharaiah, M. S. (1968). *J. Electrochem. Soc.*, **115**, 1021
49. Brewer, L., Bromley, L. A., Gilles, P. W. and Lofgren, N. L. (1958). *Paper 33—Chemistry of Uranium—Collected Papers—Rep*. TID-5290
50. Mac Wood, G. E. (1958). *Paper 58—Chemistry of Uranium—Collected Papers—Rep. TID-5290*
51. Egan, J. J., McCoy, W. and Bracker, J. (1962). *Symp. Thermodynamics of Nuclear Materials*, 163 (Vienna: IAEA)
52. Zmbov, K. F. (1970). *J. Inorg. Nucl. Chem.*, **32**, 1378
53. Blue, G. D., Green, J. W., Bautista, R. G. and Margrave, J. L. (1963). *J. Phys. Chem.*, **67**, 877
54. Hildebrand, D. L. and Murad, E. (1966). *J. Chem. Phys.*, **44**, 1524
55. Darnell, A. J. and Keneshea, F. J. (1958). *J. Phys. Chem.*, **62**, 1143
56. Stull, D. R., Editor (1965). *Janaf Thermochemical Tables* (Midland, Mich.—Dow Chemical Company)
57. Van Deventer, E. H., Rudzitis, E. and Hubbard, W. N. (1970). *J. Inorg. Nucl. Chem.*, **32**, 3233
58. Hultgren, R., Orr, R. L., Anderson, P. D. and Kelly, K. K. (1963). *Selected Values of Thermodynamic Properties of Metals and Alloys*, 279 (New York: Wiley)
59. Kent, R. A. (1968). *J. Am. Chem. Soc.*, **90**, 5657
60. Carniglia, S. C. and Cunningham, B. B. (1955), *J. Am. Chem. Soc.*, **77**, 1451
61. Phipps, T. E., Sears, G. W., Seiffert, R. L. and Simpson, O. C. (1950). *J. Chem. Phys.*, **18** 713
62. Feber, R. C. and Herrick, C. C. (1965). *Rep*. LA-3184
63. Rand, M. H., *Assessed data, to be published—Cited by Brown*, ref. *10*
64. Laser, M. and Merz, E. (1969). *J. Inorg. Nucl. Chem.*, **31**, 349
65. Maslov, P. G. and Maslov, Y. P. (1965). *Zh. Obshch. Khim.*, **35**, 2112
66. Fuger, J. and Brown, D., unpublished results
67. Vdovenko, V. M., Romanov, G. A. and Solntseva, L. V. (1969). *Radiokhimiya*, **11**, 466

68. Vdovenko, V. M., Romanov, G. A. and Solntseva, L. A. (1970). *Radiokhimiya*, **12**, 764
69. Fuger, J. and Brown, D. (1971). To appear in *J. Chem. Soc. (A)*
70. Martynova, M. S., Kudryashova, Z. P. and Vasil'kova, I. V. (1968). *At. Energ.*, **25**, 266
71. Darnell, A. J. (1960). *J. Inorg. Nucl. Chem.*, **15**, 359
72. Heus, R. J. and Egan, J. J. (1966). *Z. Phys. Chem.*, **49**, 38
73. Bilz, W. and Fendius, C. (1928). *Z. Anorg. Allgem. Chemie*, **176**, 49
74. Smith, B. C., Thacker, L. and Wassef, M. A. (1969). *Ind. J. Chem.*, **7**, 1154
75. Westrum, E. F., Jr. and Robinson, H. P. (1949). *Paper 6.50—The Transuranium Elements*, 887 (New York: McGraw-Hill)
76. Brewer, L., Bromley, L., Gilles, P. W. and Lofgren, L. F. (1949). *Paper 6.40—The Transuranium Elements*, 861 (New York: McGraw-Hill)
77. Gruen, D. M. and DeKock, C. W. (1967). *J. Inorg. Nucl. Chem.*, **29**, 2569
78. Connick, R. E. and McVey, W. H. (1951). *J. Am. Chem. Soc.*, **73**, 1798
79. Gagarinskii, Yu. V. and Khanaev, E. I. (1967). *Russ. J. Inorg. Chem.*, **12**, 54
80. Benz, R. (1962). *J. Inorg. Nucl. Chem.*, **24**, 1191
81. Bjorklund, C. W., Reavis, J. C., Leary, J. A. and Walsh, K. A. (1959). *J. Phys. Chem.*, **63**, 1774
82. Chudinov, E. G. and Choporov, D. Y. (1970). *At. Energ.*, **28**, 62
83. Mandelberg, C. and Davies, D. (1961). *J. Chem. Soc.*, 2031
84. Berger, R. and Gaumann, T. (1961). *Helv. Chim. Acta*, **44**, 1084
85. Chudinov, E. G. and Choporov, D. Ya. (1958). *Rep.* NP-7516; AEC-tr 3871
86. Fried, S. and Davidson, N. R. (1949). *Paper 6.11—The Transuranium Elements*, 784 (New York: McGraw-Hill)
87. Dawson, J. K., Elliott, R. M., Hurst, R. and Truswell, A. E. (1954). *J. Chem. Soc.*, 558
88. Chudinov, E. G. and Choporov, D. Ya. (1970). *At. Energ.*, **28**, 62
89. Settle, J. L., Feder, H. M. and Hubbard, W. N. (1963). *J. Phys. Chem.*, **67**, 1892
90. Osborne, D. W., Weinstock, B. and Burns, J. H. (1970). *J. Chem. Phys.*, **52**, 1803
91. Wolf, A. S., Posey, J. C. and Rapp, K. E. (1965). *Inorg. Chem.*, **4**, 751
92. Stein, L. (1964). *Inorg. Chem.*, **3**, 995
93. Weigel, F., Hoffmann, G. and Tèr Meer, N. (1969). *Radiochimica Acta*, **11**, 210
94. Gross, P., Hayman, C. and Stuart, M. C., cited by Settle *et al.* (ref. *89*)
95. Brickwedde, H. J., Hoge, H. J. and Scott, R. B. (1948). *J. Chem. Phys.*, **16**, 429
96. Llewellyn, D. R. (1953). *J. Chem. Soc.*, 28
97. Osborne, D. W. and Westrum, E. F., Jr. (1953). *J. Chem. Phys.*, **21**, 1884
98. Jones, W. M., Gordon, J. and Long, E. A. (1952). *J. Chem. Phys.*, **20**, 695
99. Westrum, E. F., Jr., Hatcher, J. B. and Osborne, D. W. (1953). *J. Chem. Phys.*, **21**, 419
100. Sandenaw, T. A. (1963). *J. Nucl. Mater.*, **10**, 165
101. Flotow, H. E., Osborne, D. W. and Westrum, E. F. (1968). *J. Chem. Phys.*, **49**, 2438
102. Osborne, D. W., Westrum, E. F. and Lohr, H. R. (1957). *J. Am. Chem. Soc.*, **79**, 529
103. Leask, M. J. M., Roberts, L. E. S., Walter, A. J. and Wolf, W. P. (1963). *J. Chem. Soc.*, 4788
104. Westrum, E. F., Jr. and Grønvold, F. C. (1959). *J. Am. Chem. Soc.*, **81**, 1777
105. Girdhar, H. L. and Westrum, E. F. (1968). *J. Chem. Eng. Data*, **13**, 531
106. Frederickson, D. R. and Chasanov, M. G. (1970). *J. Chem. Thermodyn.*, **2**, 623
107. Moore, G. E. and Kelley, K. K. (1947). *J. Am. Chem. Soc.*, **69**, 2105
108. Engel, T. K. (1969). *J. Nucl. Mater.*, **31**, 211
109. Grønvold, F., Kveseth, N. J., Sveen, A. and Tichy. (1970). *J. Chem. Thermodyn.*, **2**, 665
110. Ogard, A. E. and Leary, J. A. (1967). *Symp. Thermodynamics of Nuclear Materials*, 651 (Vienna: IAEA)
111. Conway, J. B. and Hein, R. A. (1965). *J. Nucl. Mater.*, **15**, 149
112. Affortit, C. (1969). *High Temp.–High Press.*, **1**, 27
113. Popov, M. M., Galichenko, G. L. and Senin, M. D. (1958). *Zh. Neorg. Khim.*, **3**, 1734
114. Hein, R. A. Stodahl, L. H. and Szwarc, R. (1968). *J. Nucl. Mater.*, **25**, 99
115. Hein, R. A., Flagella, P. N. and Conway, J. B. (1968). *J. Am. Ceram. Soc.*, **51**, 291
116. Leibowitz, L., Mishler and Chasanov, M. G. (1969). *J. Nucl. Mater.*, **29**, 356
117. Grossmann, L. N. and Kaznoff, A. I. (1968). *J. Am. Ceram. Soc.*, **51**, 59
118. Affortit, C. and Marcon, J. P. (1970). *Rev. Int. Hautes Temp. et Refract.*, **7**, 236
119. Westrum, E. F., Jr., Takanashi, Y. and Grønvold, F. (1965). *J. Phys. Chem.*, **69**, 3192
120. Gotoo, K. and Naito, K. (1965). *J. Phys. Chem. Solids*, **26**, 1679

121. Khomyakov, K. G., Spitsyn, V. I. and Zhvanko, S. A. (1961). *Issled. B. Obl. Khim. Urana, Sb. Statei,* 141
122. Maglic, K. and Herak, R. (1970). *Rev. Int. Hautes Temp. et Refract.,* 7, 247
123. Ogard, A. E. (1970). *Intern. Conf. Plutonium, AIME Nuclear Metallurgy,* 17, 78
124. Kruger, O. L. and Sauvage, H. (1968). *J. Chem. Phys.,* 49, 4540
125. Huber, E. J., Jr., Holley, C. E., Jr. and Meierkord, E. H. (1952). *J. Am. Chem. Soc.,* 74, 3406
126. Rothe, W. A. and Becker, G. (1932). *Z. Phys. Chem. (A),* 159, 1
127. Popov, M. M. and Ivanov, M. I. (1957). *Soviet J. At. Energy,* 2, 439
128. Fitzgibbon, G. C., Pavone, D. and Holley, C. E., Jr. (1967). *J. Chem. Eng. Data,* 12, 122
129. Mukaibo, T., Naito, K., Sato, K. and Uchijimo, T. (1962). *Symp. Thermodynamics of Nuclear Materials,* 723 (Vienna: IAEA)
130. Vidavskii, L. M., Byakhova, N. J. and Ippolitova, E. A. (1965). *Zh. Neorg. Khim.,* 10, 1746
131. Cordfunke, E. H. P. and Ailing, P. (1965). *Trans. Faraday Soc.,* 61, 50
132. Roberts, L. E. J., Markin, T. L. and Walter, A. J. (1962). *Symp. Thermodynamics of Nuclear Materials,* 693 (Vienna: IAEA)
133. Roberts, L. E. J. and Markin, T. L. (1967). *Proc. Brit. Ceram. Soc.* 8, 202
134. Gerdanian, P. and Dodé, M. (1967). *Symp. Thermodynamics of Nuclear Materials,* 41 (Vienna: IAEA)
135. Pattoret, A., Drowart, J. and Smoes, S. (1967). *Symp. Thermodynamics of Nuclear Materials,* 613 (Vienna: IAEA)
136. Huber, E. J., Jr. and Holley, C. E., Jr. (1968). *J. Chem. Eng. Data,* 13, 545
137. Johnson, G. K., Van Deventer, E. H., Kruger, O. L. and Hubbard, W. N. (1969). *J. Chem. Thermodyn.,* 1, 89
138. Holley, C. E., Jr., Mulford, R. N. R., Huber, E. J., Jr., Head, E. L., Ellinger, F. H. and Bjorklund, C. W. (1958). *2nd Conf. Peaceful Uses Atomic Energy,* 6, 215 (Geneva: United Nations)
139. Markin, T. L. and Rand, M. H. (1965). *Symp. Thermodynamics with Emphasis on Nuclear Materials,* 145 (Vienna: IAEA)
140. Dean, G., Boivineau, J. C., Chereau, P. and Marcon, J. P. (1970). *Intern. Conf. Plutonium, AIME-Nuclear Metallurgy,* 17, 753
141. Chikalla, T. D. and Eyring, L. (1967). *J. Inorg. Nucl. Chem.,* 29, 2281
142. King, E. G. and Weller, W. W. (1959). *U.S. Bur. of Mines Report Invest.* 5485
143. Westrum, E. F., Jr. and Grønvold, F. (1962). *Symp. Thermodynamics of Nuclear Materials,* 3 (Vienna: IAEA)
144. Aronson, S. and Ingraham, A. (1967). *J. Nucl. Mater.,* 24, 74
145. Westrum, E. F., Jr., Walters, R. R., Flotow, H. E. and Osborne, D. W. (1968). *J. Chem. Phys.,* 48, 155
146. Westrum, E. F., Jr. and Grønvold, F. (1970). *J. Inorg. Nucl. Chem.,* 32, 2169
147. Grønvold, F. and Westrum, E. F., Jr. (1968). *J. Inorg. Nucl. Chem.,* 30, 2127
148. Slater, R. C. L. M. (1964). *Z. Kristallogr.,* 120, 278
149. MacLeod, A. C. and Hopkins, S. W. J. (1967). *Proc. Brit. Ceram. Soc.,* 8, 15
150. Moser, J. B. and Kruger, O. L. (1967). *J. Appl. Phys.,* 38, 3215
151. Moser, J. B. and Kruger, O. L. (1968). *J. Am. Ceram. Soc.,* 51, 369
152. Eyring, L. and Westrum, E. F., Jr. (1953). *J. Am. Chem. Soc.,* 75, 4802 (based on work done in 1946)
153. Bear, J. and McTaggart, F. K. (1958). *Australian J. Chem.,* 11, 458
154. Eastman, E. D., Brewer, L., Bromley, L. A., Gilles, P. W. and Lofgren, N. L. (1950). *J. Am. Chem. Soc.,* 72, 4019
155. Aronson, S. (1967). *J. Inorg. Nucl. Chem.,* 29, 1611
156. O'Hare, P. A. G., Settle, J. L., Feder, H. M. and Hubbard, W. N. (1967). *Symp. Thermodynamics of Nuclear Materials,* 265 (Vienna: IAEA)
157. O'Hare, P. A. G., Settle, J. L. and Hubbard, W. N. (1966). *Trans. Faraday Soc.,* 62, 558
158. Cater, E. D., Rauh, E. G. and Thorn, R. J. (1966). *J. Chem. Phys.,* 44, 3106
159. Drowart, J., Pattoret, A. and Smoes, S. (1964). *J. Nucl. Mater.,* 12, 319
160. Wagman, D. D., Evans, W. H., Parker, V. B., Halow, I., Bailey, S. M. and Schumm, R. H. (1968). *NBS Technical Note,* 270–3 (Washington: National Bureau of Standards)
161. O'Hare, P. A. G. and Settle, J. L. (1968). *Rep.* ANL-7575, 112

162. Kolar, D., Komac, M., Drofenic, M., Bohinc, M., Marinković, V. and Vene, N. (1967). *Symp. Thermodynamics of Nuclear Materials*, 279 (Vienna: IAEA)
163. Takahashi, Y. and Westrum, E. F. Jr. (1965). *J. Phys. Chem.*, **69**, 3618
164. Satoh, H. (1938). *Sci. Papers Inst. Tokyo*, **35**, 182
165. Counsell, J. F., Dell, R. M. and Martin, J. F. (1968). *Trans. Faraday. Soc.*, **62**, 1736
166. Westrum, E. F., Jr. and Barber, C. M. (1966). *J. Chem. Phys.*, **45**, 635
167. Speidel, E. D. and Keller, D. L. (1963). *Rep.* BMI 1633
168. Affortit, C. (1970). *J. Nucl. Mater.*, **34**, 105
169. Conway, J. B. and Flagella, P. N. (1969). *Rep.* GEMP-1012
170. Spear, K. E. and Leitnaker, J. M. (1968). *J. Am. Ceram. Soc.*, **51**, 706
171. Neuman, B., Kroeger, C. and Haebler, H. (1932). *Z. Anorg. Allgem. Chemie*, **207**, 145
172. Neuman, B., Kroeger, C. and Kunz, H. (1934). *Z. Anorg. Allgem. Chemie*, **218**, 379
173. Brewer, L., Bromley, L. A., Gilles, P. W. and Lofgren, L. N. (1950). *The Chemistry and Metallurgy of Miscellaneous Materials—Thermodynamics*, 42 (New York: McGraw-Hill)
174. Olson, W. M. and Mulford, R. N. R. (1965). *J. Phys. Chem.*, **69**, 1223
175. Aronson, J. and Auskern, A. B. (1966). *J. Phys. Chem.* **70**, 3937
176. Frederickson, D. R., cited in ref. 156
177. Gross, P., Hayman, C. and Clayton, H. (1962). *Symp. Thermodynamics of Nuclear Materials*, 653 (Vienna: IAEA)
178. Inouye, H. and Leitnaker, J. M. (1968). *J. Am. Ceram. Soc.*, **51**, 6
179. Gingerich, K. A. (1969). *J. Chem. Phys.*, **51**, 4433
180. Lapat, P. E. and Holden, R. B. (1964). *Intern. Symp. Compounds of Interest in Nuclear Reactor Technology—AIME Nuclear Metallurgy*, **10**, 225
181. Naoumidis, A. and Stöcker, H. J. (1967), *Proc. Brit. Ceram. Soc.*, **8**, 193
182. Olson, W. M. and Mulford, R. N. R. (1966). *J. Phys. Chem.*, **70**, 2932
183. Lapage, R. and Bunce, J. L. (1967). *Trans. Faraday Soc.*, **63**, 1889
184. Campbell, G. M. (1969). *J. Phys. Chem.*, **73**, 350
185. Campbell, G. M. and Leary, J. A. (1966). *J. Phys. Chem.*, **70**, 2703
186. Kent, R. A. and Leary, J. A. (1969). *High Temp. Sci.*, **1**, 169
187. Marcon, J. P. and Poitreau (1970). *J. Inorg. Nucl. Chem.*, **32**, 463
188. Counsell, J. F., Dell, R. M., Junkison, A. R. and Martin, J. F. (1967). *Trans. Faraday Soc.*, **63**, 72
189. Stalinski, B., Bieganski, Z. and Troć, R. (1966). *Phys. Stat. Sol.*, **17**, 857
190. Brugger, J. E. (1966). *Rep.* ANL 7175
191. Gingerich, K. A. and Aronson, N. S. (1966). *J. Phys. Chem.*, **70**, 2577
192. Gingerich, K. A. (1964). *Intern. Symp. High Temperature Technology*, 557 (Washington D.C.: Butterworth)
193. Gingerich, K. A. (1964). *Rep.* NYO-2541-1
194. O'Hare, P. A. G. and Hubbard, W. N. (1966). *Trans. Faraday Soc.*, **62**, 2709
195. Gingerich, K. A. and Lee, P. K. (1964). *J. Chem. Phys.*, **40**, 3250
196. Gingerich, K. A. (1966). *Naturwiss.*, **53**, 525
197. Reishus, J. W. and Gundersen, G. E. (1967). *Rep.* ANL-7375
198. Dogu, A., Val, C. and Accary, A. (1968). *J. Nucl. Mater.*, **28**, 271
199. Harness, J. B., Mathews, J. C. and Morton, N. (1964), *Brit. J. Appl. Phys.*, **15**, 963
200. Westrum, E. F., Jr., Takahashi, Y. and Stout, N. D. (1965). *J. Phys. Chem.*, **69**, 1520
201. Takahashi, Y. and Westrum, E. F., Jr. (1964). *J. Chem. Eng. Data*, **10**, 128
202. Andon, R. L. J., Counsell, J. F., Martin, J. F. and Hedger, H. J. (1964). *Trans. Faraday Soc.*, **60**, 1030
203. Westrum, E. F., Jr., Suits, E. and Lonsdale, H. K. (1965). *Third Symp. Thermophysical Properties*, Lafayette, Ind., 156
204. Farr, J. D., Witteman, W. G., Stone, P. L. and Westrum, E. F., Jr. (1965). *Third Symp. Thermophysical Properties*, Lafayette, Ind., 162
205. De Combarieu, A., Costa, P. and Michel, J. C. (1963). *C. R. Acad. Sci. Paris*, **256**, 5518
206. Levinson, L. S. (1964). *Carbides in Nuclear Energy*, **1**, 429 (London: Macmillan)
207. Harrington, L. C. and Vozella, P. A. (1964). *Carbides in Nuclear Energy*, **1**, 342 (London: Macmillan)
208. Leitnaker, J. M. and Godfrey, T. G. (1967). *J. Nucl. Mater.*, **21**, 175
209. Mukaibo, T., Naito, K., Sato, K. and Uchijma, T. (1962). *Symposium on Thermodynamics of Nuclear Materials*, 645 (Vienna: IAEA)

210. Boettcher, A. and Schneider (1958). *2nd Intern. Conf. Peaceful Uses At. Energy*, **6**, 561 (Geneva: United Nations)
211. International Atomic Energy Agency (1963). *Tech. Rep. Series* No. 14 (Vienna: IAEA)
212. Levinson, L. S. (1963). *J. Chem. Phys.*, **38**, 2105
213. Sandenaw, T. A. and Gibney, R. B. (1970). *Plutonium 1970—AIME Nuclear Metallurgy*, **17**, 104
214. Akhachinskii, V. V. (1968). *Panel Meeting on Uranium Carbides and Plutonium Carbides* (Vienna: IAEA)
215. Kruger, O. L. and Savage, H. (1964). *J. Chem. Phys.*, **40**, 3324
216. Huber, C. E., Jr., Holley, C. E., Jr. and Krikorian, N. H. (1968). *J. Chem. Eng. Data*, **13**, 256
217. Egan, J. J. (1964). *J. Phys. Chem.*, **68**, 978
218. Satow, T. (1967). *J. Nucl. Mater.*, **21**, 249
219. Aronson, S. (1964). *Intern. Symp. Compounds of Interest in Nuclear Reactor Technology*, 247 (Metallurgical Soc. of AIME)
220. Rudy, E. (1962). *Symp. Thermodynamics of Nuclear Materials*, 243 (Vienna: IAEA)
221. Rudy, E. (1963). *Z. Metalkunde*, **54**, 112
222. Aronson, S. and Safodsky (1965). *J. Inorg. Nucl. Chem.*, **27**, 1769
223. Satow, T. (1967). *J. Nucl. Mater.*, **21**, 255
224. Farr, J. D., Huber, E. J., Jr. and Holley, C. E., Jr. (1959). *J. Phys. Chem.*, **63**, 1455
225. Storms, E. K., Huber, E. J., Jr. (1967). *J. Nucl. Mater.*, **23**, 19
226. Huber, E. J., Jr., Head, E. L. and Witteman, W. G. (1969). *J. Chem. Thermodyn.*, **1**, 579
227. Huber, E. J., Jr., Head, E. L. and Holley, C. E., Jr. (1963). *J. Phys. Chem.*, **67**, 1730
228. Huber, C. E., Jr. and Holley, C. E., Jr. (1962). *Symp. Thermodynamics of Nuclear Materials*, 581 (Vienna: IAEA)
229. Leitnaker, J. M. and Godfrey, T. G. (1966). *J. Chem. Eng. Data*, **11**, 392
230. Piazza, J. R. and Sinnott, M. J. (1962). *J. Chem. Eng. Data*, **7**, 451
231. Robinson, W. C. and Chiotti, P. (1964). *Rep.* TID-4500
232. Behl, W. K. and Egan, J. J. (1966). *J. Electrochem. Soc.*, 376
233. Vozella, P. A., Miller, A. D. and DeCrescente, M. A. (1968). *J. Chem. Phys.*, **49**, 876
234. Marcon, J. P., Poitreau, J. and Roullet, G. (1970). *Plutonium 1970—AIME Nuclear Metallurgy*, **17**, 799
235. Johnson, G. K., Van Deventer, E. H., Kruger, O. L. and Hubbard, W. N. (1970). *J. Chem. Thermodyn.*, **2**, 617
236. Campbell, G. M., Kent, R. A. and Leary, J. A. (1970). *Plutonium 1970—AIME Nuclear Metallurgy*, **17**, 781
237. Campbell, G. M., Mullins, L. J. and Leary, J. A. (1967). *Symp. Thermodynamics of Nuclear Materials*, 75 (Vienna: IAEA)
238. Olson, W. N. and Mulford, R. N. R. (1967). *Symp. Thermodynamics of Nuclear Materials*, 467 (Vienna: IAEA)
239. Harris, P. S., Phillips, P. A., Rand, M. H. and Tetenbaum, M. (1967). *Rep.* AERE-R-5353
240. Battles, J. E., Shinn, W. A., Blackburn, P. E. and Edwards, R. K. (1970). *High Temp. Sci.*, **2**, 80
241. Kent, R. A. (1969). *High Temp. Sci.*, **1**, 169

6
Actinide Chalcogenides and Pnictides

R. M. DELL and N. J. BRIDGER

Atomic Energy Research Establishment, Harwell

6.1   INTRODUCTION

The decade 1960–1970 saw the development of a world-wide interest in the compounds formed between the actinide metals, especially uranium and plutonium, and the non-metals of Groups Vb and VIb of the Periodic Table. During this period several hundred papers were published describing the preparation of the compounds, their crystallography and ranges of composition and their chemical, thermodynamic and physical properties. The past 1–2 years has, however, seen a decline in the amount of research being undertaken in this field. Although different workers were motivated by different considerations, two general themes run through much of this research: (1) the desire to understand the nature of the chemical bonding and, for the metallic conductors, the electronic band structure in these compounds and (2) the technological interest in the possible application of appropriate compounds, particularly the nitrides, as nuclear fuels.

The title selected for this review deserves some explanation. The use of the term 'chalcogenide' to describe metallic sulphides, selenides and tellurides leaves something to be desired, for oxides are not included in this review, while the acronym 'pnictide' for the corresponding Group Vb compounds (*phosphorus, nitrogen, arsenic* and *antimonide*) is even less pleasing and we have included bismuthides in this group. Nevertheless, when discussing the compounds as a class it is necessary to adopt some collective noun and it seems a lesser evil to continue with such terms as are already widely employed in the literature than to invent new ones.

Of the seven actinide elements from actinium to americium, only two (Th and U) can be handled without the requirement for glovebox containment and only two others (Np and Pu) are readily available in reasonable quantities. These factors, together with the technological interest in U and Pu compounds, explain why almost all of the work to date has been on the compounds of Th, U and Pu; quite recently one or two papers on the compounds of higher actinides and of neptunium have appeared.

It is intended in this review to deal only with the fundamental chemical and physical properties of these substances. There exists also an extensive technological literature, particularly for the nitrides, on such subjects as fabrication and sintering of compacts and ceramic shapes, mechanical properties (tensile strength, hardness, creep, etc.), compatibility with reactor components (steel, sodium etc.) and behaviour under irradiation. These and related subjects are deemed to be beyond the scope of the present article. Thermodynamic properties are also excluded as these are treated in Professor Fuger's review. (Chapter 5.)

6.2   GENERAL PREPARATIVE PROCEDURES

These compounds are commonly made by reaction of the metal with the non-metal or one of its gaseous derivatives, generally the hydride (H_2S, NH_3 etc.). Direct reaction with the massive metal presents kinetic problems and usually results in the formation of a protective surface coating of product. It is therefore conventional first to convert the metal to a fine powder by taking it through an intermediate hydriding-dehydriding cycle, in which the actinide metals react with hydrogen at moderate temperatures to form finely divided metallic hydrides; on raising the temperature and outgassing, these hydrides decompose to yield reactive metal powders, uncontaminated by the oxide layer normally present on metal powders which have been exposed to air. In some cases the non-metal will react with the metallic hydride directly.

Thorium reacts with hydrogen at 650 °C to form ThH_2, but without the desired disintegration of the structure; on cooling to 250 °C, further hydridation occurs, with disintegration, to yield Th_4H_{15} powder. This decomposes in vacuum at 700 °C to give finely divided thorium metal. Uranium metal, in contrast, reacts with hydrogen at 250 °C to form UH_3 powder in one step; this hydride subsequently decomposes when heated above 300 °C *in vacuo*. Plutonium metal reacts with hydrogen below 200 °C to form PuH_{3-x}, which loses hydrogen with difficulty when heated to 400 °C in vacuum[1]; at higher temperatures metal sintering occurs (m.p. Pu = 639 °C). Although this method of producing reactive plutonium metal is not entirely successful, some improvement can be achieved by repeating the hydriding cycle with intermediate crushing of the powder. The actinide hydrides and the finely divided metals are both reactive to air at room temperature and may ignite spontaneously on exposure to the atmosphere. Phase rule data on the $Np—H_2$ system and the $Am—H_2$ system have been reported by Mulford and co-workers[2, 3].

6.2.1 General methods of preparing the higher chalcogenides

6.2.1.1 Gaseous reaction

The reaction of H_2S with finely divided uranium or UH_3 is the preferred route to US_2. With plutonium, the analogous reaction is more difficult to effect quantitatively because of the greater stability of plutonium hydride. Aronson[4] reports the ready reaction of thorium with H_2S to give ThS_2. The use of H_2Se to give higher selenides is less favoured, although USe_2 has been prepared in this way[5]. There is no report of H_2Te being used to form tellurides.

6.2.1.2 Direct reaction between the solid elements

The direct reaction between the solid elements is usually strongly exo-thermic and the heat generated can both sinter the reactive metal and volatilise the chalcogen; to compensate for this, excess of the chalcogen is generally employed. An extreme instance of the direct reaction method is when the elements are arc-melted together. This route is generally used for tellurides, antimonides and bismuthides.

6.2.1.3 The Faraday method

This method is widely used for the preparation of higher chalcogenides. Weighed quantities of the finely divided metal and the chalcogen are placed at opposite ends of a silica tube which is evacuated and sealed. The tube is then heated to the temperature required to give one atmosphere vapour pressure of the chalcogen and reaction takes place between the metal and the vapour. This avoids the rapid evolution of heat as in direct reaction of the mixed elements which can lead to pyrophoric conditions. The method has the advantage that the overall stoichiometry (x/M ratio) can be pre-determined without loss of either element and, with careful selection of the temperatures employed, intermediate compositions can be formed. The practical upper limit of temperature is around 900 °C since above this temperature there may be contamination by reaction with the silica container. Examples of this method are the preparation of βUS_2 by the reaction of finely divided uranium with sulphur vapour at 400 °C[6] and the preparation of USe_2[7] and $ThTe_2$[8] at 500 °C and 600 °C respectively.

6.2.2   General methods of preparing the pnictides

6.2.2.1   Gaseous reaction

The reaction between nitrogen on ammonia and finely divided thorium, uranium or plutonium (or their hydrides) is the most widely used method for the preparation of the higher nitrides on the laboratory scale. Reaction occurs readily below 700 °C to give Th_3N_4, $UN_{1.7}$ or PuN. Phosphine reacts readily with the finely divided metals or hydrides[6] to give, for example, Th_3P_4 or UP_2. However, phosphine presents purification problems, and there are potential explosion hazards[9] which make it a dangerous reagent for preparing phosphides of radioactive elements. There is only one recorded instance of the use of arsine to prepare uranium arsenide[10].

6.2.2.2   Direct reaction of the elements

The Faraday method is used extensively to prepare the higher phosphides, while the remaining pnictides are generally prepared by direct reaction of the elements either in a sealed tube or even in an open crucible for the bismuthides. Arc melting has also been employed successfully.

6.2.3   Degradation

The methods described so far usually yield compounds of high x/M ratio; in order to prepare phases of lower x/M ratio, these compounds can be degraded in various ways.

6.2.3.1   Vacuum and gaseous degradation

When the higher composition phases are heated *in vacuo* or in a flow of argon, they decompose with vaporisation of the non-metal. This method is useful in preparing compounds of intermediate composition and, in some cases, the mono-compound can be formed. For instance, $UN_{1.7}$ degrades in flowing argon at 1450 °C to give $UN_{1.00}$, $ThN_{1.33}$ degrades in a vacuum at 1500 °C to give $ThN_{1.00}$ and UP_2 can be successively degraded to U_3P_4 and UP in vacuum. Some workers find degradation in a hydrogen flow to be beneficial, as this can lower the temperature necessary to cause the required decomposition.

6.2.3.2 Homogenisation

In some cases the preceding method of degradation is not suitable for the preparation of intermediate compounds or, especially, of the mono-phases. High temperatures are often necessary and there is the risk of forming a substoichiometric compound. In other cases degradation may result in a stable phase of higher composition than that required; PuS_2, for example, cannot be degraded below Pu_3S_4 in a vacuum without vaporising the entire specimen. Reduction of the higher composition phase with the parent metal or its hydride is then a useful alternative method.

$$MX_2 + M \rightarrow 2MX$$

Temperatures up to 1800 °C may be required, depending upon the reaction. Such temperatures are conveniently attained using a radio frequency induction heater, the reaction being conducted in a tantalum or tungsten crucible. This method gives good control over the stoichiometry of the phase being prepared. Some workers have used aluminium as the reductant but this can lead to contamination with aluminium and is not a reliable alternative for preparing pure phases.

6.2.4 Oxygen impurities

Throughout this field of research the role of oxygen is crucial. Once introduced during the preparative stages it cannot be removed; every care should therefore be taken to select starting materials which do not contain oxygen as a gross impurity and to exclude air and water vapour during the preparation. Some of the compounds react with air at room temperature and have to be handled in an inert atmosphere. The most reactive compounds, for example PuN, behave as strong desiccants even at room temperature.

Oxygen impurities are found generally as 'oxy'-compounds in the chalcogenides (e.g. UOS, Pu_2O_2S) and as impurity oxide phases in the pnictides. These second phase impurities are frequently located at the grain boundaries in sintered specimens. There is x-ray evidence that some compounds may dissolve oxygen into the lattice with expansion of the unit cell.

6.2.5 Other preparative routes

The production and purification of the actinide metals is difficult on the laboratory scale, particularly for the rarer actinides, and industrially the use of the metal would be expensive in the manufacture of nuclear fuels. Preparative routes which avoid the use of the metals include the reduction of uranium and plutonium oxides with carbon and the simultaneous reaction with nitrogen or phosphorus. It is difficult to obtain stoichiometric compounds this way and residual oxygen and carbon contamination is usually higher than desired.

6.2.6   Single crystal growing

For many physical property measurements it is desirable to have single crystals of these compounds; these are grown from both the melt and the vapour. Large crystals are best prepared from the melt. Aside from the usual slow cooling techniques, two variants have been developed with considerable success for particular compounds.

(a) Large single crystals of UN, suitable for neutron diffraction studies, were extracted from an ingot of this material prepared by an arc-casting technique, using consumable uranium electrodes under 20 atm. pressure of nitrogen gas[11]. Single crystals of ThN have recently been prepared in a similar way[12].

(b) Uranium monosulphide US, unlike UN, melts without decomposing (m.p. 2460 °C). Single crystals of this phase have been prepared by taking a melt of composition $US_{1.10-1.15}$ at 1700 °C and allowing sulphur to vaporise slowly. Crystals of US then nucleate and grow in the melt as the overall S/U ratio slowly decreases[13].

The more general method of crystal growing for these compounds, albeit yielding smaller crystals, is from the vapour phase by a chemical transport technique. The polycrystalline compound is sealed in a silica tube along with a small quantity of halogen (generally iodine, sometimes bromine). A temperature gradient of, say, 50 °C is set up along the tube. The compound reacts with the halogen to yield an equilibrium vapour containing actinide metal halides as well as the non-metal species. At the far end of the tube, where the temperature is different, this vapour composition is no longer in equilibrium and reacts back to nucleate the solid phase which grows as small single crystals. The composition of the crystals seems to depend more upon the experimental conditions than upon the composition of the initial powder. Single crystals of, for example, U_3P_4, U_3As_4, U_3Se_5, UTe_2 have all been made by this technique. The method is not normally suitable for the mono-compounds MX because silica cannot be used at the high temperatures where these are the stable phases.

6.2.7   General analytical procedures

Most of the published work on these compounds includes little or no accurate analytical results. Many workers rely on weight changes recorded during preparation of their samples, or ignition of the compounds to oxide to determine the metal content. These are adequate procedures if a single phase is present and there is no oxygen contamination, but are not adequate for more complex situations or for accurate determinations of stoichiometry.

In the analysis of the nitrides of thorium, uranium and plutonium[14] the nitrogen contents were measured by the Dumas Method and by a modified form of the Kjeldhal method and the metal was determined by ignition to the oxide and, in the case of plutonium, by controlled potential coulometry[15]. Accurate procedures for the analysis of uranium phosphides[16] and uranium sulphides[17] have been developed in connection with the characterisation of specimens for heat capacity measurements.

Oxygen impurity in these compounds is determined absolutely by vacuum or inert gas fusion analysis. Where solubility of oxygen in the compound is small (e.g. UN, UP) some idea of the oxygen content (above 0.2 wt %) is obtained by estimating the UO_2 content from x-ray diffraction measurements.

6.3 CHALCOGENIDES: PREPARATION, CRYSTAL CHEMISTRY AND STOICHIOMETRY

The crystal structures of the sulphides and selenides of any one actinide element are generally similar, with a rather greater distinction existing between these and the tellurides. On the other hand, the resemblance between corresponding compounds of one actinide metal and the next is much less marked than in the rare earth series. It is therefore convenient, from a crystallographic viewpoint, to consider in turn the compounds formed by each of the principal actinide elements.

6.3.1 Thorium

There have been surprisingly few structural investigations of the thorium chalcogenides and most of these are prior to 1960. In early work[18, 19], thorium was reported to form a disulphide, ThS_2, and three lower sulphides, Th_7S_{12}, Th_2S_3 and ThS. Subsequently[20] a polysulphide, $ThS_{2.5}$ was obtained by heating the metal with excess sulphur at 400 °C; this decomposed to ThS_2 on heating in vacuum to 900 °C. The crystallographic data are summarised in Table 6.1 together with values for the corresponding selenides. The existence of the polysulphide $ThS_{2.5}$ has been confirmed recently[21].

Graham and McTaggart[20] believed that the lower sulphides, which they were unable to prepare using high purity crystal bar thorium, were oxygen-stabilised phases. While it is true that the difficulties of avoiding contamination by oxygen have only gradually become fully appreciated and the earlier data must be treated with reservation, it seems unlikely that the lower phases are all oxygen-stabilised. In a recent thermochemical study of the Th—S system[4], the phases Th_7S_{12}, Th_2S_3 and ThS were all observed, while ThS has been synthesised in a pure state, its lattice parameter being a = 5.68 ±0.01 Å[4, 22, 23]. It seems likely that under appropriate conditions the lower thorium sulphides can be prepared, even in the absence of oxygen. The range of homogeneity of ThS has not been established with certainty, although the lattice parameter is reported to vary from 5.673 to 5.690[18].

In a detailed x-ray powder diffraction study of the selenides, D'Eye[25] determined the structures and space groups of $ThSe_2$ and Th_7Se_{12}. There remains some doubt as to whether Th_7Se_{12} and Th_2Se_3 are oxygen stabilised phases and also the ranges of composition over which they exist.

The thorium tellurides do not appear to be isostructural with the other chalcogenides. From x-ray and chemical analysis, phase rule considerations and tensimetric studies D'Eye and Sellman[8] reported a polytelluride $ThTe_{2.66}$, a ditelluride $ThTe_2$ and a monotelluride $ThTe$. The latter possessed the b.c.c. CsCl structure with $a = 3.83$Å. In contrast to ThS and ThSe which melt above 1800 °C, ThTe may be degraded well below 1000 °C *in vacuo*, a

Table 6.1   Crystallographic data for the thorium sulphides and selenides

Type	Sulphides		Selenides	
	Eastman et al.[18] Zachariasen[19]	Graham and McTaggart[20]	D'Eye et al.[24, 25]	Graham and McTaggart[20]
Poly-		$ThS_{2.5}$ tetragonal $a = $ 5.43 Å $c = 10.15$ Å	$ThSe_{2.33}$	$ThSe_{2.5}$ tetragonal $a = $ 5.629 Å $c = 10.764$ Å
ThX_2	ThS_2, orthorhombic $PbCl_2$ type $a = 4.268$ Å $b = 7.264$ Å $c = 8.617$ Å	ThS_2, orthorhombic $PbCl_2$ type $a = 4.283$ Å $b = 7.275$ Å $c = 8.617$ Å	$ThSe_2$, orthorhombic $PbCl_2$ type $a = 4.42$ Å $b = 7.61$ Å $c = 9.07$ Å	$ThSe_2$, orthorhombic $PbCl_2$ type $a = 4.435$ Å $b = 7.629$ Å $c = 9.085$ Å
Th_7X_{12}	Th_7S_{12}, hexagonal $a = 11.063$ Å $c = 3.991$ Å	Not observed	Th_7Se_{12}, hexagonal $a = 11.57$ Å $c = 4.23$ Å	Not observed
Th_2X_3	Th_2S_3 (small range) orthorhombic Sb_2S_3 type† $a = 10.99$ Å $b = 10.85$ Å $c = 3.96$ Å	Not observed	Th_2Se_3 (small range) orthorhombic Sb_2S_3 type $a = 11.34$ Å $b = 11.57$ Å $c = 4.27$ Å	Not observed
ThX	$ThS_{1.0}$ Cubic NaCl type $a = 5.68$ Å	Observed only as minor phase	$ThSe_{1.0}$ Cubic NaCl type $a = 5.875$ Å	Observed only as minor phase

†Single Crystal Weissenberg study

property which stems from its different crystal structure. Graham and McTaggart[20] report $ThTe_3$ with structural similarities to the trichalcogenides of Ti, Zr and Hf, $ThTe_2$ (hexagonal?) and Th_2Te_3 (hexagonal, $a = 12.49$Å, $c = 4.35$Å). ThTe was stated to be stabilised by oxygen but was not prepared pure. There is evidently scope for a further investigation of the Th—Te system.

The three oxychalcogenides ThOS, ThOSe and ThOTe are isostructural with each other, crystallising with the tetragonal PbFCl type structure with two molecules per unit cell[8]. The unit cell parameters are as shown.

	ThOS[19]	ThOSe[24]	ThOTe[8]
a (Å)	3.96	4.04	4.12
c (Å)	6.75	7.02	7.56*
c/a	1.70	1.74	1.84

*In reference[8] the c parameter of ThOTe is printed, erroneously, as 9.544 KX units.

The chemical stability of the thorium chalcogenides and oxychalcogenides decreases in the order sulphides > selenides > tellurides. For example, ThS_2 melts unchanged at 1905 °C, $ThSe_2$ degrades at 1000 °C before melting, while $ThTe_2$ decomposes well below 1000 °C. This decrease in stability is attributable to increasing size of the anions.

Single crystals of the thorium oxychalcogenides have been grown by halogen transport in a silica tube[26].

6.3.2 Uranium

6.3.2.1 Uranium sulphides

In the first detailed structural study of the sulphides Zachariasen[19] reported three phases, US_2, U_2S_3 and US. The structures of U_2S_3 and US were determined, the former from a single crystal Weissenberg study on a crystal weighing only 0.2 microgramme. Later Picon and Flahaut[27] reported a trisulphide US_3, three polymorphic forms of US_2 and the lower sulphides U_3S_5, U_2S_3 and US.

The starting point for the preparation of the various uranium sulphides is generally βUS_2. This is obtained by reacting powdered uranium metal with either H_2S gas or with sulphur vapour[6, 27]. Alternative preparations are by passing H_2S gas through a fused salt solution of UCl_4 in NaCl—KCl eutectic so as to precipitate βUS_2[28]:

$$UCl_4 + 2H_2S \rightarrow US_2 + 4HCl$$

or by heating a mixture of uranium tetrachloride and aluminium powder in H_2S at 350–550 °C[29].

$$3UCl_4 + 4Al + 6H_2S \rightarrow 3US_2 + 4AlCl_3 + 6H_2$$

Uranium trisulphide may be prepared by reacting US_2 with sulphur at 600–800 °C in a sealed tube for several days[27]. As this compound is paramagnetic it is clearly a polysulphide. The degradation of US_2 in vacuum takes place only at $\geqslant 1500$ °C to yield first U_3S_5 and then U_2S_3. The monosulphide US cannot conveniently be prepared by thermal decomposition and is usually produced by annealing mixtures of $US_2 + U$ in the appropriate ratio at ~ 1800 °C[22].

The U—S system has been reinvestigated[7]; the existence and structures of US_3, U_3S_5 and US were confirmed but U_2S_3 was not observed. This study also showed that of the three so-called 'polymorphs' of US_2 only one (the β phase) truly possessed this composition. The α phase is a defective structure with a range of composition between about $US_{1.80}$ and $US_{1.93}$, while the exact composition of the low temperature γ phase remains in doubt. Studies at Harwell have confirmed the U_2S_3 diffraction pattern as a phase formed during the reaction of βUS_2 with uranium metal at $>1600\,°C$; as this temperature range was not reached by Grønvold[7] it is hardly surprising that he did not observe U_2S_3. The most reliable crystallographic data for the uranium sulphides are summarised in Table 6.2. In addition to the phases shown, there is one report[21] of a second polysulphide U_2S_5 isostructural with Th_2S_5.

A tentative phase diagram for the U—S system, based on the work of the above authors, has been published[40] (Figure 6.1). While certain points of

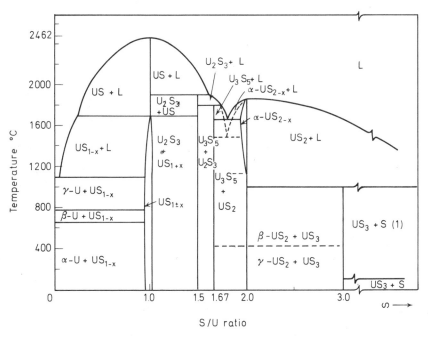

Figure 6.1 Phase diagram of the U–S system. (From Cordfunke[40], by courtesy of Elsevier Publishing Co.)

detail remain in doubt, the diagram is believed to be essentially correct. αUS_{2-x} is shown as a substoichiometric phase formed at high temperatures, while U_3S_5 and U_2S_3 are represented as stoichiometric phases decomposing at 1800 °C and 1950 °C respectively. The monosulphide is known to exhibit a narrow range of composition ($US_{0.96}$—$US_{1.01}$), with the stoichiometric compound melting at $2462 \pm 30\,°C$.

Single crystals of βUS_2 of a size suitable for physical property measurements have been made by the iodide transport method; polycrystalline US_2 was reacted with iodide in a sealed tube at 940 °C and crystals of US_2 were

Table 6.2 Crystallographic data for the uranium chalcogenides

Type	Sulphides	Selenides	Tellurides
Poly	US_3^*, monoclinic[7] $a = 5.37$ Å $b = 3.96$ Å $\left.\right\}\beta = 97.2°$ $c = 9.06$ Å (ZrSe$_3$ structure)	USe_3, monoclinic[7] $a = 5.65$ Å $b = 4.06$ Å $\left.\right\}\beta = 97.5°$ $c = 9.55$ Å	UTe_5 [35, 36] UTe_3 [37] (ZrSe$_3$ structure) U_3Te_7
γUX_2	γUS_2, hexagonal[27] $a = 7.25$ $c = 4.07$	γUSe_2, hexagonal[5] $a = 7.68$ $c = 4.21$	
βUX_2	US_2, orthorhombic[7] $a = 4.12$ $b = 7.12$ $c = 8.48$ (PbCl$_2$ structure)	USe_2, orthorhombic[5] $a = 4.26$ $b = 7.46$ $c = 8.98$ (PbCl$_2$ structure)	UTe_2, orthorhombic[37, 38] $a = 4.16$ $b = 6.13$ $c = 13.97$ Space group Immm (new structure)
UX_{2-x} ('α UX$_2$')	$US_{1.9}^*$, tetragonal[7, 30] $US_{1.80}$ $US_{1.95}$ $a = 10.30$ 10.26 $c = 6.36$ 6.34 (Fe$_2$B structure)	$USe_{1.9}$, tetragonal[7] $USe_{1.80}$ $USe_{2.0}$ $a = 10.77$ 10.70 $c = 6.67$ 6.61	UTe_{2-x}, hexagonal[37, 39] $a = 12.31$ $c = 4.24$ (Th$_7$S$_{12}$ structure)
U_3X_5	$US_{1.60}^*$, orthorhombic[31] $a = 7.43$ $b = 8.11$ $c = 11.76$	U_3Se_5, orthorhombic[5] $a = 7.76$ $b = 8.44$ $c = 12.26$	U_3Te_5 [35, 36]
U_2X_3	U_2S_3, orthorhombic[27] $a = 10.36$ $b = 10.60$ $c = 3.86$ (Sb$_2$S$_3$ structure)	U_2Se_3, orthorhombic[5] $a = 11.33$ $b = 10.94$ $c = 4.06$ (Sb$_2$S$_3$ structure)	U_2Te_3, b.c.c. $a = 9.396$[34] 9.406[37] (Th$_3$P$_4$ structure)
U_3X_4		U_3Se_4, b.c.c.[5] $a = 8.80$ (Th$_3$P$_4$ structure)	U_3Te_4, b.c.c. $a = 9.398$[34] 9.416[37] (Th$_3$P$_4$ structure)
UX	US, f.c.c.[22] $a = 5.49$ (NaCl structure)	USe, f.c.c.[5, 32, 33, 34] $a = 5.66 - 5.76$ (NaCl structure)	UTe, f.c.c.[32, 37] $a = 6.16$ (NaCl structure)

*Single crystal determinations

deposited at the cool end of the tube at 700 °C[41]. Single crystals of US have been grown from a melt of composition $US_{1.15}$ as described above.

6.3.2.2 Uranium selenides

The uranium–selenium system has been investigated by several groups of workers[5, 7, 32, 33] and there is general agreement on the existence of a poly-selenide USe_3 with monoclinic symmetry and a monoselenide USe with the rocksalt structure*. Between these end members, intermediate selenides isostructural with all the corresponding sulphides have been described by one or more authors and it seems likely that under appropriate conditions all these compounds may be formed. In addition, Khodadad[5] prepared a U_3Se_4 phase (b.c.c. Th_3P_4 structure) by degrading higher selenides in vacuum at 1350–1450 °C. Although this phase has not been reported by others and the corresponding sulphide is unknown, there seems no reason to doubt its existence. The known selenides are listed in Table 6.2 A single crystal of USe has been prepared by casting granular USe powder in a tantalum crucible under argon at 1850–2000 °C[34]. There is some disagreement concerning the lattice parameter of the cubic USe phase, values of 5.66Å[5], 5.71Å[34], 5.75Å[32] and 5.76Å[33] having been reported. This is a considerably wider range than observed for US and may indicate a wider compositional range for the mono-selenide. Recently, single crystals of U_3Se_5 have been prepared by the iodine transport method[42].

6.3.2.3 Uranium tellurides

The uranium-tellurium system is still rather confused. Originally Ferro[32] reported five phases — a polytelluride, UTe_2 (tetragonal symmetry), U_2Te_3 and U_3Te_4 (b.c.c. Th_3P_4 structure) and UTe (NaCl structure). Later investigators agree on the existence of these compounds, although recent work[38] shows that UTe_2 possesses orthorhombic symmetry and it seems likely that the compound reported by Ferro as UTe_2 was, in fact, $UOTe$. Russian authors[35, 36] have investigated the polytellurides and find a compound UTe_5 which decomposes peritectically to form UTe_3 and a molten U—Te alloy. On the basis of x-ray analysis and differential thermal analysis they present a phase diagram for the binary system extending over the composition range 58–100 atom % Te. X-ray powder data are reported for UTe_3, U_3Te_4, U_2Te_5, U_3Te_7, UTe_2, U_4Te_7, U_3Te_5 and U_2Te_3, although on the basis of other workers' results some of these must be regarded as speculative. Single crystals of UTe_3 and UTe_2 were grown by chemical transport methods[43]. Several different researchers agree that the phase with the Th_3P_4 structure extends over the compositional range U_3Te_4 to U_2Te_3 [32, 34, 37] (cf. plutonium tellurides, below). Recently Breeze et al.[37] reported U_7Te_{12} with the hexagonal Th_7S_{12} structure, and this structure has been corroborated[39]. Uranium monotelluride (UTe) has the rocksalt structure, unlike $ThTe$. It is difficult to prepare in a pure state as reaction of the higher tellurides

*Grønvold[7] alone failed to find USe, as the temperatures he employed were too low.

with uranium powder is slow below 1400 °C, while above this temperature UTe decomposes peritectically into U_3Te_4 and liquid uranium metal[39]. On freezing, recombination to UTe does not take place readily and so the preparation of crystals of UTe may prove to be difficult.

The data for the U—Te system in Table 6.2 represent the reviewers' assessment of the situation at present, although it must be admitted that some subjective judgement is involved and further investigation is required. Comparing the three chalcogenide systems it will be seen that for uranium, as for thorium, the crystal chemistry of the selenides bears a close resemblance to that of the sulphides, with the tellurides behaving rather differently.

6.3.2.4 Uranium oxychalcogenides

The oxychalcogenides UOS, UOSe, and UOTe are isostructural with each other and with the corresponding thorium compounds (tetragonal PbFCl structure). Their lattice parameters are as shown:

	UOS[27]	UOSe[44]	UOTe[45, 46]
a (Å)	3.84	3.90	4.00
c (Å)	6.68	6.98	7.49
c/a	1.74	1.79	1.87

These oxychalcogenides are readily prepared by annealing a higher chalcogenide with the appropriate quantity of UO_2, for example $US_2 + UO_2 \rightarrow 2UOS$. Single crystals have been grown by halogen transport[26]. Recently two further compounds have been reported, a new uranium oxysulphide crystallising in the cubic system[47] ($a_0 = 6.06$ Å), which is considered to be a substituted UO_2 with some of the oxide ions replaced by sulphide, and a second uranium oxytelluride of formula U_2O_2Te, crystallising in the tetragonal system ($a = 3.96$ Å, $c = 12.50$ Å) and isostructural with the rare earth oxytellurides.

6.3.3 Plutonium

The chemistry of plutonium resembles that of cerium in many respects, the similarity being most marked in the crystal chemistry of the chalcogenides. This was first noted by Zachariasen[48] who drew attention to the close structural relationship between the sulphides and oxysulphides of plutonium and cerium. More recent studies have confirmed and extended this observation.

There have been two major crystallographic investigations of the Pu—S and Pu—Se system, one by Marcon and Pascard[49-51] and the other by Allbutt et al.[52]. In addition a more restricted study of the Pu—S system was made by Kruger and Moser[1]. Broadly speaking the findings of these three investigations are in very good agreement although they differ in points of detail.

Allbutt *et al.* also made a brief study of the plutonium tellurides and showed that there were many points of structural similarity between all three Pu–chalcogen systems (Table 6.3).

The dichalcogenides PuS_2, $PuSe_2$ and $PuTe_2$ are prepared by the Faraday method. No compounds of higher X/Pu ratio than 2.0 have been prepared. The dichalcogenides tend to decompose on heating with loss of the non-metal, leaving a non-stoichiometric PuX_{2-x} phase of tetragonal Fe_2As

Table 6.3   Crystallographic data for the plutonium chalcogenides

Type	Sulphides	Selenides	Tellurides
PuX_2	PuS_2, monoclinic[50] (pseudo-tetragonal) $a = b = 7.96$ Å $c = 3.98$ $\beta = 90°$ (CeSe$_2$ structure)	$PuSe_{1.99}$[52] (pseudo-tetragonal) $a = 4.13$ Å $c = 8.34$ Å $c/a = 2.019$	$PuTe_2$[52] (pseudo-tetragonal) $a = 4.39$ $c = 8.94$ $c/a = 2.036$
PuX_{2-x}	$PuS_{1.8-1.9}$, tetragonal[49, 50] $a = 3.94$ $c = 7.96$ $c/a = 2.02$ (Fe$_2$As structure)	$PuSe_{1.8}$, tetragonal[52] $a = 4.09$ $c = 8.36$ $c/a = 2.045$ (Fe$_2$As structure)	$PuTe_{1.8}$, tetragonal[52] $a = 4.33$ $c = 8.98$ $c/a = 2.073$ (Fe$_2$As structure)
Pu_2X_3	αPu_2S_3, orthorhombic[52] $a = 7.39$ $b = 15.32$ $c = 3.98$ (αCe_2S_3 structure)	ηPu_2Se_3, orthorhombic[49] $a = 4.10$ $b = 11.10$ $c = 11.32$ (Sb$_2$S$_3$ structure)	Complex phase[52] $-$ not indexed
βPu_2X_{3-x}	Pu_2S_{3-x}, tetragonal[51] $a = 14.90$ $c = 19.78$ (Ce$_5$S$_7$ structure)		
γPu_2X_3	Pu_2S_3, cubic[51, 52] $a = 8.46$ (Th$_3$P$_4$ structure)	Pu_2Se_3, cubic[49, 52] $a = 8.80$ (Th$_3$P$_4$ structure)	Pu_2Te_3(?), cubic[52] $a = 9.355$ (Th$_3$P$_4$ structure)
Pu_3X_4	Pu_3S_4, cubic[51, 52] $a = 8.40$ (Th$_3$P$_4$ structure)	Pu_3Se_4, cubic[52] $a = 8.77$ (Th$_3$P$_4$ structure)	
PuX	PuS, cubic[51, 52] $a = 5.54$ (NaCl structure)	PuSe, cubic[51, 52] $a = 5.77$ (NaCl structure)	PuTe, cubic[52] $a = 6.15$ (NaCl structure)

type structure. This phase extends from $PuX_{2.0}$ down to $PuX_{1.8}$ and, as the X/M ratio decreases, the a parameter of the tetragonal cell contracts. For example, the c/a ratio increases from a value of 2.036 for $PuTe_2$ to 2.073 for $PuTe_{1.8}$. The detailed crystal structure of $PuS_{2.0}$ appears to be rather subtle. Originally it was indexed as cubic with $a = 7.960$ Å, but as sulphur is lost on heating the powder lines are resolved into doublets of a tetragonal structure with $c/a > 2.00$. It therefore seemed likely that $PuS_{2.00}$ was tetragonal with $c/a = 2.00$. However, a single crystal study of $CeSe_2$, which has a very similar powder diffraction pattern, has shown that this crystal is in fact monoclinic with $a = 2b$ and β fortuitously equal to 90 degrees [50]. By analogy it is probable that PuS_2, $PuSe_2$ and $PuTe_2$ all possess monoclinic symmetry, although single crystal measurements would be needed to demonstrate the point unequivocally.

Continued vacuum degradation of PuS_{2-x} above 600 °C produces a sesquisulphide, αPu_2S_3, which is stable to 1100 °C. This crystallises with an orthorhombic unit cell and appears to have a narrow range of homogeneity. At higher temperatures in vacuum Marcon and Pascard[51] found two further 'sesquisulphides', namely βPu_2S_3 (tetragonal, composition range $PuS_{1.5}$–

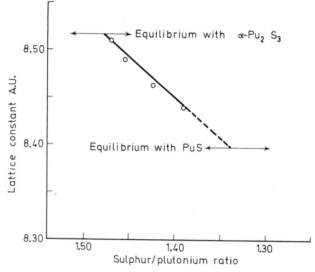

Figure 6.2   Variation of lattice constant of γPu_2S_3 with composition
(From Allbutt, Dell and Junkison[52], by courtesy of North Holland Publishing Co.)

$PuS_{1.4}$) and γPu_2S_3 (cubic, composition range $PuS_{1.5}$–$PuS_{1.33}$). Allbutt, on the other hand, observed only the γ phase which had the Th_3P_4 structure[52]. The lattice parameter of this phase varies linearly with S/Pu ratio as shown in Figure 6.2. The measured powder densities are consistent with a model in which Pu_2S_3 is a cation deficient Pu_3S_4 structure. Vacuum degradation at still higher temperatures does not lead to PuS; rather, congruent evaporation occurs at a composition near $PuS_{1.38}$.

In the sesquiselenide system Marcon and Pascard report an η phase (orthorhombic Sb_2S_3 structure), as well as γPu_2Se_3 isostructural with γPu_2S_3. Again, Allbutt et al. found only the γ phase, prepared directly by vacuum degradation of $PuSe_{2-x}$ above 700 °C. As with γPu_2S_3, a progressive loss of selenium occurred with increasing temperature and duration of heating until finally the Pu_3Se_4 composition was reached. In the telluride system, thermal degradation of $PuTe_{1.8}$ gave first a complex phase (unindexed) and then the cubic γ phase.

The monocompounds PuS, PuSe and PuTe are not easily prepared in a pure state although acceptable products have been obtained by heating a mixture of a higher chalcogenide and plutonium hydride to 1500–1800 °C. The lattice parameter of the PuS phase varies from $a = 5.528$ Å to $a = 5.540$ Å, corresponding to a composition range from $PuS_{0.95}$ to $PuS_{1.00}$ [51]. Insufficient work has been carried out to establish the composition range of PuSe and PuTe.

The plutonium oxychalcogenides Pu_2O_2X are of two different structures. Pu_2O_2S and Pu_2O_2Se are of hexagonal symmetry with the La_2O_3 structure type, isostructural with Ce_2O_2S.

$$Pu_2O_2S \quad a = 3.92 \quad c = 6.76 \quad c/a = 1.72$$
$$Pu_2O_2Se \quad a = 3.96 \quad c = 6.98 \quad c/a = 1.76$$

Pu_2O_2Te, however, was indexed as tetragonal with $a = 6.33$, $c = 11.66$, $c/a = 1.84$ [52].

These compounds are all derived from trivalent plutonium. In addition, Marcon describes two further oxysulphides stemming from tetravalent plutonium and a phase containing both Pu^{3+} and Pu^{4+} ions. These are as follows:

Composition	Symmetry	Type	Lattice parameters
PuOS	Tetragonal	UOS	$a = 3.80$, $c = 6.59$
$Pu_2O_2S_3$	Tetragonal	$PuS_{1.9}$	$a = 3.95$, $c = 7.95$
$Pu_4O_4S_3$	Monoclinic		$\begin{cases} a = 4.07, & b = 6.73 \\ c = 3.87, & \beta = 118° \end{cases}$

PuOS forms a complete range of solid solutions with UOS. It is a less stable compound, decomposing at 600 °C whereas UOS melts without decomposition. The second tetragonal oxysulphide phase was first described by Marcon as a polymorphic form of PuOS [53]. Later he inferred that this was incorrect and that the formula should be $Pu_2O_2S_3$ [51]. This compound is said to contain a polysulphide ion and may be written as $(Pu^{4+})_2 (O^{2-})_2 (S^{2-}) (S-S)^{2-}$. $Pu_4O_4S_3$, prepared by decomposing PuOS above 700 °C, may be regarded as having the formula $(Pu^{4+})_2(Pu^{3+})_2(O^{2-})_4(S^{2-})_3$. The corresponding mixed compound $(U^{4+})_2(Pu^{3+})_2(O^{2-})_4(S^{2-})_3$ has also been prepared and it is to be presumed that analogous oxyselenides exist but these have not been investigated.

6.3.4 Other actinides

The only other actinide metal–chalcogen system to have been investigated in any detail is that of Np—S [21]. The crystal chemistry of this system is intermediate between that of U—S and Pu—S, in agreement with the position of neptunium in the Periodic Table.

The highest polysulphide NpS_3 has a monoclinic lattice, isostructural with US_3, with lattice parameters $a = 5.36$ Å, $b = 3.87$ Å, $c = 18.10$ Å and $\beta = 99° 30'$ (N.B. The quoted c parameter is doubled compared to that for US_3 shown in Table 6.2; there appears to be uncertainty over this point). By analogy with US_3 which contains U^{4+} ions, it is concluded that NpS_3

is based upon Np^{4+} ions. A second polysulphide Np_2S_5 is also known. This has a tetragonal lattice, isostructural with Th_2S_5 and U_2S_5. Within the series an actinide contraction is observed[51]:

$$Th_2S_5 \qquad a = 10.80* \qquad c = 10.20$$
$$U_2S_5 \qquad a = 10.57 \qquad c = 9.88$$
$$Np_2S_5 \qquad a = 10.48 \qquad c = 9.84$$

The trisulphide NpS_3 is relatively unstable and on heating above 500 °C it decomposes directly to Np_3S_5 (Note: NpS_2 has not been reported). Np_3S_5 is isostructural with U_3S_5 (orthorhombic $a = 8.42$, $b = 8.07$, $c = 11.71$ Å) and is thought to be an ionic compound containing both Np^{4+} and Np^{3+} ions, $Np^{4+} (Np^{3+})_2 (S^=)_5$.

In all the above sulphides Np behaves structurally like U. However, on decomposing Np_3S_5 *in vacuo* above 900 °C, the sesquisulphide formed does not normally possess the U_2S_3 structure, although such a phase was reported by Zachariasen in the first investigation of the neptunium sulphides[20] $(a = 10.3$ Å, $b = 10.6$ Å, $c = 3.86$ Å). Rather, three other polymorphs of Np_2S_3 are known and these are isostructural with α, β and γ Pu_2S_3 respectively. The orthorhombic αNp_2S_3 $(a = 3.98$ Å, $b = 7.39$ Å, $c = 15.50$ Å) transforms into βNp_2S_3 near 1200 °C and γNp_2S_3 near 1500 °C. The latter compound has the b.c.c. Th_3P_4 structure and extends over the range Np_3S_4 $(a = 8.440$ Å) to Np_2S_3. Marcon concludes that in the sesquisulphides neptunium is present as Np^{3+} ions, corresponding to the Pu^{3+} ions in Pu_2S_3.

Finally, Np_2S_3 reacts with neptunium metal to yield the monosulphide, NpS, with the rocksalt structure $(a = 5.532$ Å).

The remaining data in the literature pertaining to other actinides all relate to chalcogenide phases crystallising with the b.c.c. Th_3P_4 structure or to tetragonal oxychalcogenides. As one ascends the Periodic Table beyond plutonium the quantity of material available for investigation becomes vanishingly small and the experimental difficulties corresponding greater. Berkelium sesquisulphide, Bk_2S_3, was prepared by reacting a mixture of H_2S and CS_2 vapour with Bk_2O_3 at 1100 °C [54]. Microgramme quantities of the isotope [249]Bk were employed. A similar technique was employed in the preparation of Cf_2S_3, using 1 μg of [249]Cf, a daughter product formed by the β decay of [249]Bk. The lattice parameters reported for the various cubic sesquisulphides are as follows:

Compound	a_0	Reference
Ac_2S_3	8.99 Å	48
Th_2S_3	Sb_2S_3 structure	—
Pa_2S_3	—	—
U_2S_3	Sb_2S_3 structure	—
Np_2S_3	8.44 Å	21
Pu_2S_3	8.46 Å	51, 52
Am_2S_3	8.45 Å	48
Cm_2S_3	8.44 Å	54
Bk_2S_3	8.44 Å	54
Cf_2S_3	8.39 Å	55

*Compared to the data of Graham and McTaggart (Table 6.1), Marcon considers the 'a' parameter to be multiplied by $\sqrt{2}$ (reference 21, 1967) or by 2 (reference 51, 1969).

A comparison of the lattice constants from compound to compound is not too meaningful unless one knows where the measured compound lies in the composition range M_2S_3–M_3S_4; for instance, the value for 'Np_2S_3' relates to a two phase mixture with NpS and is therefore almost certainly that of Np_3S_4.

It is of interest that Th_2S_3 and U_2S_3 do not appear to crystallise readily with this structure, if at all. This is understandable if, following Marcon[21], we assume that the M_2S_3 structure contains M^{3+} ions. Th^{3+} and U^{3+} ions are known to be very unstable and it seems likely that Th_2S_3 and U_2S_3 contain tetravalent cations which favour the Sb_2S_3 structure. On the same basis, we might predict that Pa_2S_3, when prepared, will not crystallise in the cubic M_2X_3 structure. The ions Ac^{3+}, Pu^{3+} and Am^{3+} are well known and stable.

Recently, the compounds Np_3Se_4, Np_3Te_4, Am_3Se_4 and Am_3Te_4, all crystallising in the Th_3P_4 structure, have been prepared[56]. The measured lattice parameters were:

$$Np_3Se_4 \qquad a_0 = 8.83\,\text{Å}$$
$$Np_3Te_4 \qquad a_0 = 9.40\,\text{Å}$$
$$Am_3Se_4 \qquad a_0 = 8.78\,\text{Å}$$
$$Am_3Te_4 \qquad a_0 = 9.39\,\text{Å}$$

It is of interest that the americium compounds were formed as one component of a two-phase mixture when equiatomic proportions of americium and the chalcogenide were annealed at 1000 °C. This raises the interesting question as to whether the monocompounds AmSe and AmTe do not exist or whether the temperature was not sufficiently high for reduction of Am_3X_4 to take place.

Finally, mention may be made of the oxy-sulphides. Neptunium oxysulphide, NpOS, and protactinum oxysulphide, PaOS, were both first prepared by Zachariasen and co-workers[19, 57]. They are tetragonal, isostructural with ThOS and UOS:

Compound	ThOS	PaOS	UOS	NpOS	PuOS
c	6.75 Å	6.70 Å	6.68 Å	6.65 Å	6.59 Å
a	3.96 Å	3.83 Å	3.84 Å	3.83 Å	3.80 Å
c/a	1.70 Å	1.75 Å	1.74 Å	1.73 Å	1.73 Å

According to Marcon[21] these phases all derive from the M^{4+} ion. In addition Marcon has prepared two further oxysulphides whose formulae were deduced from the fact that they are isostructural with the corresponding plutonium compounds[51], namely:

$$Np_2O_2S \quad \begin{cases} \text{hexagonal } a = 3.95, c = 6.80\,\text{Å} \\ \text{derived from } Np^{3+} \text{ ions} \end{cases}$$

$$Np_4O_4S_3 \quad \text{monoclinic } a = 4.07, b = 6.76$$
$$c = 3.89, \beta = 118°$$

derived from Np^{3+} and Np^{4+} namely:
$$(Np^{4+})_2(Np^{3+})_2(O^=)_4(S^=)_3$$

No oxyselenides or oxytellurides of neptunium have yet been reported.

6.4 PNICTIDES: PREPARATION, CRYSTAL CHEMISTRY AND STOICHIOMETRY

The nitrides are crystallographically distinct from the other pnictides, but for any particular actinide metal there is generally a close structural relationship between the various phosphides, arsenides, antimonides and bismuthides. It is convenient therefore to consider the nitrides first, as a separate class of compounds, and then to group the remaining pnictides together for each of the actinide metals.

6.4.1 Nitrides

Thorium forms two nitrides (Th_3N_4 and ThN), uranium forms three crystallographically distinct phases ($\alpha U_2N_{3 \pm x}$, βU_2N_3 and UN), while plutonium mononitride (PuN) is the only known compound in the Pu—N system. Other actinide nitrides which have been reported are PaN, NpN and AmN. All the mononitrides crystallise with the rocksalt structure and exhibit complete mutual miscibility so far as is known.

6.4.1.1 Thorium

In an early study, Zachariasen[58] reported thorium sesquinitride, Th_2N_3, with hexagonal symmetry. Later workers were unable to reproduce this composition and it has now been shown[59] that this hexagonal structure relates to an oxynitride Th_2N_2O. Th_3N_4, however, has been substantiated by several different groups[60-62] and shown to possess a hexagonal structure closely related to Th_2N_2O [59]:

Th_2N_2O	Th_3N_4
$a = 3.883$ Å	$a = 3.871$ Å
$c = 6.187 (= 2 \times 3.094$ Å$)$	$c = 27.385 (= 9 \times 3.043$ Å$)$

It is a maroon coloured powder, prepared by reacting finely divided thorium metal with nitrogen at 500–1000 °C. As the compound is diamagnetic and exhibits a high electrical resistivity, it presumably has the ionic structure $(Th^{4+})_3(N^{3-})_4$; this explains why N/Th ratios greater than 1.33 are not observed.

The phase diagram of the Th—N system (Figure 6.3) indicates that the higher nitride has no range of composition below 1500 °C. At higher temperatures the metal-rich boundary widens to $ThN_{1.29}$ at 1960 °C. The mononitride ThN (NaCl type, $a = 5.16$ Å) is formed by heating the higher nitride in vacuum above 1400 °C. It is a greenish-yellow compound, exhibiting an electrical conductivity typical of a metal and possessing no significant range of stoichiometry below 1400 °C. Both Th_3N_4 and ThN are extremely hygroscopic and reactive chemically.

The reaction of thorium metal with ammonia gas at high pressure (2500–5500 atm) and at temperatures of 400–500 °C has been reported[63] to yield a

Table 6.4 Crystal structures of thorium-nitrogen compounds[63]

Compound	Symmetry	a (Å)	b (Å)	c (Å)	β
$\alpha\,ThN_{1.3}$	orthohexagonal	6.69_6	3.86_6	27.36_8 (6.08_2)	$90°$
$ThNO_{0.5}$	orthohexagonal	6.72_0	3.88_0	(6.18_0)	$90°$
$\beta\,ThN_{1.3}$	monoclinic	6.95_2	3.83_0	6.20_6	$90.7_1°$
$ThN(NH)_{0.5}$	monoclinic	7.16_7	3.86_0	6.24_2	$92.2_1°$

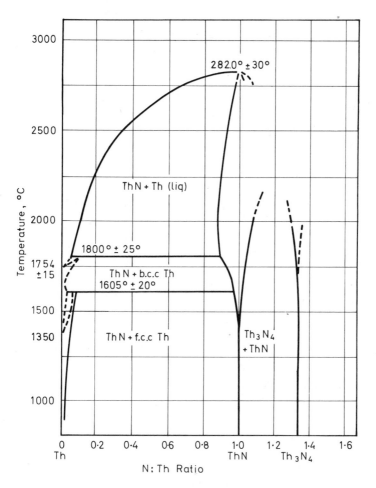

Figure 6.3 Phase diagram of Th–N system
(From Benz, Hoffmann and Rupert[61], by courtesy of the American Chemical Society)

thorium nitride–imide Th_2N_2 (NH). This compound is an analogue of Th_2N_2O with a monoclinically distorted lattice. It degrades thermally above 270 °C to yield a second crystallographic modification of Th_3N_4 (βTh_3N_4). The structural relationship between the four compounds, is shown in Table 6.4.

6.4.1.2   Uranium

Uranium nitrides are prepared by reacting finely divided uranium powder with nitrogen or ammonia gas at temperatures above 400 °C. The phase first formed, $\alpha U_2 N_{3+x}$, has the b.c.c. Mn_2O_3-type structure, with a composition usually in the range $UN_{1.60}$–$UN_{1.75}$, depending upon the precise experimental conditions employed. This phase has a wide range of stoichiometry, the lower limit lying at $UN_{1.55\pm.02}$ at 950–1000 °C and $UN_{1.57\pm.02}$

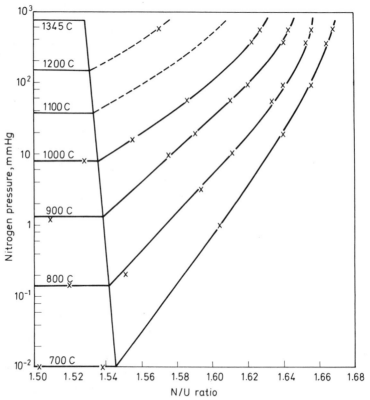

Figure 6.4   Dissociation pressure isotherms for U_2N_{3+x}
(From Bugl and Bauer[64], by courtesy of the Metallurgical Society of A.I.M.E.)

at 700 °C [64-66]. There is no evidence for stoichiometric U_2N_3 with this structure; decomposition to UN occurs below the temperature at which it would be stable, as shown in Figure 6.4.

The upper end of the composition range for this phase is still uncertain. Originally UN_2 with the fluorite structure (cf. UO_2) was reported[67, 68], but this has now been discredited. Extrapolation of the data of Figure 6.4 indicates that the N/U ratio will not exceed 1.75 even at low temperatures (700 °C) and high pressures (100 atm). This was confirmed experimentally; heating at 500 °C under 125 atm N_2 pressure gave only $UN_{1.70}$ [69]. On the other hand, several authors have reported the preparation of $UN_{1.80}$–$UN_{1.86}$ by the reaction of ammonia or nitrogen with UH_3 rather than uranium metal [69–71]. Even though the thermodynamic data indicate a practical upper limit of $UN_{1.75}$ for the $\alpha U_2 N_{3+x}$ phase, there can be little doubt about the existence of phases with $N/U > 1.8$. These may result from the very high virtual activity of nitrogen associated with ammonia gas [72].

The lattice parameter of $U_2 N_{3+x}$ as a function of composition is plotted in Figure 6.5 [71]. Above $UN_{1.8}$ the structure was reported to be fluorite rather than $Mn_2 O_3$. However, these two structures are crystallographically

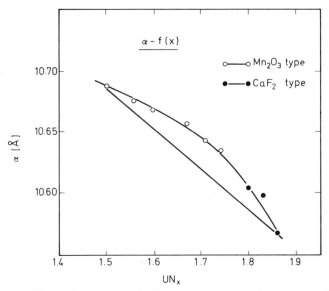

Figure 6.5   Lattice parameter of $\alpha U_2 N_{3+x}$ phase as a function of composition
(From Trzebiatowski and Troc[71], by courtesy of the Polish Academy of Science)

related, the $Mn_2 O_3$ structure being formed from eight fluorite unit cells $(2 \times 2 \times 2)$ by loss of $\frac{1}{4}$ of the anions and an associated rearrangement of the atoms. The x-ray pattern of the $Mn_2 O_3$ structure consists essentially of the fluorite pattern with additional lines resulting from the b.c.c. superlattice. As the N/U ratio increases in the $\alpha U_2 N_{3+x}$ phase, the intensities of the additional lines diminish; it is difficult to decide whether the structure is truly fluorite above $UN_{1.80}$ or whether the b.c.c. lines are simply too faint to record, although it is probable that at high N/U ratios the structure is fluorite with random nitrogen vacancies. A further possibility is that compositions above $UN_{1.80}$ are stabilised by hydrogen, although there is no evidence for this suggestion.

The 'second' uranium sesquinitride, βU_2N_3, was first described by Vaughan[73]. It has the hexagonal, La_2O_3 type structure with lattice parameters $a = 3.698\,\text{Å}$ and $c = 5.839\,\text{Å}$. The phase was first observed as an intermediate surface layer between αU_2N_{3+x} and UN on uranium metal reacted with nitrogen at 775–900 °C. Later workers prepared βU_2N_3 by quenching αU_2N_{3+x} from near 1100 °C [69, 74, 75] and, on this basis, Benz and Bowman concluded that βU_2N_3 was a high temperature phase which reverted to αU_2N_3 below 1120 ± 30 °C [75]. However, detailed annealing experiments disproved this idea[69] and Stöcker and Naoumidis reported that hexagonal uranium nitride extends over the homogeneity range U_3N_4 to U_2N_3 and is stable at temperatures as low as 900 °C provided the nitrogen pressure is sufficiently small[76]. A more recent investigation[77] shows that βU_2N_3 is, in fact, a phase of almost fixed composition, $UN_{1.44-1.45}$. When uranium is heated in nitrogen (1 atm) αU_2N_{3+x} with progressively smaller values of x is formed up to 1150 °C, then a mixture of α and β U_2N_3 from 1150° to 1350 °C, pure βU_2N_3 at 1350 °C and a mixture of βU_2N_3 and UN at 1380–1400 °C. Thus the transformation between the two 'sesquinitrides' can be represented as:

$$\text{b.c.c. } U_2N_{3+x} \rightleftharpoons \beta U_2N_{3-y} + \tfrac{1}{2}(x+y)N_2$$

the hexagonal phase being favoured at higher temperatures and lower nitrogen pressures.

Uranium mononitride (UN) has been studied very extensively because of its interest as a potential nuclear fuel. It is a near-stoichiometric compound prepared as a dark grey powder by heating αU_2N_{3+x} at temperatures above 1100 °C in vacuum or flowing argon. The powder is liable to be pyrophoric in air and, for this reason, is best prepared at 1400 °C or above so as to reduce its surface area and stabilise it towards exposure to air. UN may also be prepared from UO_2 by heating a mixture of $UO_2 + C$ in flowing nitrogen at 1450 °C:

$$UO_2 + 2C \underset{1450\,°C}{\overset{N_2}{\rightleftharpoons}} UN + 2CO$$

Since uranium monocarbide and mononitride form a complete range of solid solutions $U(C_{1-x}N_x)$, the reaction product invariably contains some carbon. Using a fluidised-bed with excess of nitrogen, products containing a nitrogen content corresponding to 96% UN have been obtained[78].

UN melts at 2850 °C with a dissociation pressure of 2.5 atm nitrogen[64, 79]. There is evidence that the stoichiometry range of the mononitride phase widens above 1500 °C; the lower phase boundary is reported as $UN_{0.96}$ at 1500 °C and $UN_{0.80}$ at 2000 °C [75]. The defects introduced into the rocksalt lattice at these temperatures cannot be quenched in and UN is essentially a line compound ($UN_{1.00\pm0.01}$) below 1000 °C with a constant lattice parameter, $a = 4.889\,\text{Å}$. The solubility of oxygen in UN is small, at least up to 1500 °C, and as little as 0.2 wt% oxygen can be detected by x-ray diffraction as impurity UO_2. This implies that the solubility of oxygen in UN is probably not more than 2 at%. The lattice parameter of UN saturated with oxygen increases to 4.891–4.892 Å [80]. It is of interest that while the solubility of oxygen in UN is so small, for uranium monocarbide the limit of solubility is as high as U $(C_{0.65}O_{0.35})$.

6.4.1.3 Plutonium

Plutonium mononitride is best prepared by heating plutonium hydride in nitrogen at 400 °C[28, 81]. No higher nitrides of plutonium have ever been reported. Attempts to prepare PuN by direct reaction between the metal and nitrogen are not particularly successful as the reaction does not go to completion even well above the melting point of plutonium (639 °C). However, the reaction of bulk metal with nitrogen may be catalysed by a preliminary surface hydriding; upon admitting nitrogen at 250 °C reaction occurs with the hydride to liberate hydrogen, which then serves to hydride further underlying plutonium metal, so leading to a cyclic catalytic reaction[82, 83].

Plutonium mononitride is a stoichiometric phase, crystallising in the rocksalt structure with a lattice parameter varying from $a = 4.903$ to $a = 4.909$ Å for various preparations; the higher values are associated with an oxygen solubility up to 10 at % of the anion[28]. Excess oxygen is present as hexagonal Pu_2O_3. PuN forms a complete range of solid solutions with UN although homogenisation of the mixed nitrides to (U, Pu)N may be effected only at temperatures of > 1700 °C. Because the lattice parameters of UN and PuN are very similar, it is difficult by x-ray techniques to distinguish between incompletely homogenised and poorly crystalline samples. The unit cell constant of PuN gradually expands at a linear rate on standing; this is attributable to lattice defects introduced by self-irradiation damage[84].

6.4.1.4 Other actinides

Neptunium mononitride NpN, first reported by Zachariasen[58], has the rocksalt structure with lattice parameter $a = 4.90$ Å. Its melting point $(2830 \pm 30$ °C) and dissociation pressure have been determined recently[85]. It is not known whether other nitrides of neptunium exist. Protactinium mononitride PaN is said to be formed by reacting ammonia with $PaCl_4$ or $PaCl_5$ at 800 °C[57] but no x-ray data are available. Recently americium mononitride (AmN, $a = 5.00$ Å) also has been prepared[86].

6.4.2 Phosphides, arsenides, antimonides and bismuthides

6.4.2.1 Thorium

The thorium pnictides, like the chalcogenides, have not been investigated as thoroughly as the corresponding uranium compounds. Despite the large change in atomic mass and ionic radius from phosphorus to bismuth, the crystal structures are analogous and only three structures have been established with certainty. These are summarised in Table 6.5.

The first striking observation is that the thorium pnictides are isostructural with the corresponding plutonium chalcogenides PuX_2, Pu_3X_4 and PuX. However, the three compounds Th_3Y_4 show no evidence for a wide range of stoichiometry, unlike the corresponding plutonium chalcogenides which extend from Pu_3X_4 to Pu_2X_3.

Table 6.5   Crystallographic data for the thorium pnictides

Type	Phosphides	Arsenides	Antimonides	Bismuthides
ThY_2*		$ThAs_2$ [91] $a = 4.09$ Å $c = 8.57$ Å	$ThSb_2$ [93] $a = 4.35$ Å $c = 9.17$ Å	$ThBi_2$ [94, 95] $a = 4.29$ Å $c = 9.30$ Å
			← Tetragonal Fe_2As structure →	
Th_3Y_4	Th_3P_4 [87] $a = 8.65$ Å	Th_3As_4 [87, 91] $a = 8.85$ Å	Th_3Sb_4 [93] $a = 9.37$ Å	Th_3Bi_4 [94, 95] $a = 9.56$ Å
			← Body Centred Cubic Th_3P_4 structure →	
ThY	ThP_{1-x} [88-90] $a = 5.83$ Å	$ThAs$ [91, 92] $a = 5.97$ Å	$ThSb$ [93] $a = 6.32$ Å	
			← Face Centred Cubic NaCl structure →	

*Throughout this paper the symbol Y denotes a general pnictide element and not yttrium

This is readily understood if we consider the pnictides Th_3Y_4 to be purely ionic; as thorium cannot assume a valence greater than 4, the Y/Th ratio will be limited to 1.33. A similar explanation was offered for Th_3N_4 (v.s.) though it is of interest that this compound is not isostructural with the other pnictides.

The existence of di-pnictides, ThY_2, may only be understood in terms of a polyanion structure, $Th^{4+}(Y-Y)^{4-}$. In this regard it should be noted that the Fe_2As structure has three distinct M—Y lengths. The Th—P system has been investigated by several authors[88-90, 96-98] and it is generally agreed that there is no diphosphide ThP_2 with the Fe_2As structure. The non-existence of this phase is surprising having regard to the relative stability of UP_2. Hulliger[99] claims to have prepared a second (low temperature) form of $ThAs_2$ with orthorhombic symmetry ($a = 7.29$ Å, $b = 9.78$ Å, $c = 4.00$ Å), isostructural with $ZrAs_2$, $HfAs_2$, TiP_2, ZrP_2 and HfP_2. This transformed to the more usual $ThAs_2$ structure on heating to 1150 °C. Attempts to prepare a corresponding form of ThP_2 invariably led to Th_3P_4, although it was claimed that faint impurity lines could be indexed as ThP_2 with the ZrP_2 structure ($a = 6.95$ Å, $b = 9.42$ Å, $c = 3.90$ Å). It is possible then that an unstable form of ThP_2 with this structure does exist.

The preparative methods for the thorium pnictides are quite straightforward. Phosphides and arsenides are generally prepared from finely divided thorium metal powder (ex-hydride) either by direct synthesis in a sealed ampoule or by reaction with the appropriate gas (PH_3 or AsH_3). The products so formed (Th_3P_4 or $ThAs_2$) may be converted to the lower compounds (ThP, Th_3As_4, ThAs) by reduction with the appropriate quantity of thorium powder at high temperature *in vacuo*. The antimonides and bismuthides are also prepared by the sealed ampoule reaction method, although it is possible that arc-melting could be used for these alloys (the melting point of thorium metal is $1770 \pm 10\,°C$ [88]). According to Price and Warren[87], Th_3P_4 decomposes above $1200\,°C$, Th_3As_4 is stable to $2050\,°C$ while Th_3Sb_4 melts (with decomposition?) at $1600\,°C$. The relatively low decomposition temperature of Th_3P_4 allows of the possibility of preparing ThP_{1-x} by direct vacuum degradation at $\sim 1500\,°C$ or by synthesis from the elements at $1600–1800\,°C$ in a high temperature modification of the Faraday method[88].

The present state of knowledge concerning ThP and ThBi is not entirely satisfactory. Thorium monophosphide, a dark blue solid, is generally regarded as being a sub-stoichiometric phase existing over a wide composition range[88, 96, 98]. A detailed phase study of the Th–ThP system[88], showed that at $1000\,°C$ the composition range of the monophosphide extended from $ThP_{0.96}$ $(a = 5.840\,Å)$ down to $ThP_{0.55}$ $(a = 5.830\,Å)$. At higher temperatures the P/Th ratio of the P-rich boundary decreased progressively, reaching $ThP_{0.73}$ at $1850\,°C$. Conversely, the P/Th ratio of the Th-rich boundary increased with increasing temperature so that at $2000\,°C$ the homogeneity range was reported to be only $ThP_{0.7}$ to $ThP_{0.75}$. This is a rather surprising result as for most compounds the compositional range widens with increasing temperature (cf. ThN, Figure 6.3).

Later papers have cast doubt upon the results of this apparently careful study. Baskin[89] prepared $ThP_{0.99}$ and $ThP_{0.95}$ with significantly lower lattice parameters than found by Gingerich[88] for these compositions. Javorsky and Benz[90] repeated the phase study of the Th–ThP system, using thermal and metallographic as well as x-ray techniques, and present a phase diagram in which the ThP phase has a minimum P/Th ratio of 0.98 at all temperatures. The ThP phase is reported to melt congruently with the P/Th ratio of 0.99 ± 0.02 at $2990\,°C$. As the results of these two phase studies stand in such marked contrast, it will be necessary to conduct an independent referee investigation to clarify the true situation. Benz has also made a phase study of the Th—ThAs system[92].

The position regarding ThBi is similarly confused. As UBi exists with the rocksalt structure, a thorium analogue might be expected. Ferro[94] found a complex x-ray situation at the 50/50 mole % composition with evidence for two phases, neither of which had the NaCl or CsCl structures. Dahlke et al.[95] measured the dissociation pressure of Bi in the Th—Bi system and found evidence for a mono-compound at $1000\,°C$, but on cooling the specimen of composition Bi/Th = 1.0 to room temperature its x-ray pattern was that of Th_3Bi_4. A further phase study of this system is also required.

6.4.2.2   Uranium

The uranium pnictides have been fairly extensively studied, particularly the phosphides. Generally they are isostructural with their thorium analogues (Table 6.6). The uranium phosphides may be prepared either from the elements in a sealed ampoule or by passing phosphine gas over finely divided uranium powder (ex uranium hydride). The latter reaction occurs readily at $500\,°C$ to yield UP_2, which decomposes at $800\,°C$ in argon to U_3P_4. The latter may be degraded further to UP, a fine grey powder, by heating in vacuum at 1300–1400 °C. There is evidence that U_3P_4 has a certain range of composition, but the width of this has not been well defined[16]. Single crystals of U_3P_4 and UP_2 have been grown by the iodine transport method[112, 113]; an x-ray study of the latter has shown the tetragonal unit cell to be a pseudo cell up to 356 K [114].

UP is invariably found to be very well crystalline with sharp x-ray diffraction lines and a lattice constant of $a = 5.589 \pm .001$Å. It appears to be an almost stoichiometric compound, at least up to 1500 °C. At temperatures

Table 6.6   Crystallographic data for the uranium pnictides

Type	Phosphides	Arsenides	Antimonides	Bismuthides
UY_2	UP_2 [100–104] $a = 3.81$ Å $c = 7.76 - 7.78$ Å	UAs_2 [104, 105] $a = 3.95$ Å $c = 8.12$ Å	USb_2 [105, 108] $a = 4.27$ Å $c = 8.75$ Å	UBi_2 [110, 111] $a = 4.45$ Å $c = 8.92$ Å
		← Tetragonal Fe_2As structure →		
U_3Y_4	U_3P_4 [16, 100–103] $a = 8.21$ Å	U_3As_4 [105, 106] $a = 8.51$ Å	U_3Sb_4 [105, 108] $a = 9.11$ Å	U_3Bi_4 [110, 111] $a = 9.36$ Å
		← B.C.C. Th_3P_4 structure →		
UY	UP [16, 100–103] $a = 5.589$ Å	UAs [10, 106, 107] $a = 5.78$ Å	USb [105, 108, 109] $a = 6.19$ Å	UBi $a = 6.36$Å[109,110] $a = 6.40$Å[111]
		← F.C.C. NaCl structure →		

above 1800 °C in vacuum UP loses phosphorus by decomposition and liquid uranium metal is formed at the grain boundaries. Metallographic examination of quenched compacts suggests that the free metal is present at the sintering temperature and is not formed by disproportionation of UP_{1-x} on cooling[6]. It is not possible to cast UP by arc-melting without significant decomposition occurring. The melting point of stoichiometric UP, determined directly in sealed tungsten vessels, was found to be 2610 ± 20 °C[115] although Benz and Ward, in a phase study of the U—UP system, found a temperature of 2850 °C for the highest melting point[116]. The solubility of oxygen in uranium monophosphide is extremely small; as little as 0.2 wt % oxygen present in a preparation of UP can be detected by x-ray diffraction as a separate UO_2 phase[16]. No oxy-phosphide is found and the UP—UO_2 system shows a simple eutectic at 38 wt % UP—62 wt % UO_2 (2390 ± 30 °C)[117].

The first systematic study of the uranium arsenides established the crystal structures of UAs_2, U_3As_4 and UAs[104, 106]. No evidence for non-stoichiometric compounds was observed up to 900 °C. These compounds have also been prepared by Trzebiatowski et al.[105], employing the Faraday method at 800 °C to produce UAs_2 and U_3As_4. The latter was converted to UAs by thermal decomposition in vacuum at 1300–1400 °C. Uranium mono-arsenide has also been prepared by reacting uranium powder with the stoichiometric quantity of AsH_3 gas at 300 °C, followed by homogenisation in vacuum at 1200–1400 °C[10]. Single crystals of UAs_2 and U_3As_4 have been grown from the powder by iodine transport[112, 113].

The U—UAs portion of the U—As phase diagram has been investigated[107] using thermal, x-ray and metallographic techniques on quenched samples. Uranium solubility in the UAs phase was found not to exceed 0.15% up to 2200 °C. The mono-arsenide decomposes at high temperature before melting; the highest melting point observed experimentally was 2705 °C.

The principal structural studies of the uranium antimonides and bismuthides were carried out by Ferro[108, 110]. In each system three compounds were found, isostructural with the corresponding phosphides and arsenides. The compounds are made by heating stoichiometric mixtures of the powdered elements in evacuated and sealed tubes at 800–900 °C. Trzebiatowski and co-workers[105] prepared USb_2 and U_3Sb_4 in a similar fashion and USb by decomposing the higher antimonides in vacuum at 1300–1400 °C. The crystal structures were confirmed by these authors and also by Beaudry and Daane[118] who studied the U—Sb phase diagram. USb_2 and U_3Sb_4 undergo peritectic decomposition at 1355 °C and 1695 °C respectively. A subantimonide (U_4Sb_3) was reported and found to melt congruently at 1800 °C. Single crystal studies of U_4Sb_3 showed it to have a hexagonal unit cell with $a_0 = 9.27$Å and $c_0 = 6.20$Å. The melting point observed for USb (1850 °C) has been confirmed as 1850 ± 30 °C[119].

The three uranium bismuthides were prepared also by Trzebiatowski and Zygmunt[111], who investigated their crystal structures and magnetic properties. The x-ray data were in good agreement with those of Ferro, except for UBi (see Table 6.5). The U—Bi phase diagram has been determined by Teitel[120] and by Cotterill and Axon[121]. According to Teitel, UBi_2 and U_3Bi_4 decompose peritectically at 1010 °C and 1150 °C respectively, while UBi decomposes to two liquid phases between 1400 °C and 1450 °C.

6.4.2.3   Plutonium

From the comparatively limited investigations carried out so far, plutonium appears to form only monopnictides of rocksalt structure with N, P, As, Sb and Bi. The lattice parameters are as follows:

Compound	PuN	PuP	PuAs	PuSb	PuBi
a_0 (Å)	4.90	5.66	5.86	6.24	6.35

Plutonium monophosphide has been prepared[1] by reacting phosphine with partially decomposed plutonium hydride at 400 °C. The reaction ceased after a while, as further PuH_3 was formed, and it was necessary to decompose this hydride in vacuum at 500 °C before continuing the reaction with PH_3. By multiple cycles of reaction/decomposition at increasing temperature up to 600 °C pure PuP was ultimately formed. This was vacuum annealed at 1400 °C. The method is not particularly convenient and it seems likely that the Faraday technique, using powdered plutonium (made from the hydride) and red phosphorus will prove more satisfactory. PuP has also been made by reaction of the powdered elements mixed together, but this requires the use of a pressure vessel at 550–650 °C[122]. PuP melts with decomposition at 2600 °C in argon at 2 atm pressure[1]. For this reason it cannot be fabricated by arc-casting. The sintering behaviour of PuP powder has been described[123]. The lattice parameter of PuP in equilibrium with plutonium (a_0 = 5.6562 Å) differs significantly from that in equilibrium with phosphorus (a = 5.644 Å) suggesting a measurable range of stoichiometry.

Plutonium arsenide was first prepared by Pardue et al. from the elements[124]. They reported three distinct compounds, $PuAs_{1-x}$, PuAs and $PuAs_{1+x}$. This seems very surprising and must be regarded as dubious in the light of the fact that Kruger and Moser, reacting PuH_{2-x} with AsH_3, found only the compound PuAs[125]. This compound melted with decomposition at 2420 ± 30 °C in a flow of argon at 3 atm pressure.

The same authors also prepared PuSb by arc-melting together the elements[125]. The observed melting point under similar conditions was 1980 ± 30 °C. Recently PuSb has again been prepared by Lam[126] who finds a lattice constant a_0 = 6.2396 $\pm$ .0001 Å.

PuBi was prepared by melting the elements together[127]. The compound, which was pyrophoric in air, had a lattice constant a_0 = 6.350 Å.

6.4.2.4   Other actinides

The only other compounds which have been reported are neptunium monoantimonide, NpSb[126] and americium monoantimonide, AmSb[56]. These were prepared by arc-melting equiatomic proportions of the elements. Both compounds have the rocksalt structure with lattice parameters as shown:

$$NpSb \quad a_0 = 6.2485 \pm .0007 \text{ Å}$$
$$AmSb \quad a_0 = 6.2426 \pm .0002 \text{ Å (as cast)}$$
$$6.2380 \pm .0002 \text{ Å (annealed at 1000 °C)}$$

It is not known whether there are any other compounds in the Np—Sb or Am—Sb systems.

6.5   CHEMICAL PROPERTIES

6.5.1   Introduction

Detailed study of the chemical properties of these compounds has been confined to materials having potential as nuclear fuel, in particular nitrides, phosphides and sulphides. Systematic investigations have centred largely on their behaviour in different gaseous environments over a wide temperature range, with little information being available on their reactions with liquids. There is a fairly extensive coverage of chemical compatibility with various reactor materials and their behaviour under irradiation, but this is not discussed here.

6.5.2   Thorium compounds

No systematic studies have been made of the chemical reactivity of the thorium compounds, but some qualitative information has been published from the Argonne National Laboratory. Shalek[22] followed the reaction of ThS with air using a Differential Thermal Analyser (DTA) and found two strong exothermic peaks in the range 460–510 °C, indicating ignition. He suggested the first peak to be intermediate oxidation to ThOS followed by oxidation to ThO_2.

Using DTA also, Baskin[89] followed the oxidation of thorium phosphide in air and found oxidation to occur rapidly between 650 °C and 680 °C. Thermo-gravimetric-analysis (TGA) studies confirmed this and the weight change corresponded to complete oxidation of the ThP to ThO_2 and P_2O_5. Subsequent heating to 950 °C failed to remove the P_2O_5 which was suggested to be in the form of a vitreous phosphate.

Finally, the oxidation of thorium mononitride has been followed (both in dry air and moist oxygen) using TGA[62]. With a heating rate of 25 °C/min, it was observed that ThN powder ignited in dry oxygen at 650 °C to give ThO_2. Although ThN was found to be stable in dry oxygen at room temperature, it oxidised in moist air at 25 °C to give a hydrated oxide, probably $ThO_2 \cdot 2H_2O$; complete oxidation occurred in under 5 days. This behaviour is very similar to that found in PuN but is in marked contrast to that of UN.

6.5.3 Uranium monosulphide

The reaction of US with oxygen was first studied by Shalek[22] using a DTA apparatus. The compound ignited between 360 °C and 375 °C, where three very sharp exothermic peaks were found in quick succession. These were followed by a weak exothermic peak at 540 °C and a weak endothermic peak at 765 °C. X-ray analysis of the product at 650 °C indicated the presence of U_3O_8 and the peak at 540 °C was identified as the oxidation of UO_{2+x} to U_3O_8. Separation of the initial three peaks was difficult since they were found to be self sustaining. TGA measurements have since been carried out

Table 6.7 Oxidation processes of US in oxygen

Temperature range (°C)		Reaction	
20–220		None	
220–350		US	$\to$ US(O)
350–400 $\begin{cases} 350 \\ 365 \\ 395 \end{cases}$		US(O)	$\to \gamma US_2$, UOS, UO_2
		γUS_2	$\to$ UOS, UO_2
		UO_2	$\to UO_{2+x}$
400–600		UOS	$\to UO_2SO_4$
540		UO_{2+x}	$\to U_3O_8$
810		UO_2SO_4	$\to U_3O_8$

by Ban[255] who noted weight changes corresponding to the peaks found by Shalek and interpreted them by x-ray and chemical analysis. The initial peaks were attributed to intermediate formation of compounds having formulae $U_2S_yO_8$.

The TGA and DTA measurements have been repeated by Nakai et al.[128] whose TGA measurements agree with those of Shalek. Consideration of these results, together with extensive sulphur analysis and x-ray diffraction data, leads to a more likely interpretation of the mechanism. By interrupting the DTA between the first and second, and the second and third peaks, these were found to correspond to (1) the ignition of US to UOS and γUS_2 and, (2) the subsequent oxidation of γUS_2 to give more UOS and UO_2. The third peak was attributed to the oxidation of UO_2 to UO_{2+x}. Between 400 °C and 600 °C the UOS is oxidised to UO_2SO_4, the decomposition of which accounts for the peak at 765 °C (Table 6.7).

βUS_2 and UOS appear as impurities in US[129, 130], possibly as a result of partial oxidation. Using an electron microscope, Sole[130] found that UOS and βUS_2 precipitates in US single crystals and described their orientation relationships.

6.5.4 Uranium nitrides

Besson et al.[131], using TGA, studied the isothermal oxidation of sintered components of polycrystalline UN over the temperature range 350–480 °C. They found the products of oxidation to be gaseous nitrogen and an oxide of composition between U_3O_8 and UO_3. Bugl and Bauer[132] again followed the weight change during oxidation of UN in various gaseous environments using nitride compacts of variable composition and density. They found that stoichiometric single crystals of UN reacted in O_2, CO_2 and wet air above 300 °C and rapidly at 500 °C. Powder-compacted UN, 96% dense, was found to react thirty times as fast as the single crystals.

Dell et al.[133] have investigated quantitatively the oxidation of UN powder and single crystals in pure oxygen using isothermal gravimetric techniques and a variety of subsidiary analytical methods. They found that powdered UN reacted isothermally with oxygen between 230 °C and 270 °C, the kinetics of the reaction being independent of oxygen pressure and the activation energy 31 ± 1 kcal mol^{-1}. Complete oxidation at these temperatures gave a product oxide containing nitrogen of composition $UO_3N_{0.2-0.4}$. Subsequent heating to 500 °C was needed to remove the remaining nitrogen, the final product being αUO_3. It was suggested that the excess nitrogen was trapped in the oxide lattice as a result of the mechanism of the reaction. X-ray analysis of partially oxidised UN revealed the presence of αU_2N_3 as well as UO_2 and UN. Since molecular nitrogen does not react with UN to form αU_2N_3 at the temperatures studied, the formation of the αU_2N_3 was considered to be an intrinsic part of the reaction mechanism and not a by-product.

Single crystal studies of the oxidation of UN using x-ray techniques[133] and electron microscopy[134] confirm the presence of αU_2N_3 as a definite intermediate in the reaction. The oxidation products present on the surface of the UN consist of a layered structure in which the αU_2N_3 is sandwiched between the UN crystal and an outer layer of UO_2. Both the U_2N_3 and the UO_2 are epitaxially orientated with respect to the underlying UN. According to the mechanism of the oxidation proposed by Dell et al.[133], (Figure 6.6) oxygen is first chemisorbed at the outer oxide surface, then diffuses interstitially through the UO_2 lattice to the UO_2/U_2N_3 interface where it reacts to release nitrogen and form further UO_2. Some of the released nitrogen is liberated as gas, some dissolves in the oxide and some reacts with the underlying U_2N_3 to give U_2N_{3+x}. This dissolved nitrogen next diffuses to the U_2N_3/UN interface under the influence of a nitrogen concentration gradient, and there converts further UN to U_2N_3. Apart from the first step the entire process is a solid–solid reaction, so accounting for the epitaxial orientation of the U_2N_3 and UO_2 layers with respect to the substrate UN crystal.

The same authors have also studied the ignition of UN powder in the temperature range 280–400 °C [135]. They list more than twelve factors which

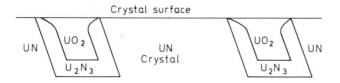

Initial stages of oxidation-product nuclei on surface of UN crystal

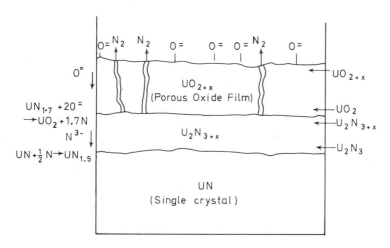

Later stages of oxidation - sandwich structure

Figure 6.6 Mechanism of oxidation of UN

(From Dell, Wheeler and McIver[133], by courtesy of the Faraday Society)

influence the ignition temperature. Ignition of UN was found to give U_3O_8 as the pure oxide containing no residual nitrogen; this is a consequence of the high temperatures reached momentarily during ignition.

Uranium mononitride is quite stable in boiling water at 100 °C. The hydrolysis of UN has therefore been followed at higher temperatures in water vapour/argon mixtures of controlled composition[136]. Using powdered

UN the reaction rates were measured between 340 °C and 420 °C and the reaction was found to proceed broadly as:

$$UN + 2H_2O \rightarrow UO_2^{\cdot} + NH_3 + \tfrac{1}{2}H_2,$$

although some specimens yielded products containing significant quantities of αU_2N_3 and dissolved nitrogen. A similar epitaxial relationship was found between the U_2N_3, UO_2 and UN in partially reacted single crystals, indicating a similar mechanism to that for the oxidation of UN.

The relative stability of UN to water compared to UC and ThN is attributed to the presence of this coherent layer of αU_2N_3. Sole found U_2N_3 to be the initial product of oxidation[134] and the αU_2N_3 is probably present as a protective layer on pure UN.

Further work by Sugihara and Imoto[137] has been carried out on the reaction of UN and U_2N_3 with steam in inert and nitrogen atmospheres. The hydrolysis of U_2N_3 was found to be complex, the products (as in the case of UN) being UO_2, U_2N_{3+x}, NH_3 and UO_2 with dissolved nitrogen.

A systematic study of the reaction of UN with acids has also been carried out[138]. UN was found to be inert to HCl and H_2SO_4 but the reaction with nitric acid yielded uranyl nitrate and gaseous products (nitrogen and nitrogen oxides). The rate was dependent on the acid concentration.

6.5.5   Uranium monophosphide

Baskin[139] found UP to exhibit a greater resistance to oxidation than UC, UN, US and UAs. It is the only compound which, in the powdered state, does not ignite on being heated fairly rapidly in oxygen. DTA measurements revealed three broad peaks in the oxidation of UP powder, at 440 °C, 560 °C and 700 °C. X-ray analysis of the partially oxidised UP showed the first two peaks to correspond to the oxidation of UP to amorphous UO_2 and P_2O_5. The higher phosphides UP_2 and U_3P_4 were also present, this being analogous to the formation of US_2 and U_2N_3 during the oxidation of US and UN respectively. The DTA peak at 700 °C represents the oxidation of UO_2 to U_3O_8. Above 700 °C the U_3O_8 reacted with the P_2O_5 to give $(UO_2)_2P_2O_7$ as the final product. The stability of UP to oxidation is attributed to the surface formation of the amorphous phase containing P_2O_5 and UO_2. Being a vitreous phase this effectively excludes gases and limits further oxidation. UP is also very resistant to hydrolysis and forms a hydrated uranyl phosphate which gives similar protection to the underlying UP.

6.5.6   Plutonium nitride

The oxidation and hydrolysis of PuN have been studied by Dell and Bridger[254]. PuN powder reacts with dry oxygen below 280 °C to give $PuO_{2.10}N_{0.10}$; the excess oxygen may be chemisorbed on the finely divided oxide and the nitrogen dissolved in the oxide, as in the case of UN. The kinetics obeyed a para-linear rate law, the linear part having an activation energy of 10 kcal mol^{-1}. Above 280 °C the PuN ignites in dry oxygen to PuO$_2$. Small amounts of water vapour in the oxygen markedly increased the oxidation rates, while wet oxygen ($P_{H_2O} = 20$ Torr) caused the powder to ignite at a temperature as low as 166 °C. A study of the hydrolysis of PuN using TGA confirmed the sensitivity of PuN to water vapour. The reaction proceeds as:

$$PuN + (2+x)H_2O \longrightarrow PuO_2 \cdot xH_2O + NH_3 + \tfrac{1}{2}H_2$$

The value of x depended entirely on the temperature of the hydrolysis and was independent of the partial pressure of the water vapour. Values of $x = 2$ at 20 °C, $x = 0.45$ at 60 °C decreasing to $x = 0$ above 350 °C were found. In the temperature range 50–350 °C the apparent activation energy is 1–2 kcal, indicating a diffusion controlled reaction. It was concluded that diffusion of gas into and out of the pores of the material controlled the reaction rate, the oxide product being completely non-protective. UN and PuN stand in striking contrast as regards their reactivity to water vapour, the former being protected by an epitaxial surface layer of U$_2$N$_3$ while plutonium has no higher nitride. The reactivity of ThN may be ascribed to the fact that the higher nitride of thorium (Th$_3$N$_4$) is not epitaxial on ThN and therefore is also non-protective.

6.6 MAGNETIC AND ELECTRICAL PROPERTIES

Magnetic property measurements have largely been confined to uranium compounds. This is because thorium compounds are generally diamagnetic, and therefore of limited interest, while the determination of magnetic properties for radioactive compounds necessitates the use of special equipment and handling techniques. A few measurements on plutonium compounds have been reported recently. Electrical property measurements (resistivity, thermoelectric effect, Hall coefficient) have been made on a limited scale, particularly for the monocompounds of rocksalt structure.

6.6.1 Thorium compounds

The magnetic susceptibility of ThN is independent of temperature with a value of $\sim 35 \times 10^{-6}$ e.m.u. mol^{-1}[140, 141]. Allowing for the diamagnetic contributions of the Th^{4+} ion and the N^{3-} ion, the paramagnetic susceptibility of ThN is about 70×10^{-6} e.m.u. mol^{-1}. This small value confirms that no unpaired electrons are present in localised states; the temperature-independent paramagnetism stems from the Pauli paramagnetism of free electrons in the conduction band (cf. TiN, YN etc.). Th$_3$N$_4$ has an even smaller susceptibility, $\sim 1 \times 10^{-6}$ e.m.u. mol^{-1}, in agreement with its high resistivity (see below).

Comparatively few other thorium pnictides or chalcogenides have been studied in any detail. All the sulphide phases are diamagnetic according to Eastman et al.[18], indicating that the oxidation state of thorium is +4 even in the sub-sulphides. Later investigations have confirmed that ThS, like ThN, exhibits a small, temperature-independent susceptibility, again associated with conduction electrons[68, 100]. Th_3P_4 and ThP_{1-x} behave similarly[98]. Recent work by Russian and Japanese authors has been concerned with the magnetic properties of solid solutions of US—ThS[142, 143], UP—ThP[101, 144], UP_2—ThP_2[144] and USb_2—$ThSb_2$[145] (see Section 6.7).

The nuclear magnetic resonance of ^{14}N in ThN and ^{31}P in $ThP_{0.95}$ has been studied by Kuzneitz[146]. The shapes and widths of the n.m.r. lines were independent of temperature (77–300 K) and applied magnetic field, as expected for diamagnetic compounds. Measurements of the Knight Shift

$$K = \frac{H_{ref} - H_0}{H_0}$$

where H_0 is the resonant field for the specimen and H_{ref} that of the nuclide in a reference substance, gave positive values of K for ThN and ThP and negative values for ThC and ThC_2. This is taken by the authors as evidence for the ionic character of the pnictides. The nuclear spin-lattice relaxation time (T_1) and free inductive decay time constant (T_2) for ^{31}P in $ThP_{0.95}$ have also been measured using pulsed n.m.r. techniques[147].

Th_3N_4 has a high resistivity[148] (10^5–10^8 Ω cm at 20 °C) and a temperature-coefficient typical of an ionic semi-conductor, confirming that Th_3N_4 is a purely ionic compound. In contrast, ThN has a resistivity of only 2×10^{-5} Ω cm at 20 °C, a value almost identical to that of thorium metal. The resistivity increases linearly with temperature throughout the range 4–850 K. The Hall coefficient of ThN has a value of -1.6×10^{-4} cm^3 C^{-1}, independent of temperature, indicating electronic conduction in a single band. From this value, the number of current carriers is estimated as 1.47 electrons per thorium atom. We may therefore represent ThN as $Th^{4+}N^{2.53-}(1.47\varepsilon)$, where the formal charge on the nitride ion is less than $3-$. Auskern and Aronson[148] also report thermoelectric power measurements on ThN.

Both ThP and ThS are excellent metallic-type conductors, the measured resistivities at 20 °C being 34 $\mu\Omega$ cm for ThP[149] and 16–70 $\mu\Omega$ cm for ThS[22, 150]. Neither specimen was of particularly high purity or density, so that the high conductivities are even more remarkable. Similar results for ThP were obtained by Adachi and Imoto[101]. The Seebeck coefficient of ThS has been measured over the temperature range 20–1000 °C[150]; as with ThN[148], it has a small negative value which increases (negatively) with increasing temperature. This is consistent with the metal-type conductivity observed in these compounds.

Thermoelectric power measurements on Th_3P_4, Th_3As_4, Th_3Sb_4 and solid solutions $Th_3(As, Sb)_4$ have also been reported[87]. The theoretical significance of these results has not been discussed.

6.6.2 Uranium chalcogenides

The magnetic data which have been reported for the uranium chalcogenides are summarised in Table 6.8. In general the agreement between the results of different investigators is rather poor; this may be attributed to the difficulty of preparing these compounds as pure single phases and the disproportionate effect that impurity phases, particularly ferromagnetic compounds, can exert upon magnetic behaviour.

In most cases the measurements have only been made down to 78 K and magnetic ordering which occurs at lower temperatures has to be inferred from the sign of the Weiss constant. It seems likely that the higher phases UX_3, UX_2 (and U_3S_5) will order antiferromagnetically with Néel temperatures below 78 K. Support for this view is provided, in the case of USe_2, by heat capacity measurements[151] which reveal a λ-type anomaly in the C_p

Table 6.8 Magnetic properties of uranium chalcogenides

Compound	Paramagnetic moment μ_{eff} (BM)	Weiss constant θ_p (°)	Curie temperature Tc (K)	Ferromagnetic moment μ_{ferro} (BM)	References
US$_3$	3.42, 3.0	$-150°$, $-100°$			7, 153
US$_2(\beta)$	2.83, 3.14	$-30°$			7, 153
US$_{1.9}(\alpha)$	1.83, 3.07	$-30°$, $-50°$			7, 153
U$_3$S$_5$	3.42, 3.10	$-20°$, $0°$, $-16°$			7, 153, 154
U$_2$S$_3$	2.53	$+27°$			153
US	2.25, 2.22	$+173°$, $+190°$	178, 180	1.1, 1.05, 1.55	68, 100, 153-7
USe$_3$	3.05	$-40°$, $-120°$			7, 33
USe$_2$	3.05, 3.2	$-40°$, $-10°$			7, 33, 154
U$_2$Se$_3$	3.2	$-10°$	180		154
U$_3$Se$_4$	2.45, 3.06	$+164°$	160, 130	0.5	154, 158
USe	1.8, 2.51	$+182°$, $+188°$	180, 185	0.7, 1.0, 1.3	33, 154, 158
UTe$_3$	3.16	$-56°$			159
UTe$_2$	3.12, 3.35	$-54°$, $-80°$			46, 159
U$_2$Te$_3$		$+123°$	122		160
U$_3$Te$_4$	3.12		105, 120	0.44	158, 160
UTe	2.36, 2.84	$+104°$	103	1.10	160, 161

Table 6.9 Calculated magnetic moments for uranium ions in various configurations

Configuration	Ground state	Free ion moment (BM)	Spin-only moment (BM)
5f^1	$^2F_{5/2}$	2.54	1.73
5f^2	$^3H_4$	3.57	2.83
5f^3	$^4I_{9/2}$	3.61	3.88
5f^{1}6d^1	$^3H_4$	3.57	2.83
6d^2	$^3F_2$	1.63	2.83

curve at 13 K. This anomaly, which is associated with an entropy increment of 0.16 cal mol^{-1}K^{-1}, presumably results from magnetic ordering. However, no such anomaly is observed in the C_p curves for US_2 and US_3, at least down to 5 K (Figure 6.7)[152]. Possibly ordering occurs at a still lower temperature.

The paramagnetic moments of the higher chalcogenides all lie in the range 3.1 ± 0.3 BM. This closeness of the moments for UX_3 and UX_2 is remarkable and suggests that the uranium ions have the same configuration in all six compounds. By comparison with the calculated magnetic moments of uranium ions in various ground states (Table 6.9) it seems probable that

U^{4+} ions are present with either the $5f^2$ or $5f^16d^1$ configuration and with a magnetic moment intermediate between the spin-only and free-ion values. Some indirect support for this conclusion stems from the observation that the UX_3 compounds crystallise in the $ZrSe_3$ structure which is known only for tetravalent cations (Ti, Zr, Hf, Th). Also, the oxidation of US_3 in air leads initially to UO_2 rather than UO_3. In order to explain the occurrence

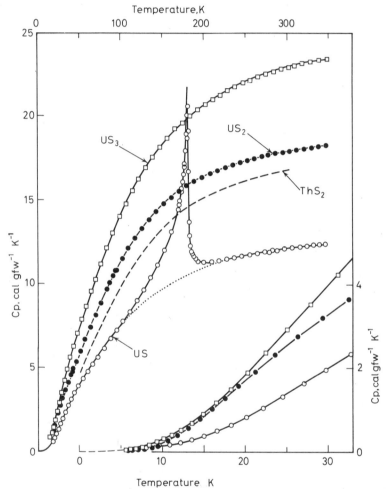

Figure 6.7   Molar heat capacities of US_3, US_2, US and ThS_2
(From Westrum and Grønvøld[152], by courtesy of the International Atomic Energy Agency)

of U^{4+} ions in US_3, USe_3 and UTe_3 it is necessary to postulate that these compounds contain polychalcogenide ions in which two anion valency electrons per formula are saturated by anion–anion bonds.

With increasing uranium content in the chalcogenides, the ferromagnetic interactions become stronger than the antiferromagnetic interactions and most of the compounds U_2X_3, U_3X_4 and UX order ferromagnetically. The

Curie temperatures, in the range 100–180 K, are remarkably high when compared to those of the corresponding rare-earth chalcogenides. This may be understood in terms of the fact that 5f orbitals are more spatially extended than 4f, leading to stronger magnetic interactions between neighbouring ions. U_2Se_3 is an uncertain intermediate case; Checkernikov *et al.*[154] reported a small magnetisation up to a Curie temperature of 180 K, with a superimposed strong antiferromagnetic interaction leading to $\theta_p = -10°$. They also observed that the compounds U_3Se_4, USe[154], U_3Te_4 and UTe[160] in the ferromagnetic region exhibited a magnetisation curve $\sigma(T)$ containing a maximum. With increasing applied field the maximum in $\sigma(T)$ shifted towards lower temperatures. This behaviour has been

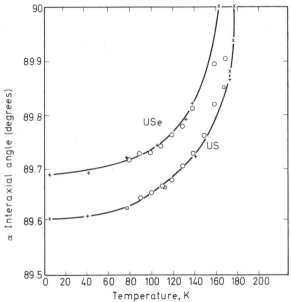

Figure 6.8 Variation of inter-axial angle with temperature for US and USe
(From Marples[166], by courtesy of *J. Phys. Chem. Solids*)

re-investigated and broadly confirmed[158]. A related phenomenon is observed in $UP_{0.75}S_{0.25}$ (see Section 6.7) where there are known to be transitions from a ferro- to a ferri-magnetic state and then to an antiferromagnetic state as the temperature is lowered.

Uranium monosulphide, in contrast, appears to behave as a simple ferromagnet. The spontaneous magnetisation as a function of temperature has been investigated and found to decline monotonically with increasing temperature, there being no evidence of a maximum[154–157]. The ferromagnetic Curie temperature (178–180 K) agrees excellently with the temperature of a sharp peak in the $C_p(T)$ curve (Figure 6.7). The saturation ferromagnetic moment measured for polycrystalline samples in fields of 10 koersted was found to be 1.05–1.25 BM[154, 156]. However, measurements on a single crystal sphere of US in the direction of easy magnetisation, in fields up to 40 koersted, yielded a value for the saturation magnetisation at 0 K of 1.55 BM[157].

This value is still substantially smaller than the paramagnetic moment (2.25 BM) which, in turn, is less than the paramagnetic moment in the higher chalcogenides UX_2 and UX_3. This raises the question whether US has the $5f^2$ (U^{4+}) configuration or whether it has a different electronic structure. This has been debated by several authors[157, 162−164] who arrive at different conclusions; at present the situation remains unresolved. Finally, neutron diffraction experiments on a single crystal of US have shown that in the ferromagnetic state the easy direction of magnetisation is along the (111) axis[165]. The internal anisotropy field was estimated to be 35 koersted.

Further indirect evidence relating to ferromagnetic ordering in US and USe has been obtained from low temperature x-ray diffraction studies. Marples has shown that below T_c (180 K) US transforms from cubic to rhombohedral symmetry[166]. The interaxial angle α, initially 90 degrees, decreases at first rapidly and then more slowly with falling temperature, to reach 89.6 degrees at 4 K (Figure 6.8). Similar behaviour was observed for USe, except that the transformation began at 160 K rather than at the Curie temperature (see above). These rhombohedral distortions are in agreement with the observation, cited above, that ferromagnetic US has unpaired spins which are aligned along the (111) directions.

The electrical properties of the uranium chalcogenides have been investigated only in broad outline. Single crystals of βUS_2 have been shown to be semi-conductors with a resistivity of $\sim 25 \, \Omega$ cm at room temperature[41]; this is in marked contrast to ThS_2, which is an insulator of resistivity $\sim 10^{10}$ Ω cm[167]. The other uranium dichalcogenides USe_2 and UTe_2 have resistivities of $\sim 10^{-2} \, \Omega$ cm[34, 160]. It is evident that whereas ThS_2 is a simple ionic compound, the uranium dichalcogenides possess a conduction band. Assuming that f electrons from the uranium contribute to this band, this implies that less than two electrons remain localised on the uranium ion.

Electrical measurements have also been made on the monocompounds $US^{22, 150}$, USe^{34} and $UTe^{34, 160}$. The reported data are summarised in Table 6.10.

Table 6.10   Electrical properties of uranium monochalcogenides at 300 K

	US	USe	UTe
Electrical resistivity ($\mu\Omega$ cm)	286	280	1300
Hall coefficient (cm³ C⁻¹)	—	+0.008	—
Thermoelectric power (μV/K)	+50	+38	—

The resistivity of US shows only a very weak dependence on temperature over the range 20–900 °C and has a numerical value several times larger than that of ThS. ThS is a typical metal, while US and USe have positive thermoelectric power coefficients indicating semi-metallic behaviour and positive hole conduction. Recent photoelectron emission studies on US[168] have provided valuable information on the band structure of this compound.

6.6.3   Uranium oxychalcogenides

Each of the three uranium oxychalcogenides has been investigated by magnetic and neutron diffraction techniques. The compounds all order antiferromagnetically on cooling. In the paramagnetic region, susceptibility measurements on UOS[169] and UOSe[44] are found to yield a plot of $1/\chi$ versus T which consists of two linear portions intersecting at a transition temperature T_t. For UOTe no such transition was observed and the Curie-Weiss plot was linear[46]; however, in this case the measurements were made only to 400 K and it is possible that the transition occurs at a higher temperature. The experimental results are summarised in Table 6.11. Further data, not in particularly good agreement with these, are reported briefly by Boelsterli and Hulliger[26].

Table 6.11   Magnetic properties of uranium oxychalcogenides

Property	UOS[169]	UOSe[44]	UOTe[46, 170]
Néel temperature T_N	55 K	90 K	162 K
Transition temp. T_t	450 K	300 K	?
Magnetic moment ⎱ below T_t	2.78 BM	2.87 BM	3.26 BM
Weiss constant ⎰	$-51°$	$-10°$	$-56°$
Magnetic moment ⎱ above T_t	3.13 BM	3.30 BM	
Weiss constant ⎰	$-108°$	$-120°$	
Ordered magnetic moment (neutrons)	1.9	2.21	2.0
Type of magnetic order	+ + − −	+ + − −	+ −

These three compounds all crystallise with the tetragonal PbFCl structure. The neutron diffraction measurements show that in the antiferromagnetic state the uranium ions are aligned parallel to each other within the (001) planes, giving ferromagnetic sheets perpendicular to the c axis. The direction of the spins is along the c axis. In the case of UOS and UOSe the magnetic unit cell is double the crystallographic cell in the c-direction. There are two possible models for this magnetic structure corresponding, respectively, to alignment of (001) planes of uranium ions in the sequence + + − − or + − − +. From a consideration of the intensities of the magnetic reflections it was concluded that the former type of ordering is present in both UOS and UOSe. With UOTe the situation appears to be different as the magnetic unit cell is the same as the crystallographic cell and the ordering is of the type + −[170]. The magnitude of the magnetic moment on the uranium ions in the anti-ferromagnetic state is calculated from the intensities of the magnetic reflections in the neutron diffraction pattern.

From a consideration of bond distances it seems certain that these compounds contain U^{4+} ions with the $5f^2$ configuration (ground state $^3H_4$). In a ligand field of cubic symmetry the ($J = 4$) level splits into a singlet Γ_1, a doublet Γ_2 and two triplets Γ_4 and Γ_5. If the Γ_5 triplet is the lowest level, the effective saturation moment is 2 BM in reasonable agreement with the values deduced from the neutron data at 4.2 K. At higher temperatures the Γ_5 triplet may be expected to be split, giving rise to higher magnetic moments. Similarly, the transition T_t may be associated with the population of an excited state at higher temperatures.

Heat capacity measurements have been made on UOTe over the temperature range 21–362 K[171]. A pronounced λ-type anomaly is observed at 160 K, corresponding to the antiferromagnetic ordering process. The excess entropy increment associated with this peak is 1.07 cal mol^{-1}K^{-1}. No heat capacity data for UOS or UOSe have been reported; these would be of interest on account of the paramagnetic transition observed at T_t. Electrical resistivity measurements on UOSe[44] reveal a maximum in resistivity at the Néel temperature (90 K) and a minimum at 40 K. Metallic-type conductivity is observed in the range 40–90 K and activated conduction below 40 K and above 90 K. Further physical property measurements on UOSe would clearly be desirable to help explain these results.

6.6.4 Uranium pnictides

6.6.4.1 Uranium nitrides

The magnetic susceptibilities of the three uranium nitride phases were first measured by Trzebiatowski et al.[74] over the temperature range 80–300 K. UN and αU_2N_{3+x} showed paramagnetic behaviour, while βU_2N_3 was ferromagnetic below a Curie point of 186 K. The data for UN and βU_2N_3 are summarised in Table 6.12, together with those obtained by Allbutt et al.[100]. There is good agreement between the two investigations for UN, although for βU_2N_3 there is a discrepancy in the observed Curie temperatures. The large negative value of the Weiss Constant for UN indicates antiferromagnetic ordering below 80 K.

Table 6.12   Magnetic properties of UN and βU_2N_3

Compound	Paramagnetic moment (BM)	Weiss constant $\theta p(°)$	Curie temp. T_c(K)	References
UN	3.0	−310°		74
	3.11	−325°		100
βU_2N_3			186	74
	1.84	162°	235	100

Subsequently, Trzebiatowski and Troc studied the magnetic behaviour of the αU_2N_{3+x} phase as a function of composition[71]. The susceptibility falls with increasing N/U ratio as shown in (Figure 6.9); extrapolation to the composition UN$_2$ indicates that this (hypothetical) compound would be diamagnetic, suggesting the presence of U^{6+} ions. This is in marked contrast to the situation for the other dipnictides and dichalcogenides which contain U^{4+} ions and polyanions. The U_2N_{3+x} phase in equilibrium with UN showed a susceptibility maximum (antiferromagnetic ordering) at 96 K. This temperature agrees well with that deduced from heat capacity measurements on UN$_{1.59}$ where there was a small thermal anomaly at 94 K[66]. Similar Cp measurements on a sample of higher N/U ratio (UN$_{1.73}$) revealed that the transition lay at a much lower temperature (33 K) with a greatly reduced enthalpy and entropy of ordering. The increasing proportion of U^{6+} ions present is clearly responsible for weakening the magnetic exchange interactions, so reducing both the Néel temperature and the entropy increment.

Magnetic measurements on UN have been extended down to liquid helium temperatures[141], the observed Néel temperature being 52 K. This value agrees exactly with the temperature of the λ-anomaly in the specific heat curve [66, 172] and also with the antiferromagnetic ordering temperature observed directly by neutron diffraction[173]. The entropy increment associated with antiferromagnetic ordering in UN (0.15 cal mol^{-1} K^{-1}) is considerably

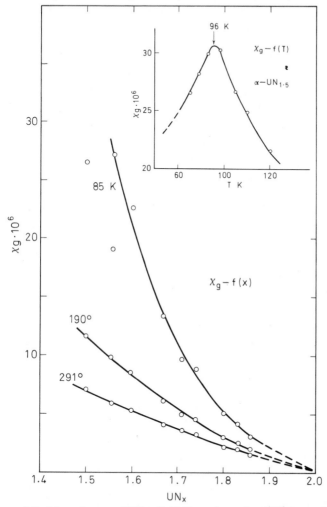

Figure 6.9 Magnetic susceptibility of $\alpha U_2 N_{3+x}$ phase of variable composition (From Trzebiatowski and Troc[71], by courtesy of the Polish Academy of Sciences)

smaller than that in UP (0.74 cal mol^{-1} K^{-1}) and also smaller than the entropy of ferromagnetic ordering in US (1.6 cal mol^{-1} K^{-1}) [174] and USe (1.05 cal mol^{-1} K^{-1}) [151]. In fact, it corresponds to the disordering of only 0.08 spins[66]. The explanation of this small entropy increment is not yet entirely clear, although similar effects in transition metals (e.g. chromium) are attributed to spin density waves.

Neutron diffraction experiments[173] show that UN undergoes Type I antiferromagnetic ordering; that is, uranium ions in individual (001) planes are ordered ferromagnetically, with antiferromagnetic coupling of adjacent planes (Figure 6.10) The magnetic unit cell is the same size as the chemical cell, but is centred on one face only so that the symmetry of the structure is reduced from cubic to tetragonal. The magnetic moments of the uranium atoms are perpendicular to the ferromagnetic sheets. An identical structure is assumed by the other uranium monopnictides in the antiferromagnetic state. From the intensities of the magnetic reflections, the magnetic moment of the uranium atom in antiferromagnetic UN at 12 K was calculated to be 0.75 BM. This is a considerably smaller value than that found for UP (1.7 BM)[16], in qualitative agreement with the smaller entropy of disordering. It is evident that in the ordered state the uranium atoms in UN have few unpaired electrons. This may be understood if the ground energy level is

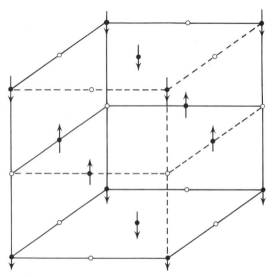

Figure 6.10 Antiferromagnetic structure of uranium monopnictides
(From Curry[173], by courtesy of the Physical Society)

assumed to be split in the cubic crystal field and tetragonal uranium exchange field, so that at low temperatures most of the electrons will be paired in the lower energy state thereby reducing the effective magnetic moment.

Further information on the paramagnetic state of UN stems from n.m.r. measurements on the ^{14}N isotope[175]. The Knight shift (K) is found to be a linear function of the paramagnetic susceptibility, $K = K + \alpha\chi_m$. The temperature independent term K has a negative value for UN (-32×10^{-4}) but a positive value for ThN ($+11 \times 10^{-4}$). This difference is ascribed to temperature independent (Van Vleck) paramagnetism in UN, with a magnitude of χ_0 1000×10^{-6} e.m.u. mol^{-1}. This is comparable to the temperature-independent susceptibility of UC; making allowance for it would significantly change the slope of the Curie-Weiss plot and therefore the derived paramagnetic moment of UN. Raphael and de Novion have extended their susceptibility measurements on UN to 1000 K [141] and find a value of $\chi_0 = \cdot 500 \times 10^{-6}$ e.m.u. mol^{-1}. Applying the Curie-Weiss law in the modified form

$$\chi = \chi_0 + \frac{C}{T - \theta_p},$$

they then obtain values of 2.06 BM for the paramagnetic moment of UN and $-160°$ for the Weiss constant (cf. Table 6.12). This study emphasises the importance of making magnetic measurements at high as well as low temperatures when there is any possibility of a temperature-independent component to the susceptibility.

A number of other physical properties of UN undergo transition at the Néel temperature. The expansion coefficient, as determined by low temperature x-ray diffraction measurements, changes sign at 52 K, there being a small lattice expansion on cooling from 52 K to 4 K [166]. This is attributed to a stabilisation of the 5f band in the antiferromagnetic state, with consequent electron transfer from the valence band to the 5f band. Similarly, the curve of thermoelectric power versus temperature shows an abrupt change of slope at 52 K [176]. The electrical resistivity of UN shows a metallic-type temperature dependence with a room temperature value of 150–250 $\mu\Omega$cm[176-177]. A comprehensive study of the physical properties of UN is in progress at Oak Ridge National Laboratory[178].

6.6.4.2   Uranium phosphides

The paramagnetic properties of the three uranium phosphides have been reported by several different authors (Table 6.13). The observed values are not in good agreement, possibly[179] because of variations in purity and stoichiometry from sample to sample.

The magnetic moment increases in the order $UP_2 < U_3P_4 < UP$. On the assumption that each compound contains U^{4+} ions (5f^2, $^3H_4$) the value for

Table 6.13   Paramagnetic properties of the uranium phosphides

Compound	Paramagnetic moment (BM)	Weiss constant θ_p	Ordering temperature (K)		References
			T_N	T_C	
UP_2	2.24–2.30	$+77$–$86°$	199–206°		100, 101, 103, 144
U_3P_4	2.61–2.77	$+134$–$151°$		134–165°	100, 101, 103
UP	3.17–3.56	-15–$49°$	116–130°		100, 101, 103, 144, 179, 180

UP (Table 6.13) is nearest to that calculated for the free ion. Crystal field splitting of the ground state plays an increasing role in U_3P_4 and UP_2. Troc et al. have discussed recently the values of the magnetic moments of U_3P_4 and similar Th_3P_4-type compounds[181]. Further information on the paramagnetic state of uranium phosphides is derived from n.m.r. measurements on the ^{31}P isotope[182-186]; these experiments, together with spin-lattice relaxation time measurements[147], provide evidence of an indirect coupling between the ^{31}P nuclei and the localised uranium moments via the conduction electrons.

On cooling, UP and UP_2 both order antiferromagnetically while U_3P_4 orders ferromagnetically. The Néel and Curie temperatures deduced from the magnetic data are confirmed by the positions of the λ point anomaly in the specific heat curves. Low temperature heat capacities and thermodynamic functions have been determined for all three phosphides. The reported ordering temperatures are UP 121 K [16], U_3P_4 136.5 K [16] (138.3 K [253]) and UP_2 203.2 K [187]. The entropy increments associated with the ordered → paramagnetic transition are generally smaller than predicted from the value of the magnetic moment. Uranium monophosphide is unique in possessing a second, exceedingly sharp C_p anomaly at 22.5 K. This unexpected observation has been further investigated by low temperature neutron diffraction and magnetic susceptibility studies.

Two independent neutron diffraction investigations have shown that UP, like UN, assumes type I antiferromagnetic order below T_N [188−189]. On cooling further, a sudden, reversible increase in the intensity of the magnetic diffraction peaks is observed at 25–30 K [188, 190]. This may be explained by a change in the ground state of the uranium ion, leading to an increase in the magnitude of the ordered magnetic moment from 1.7 BM to 1.9 BM[188]. The magnetic ordering remains type I antiferromagnetic down to 4 K and it is only the numerical value of the magnetic moment which changes. Magnetic susceptibility measurements show a peak at 22.5 K with a sharp drop in susceptibility below this temperature[180, 191]. Finally, n.m.r. studies of the ^{31}P nuclide in UP have revealed an abrupt transition in the frequency of the resonance line at 22.9 K [192]. This too may be explained by a change in the hyperfine field at the ^{31}P nucleus.

Unlike UN, the curve of lattice parameter against temperature for UP passes smoothly through T_N, though there is evidence of a small expansion anomaly below 22.5 K. This rather surprising[166] difference in behaviour is attributed to a lower density of states of UP at the Fermi surface compared to UN, resulting in the absence of interband electron transfer in UP.

Neutron diffraction studies have also been made on ferromagnetic U_3P_4 [193] and antiferromagnetic UP_2 [194]. For U_3P_4, the magnitude of the ordered magnetic moment so calculated (1.5 ± 0.1 BM) is in fair agreement with that from specific magnetisation measurements (1.25 ± 0.5 BM). No information on the alignment of the spins with respect to the crystallographic axes can be deduced from powder data on a cubic ferromagnetic substance, although single crystal measurements on U_3P_4 at 73 K have shown a strong magnetic antisotropy with the easy direction along [111][112] (cf. US, USe). For UP_2, which is tetragonal, the antiferromagnetic unit cell is twice the chemical one in the c axis direction. The magnetic structure is obtained by stacking ferromagnetic sheets of uranium ions along the c axis in the order $+ - - +$ (cf. UOS). The magnetic moments of the U ions are pointing in the direction of the c axis and of magnitude 1.0 ± 0.1 BM.

On cooling, the resistivity of UP rises from its room temperature value (500 $\mu\Omega$cm) to a maximum of 600 $\mu\Omega$cm at the Néel temperature. In the antiferromagnetic state metal-type conduction is observed, the resistivity falling with decreasing temperature. No measurements at the 22.5° transition have been reported. Electrical property measurements have also been carried out recently on single crystals of U_3P_4 [195] and UP_2 [196]. The latter compound exhibits a strong anisotropy of the transport properties along the a and c axes of the tetragonal structure; for example, the resistivity is $\sim 700 \, \mu\Omega$cm along the c axis and $\sim 200 \, \mu\Omega$cm perpendicular to the c axis. Hall coefficient and thermoelectric power measurements are also reported.

6.6.4.3 Uranium arsenides, antimonides and bismuthides

The magnetic data which have been reported for the uranium arsenides, antimonides and bismuthides are summarised in Table 6.14.

Table 6.14   Magnetic properties of uranium arsenides, antimonides and bismuthides

Compound	Paramagnetic moment (BM)	Weiss constant θ (°)	Ordering temperature (K)		Ordered moment (BM)	References
			T_N	T_C		
UAs_2	2.94	+34	283°		1.61	105,197
USb_2	3.04	+18	206°		0.94(78 K)	105,198
UBi_2	3.40	−53	183°		2.1 (78 K)	111,198
U_3As_4	2.94, 2.81	+200		198°	1.71	105,181,112
U_3Sb_4	3.04, 3.01	+155, +148		146°		105,181
U_3Bi_4	3.14	+110		108°		111
UAs	3.54	+32	128°		1.89, 1.93 (78 K) 2.20 (4 K)	105,199,200
USb	3.85	+95	213° ⎱ 246° ⎰		2.08, 2.78 (78 K) 2.16, 2.85 2.64 (4 K)	105,109, 200,201
UBi	4.06	+105	290° ⎱ 285° ⎰		3.0	111,109

Within each of the three series UY_2, U_3Y_4 and UY there is a progressive increase in the paramagnetic moment as the non-metal increases in size from phosphorus to bismuth. No simple correlation exists between the measured paramagnetic moments and calculated values for various uranium ion configurations (Table 6.9).

On cooling, the UY_2 and UY compounds order antiferromagnetically and the U_3Y_4 compounds ferromagnetically. Saturation magnetisation measurements on single crystals of U_3As_4 in the ferromagnetic state have been reported[112]. For the monocompounds the Néel temperature increases progressively from UP ($T_N \sim 121$ K) to UBi ($T_N \sim 290$ K). However, for the dipnictides there is a sharp maximum in T_N at UAs_2, with the Néel temperature of UP_2 being close to that of USb_2 (Tables 6.13 and 6.14). A similar effect is seen in the Curie temperatures of the ferromagnetic U_3Y_4 series. In the ordered state, the magnetic moment of U_3As_4 (1.71 BM) is greater than that of U_3P_4. The magnetic moments of the antiferromagnetically ordered compounds are derived from neutron diffraction measurements; these values are inevitably rather imprecise as they involve the assumption of a particular uranium ion configuration and the corresponding form factor curve. For instance, the difference in the reported values for the ordered moment of USb may arise from the assumption of a $5f^3$ configuration by Kuzneitz et al.[109] for this compound and UBi rather than the more usual $5f^2$. On the other hand, it is equally possible that the difference originates in stoichiometry problems, for the spread of Néel temperatures between 213 and 246 K has been correlated with compositional differences in USb specimens[201].

At 77 K the monocompounds UAs, USb and UBi have all assumed the type I antiferromagnetic structure favoured also by UN and UP (Figure 6.10). There is no evidence of a further magnetic transition comparable to that

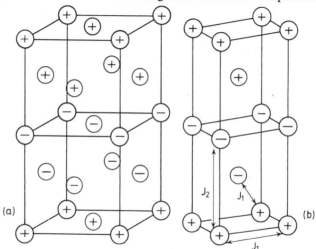

Figure 6.11   Magnetic structure of UAs at 4 K
(From Leciejewicz, Mrasik and Troc[200], by courtesy of Academic Press)

observed in UP at 22.5 K. However, UAs is exceptional in that its magnetic structure changes on cooling to 4 K to produce a new magnetic unit cell, doubled in one direction with the moments aligned along the tetragonal unique axis[200, 202]. The sequence of ferromagnetic sheets which are perpendicular to the tetragonal axis is now $+ - - +$ rather than $+ -$ as found at 78 K (Figure 6.11). This structure bears a close formal resemblance to the magnetic ordering scheme in UP_2 (Section 6.6.4.2). The transformation temperature has not been determined, either directly by monitoring the neutron diffraction pattern as a function of temperature, or indirectly by specific heat or electrical resistivity measurements. It is of interest that this change of magnetic structure does not occur in UN, UP, USb or UBi. In the usual type I magnetic structure the nearest neighbour U–U interaction (J_1) is two thirds negative and one third positive while the next nearest neighbour interaction (J_2) is positive. In UAs at 4 K, both J_1 and J_2 are two thirds positive and one third negative. The stability of the latter structure presumably results from the particular combination of magnetic moment and lattice parameter existing in UAs at low temperatures.

The antiferromagnetic structures of USb_2 and UBi_2 have also been determined by neutron diffraction at 78 K [198]. Again it is found that the magnetic cell of USb_2 is twice the size of the chemical cell along the c axis (ferromagnetic sheet sequence $+--+$) whereas for UBi_2 the magnetic and chemical unit cells are identical ($+-$). In both cases the magnetic moments are parallel to the c axis with the values listed in Table 6.14.

Few other investigations of the higher uranium pnictides have been made. The Mössbauer spectrum of USb indicates [203] that the saturation hyperfine field is 170 ± 5 koersted at the Sb nucleus at 4 K. The presence of this unique field value at all the Sb anions is consistent with the type I antiferromagnetic structure determined by neutron diffraction. The hyperfine field at the U nucleus is 4500 ± 200 koersted. Finally, a few values of electrical resistivity and thermoelectric power for rather poorly defined specimens of UAs and USb have been reported[204].

6.6.5 Plutonium compounds

Our knowledge of the magnetic properties of plutonium compounds stems largely from the work of Raphael in France[141, 205, 206], although some measurements have been made also in England and U.S.A. The first remarkable observation is that PuO_2, which is known to be an ionic compound containing the Pu^{4+} ($5f^4$) ion, has a small susceptibility ($\sim 500 \times 10^{-6}$ e.m.u. mol^{-1}) independent of temperature from 4° to 1000 K*. This compares with Curie-Weiss behaviour in $Pu(SO_4)_2 \cdot 2H_2O$ (2.9 BM), NpO_2 (2.95 BM) and UO_2 (3.28 BM). The behaviour of PuO_2 is explained in terms of the influence of the strong crystal field of the eight O^{2-} ions surrounding the Pu^{4+} on the energy levels of the $5f^4$ configuration. The crystal field raises the degeneracy of the fundamental multiplet $J = 4$ and leads to a ground state Γ_1 which is non-magnetic. The first excited magnetic state is then > 2300 cm^{-1}.

Raphael has also found that PuN, PuS, Pu_3S_4, αPu_2S_3 and PuS_2 all have susceptibilities of $500-700 \times 10^{-6}$ e.m.u. mol^{-1} at 300 K and show little or no temperature dependence up to 1000 K[141, 205, 206]. At lower temperatures the susceptibility does increase and, for the sulphides, reaches a value of $(1000-4000) \times 10^{-6}$ e.m.u. mol^{-1} at 4 K, depending on the compound. The experimental magnetic moment is considerably less than that calculated for the free Pu^{3+} ion, considered by Marcon (on the basis of crystallographic data) to be present in PuS_2 and αPu_2S_3. It is concluded that in the sulphides crystal field quenching is incomplete; a configuration is deduced with half integral J. For PuS, a basic configuration $J = 5/2$ with a fundamental level Γ_7 and a first excitation level Γ_8 separated by 1000 K fits the experimental data. For PuN the measured susceptibility rises from $\sim 600 \times 10^{-6}$ at 300 K to a maximum of $\sim 750 \times 10^{-6}$ at 13 K and then falls abruptly[141]. It is postulated that antiferromagnetic ordering occurs at 13 K.

*This result is in conflict with earlier data of Dawson[207].

The magnetic susceptibilities of PuN, PuS_2 and PuS were measured also by Allbutt and Dell over the temperature range 100–300 K [129], with results in general agreement with those of Raphael. Further measurements on PuSe and PuTe showed similar behaviour to that of PuS, though with the susceptibility increasing somewhat with increasing atomic number of the non-metal[52].

PuC behaves differently from the other plutonium compounds in showing a large, temperature dependent susceptibility, following the Curie-Weiss law with a paramagnetic moment $\mu \sim 1.7$ BM[208]. Even so, in view of the behaviour of PuN it is rather surprising to find that PuP is similar to PuC. A detailed study of PuP[209] showed Curie-Weiss behaviour with a paramagnetic moment of 1.06 BM and ferromagnetic ordering below a Curie temperature of 126 K. Saturation magnetisation measurements gave a ferromagnetic moment of 0.42 BM. N.M.R. studies were also made on PuP, including measurements of the ^{31}P Knight-shift and spin-lattice relaxation times[209–210]. An effective hyperfine field of 51 koersted at the ^{31}P nucleus was deduced.

Electrical measurements on the plutonium compounds are limited. PuN is reported to possess a negative thermoelectric power which has a maximum value at ~ 120 K [176]. Its electrical resistivity is metal-like, increasing from $\sim 10\ \mu\Omega$cm at 4 K to $\sim 650\ \mu\Omega$cm at 300 K, with a change of slope near 120 K. From these measurements it was suggested that PuN orders below 120 K [176], but the magnetic measurements[141] do not support this suggestion.

The electrical resistivity of PuS and Pu_3S_4 has been measured by Marcon[51], while Kruger and Moser have reported on the resistivity and thermoelectric power of PuS and PuP[211]. Each of the sulphides is a semiconductor with activation energies for conduction of 0.49 eV(Pu_3S_4) and 0.21–0.24 eV(PuS). A plot of thermoelectric power shows that PuS is an n-type semiconductor with a peak in the thermal e.m.f. curve at about 475 K.

PuP is intermediate between PuN and PuS in its electrical behaviour. On heating above 300 K, its resistivity rises to pass through a shallow maximum at ~ 700 K and then falls at higher temperatures. Thermoelectric measurements reveal a change from positive hole conduction below 740 K to n-type conduction above 740 K.

6.6.6   Other actinides

Very little has been reported on the magnetic and electrical properties of the other actinide compounds except for studies on NpN[212]. This compound is ferromagnetic below the Curie temperature ($(T_c = 82$ K). Above T_c, $1/\chi(T)$ is not a straight line, but the susceptibility can be written in the form

$$\chi = \chi_0 + \frac{C}{T-82},$$

where the temperature-independent term $\chi_0 = 400 \times 10^{-6}$ e.m.u. mol^{-1} and $\mu_B = 2.13$ BM. The electrical resistivity of NpN shows a very pronounced anomaly at the Curie temperature. The fact that NpN orders ferromagnetically while UN orders antiferromagnetically is clearly significant when considering the theory of magnetic exchange interactions in these compounds.

It will be interesting to discover whether NpP, NpAs, NpSb and NpBi also order ferromagnetically.

6.7 TERNARY SYSTEMS

Ternary systems are of two general types, (1) compounds with a strictly limited range of composition and (2) solid solutions formed between two binary phases with a common anion or cation. Examples of the first type are the nitrihalides ThNBr, UNCl and the nitriphosphide U_2N_2P. These are discrete compounds and in no sense solid solutions.

Solid solutions are formed between two binary compounds with the same crystal structure and with appropriately close lattice parameters, for example, USb_2 and $ThSb_2$, which have a common anion, are miscible in all proportions. The solid solutions formed between these end members may be written as USb_2-ThSb_2 or $(U_{1-x}Th_x)Sb_2$; the latter notation is often shortened to $(U,Th)Sb_2$. Similarly, considering a ternary system with a common cation, UP_2 and UAs_2 are miscible in all proportions to give $U(P_{1-x}As_x)_2$.

Most work on solid solutions has been concerned with the mono-compounds of rocksalt structure. The mutual solubility of these is determined by the proximity of the lattice constants as shown in Table 6.15.

Table 6.15   Mutual solubility of compounds A and B

Compound A	Lattice parameter a_A	Compound B	Lattice parameter a_B	$\left\{\dfrac{a_A - a_B}{a_A}\right\}100$	Molar solubility A in B	Molar solubility B in A
US	5.49	UN	4.89	11%	nil	10%
		UC	4.96	9.7%	4%	40%
		ThS	5.68	-3.5%	Complete	
UP	5.59	UN	4.89	13%	0.3%	0.6%
		UC	4.96	11%	nil	4%
		US	5.49	1.8%	Complete	
UC	4.96	UN	4.89	1.4%	Complete	

When the lattice parameters of the end members differ by more than a few per cent, there is no longer total miscibility. Since the cation radii of the actinide elements in a given valence state are close together, it follows that binary phases with a common anion (e.g. US–ThS) are usually fully miscible. On the other hand, it is rare to find solid solutions with the NaCl structure formed between two chalcogenides (e.g. US and USe) or two pnictides (e.g. PuN and PuP) since the anion radii of successive members of the same series differ markedly. Complete miscibility is observed, however, between chalcogenides and pnictides of similar lattice constant (e.g. US–UP, USe–UAs).

The homogenisation of two binary phases to form a ternary solid solution takes place much more readily with a common cation than with a common anion. For example, mixtures of US and UP powder, fabricated into a pellet, will homogenise to $U(P_{1-x}S_x)$ readily at 1500–1600°C whereas mixtures of

US + ThS or UN + PuN react more sluggishly and require temperatures of 1800–1900°C to effect homogenisation. This is in line with diffusion co-efficient measurements which show that anions have a greater mobility than cations through these phases. This may be attributed to the greater polaris-ability and ease of deformation of the anions, which more than offsets their larger size.

6.7.1 Mixed cation systems

6.7.1.1 Chalcogenides

The US–ThS system has been shown to consist of a single homogeneous phase whose lattice parameter a_0 increases linearly with mole fraction of ThS[22, 23, 100, 143]. Any impurity oxygen present in this phase precipitates as ThOS at the grain boundaries, demonstrating that ThOS is more stable than UOS[22]. No UO_2 or UOS was found even in a $(U_{0.75}Th_{0.25})S$ compact. The chemical properties[22], magnetic properties[100, 143] and thermoelectric properties[150] of the $(U_{1-x}Th_x)S$ phase have all been studied. Alloys contain-

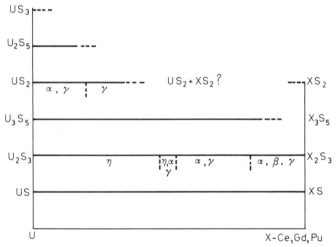

Figure 6.12 Schematic representation of various sulphides existing in the system U—X—S
(where X = Ce, Gd, Pu)
(From Marcon and Pascard[50], by courtesy of Masson et Cie)

ing up to 60 mole % ThS are ferromagnetic at low temperatures. In the para-magnetic region the effective magnetic moment$_{\mu eff}$ increases with increasing ThS content up to 3.6 BM for alloys containing 90% ThS. This value agrees well with the theoretical value of 3.58 BM for L–S coupling of the $5f^2$ electron configuration of the uranium ion (ground state $^3H_4$).

Inter-diffusion of US + PuS to form alloys of general composition $(U_{1-x}Pu_x)S$ requires temperatures approaching 2000°C for complete homogenisation to take place. This has been overcome by blending a mixture of US_2, UH_3, PuS_2 and PuH_2 powders, pelleting and reaction sintering[28]. Homogeneous phases were then prepared after 1 h at 1600°C. The magnetic properties and the electrical conductivities of (U,Pu)S solid solutions have been measured by Raphael[206]. We have seen (Tables 6.2 and 6.3) that the higher sulphides of uranium are crystallographically distinct from those of plutonium. Marcon and Pascard have made a structural investigation of the U—Pu—S system to determine the phases formed for $S/(U + Pu) > 1$ [50, 51]. The isostructural systems U—Ce—S and U—Gd—S have also been investigated (Figure 6.12). The solubility of uranium in PuS_2 and of plutonium in US_3 is small in each case, as expected if the ions involved are U^{4+} and Pu^{3+}. The solubility of Pu in U_3S_5 is very large. Although Pu_3S_5 itself could not be prepared, the compounds UPu_2S_5, UCe_2S_5, UGd_2S_5 and $ThPu_2S_5$ were all produced, from which we conclude that the M_3S_5 phase has the ionic configuration $M^{4+}(M^{3+})_2(S)_5^{2-}$. As Pu^{4+} is not known in sulphides, one can see why Pu_3S_5 has not been prepared.

In the sesquisulphide system, for uranium contents $< 20\%$, the three allotropic forms α, β and γ exist. Between 20 and 50% uranium only the α and γ phases exist. Around 50%U there is a narrow region in which the low temperature η phase coexists with the high temperature γ phase. Finally, for more than 50%U, the sesquisulphide is found only as the η (U_2S_3) form. Further details available[51].

A number of ternary uranium sulphides containing alkali and alkaline earth metals have been described. When KUS_2 is heated with sulphur in a sealed tube at 750°C, oxidation to KUS_3 (U^V) takes place[213], the magnetic behaviour of which indicates a $5f^1$ configuration. The compounds CaU_2S_5, SrU_2S_5 and BaU_2S_5 have been prepared by reacting βUS_2 with CaS, SrS and BaS respectively[214]. These compounds are isostructural with U_3S_5, although in this case the ionic structure must be $M^{2+}(U^{4+})_2(S^=)_5$, a fact confirmed by magnetic studies.

Ternary uranium chalcogenides, USiS, UGeS, USnTe, USbS, USbTe, USiSe, UGeSe and UGeTe have also been reported[215]

6.7.1.2 Pnictides

The compounds UN and ThN form a complete range of solid solution $(U_{1-x}Th_x)N$[216]. The phase relations existing in the U—Th—N ternary at 1000°C have been investigated recently[217]. At nitrogen contents lower than correspond to the monocompound, it is found that (Th,U)N is in equilibrium with either uranium, thorium or uranium + thorium, depending upon the overall Th/U ratio.

The (U,Pu)N ternary system is a possible fuel for fast breeder reactors and considerable technological research has been undertaken in the U.S.A., France and Britain, possibly the most definitive scientific paper being that by Anselin[218]. Because of low diffusion coefficients the solid solution is not easy to form and hot pressing techniques may be necessary. Attempts to nitride a U—Pu alloy have been only partially successful and at high nitrogen pressures disproportionation to U_2N_{3+x} and PuN occurs. On heating (U,Pu)N to 1900°C in vacuum, preferential volatilisation of plutonium occurs.

The analogous (U,Th)P system has been described[89, 101, 144, 219] and the magnetic and electrical properties have been measured as a function of composition. In the paramagnetic region, Curie–Weiss behaviour was observed for uranium concentrations greater than $(U_{0.67}Th_{0.33})P$[101] or $(U_{0.5}Th_{0.5})P$[144].

Checkernikov and co-workers have investigated the magnetic properties of the $(U,Th)P_2$ and $(U,Th)Sb_2$ solid solutions[144, 145]. In the former system, the Néel temperature of $UP_2(\sim 200$ K$)$ is hardly changed by the dissolution of $\leqslant 20\%$ ThP_2. However, in the diantimonides the Néel temperature decreases from 168 K for $(U_{0.8}Th_{0.2})Sb_2$ to 104 K for $(U_{0.5}Th_{0.5})Sb_2$; at lesser uranium contents antiferromagnetic ordering is not observed.

Finally, there have been a number of investigations of ternary pnictides containing transition metals, especially the nitrides. A recent example is the work of Benz and Zachariasen[220] who prepared and characterised the compounds Th_2CrN_3, Th_2MnN_3, U_2CrN_3 and U_3MnN_3.

6.7.2   Mixed anion systems

Oxygen is invariably present, to a greater or lesser extent, as an impurity in the binary systems. As already seen, the chalcogenides preferentially form ternary oxy-chalcogenides, whereas the solubility of oxygen in most of the pnictides is limited and there is a tendency for ThO_2, UO_2, PuO_2 etc. to precipitate as a second phase. As we have considered the oxy-compounds in some detail already we shall not deal further with ternary phases containing oxygen.

6.7.2.1   Higher anion phases

$$\left(\frac{X+Y}{M} > 1\right)$$

The second most common non-metallic impurity is nitrogen which also may be introduced from the atmosphere. Allbutt and Dell[28] showed that when UP or US were heated in a flow of nitrogen above 900°C the mixed anion compounds U_2N_2P and U_2N_2S were formed respectively. These possess the hexagonal La_2O_3 – type structure and are isostructural with Ce_2O_2S, Pu_2O_2S and βU_2N_3. They may be regarded formally as derived from the βU_2N_3 by substituting a phosphorus or sulphur atom for one of the nitrogen atoms. U_2N_2P and U_2N_2S are both stable in a vacuum up to almost 1500°C. By nitriding $U(P_{1-x}S_x)$ solid solutions it was found that U_2N_2P and U_2N_2S exhibit complete mutual solubility[129]. It was also possible to react ThS and ThP separately with nitrogen to give multiphase products in which the hexagonal cells of (presumably) Th_2N_2S and Th_2N_2P could be identified and indexed[129]

This study was extended by Benz and Zachariasen[221] who prepared a number of analogous hexagonal ternary and quaternary compounds, including those containing selenium and arsenic in place of sulphur and phosphorus, and others in which one of the two nitrogen atoms is replaced by oxygen. Crystallographic data for these phases are given in Table 6.16. The c/a ratio observed in these compounds is close to that in the ionic compounds Ce_2O_2S ($c/a = 1.698$), Pu_2O_2S ($c/a = 1.724$) but much larger than in βU_2N_3 or Th_2N_2O. The substitution of S or P leads to a larger expansion in the c direction than the a direction. Another unusual compound with this crystal structure is Th_2H_2C [222].

Table 6.16   Hexagonal ternary phases containing nitrogen

	a_0	c_0	c/a	Reference
U_2N_2S	3.826 / 3.828	6.587 / 6.587	1.722	28 / 221
U_2N_2P	3.803 / 3.802	6.556 / 6.552	1.724	28 / 221
U_2N_2Se	3.862	6.856	1.775	221
U_2N_2As	3.833	6.737	1.758	221
Th_2N_2S	3.996 / 4.008	6.912 / 6.920	1.730	28 / 221
Th_2N_2P	4.006	6.846	1.709	28
Th_2N_2Se	4.029	7.156	1.776	221
$Th_2(N,O)_2P$	4.029	6.835	1.696	221
$Th_2(N,O)_2As$	4.041	6.979	1.727	221
Th_2N_2O	3.883	6.187	1.593	59
βU_2N_3	3.698	5.839	1.579	

Attempts to prepare U_2N_2S by the converse reaction, that is between UN and excess of H_2S gas, were not successful. The more stable compound βUS_2 was found, with elimination of nitrogen.

Another class of mixed chalcogenide–pnictide compounds of U and Th has been described by Hulliger [223]. These compounds, of general formula ThXY and UXY (where Y is any pnictide except nitrogen), crystallise with the tetragonal PbFCl structure. Single crystals were grown by the halogen transport technique in sealed tubes. Eight of the twelve possible thorium compounds were prepared (ThPS, ThAsS, ThPSe, ThAsSe, ThSbSe, ThAsTe, ThSbTe, ThBiTe) and nine of the uranium compounds (UPS, UAsS, USbS, UPSe, UAsSe, USbSe, UAsTe, USbTe, UBiTe). The uranium compounds all order ferromagnetically on cooling, with Curie temperatures ranging from 90 to 150 K. Lattice constants and magnetic data are given in the reference.

The nitrihalides of thorium and uranium (ThNZ and UNZ, where Z is a halogen atom) crystallise mostly in the same tetragonal (PbFCl) structure. These may be regarded as the analogous nitrogen-containing compounds, except that the chalcogen atom is replaced by a halogen. Juza and co-workers[224–226] have prepared UNCl, UNBr, UNI and ThNCl, ThNBr and ThNI, all with this structure, and ThNF which has the rhombohedral LaOF structure. The compounds are formed by reacting the actinide tetrahalide with ammonia or the corresponding metal nitride. The structure of these compounds is presumably $M^{4+}N^{3-}Z^-$ and we may understand the non-existence of the nitrichalcogenides of this type in terms of the instability of N^{3-} and X^- ions.

The magnetic properties of UP_2-UAs_2 [227] and $U_3P_4-U_3As_4$ [228] solid solutions have also been studied. The former system shows a complete range of miscibility, with positive deviation from Vegard's law. All compositions order antiferromagnetically on cooling, the Néel temperature rising non-linearly with increasing arsenic content. The $U_3P_4-U_3As_4$ system also shows complete miscibility, although there is no solubility of U_3P_4 in U_3Sb_4 and only limited solubility of U_3As_4 in U_3Sb_4. The $U_3(P,As)_4$ phase is ferromagnetic for all compositions, the Curie temperature rising with increasing arsenic content.

6.7.2.2 Monocompounds

Among the mixed anion compounds of rocksalt structure by far the most attention has been directed towards the carbonitrides and the sulpho-phosphides. The carbonitrides are of technological interest as possible reactor fuels, while the sulphophosphides present many points of scientific interest in connection with magnetic interactions and ordering phenomena.

(a) Carbonitrides. The magnetic [140] and electrical [229] properties of the thorium compounds have recently been investigated. All compositions in the $Th(N_{1-x}C_x)$ phase show a weak, temperature-independent paramagnetism. The substitution of C for N in ThN leads to a progressive increase in resistivity, Hall coefficient and thermoelectric power.

The uranium carbonitrides $U(N_{1-x}C_x)$ were first investigated metallographically and by x-ray diffraction by Williams and Sambell [230] and subsequently by Anselin et al.[80]. The latter have also studied the quaternary system $(U_{1-x}Pu_x)(C_{1-x}N_x)$ made by reacting UC with PuN. The $U(N_{1-x-y}C_xO_y)$ system, formed by dissolving oxygen in the uranium carbonitrides has also been investigated[80, 231]. Recently de Novion and Costa[232] have measured the magnetic susceptibility (4–900 K), electrical resistivity (4–300 K) and specific heat (from 1 K upwards) of the uranium carbonitrides as a function of composition. They find that dissolution of carbon in UN leads to a lowering of the Néel temperature and that the antiferromagnetic behaviour of UN extends across the $U(N_{1-x}C_x)$ phase only to $x \sim 0.1$; at this point the electronic specific heat shows a sharp maximum. This result conflicts with magnetic measurements on $U(N_{1-x}C_x)$ by Ohmichi and Nasu [233] who claim to have observed antiferromagnetic behaviour for C contents up to $U(N_{0.5}C_{0.5})$. However, Kuznietz [234] has criticised this work and points out that it conflicts not only with the magnetic measurements of de Novion and Costa but also with neutron diffraction measurements on $UN_{0.9}C_{0.1}$ and $UN_{0.74}C_{0.26}$ where no magnetic ordering was observed [235].

(b) Sulphophosphides. The $U(P_{1-x}S_x)$ system has attracted much interest recently because UP is an antiferromagnet while US is a ferromagnet at low temperatures; one is therefore led to enquire what type of magnetic interaction is present at intermediate compositions. The first experiments along these lines[100], showed a linear variation in paramagnetic moment from UP to US, while the variation of Weiss constant with composition indicated that the ferromagnetic interactions of the US lattice dominated over the antiferromagnetic interactions of the UP lattice. Later work [179, 236] confirmed this conclusion and revealed that the changeover from antiferromagnetic to ferromagnetic ordering lay between 10% and 20% US in UP.

In a later study of this system the magnetic phase diagram for compositions in the range $U(P_{1-x}S_x)$ was determined by means of neutron diffraction [237-240]. A complex state of affairs was revealed. For $x \geqslant 0.33$ the ordering is ferromagnetic. For $x = 0$ (UP) and 0.02 the ordering is type I antiferromagnetic (sheets arranged $+ - + -$). For $0.05 \leqslant x \leqslant 0.10$ type I antiferromagnetic ordering occurs first, but on cooling further a transition to type I_A ($+ + - -$) is found. For $x = 0.15$ and 0.20 the ordering is type I_A directly. Finally, for samples with $x = 0.25$ and 0.28 the following transitions are observed:

At 100 K paramagnetic $\rightarrow$ Ferromagnetic
At   70 K ferromagnetic $\rightarrow$ Metamagnetic
At   20 K metamagnetic $\rightarrow$ Antiferromagnetic, type I_A.

This complex behaviour is explained in terms of long range magnetic interactions via the conduction electrons [239].

Further evidence for the intermediate ordered state in $U(P_{0.75}S_{0.25})$ is provided by magnetisation measurements [241]. The Curie temperature is around 100 K and the lower transition to type I_A antiferromagnetism at 20 K. At temperatures of 57 and 67 K the specimen was observed to be metamagnetic, that is, the application of a field stronger than a critical value caused the specimen to become ferromagnetic. This is compatible with a ferrimagnetic structure. Further evidence for the low temperature transition to the type I_A antiferromagnetic state was obtained from lattice parameter measurements, there being a small but significant discontinuity at 20 K [179], and from specific heat measurements [17] which showed a small deviation from the smoothed curve at 17–21 K [17].

No neutron diffraction studies were carried out on $U(P_{0.25}/S_{0.75})$ as it was assumed to remain ferromagnetic from the Curie temperature (167 K) down to 4 K. This now seems unlikely as x-ray diffraction studies [179] show that this substance, like US, becomes rhombohedral below T_c, but (unlike US) reverts back to cubic at 45–50 K. This is evidence, supported also by specific heat data [17], that there is a magnetic transition at this temperature. A further investigation of this substance is warranted.

The nuclear magnetic resonance and Knight shift of ^{31}P has been observed in the paramagnetic state of UP—US solid solutions [186, 242-244]. For all compositions studied the Knight shift (K) could be expressed in the form $K = K_o + \alpha\chi_m$ where α had a constant value. This constancy of α with varying concentration of conduction electrons in passing from UP to US implies that the long range electronic interaction, responsible for the ordered magnetic structure in these compounds, cannot explain the Knight shift[244]. Rather, the U—P magnetic coupling must be through localised electron bonds[147]. Finally, it should be noted that the n.m.r. line widths reveal a certain degree of inhomogeneity in all the $U(P_{1-x}S_x)$ powder specimens[245].

(c) Selenoarsenides. Just as US and UP form a complete range of solid solutions, so also do USe and UAs. Again there is a transition from ferromagnetic to antiferromagnetic ordering across the compositional range. Magnetic susceptibility measurements[246] on $U(As_{1-x}Se_x)$ indicate antiferromagnetic behaviour up to $x = 0.40$, with approximately constant Néel temperature $(130 \pm 5$ K), followed by ferromagnetic behaviour for $x \geqslant 0.45$. However, the observed field dependence at intermediate compositions implies more complex ordered phases as found in $U(P_{1-x}S_x)$. A preliminary neutron study of this system has recently been made[202].

(d) Other Solid Solutions. There are a number of isolated studies of other solid solutions with the rocksalt structure. For example, in the UC—US system[247] there are two rocksalt phases, one extending from UC to $UC_{0.96}S_{0.04}$ and the other from US to $US_{0.60}C_{0.40}$. Again, Baskin[248] carried out phase studies for the UN—UP system and found $<1\%$ mutual solubility. A eutectic occurs at 2300°C and ~ 40 wt % UP.

Aside from the carbonitride system, very little work on mixed anion solid solutions of thorium or plutonium has been undertaken.

6.8 CONCLUDING REMARKS

From the foregoing review it will be evident that this class of compound has stimulated a great deal of interest and a large number of experimental studies have been carried out to establish crystallographic structures and compositions and to measure physical and chemical properties. In addition there have been a number of thermodynamic investigations and many technological studies, neither of which we have reviewed. Most of the research to date has been concerned with Th, U and Pu and there is ample scope to extend this area of work to other actinides, particularly Pa, Np and Am.

Despite the number of experimental studies, the theory of chemical bonding in these semi-metallic compounds is still confused. There is general agreement on the existence of a conduction band, although the orbitals from which it is built up have not been identified for certain. Recent photoelectric emission experiments suggest that the conduction band in US is a narrow f–d band[168]. Most authors favour the view that localised, unpaired electrons on the actinide ions are f electrons, although here again the neutron diffraction data cannot be regarded as unequivocal. Certainly, the number of unpaired electrons on the uranium ion in compounds as simple as US and UN is not universally agreed.

Theories of chemical bonding which have been advanced vary widely. Simple models based on the idea of an ionic lattice with a superimposed conduction band have had some success in explaining observed lattice parameters[163, 249] and crystal structures[50, 51]. On the other hand, more sophisticated energy level calculations have been useful in explaining observed electrical and magnetic properties [209, 250]. Evidence is now accumulating to support the view that magnetic ordering at low temperatures is determined by long range interactions via the conduction electrons (Ruderman–Kettel–Kasuya–Yosida coupling mechanism) [164, 251]. Some elegant experiments pointing in this direction are to be found in a recent paper of Kuznietz concerning the influence of O and C substituents on magnetic ordering in UN and GdN [252]. Finally we may note that Friedman and Grunzweig–Genossar, in their latest paper on n.m.r. line shifts of ^{31}P in US—UP solid solutions [186], have found it necessary to postulate covalent U—P bonds to explain the line shift and conclude that UP has a hybrid mixture of ionic, metallic (conduction electrons) and covalent bonding.

In this confused situation, a detailed review of theories of bonding in these compounds is hardly timely. Rather, there is a need for further experimental work, particularly on neptunium compounds as these are intermediate between uranium, whose most stable valence states are $+4$, $+5$ and $+6$, and plutonium, whose most stable valencies are $+3$ and $+4$. A comparison of the behaviour of neptunium chalcogenides and pnictides with those of U and Pu will provide facts which must be explained in a satisfactory theory of bonding and a start in this direction has been made by Marcon[21]. Much fascinating chemistry remains to be unravelled.

References

1. Kruger, O. L. and Moser, J. B. (1966). *J. Inorg. Nucl. Chem.*, **28,** 825
2. Mulford, R. N. R. and Wiewandt, T. A. (1965). *J. Phys. Chem.*, **69,** 1641
3. Olsen, W. M. and Mulford, R. N. R. (1966). *J. Phys. Chem.*, **70,** 2934
4. Aronson, S. (1967). *J. Inorg. Nucl. Chem.*, **29,** 1611
5. Khodadad, P. (1961). *Bull. Soc. Chim. France,* 133
6. Allbutt, M., Junkison, A. R. and Carney, R. F. (1967). *Proc. Brit. Ceram. Soc.*, **7,** 111
7. Grønvold, F., Haraldsen, H., Thurmann-Moe, T. and Tufte, T. (1968). *J. Inorg. Nucl. Chem.*, **30,** 2117
8. D'Eye, R. W. M. and Sellman, P. G. (1954). *J. Chem. Soc.*, 3760
9. McKay, H. A. C. (1964). *Chem. Ind. (London).,* 1978
10. Baskin, Y. (1967). *J. Inorg. Nucl. Chem.*, **29,** 2480
11. Endebrock, R. W., Foster, E. L. and Keller, D. L. (1964). *Compounds of Interest in Nuclear Reactor Technology, Proceedings of a Symposium, A.I.M.E. Nucl. Met.*, **10,** 557
12. *K.F.K. Report,* 1023
13. Van Lierde, W. and Bressers, J. (1966). *J. Appl. Phys.*, **37,** 444
14. Milner, G. W. C. (1964, 1965). *A.E.R.E. Reports* R.4713 and R.5002
15. Milner, G. W. C., Phillips, G. and Fudge, A. J. (1968). *Talanta,* **15,** 1241
16. Counsell, J. F., Dell, R. M., Junkison, A. R. and Martin, J. F. (1967). *Trans. Faraday Soc., **63,** 72
17. Counsell, J. F., Martin, J. F., Dell, R. M. and Junkison, A. R. (1968). *Thermodyn. Proc. Symp. I.A.E.A. (Vienna),* 385
18. Eastman, E. D., Brewer, L., Bromley, L. A., Gillies, P. W. and Lofgren, N. L. (1950). *J. Amer. Chem. Soc.*, **72,** 4019
19. Zachariasen, W. H. (1949). *Acta Crystallogr.*, **2,** 288, 291
20. Graham, J. and McTaggart, F. K. (1960). *Aust. J. Chem.*, **13,** 67
21. Marcon, J. P. (1967). *Compt. Rend. Ser. C.*, **265,** 235
22. Shalck, P. D. (1963). *J. Amer. Ceram. Soc.*, **46,** 155
23. Cater, E. D., Thorn, R. J. and Walters, R. R. (1964). *Compounds of Interest in Nuclear Reactor Technology, Proceedings of a Symposium, A.I.M.E. Nucl. Met.*, **10,** 237
24. D'Eye, R. W. M., Sellman, P. G. and Murray, J. R. (1952). *J. Chem. Soc.*, 2555
25. D'Eye, R. W. M. (1953). *J. Chem. Soc.*, 1670

26. Boelsterli, H. U. and Hulliger, F. (1968). *J. Mater. Sci.*, **3**, 664
27. Picon, M. and Flahaut, J. (1958). *Bull. Soc. Chim. Fr.*, 772
28. Allbutt, M. and Dell, R. M. (1967). *J. Nucl. Mater.*, **24**, 1
29. Yoshihara, K., Kanno, M. and Mukaibo, T. (1967). *J. Nucl. Sci. Technol.*, **4**, 578
30. Mooney-Slater, R. C. L. (1964). *Z. Kristallogr.*, **120**, 278
31. Mazurier, A. and Khodadad, P. (1968). *Bull. Soc. Chim. Fr.*, 4058
32. Ferro, R. (1954). *Z. Anorg. Chem.*, **275**, 320
33. Trzebiatowski, W. and Suski, W. (1961). *Bull. Acad. Polon. Sci. Ser. Sci. Chim.*, **9**, 277 and (1962), **10**, 399
34. Matson, L. K., Moody, J. W. and Himes, R. C. (1963). *J. Inorg. Nucl. Chem.*, **25**, 795
35. Slovyanskikh, V. K., Ellert, G. V. and Yarembash, E. I. (1967). *Inorg. Mater. (U.S.S.R.)*, **3**, 969
36. Slovyanskikh, V. K., Yarembash, E. I., Ellert, G. V. and Eliseev, A. A. (1968). *Inorg. Mater. (U.S.S.R.)*, **4**, 543
37. Breeze, E. W., Brett, N. H. and White, J.—To be published in *J. Nucl. Mater*
38. Klein Haneveld, A. J. and Jellinek, F. (1970). *J. Less-Common Metals*, **21**, 45
39. Bridger, N. J. and Moseley, P. T.—Unpublished results
40. Cordfunke, E. H. P. (1969). *The Chemistry of Uranium*, 207 (Elsevier)
41. Smith, P. K. and Cathey, L. (1967). *J. Electrochem. Soc.*, **114**, 973
42. Wilkinson, K. and Moseley, P. T.—Private communication
43. Slovyanskikh, V. K., Ellert, G. V., Yarembash, E. I. and Korsakova, M. D. (1966). *Inorg. Mater. (U.S.S.R.)*, **2**, 827
44. Murasik, A., Suski, W., Troc, R. and Leciejewicz, J. (1968). *Phys. Status Solidi*, **30**, 61
45. Klein Haneveld, A. J. and Jellinek, F. (1964). *J. Inorg. Nucl. Chem.*, **26**, 1127
46. Trzebiatowski, W., Niemiec, J. and Sepichowska, A. (1961). *Bull. Acad. Polon. Sci. Ser. Sci. Chim.*, **9**, 373
47. Petitjean, G. and Accary, A. (1970). *J. Nucl. Mater.*, **34**, 59
48. Zachariasen, W. H. (1949). *Acta Crystallogr.*, **2**, 57, 60
49. Marcon, J. P. and Pascard, R. (1966). *J. Inorg. Nucl. Chem.*, **28**, 2551
50. Marcon, J. P. and Pascard, R. (1968). *Rev. Int. Hautes Temp. Refract.*, **5**, 51
51. Marcon, J. P. (1969). *Thesis, University of Paris*
52. Allbutt, M., Dell, R. M. and Junkison, A. R. (1970). *The Chemistry of Extended Defects in Non-Metallic Solids*, 124 (North Holland)
53. Marcon, J. P. (1967). *Compt. Rend., Ser. C.*, **264**, 1475
54. Cohen, D., Fried, S., Siegel, S. and Tani, B. (1968). *Inorg. Nucl. Chem. Lett.*, **4**, 257
55. Fried, S., Cohen, D., Siegel, S. and Tani, B. (1968). *Inorg. Nucl. Chem. Lett.*, **4**, 495
56. Mitchell, A. W. and Lam, D. J. (1970). *J. Nucl. Mat.*, **37**, 349
57. Sellars, P. A., Fried, S., Elson, R. E. and Zachariasen, W. H. (1954). *J. Amer. Chem. Soc.*, **76**, 5935
58. Zachariasen, W. H. (1949). *Acta Crystallogr.*, **2**, 388
59. Benz, R. and Zachariasen, W. H. (1966). *Acta Crystallogr.*, **21**, 838
60. Aronson, S. and Auskern, A. B. (1966). *J. Phys. Chem.*, **70**, 3937
61. Benz, R., Hoffmann, C. G. and Rupert, G. N. (1967). *J. Amer. Chem. Soc.*, **89**, 191
62. Bridger, N. J. and Dell, R. M.—Unpublished results
63. Juza, R. and Gerke, H. (1968). *Z. Anorg. Allg. Chem.*, **363**, 245
64. Bugl, J. and Bauer, A. A. (1964). *Compounds of Interest in Nuclear Reactor Technology, Proceedings of a Symposium, A.I.M.E. Nucl. Met.*, **10**, 215
65. Lapat, P. E. and Holden, R. B. (1964). *Compounds of Interest in Nuclear Reactor Technology, Proceedings of a Symposium, A.I.M.E. Nucl. Met.*, **10**, 225
66. Counsell, J. F., Dell, R. M. and Martin, J. F. (1966). *Trans. Faraday Soc.*, **62**, 1736
67. Rundle, R. E. (1948). *Acta Crystallogr.*, **1**, 180
68. Didchenko, R. and Gortsema, F. P. (1963). *Inorg. Chem.*, **2**, 1079
69. Dell, R. M. and Carney, R. F. A.—Unpublished results
70. Price, C. E. and Warren, I. H. (1965). *Inorg. Chem.*, **4**, 115
71. Trzebiatowski, W. and Troc, R. (1964). *Bull. Acad. Polon. Sci. Ser. Sci. Chim.*, **12**, 681
72. Kemball, C. (1966). *Discuss. Faraday Soc.*, **41**, 190
73. Vaughan, D. A. (1965). *J. Metals*, **206**, 78
74. Trzebiatowski, W., Troc, R. and Leciezewicz, J. (1962). *Bull. Acad. Polon. Sci. Ser. Sci. Chim.*, **10**, 395
75. Benz, R. and Bowman, M. G. (1966). *J. Amer. Chem. Soc.*, **88**, 264

76. Stöcker, H. J. and Naoumidis, A. (1966). *Ber. Deut. Keram. Ges.*, **43**, 724
77. Sasa, Y. and Atoda, T. (1970). *J. Amer. Ceram. Soc.*, **53**, 102
78. Hyde, K. R., Landsman, D. A., Morris, J. B., Seddon, W. E. and Tulloch, H. J. C. (1965). *Special Ceramics, Proceedings of a Symposium*, 1, (Academic Press)
79. Olsen, W. M. and Mulford, R. N. R. (1963). *J. Phys. Chem.*, **67**, 952
80. Anselin, F., Dean, G., Lorenzelli, R. and Pascard, R. (1964). *Carbides in Nuclear Energy, Proceedings of a Symposium*, **1**, 113, (McMillan and Co.)
81. Lapage, R. and Bunce, J. L. (1967). *Trans Faraday Soc.*, **63**, 1889
82. Anselin, F. (1963). *Compt. Rend.*, **256**, 2616
83. Leary, J. A., Pritchard, W. C., Nance, R. L. and Shupe, M. W. (1965). *3rd Int. Conf. on Plutonium, Institute of Metals*, 639 (London)
84. Rand, M. H., Fox, A. C. and Street, R. S. (1962). *Nature (London)*, **195**, 567
85. Olsen, W. M. and Mulford, R. N. R. (1966). *J. Phys. Chem.*, **70**, 2932
86. Akimoto, Y. (1967). *J. Inorg. Nucl. Chem.*, **29**, 2650
87. Price, C. E. and Warren, I. H. (1965). *J. Electrochem. Soc.*, **112**, 510
88. Gingerich, K. A. and Wilson, D. W. (1965). *Inorg. Chem.*, **4**, 987
89. Baskin, Y. (1969). *J. Amer. Ceram. Soc.*, **52**, 54
90. Javorsky, C. A. and Benz, R. (1967). *J. Nucl. Mater.*, **23**, 192
91. Ferro, R. (1955). *Acta Crystallogr.*, **8**, 360
92. Benz, R. (1968). *J. Nucl. Mater.*, **25**, 233
93. Ferro, R. (1956). *Acta Crystallogr.*, **9**, 817
94. Ferro, R. (1957). *Acta Crystallogr.*, **10**, 476
95. Dahlke, O., Gans, W., Knacke, O. and Müller, F. (1969). *Z. Metallk.*, **60**, 465
96. Strotzer, E. F., Biltz, W. and Meisel, K. (1938). *Z. Anorg. Chem.*, **238**, 69
97. Meisel, K. (1939). *Z. Anorg. Chem.*, **240**, 300
98. Adachi, H. and Imoto, S. (1968). *Tech. Report. Osaka University*, **18**, 377
99. Hulliger, F. (1966). *Nature (London)*, **209**, 499
100. Allbutt, M., Junkison, A. R. and Dell, R. M. (1964). *Compounds of Interest in Nuclear Reactor Technology, Proceedings of a Symposium, A.I.M.E. Nucl. Met.*, **10**, 65
101. Adachi, H. and Imoto, S. (1969). *J. Nucl. Sci. Technol.*, **6**, 531
102. Baskin, Y. (1966). *J. Amer. Ceram. Soc.*, **49**, 541
103. Trzebiatowski, W. and Troc, R. (1963). *Bull. Acad. Polon. Sci. Ser. Sci. Chim.*, **11**, 661
104. Iandelli, A. (1952). *Atti Acad. Naz. Lincei., Cl. Sci. Fis. Mat. Nat., Rend.*, **13**, 144
105. Trzebiatowski, W., Sepichowska, A. and Zygmunt, A. (1964). *Bull. Acad. Polon. Sci. Ser. Sci. Chim.*, **12**, 687
106. Iandelli, A. (1952). *Atti Acad. Naz. Lincei, Cl. Sci. Fis. Mat. Nat., Rend.*, **13**, 138
107. Benz, R. and Tinkle, M. C. (1968). *J. Electrochem. Soc.*, **115**, 322
108. Ferro, R. (1952). *Atti Acad. Naz. Lincei., Cl. Sci. Fis. Mat. Nat., Rend.*, **13**, 53 and 151
109. Kuznietz, M., Lander, G. H. and Campos, F. P. (1969). *J. Phys. Chem. Solids*, **30**, 1642
110. Ferro, R. (1952). *Atti Acad. Naz. Lincei., Cl. Sci. Fis. Mat. Nat., Rend.*, **13**, 401 and (1953), **14**, 89
111. Trzebiatowski, W. and Zygmunt, A. (1966). *Bull. Acad. Polon. Sci. Ser. Sci. Chim.*, **14**, 495
112. Buhrer, C. F. (1969). *J. Phys. Chem. Solids*, **30**, 1273
113. Henkie, Z. (1968). *Rocz. Chem.*, **42**, 363
114. Lukaszewicz, K. and Guzy, D. (1969). *Acta Crystallogr., Sect. A.*, **25**, S.47
115. Baskin, Y. (1966). *Nucl. Sci. Eng.*, **24**, 332
116. Benz, R. and Ward, C. H. (1968). *J. Inorg. Nucl. Chem.*, **30**, 1187
117. Baskin, Y. (1965). *J. Amer. Ceram. Soc.*, **48**, 652
118. Beaudry, B. J. and Daane, A. H. (1959). *Trans. A.I.M.E.*, **215**, 199
119. Kruger, O. L. and Moser, J. B. (1967). *Chemical Engineering Progress Symposium Series*, **63**, 1
120. Teitel, R. J. (1957). *J. Metals.*, **9**, 131
121. Cotterill, P. and Axon, H. J. (1959). *J. Inst. Metals*, **87**, 159
122. Moser, J. B. and Kruger, O. L. (1966). *J. Less-Common Metals*, **10**, 402
123. Kruger, O. L. and Moser, J. B. (1966). *J. Amer. Ceram. Soc.*, **49**, 661
124. Pardue, W. M., Storhok, V. W., Smith, R. A. and Keller, D. L. (1964). *U.S.A.E.C. Report, B.M.I.*—1698
125. Kruger, O. L. and Moser, J. B. (1967). *J. Phys. Chem. Solids*, **28**, 2321
126. Lam, D. J., Private communication

127. Ellinger, F. H., Land, C. C. and Gschneidner, Jr. (1967). *Plutonium Handbook*, **1**, 191, (Gordon and Breach, Science Publishers)
128. Nakai, E., Kanno, M. and Mukaibo, T. (1969). *J. Nucl. Sci. Technol.*, **6**, 138
129. Allbutt, M. and Dell, R. M.—Unpublished results
130. Sole, M. J. Van Der Walt, C. M. and Truzenberger, M. (1968). *Acta Met.*, **16**, 667
131. Besson, J., Moreau, C. and Philippot, J. (1964). *Bull. Soc. Chim. Fr.*, 1069
132. Bugl, J. and Bauer, A. A. (1964). *Compounds of Interest in Nuclear Reactor Technology, Proceedings of a Symposium, A.I.M.E. Nucl. Met.*, **10**, 463
133. Dell, R. M., Wheeler, V. J. and McIver, E. J. (1966). *Trans. Faraday Soc.*, **62**, 3591
134. Sole, M. J. and Van Der Walt, C. M. (1968). *Acta Met.*, **16**, 501
135. Dell, R. M. and Wheeler, V. J. (1967). *J. Nucl. Mater.*, **21**, 328
136. Dell, R. M., Wheeler, V. J. and Bridger, N. J. (1967). *Trans. Faraday Soc.*, **63**, 1286
137. Sugihara, S. and Imoto, S. (1969). *J. Nucl. Sci. Technol.*, **6**, 237
138. Ferris, L. M. (1968). *J. Inorg. Nucl. Chem.*, **30**, 2661
139. Baskin, Y. (1965). *J. Amer. Ceram. Soc.*, **48**, 153
140. Aronson, S. and Auskern, A. B. (1968). *J. Chem. Phys.*, **48**, 1760
141. Raphael, G. and de Novion, C. (1969). *Solid State Commun.*, **7**, 791
142. Chechernikov, V. I. (1968). *Vestn. Mosk. Univ. Ser. III, Fiz. Astron.*, **23**, 115 (N.S.A., **23**, 12989)
143. Chechernikov, V. I., Shavishvili, T. M., Pletyushkin, V. A. Slovyanskikh, V. K. and Ellert, G. V. (1968). *Sov. Phys. J.E.T.P.*, **27**, 921
144. Chechernikov, V. I., Shavishvili, T. M., Pletyushkin, V. A. and Slovyanskikh, V. K. (1969). *Sov. Phys. J.E.T.P.*, **28**, 81
145. Chechernikov, V. I., Kuz'man, R. N., Chachkhiani, L. G., Golovnin, V. A., Irkaev, S. M., Slovyanskikh, V. K. and Shavishvili, T. M. (1970). *Soviet Physics J.E.T.P.*, **30**, 73
146. Kuznietz, M. (1968). *J. Chem. Phys.*, **49**, 3731
147. Kuznietz, M. and Matzkanin, G. A. (1969). *Phys. Rev.*, **178**, 580
148. Auskern, A. B. and Aronson, S. (1967). *J. Phys. Chem. Solids*, **28**, 1069
149. Baskin, Y. (1964). *U.S.A.E.C. Report* A.N.L.—7000, 90
150. Tetenbaum, M. (1964). *J. Appl. Phys.*, **35**, 2468
151. Westrum, E. F. (1965). *Thermodyn. Proc. Symp. I.A.E.A. (Vienna)*, **2**, 497
152. Westrum, E. F. and Grønvold, F. (1962). *Thermodyn. Proc. Symp. I.A.E.A. (Vienna)*, **1**, 3
153. Suski, W. and Trzebiatowski, W. (1964). *Bull. Acad. Polon. Sci. Ser. Sci. Chim.*, **12**, 277
154. Chechernikov, V. I., Pechennikov, A. V., Yarembash, E. I., Martynova, L. F. and Slavyanskikh, V. K. (1968). *Sov. Phys. J.E.T.P.*, **26**, 328
155. Trzebiatowski, W. and Suski, W. (1963). *Rocz. Chem.*, **37**, 117
156. Kaznierouricz, C. W. (1963). *A.N.L. Report*—6731
157. Gardner, W. E. and Smith, T. F. (1968). *11th Int. Conf. Low Temperature Physics.*, **2**, 1377
158. Suski, W., Rao, V. U. S. and Wallace, W. E.—Private communication
159. Pechennikov, A. V. Chechernikov, V. P., Barykin, M. E., Ellert, G. V., Slovyanskikh, and Yarembash, E. I. (1968). *Inorg. Mater. (U.S.S.R.)*, **4**, 1176
160. Chechernikov, V. I., Pechennikov, A. V., Barykin, M. E., Slovyanskikh, V. K., Yarembash, E. I. and Ellert, G. V. (1967). *Sov. Phys. J.E.T.P.*, **25**, 560
161. Trzebiatowski, W. and Sephichowska, A. (1960). *Bull. Acad. Polon. Sci. Ser. Sci. Chim.*, **7**, 181, and (1960), **8**, 457
162. Fisk, Z. and Coles, B. R. (1970). *J. of Physics C (Solid State Physics)*, **3**, L104
163. Allbutt, M. and Dell, R. M. (1968). *J. Inorg. Nucl. Chem.*, **30**, 705
164. Grunzweig-Genossar, J., Kuznietz, M. and Friedman, F. (1968). *Phys. Rev.*, **173**, 562
165. Wedgewood, F. A.—Private communication
166. Marples, J. A. C. (1970). *J. Phys. Chem. Solids*, **31**, 2431
167. McTaggart, F. K. (1958). *Aust. J. Chem.*, **11**, 471
168. Eastman, D. E. and Kuznietz, M. (1970). *Proc. 16th Annual Conference on Magnetism*, Miami Beach
169. Ballestracci, R., Bertaut, E. F. and Pauthenet, R. (1963). *J. Phys. Chem. Solids*, **24**, 487
170. Murasik, A., Suski, W. and Leciejewicz, J. (1969). *Phys. Status Solidi*, **34**, K.157
171. Stalinski, B., Niemiec, J. and Bieganski, Z. (1963). *Bull. Acad. Polon. Sci. Ser. Sci. Chim.*, **11**, 267

172. Westrum, E. F. and Barber, C. M. (1966). *J. Chem. Phys.*, **45**, 635
173. Curry, N. A. (1965). *Proc. Phys. Soc.*, **86**, 1193
174. Westrum, E. F., Walters, R. R., Flotow, H. E. and Osborne, D. W. (1968). *J. Chem. Phys.*, **48**, 155
175. Kuznietz, M. (1969). *Phys. Rev.*, **180**, 476
176. Costa, P., Lallemont, R., Anselin, F. and Rossignol, D. (1964). *Compounds of Interest in Nuclear Reactor Technology, Proceedings of a Symposium, A.I.M.E. Nucl. Met.*, **10**, 83
177. Moore, J. P., Fulkerson, W. and McElroy, D. L. (1970). *J. Amer. Ceram. Soc.*, **53**, 76
178. (1969). *U.S.A.E.C. Report, O.R.N.L.—4470*
179. Allbutt, M., Dell, R. M., Junkison, A. R. and Marples, J. A. (1970). *J. Inorg. Nucl. Chem.*, **32**, 2159
180. Gulick, T., Long, C. E. and Moulton, W. G. (1970). *Bull. Amer. Phys. Soc.*, **15**, 318
181. Troc, R., Mulak, J. and Suski, W. (1970). *Phys. Status Solidi.*, (Preprint)
182. Scott, B. A., Kingerich, K. A. and Bernheim, R. A. (1967). *Phys. Rev.*, **159**, 387
183. Jones, E. D. (1967). *Phys. Lett.*, **25A**, 111
184. Easwaran, K. R. K., Rao, V. U. S., Vijayaraghavan, R. and Rao, V. R. K. (1967). *Phys. Lett.*, **25A**, 683
185. Friedman, F., Grunzweig-Genossar, J. and Kuznietz, M. (1967). *Phys. Lett.*, **25A**, 690
186. Friedman, F. and Grunzweig-Genossar, J.—Private communication
187. Stalanski, B., Bieganski, Z. and Troc, R. (1967). *Bull. Acad. Polon. Sci. Ser. Sci. Chim.*, **15**, 257
188. Curry, N. A. (1966). *Proc. Phys. Soc.*, **89**, 427
189. Sidhu, S. S., Vogelsang, W. and Anderson, K. D. (1966). *J. Phys. Chem. Solids*, **27**, 1197
190. Heaton, L., Müller, M. H., Anderson, K. D. and Zauberis, D. D. (1969). *J. Phys. Chem. Solids*, **30**, 453
191. Gardner, W. E.—Private communication
192. Carr, S. L., Long, C., Moulton, W. G. and Kuznietz, M. (1969). *Phys. Rev. Lett.*, **23**, 786
193. Ciszewski, R., Murasik, A. and Troc, R. (1965). *Phys. Status Solidi*, **10**, K.85
194. Troc, R., Leciejewicz, J. and Ciszewski, R. (1966). *Phys. Status Solidi*, **15**, 515
195. Trzebiatowski, W. and Henkie, Z.—To be published
196. Henkie, Z. and Trzebiatowski, W. (1969). *Phys. Status Solidi*, **35**, 827
197. Oles, A. (1965). *J. de Physique*, **26**, 561
198. Leciejewicz, J., Troc, R., Murasik, A. and Zygmunt, A. (1967). *Phys. Status Solidi*, **22**, 517
199. Williams, J. M., Heaton, L. and Campos, F. (1968). *J. Phys. Chem. Solids*, **29**, 1702
200. Leciejewicz, J., Murasik, A. and Troc, R. (1968). *Phys. Status Solidi*, **30**, 157
201. Olsen, C. E. and Koehler, W. C. (1969). *J. Appl. Phys.*, **40**, 1135
202. Leciejewicz, J., Murasik, A., Palewski, T. and Troc, R. (1970). *Phys. Status Solidi*, **38**, K.89
203. Shenoy, G. K., Kalvius, G. M., Ruby, S. L., Dunlap, B. D., Kuznietz, M. and Campos, F. P. (1970). *Int. J. Magnetism*, **1**
204. Warren, I. H. and Price, C. E. (1964). *Can. Met. Quart.*, **3**, 183
205. Raphael, G. and de Novion, C. (1969). *J. de Physique*, **30**, 261
206. Raphael, G. (1969). *C.E.A. Report—R.3912*
207. Dawson, J. K. (1952). *J. Chem. Soc.*, 1882
208. Lam, D. J., Nevitt, M. V., Ross, J. W. and Mitchell, A. W. (1965). *3rd Int. Conf. on Plutonium, Institute of Metals, (London)* P.274
209. Lam, D. J., Fradin, F. Y. and Kruger, O. L. (1969). *Phys. Rev.*, **187**, 606
210. Fradin, F. Y. (1969). *Solid State Commun.*, **7**, 759
211. Kruger, O. L. and Moser, J. B. (1967). *J. Chem. Phys.*, **46**, 891
212. de Novion, C. H. and Lorenzelli, R. (1968). *J. Phys. Chem. Solids* **29**, 1901
213. Padiou, J. and Guillevic, J. (1969). *Compt. Rend., Ser. C.*, **268**, 822
214. Brochu, R., Padiou, J. and Prigent, J. (1970). *Compt. Rend., Ser. C.*, **270**, 809
215. Klein Haneveld, A. J. and Jellinek, F. (1969). *J. Less-Common Metals*, **18**, 123
216. Venard, J. T., Spruiell, J. E. and Cavin, O. B. (1967). *J. Nucl. Mater.*, **24**, 245
217. Venard, J. T. and Spruiell, J. E. (1968). *J. Nucl. Mater.*, **27**, 257
218. Anselin, F. (1963). *J. Nucl. Mater.*, **18**, 301
219. Adachi, H. and Imoto, S. (1967). *J. At. Energy Soc. Jap.*, **9**, 381
220. Benz, R. and Zachariasen, W. H. (1970). *J. Nucl. Mater.*, **37**, 109
221. Benz, R. and Zachariasen, W. H. (1969). *Acta Crystallogr., Sect. B.*, **25**, 294

222. Peterson, D. T. and Rexer, J. (1962). *J. Inorg. Nucl. Chem.*, **24**, 519
223. Hulliger, F. (1970). *U.S. Patents* 3 505 410 and 3 510 274
224. Juza, R. and Sievers, R. (1968). *Naturwissenschaften*, **52**, 538
225. Juza, R. and Sievers, R. (1968). *Z. Anorg. Allg. Chem.*, **363**, 258
226. Juza, R. and Meyer, W. (1969). *Z. Anorg. Allg. Chem.*, **366**, 43
227. Trzebiatowski, W. and Misiuk, A. (1968). *Rocz. Chem.*, **42**, 163
228. Trzebiatowski, W. and Misiuk, A. (1970). *Rocz, Chem.*, **44**, 695
229. Auskern, A. B. and Aronson, S. (1970). *J. Appl. Phys.*, **41**, 227
230. Williams, J. and Sambell, R. A. J. (1959). *J. Less-Common Metals*, **1**, 217
231. Imoto, S. and Stöcker, H. J. (1965). *Thermodyn. Proc. Symp. I.A.E.A. (Vienna)*, **2**, 533
232. de Novion, C. H. and Costa, P. (1970). *Compt. Rend., Ser. B.*, **270**, 1415
233. Ohmichi, T. and Nasu, S. (1970). *J. Nucl. Sci. Technol.*, **7**, 268
234. Kuznietz, M.—To be published in *J. Nucl. Sci. Technol.*
235. Koehler, W. C.—Private communication quoted by M. Kuznietz, (Ref. 205)
236. Trzebiatowski, W. and Palewski, T. (1969). *Phys .Status Solidi*, **34**, K.51
237. Lander, G. H., Kuznietz, M. and Baskin, Y. (1968). *Solid State Commun.*, **6**, 877
238. Kuznietz, M., Lander, G. H. and Baskin, Y. (1969). *J. Appl. Phys.*, **40**, 1130
239. Kuznietz, M. and Grunzweig-Genossar, J. (1970). *J. Appl. Phys.*, **41**, 906
240. Lander, G. H., Kuznietz, M. and Cox, D. E. (1969). *Phys. Rev.*, **188**, 963
241. Crangle, J., Kuznietz, M., Lander, G. H. and Baskin, Y. (1969). *J. Physics., C.*, **2**, 925
242. Kuznietz, M., Matzkanin, G. A. and Baskin, Y. (1968). *Phys. Lett.*, **28A**, 122
243. Kuznietz, M., Baskin, Y. and Matzkanin. (1969). *Phys. Rev.*, **187**, 737
244. Kuzneitz, M., Baskin, Y. and Matzkanin, G. A. (1970). *J. Appl. Phys.*, **41**, 1111
245. Kuzneitz, M., Campos, F. P. and Baskin, Y. (1969). *J. Appl. Phys.*, **10**, 3621
246. Trzebiatowski, W., Musiuk, A. and Palewski, T. (1967). *Bull. Acad. Polon. Sci. Ser. Sci. Chim.*, **15**, 543
247. Shalak, P. D. and White, G. D. (1964). *Carbides in Nuclear Energy, Proceedings of a Symposium, Harwell*, **1**, 266 (McMillan and Co.)
248. Baskin, Y. (1967). *J. Amer. Ceram. Soc.*, **50**, 74
249. Baskin, Y. (1967). *Trans. A.I.M.E.*, **239**, 1708
250. Chan, S. K. and Lam, D. J. (1970). *Fourth Int. Conf. on Plutonium and Other Actinides* (Santa Fe)
251. Grunzweig, J. and Kuznietz, M. (1968). *J. Appl. Phys.*, **39**, 905
252. Kuznietz, M. (1970). *Proc. 16th Int. Conf. on Magnetism* (Miami)
253. Stalinski, B., Bieganski, Z. and Troc, R. (1966). *Phys. Status Solidi*, **17**, 837
254. Bridger, N. J., Dell, R. M. and Wheeler, V. J. (1968). *6th International Symposium on Reactivity of Solids*, 389 (New York, Wiley)
255. Ban, Z., Despotovic, Z. and Tudja, M. (1968). *J. Nucl. Mater.*, **25**, 106

7
Complexes of the Lanthanides

THERALD MOELLER

Arizona State University

7.1 INTRODUCTION

The development of coordination chemistry as a major area of investigative effort in inorganic chemistry has been paralleled by the emergence of studies involving complexes of the lanthanide ions as a major area of interest in the chemistry of these 4f species. Prior to 1940, information relating to lanthanide complexes was restricted to a limited number of double salts and a few accounts of the syntheses of β-diketone derivatives and a few other species[1]. During the 1940s, the development of methods for the ion-exchange separation of lanthanide fission products focused particular attention upon complexation. In subsequent years, improved separations, energy absorption-emission systems, and improved understanding of bonding and bond-related properties have combined to increase publication in this area essentially at an exponentially definable rate. Although it was possible to describe the significant details of complexation of the lanthanide ions in 1953 in only some 60 references[2], a summary published in 1965, but covering published data only through 1962, cited over 500 references[3]. A very restrictive summation for the period 1963–1965 listed nearly 300 references and made no attempt[4] to cover all topics being investigated[4]. Subsequent publication has been at such an increased rate that complete coverage is no longer either possible or desirable. This article is, therefore, restricted largely to the broad topics of coordination number, molecular geometry, thermodynamic stability in solution, kinetics and mechanism of complexation, bonding, and applications, with primary emphasis upon the interpretation of physical data.

It is apparent that, although many complex derivatives of the lanthanide ions have been described, the total number and types of known and probably possible species are much less than those characteristic of the d transition metal ions. Among the factors that appear to limit the number of lanthanide complexes and to mitigate against their formation are at least the following:

1. Electronic configuration. In its ground state, each Ln^{n+} ion (Ln = any element in the series La–Lu) presents to any approaching ligand a completely paired, noble-gas atom type of outermost electronic configuration, with both the distinguishing 4f orbitals and the electrons occupying them largely or completely unavailable for bond formation. Higher-energy orbitals alone can thus participate in covalent linkages. Ligand-field stabilisation energies are only of the order of 1 kcal mol^{-1}. By contrast, the distinguishing nd orbitals of the d-transition metal species are valency-shell orbitals, and ligand-field stabilisation may amount to 100 kcal mol^{-1} or more.

2. Ionic size. Lanthanide complexes are derived very largely from the tripositive ions, Ln^{3+}. A limited number are formed by tetrapositive ions, Ln^{4+} (nearly always Ce^{4+}), and dipositive ions, Ln^{2+} (limited to Eu^{2+}, Yb^{2+}, and Sm^{2+}). Each of these ions is large for species of the charge type in question. Electrostatic attractions for ligands are thus minimised, and the strengths of bonds of this type are reduced accordingly.

3. Ligand lability. The results of synthesis, the observations of ion exchange and solvent extraction processes, and the limited studies of kinetics thus far described all indicate that ligand exchange reactions involving the lanthanide species in solution are extremely rapid. Both the number of isolable complexes and the possibilities of isomerism are thus substantially limited. Furthermore, a composition found for a solid may not persist in solution, and a given complex often cannot be recovered as such once it has been dissolved.

4. Competition by water and hydroxide ion. The water molecule is a particularly strong ligand toward the lanthanide ions and thus competes effectively for coordination sites. In a neutral or acidic aqueous environment, this competition precludes coordination by any but the strongest donors and restricts isolable complexes very largely to those of chelating ligands. Under alkaline conditions, the OH^- ion is an even stronger ligand than the water molecule. The solubilities of the hydrous lanthanide (III or IV) hydroxides or oxides are sufficiently small that precipitation of these compounds effectively prevents complexation by many ligands. Indeed, highly basic ligands such as amines, that might be expected to be strong donors, form sufficient concentrations of hydroxide ion by interaction with water to precipitate these cations.

5. Donor atoms. The lanthanide ions are typical 'A' type cations in the Ahrland-Chatt-Davies sense, or 'hard' acids in the Pearson sense. The majority of the complexes that can be isolated from aqueous systems contain ligands with oxygen donor sites. These ligands may be pure oxygen donors (e.g., oxalate ion, β-diketonate ions) or mixed oxygen-nitrogen donors (e.g., polyaminepolycarboxylate ions). Ligands with only nitrogen donor sites are limited (e.g., 1,10-phenanthroline), and those containing other donor atoms essentially unknown. The implication that interaction between an Ln^{n+} ion and a donor atom other than oxygen is weak has been shown to be in error for at least pure nitrogen donors, providing competition is reduced by using a weak donor solvent[5-9].

These factors indicate clearly that complexation of the lanthanide ions is a wholly different phenomenon from complexation of d-transition metal ions in general and that the species obtained for one type of ion cannot be compared with those for the other type. In many ways, the complexes of the lanthanides resemble more closely those of the alkaline earth metal ions (Ca^{2+}, Sr^{2+}, Ba^{2+}) than those of other types of cations. Indeed, in biological systems where complexing is involved, the lanthanide ions serve as very useful probes for calcium ion behaviour[10]. The alkaline earth metal ions resemble the lanthanides in both size and noble-gas type valency shell configurations. The resemblance is nearly exact for the Ln^{2+} ions and is only slightly modified by charge difference for the Ln^{3+} ions.

That the radius of the Y^{3+} ion (0.88 Å) lies close to those of the Ho^{3+} ion (0.894 Å) and the Er^{3+} ion (0.881 Å) [11], suggests a close resemblance among complexes of these cations. An overall general resemblance involving all of the Ln^{3+} ions is suggested by their radii (from 1.061 Å for La^{3+} to 0.848 Å for Lu^{3+}) [11]. These similarities do exist and are of particular importance as a consequence of the inevitable occurrence of yttrium with the lanthanides. Yttrium complexes differ from those of the Ln^{3+} ions only in those properties that depend upon the presence of incompletely occupied 4f orbitals, i.e., in magnetic properties, in light absorption, and in energy absorption-emission phenomena.

7.2 COORDINATION NUMBER AND MOLECULAR GEOMETRY

Contrary to a generally held and widely quoted belief, that was apparently based upon the stoichiometric compositions of certain complexes described in the early literature, the coordination number of an Ln^{n+} ion is only rarely six, and higher coordination numbers appear to the rule[1,3,4,12-14]. This situation is a reasonable consequence of the comparatively large sizes of the lanthanide ions. It is of particular interest as a consequence of the number and kinds of molecular geometry that result. At least two significant conclusions, based upon examination of available data, are apparent at the outset, namely (1) variability in coordination number either for a given lanthanide ion or for a series of lanthanide ions of the same charge is determined more by spatial accommodation of ligands than by the bonding characteristics of ligands, and (2) coordination number in solution may, and commonly does, differ from coordination number in a crystal, given the same lanthanide ion and the same ligand. Definitive data for coordination numbers and molecular geometry are available only for crystals. Data for solutions are usually suggestive rather than definitive.

7.2.1   Indicative physical data

7.2.1.1   Vibrational spectra

Infrared data have been used primarily to indicate bonding sites in ligands and to distinguish between coordinated and non-coordinated ligands in terms of their indicated site symmetries. Examples of the first of these applications include establishment of oxygen as the donor atom (1) in triphenylphosphine oxide[15], trimethyl[16,17] and methylethylenephosphates[16], di-isopropyl-N,N-diethylcarbamylmethylenephosphonate[18], tri-n-butylphosphine oxide[19], and hexamethylphosphoramide[20,21] in terms of displacement of the P=O stretching vibration to lower frequencies; (2) in triphenyl- and tribenzylarsine oxides[15,22] by similar observations of the As=O stretching vibration; and (3) in di-isopropyl-N,N-diethylcarbamylmethylenephosphonate[18], pyramidone[23], and N,N-dimethylacetamide[24-26] by corresponding observations of the C=O stretching vibration.

Differences in site symmetry have been particularly useful in distinguishing between coordinated and non-coordinated nitrate, perchlorate, and thiocyanate groups. The uncoordinated nitrate group (i.e., NO_3^- ion) has D_{3h} symmetry, with only four allowed infrared bands in the 700–1800 cm^{-1} region. Coordination of this group reduces the site symmetry to C_{2v}, with the displacement of the four original bands and the appearance of two new bands[27]. On this basis, i.r. data indicate coordination by all nitrate groups present in complexes of the types $Ln(phen)_2(NO_3)_3$(phen = 1,10-phenanthroline)[7,28], $Ln(bipy)_2(NO_3)_3$ (bipy = 2,2'-bipyridine)[28], and $Ln(terpy)(NO_3)_3$ (terpy = 2,2',6',2''-terpyridine)[29]. On the other hand, both coordinated and ionic nitrate groups are similarly indicated in the complexes $Ln(en)_3(NO_3)_3$ (en = ethylenediamine)[5], $Ln(en)_4(NO_3)_3$ (Ln = La–Sm)[5], $Ln(pn)_3(NO_3)_3$ (pn = 1,2-propanediamine)[8], $Ln(pn)_4(NO_3)_3$ (Ln = La, Nd)[8], and $Ln(dien)_2(NO_3)_3$ (dien = diethylenetriamine; Ln = La-Dy, Er, Yb, Y)[9], but only ionic nitrate in the species $Ln(en)_4(NO_3)_4$ (Ln = Eu–Yb)[5] and $Ln(dien)_3(NO_3)_3$ (Ln = La–Gd)[9].

The uncoordinated perchlorate group (i.e., ClO_4^- ion) has T_d symmetry, which symmetry is lowered to C_{3v}, as evidenced by splitting of the infrared band at 1111 cm^{-1} and the intensification of a band around 927 cm^{-1}, when coordination occurs. The presence of both ionic and coordinated groups is indicated in such lanthanide complexes as $Ln(TMP)_{5-7}(ClO_4)_3$ (TMP = trimethylphosphonate) [16], $Ln(HMPA)_4(ClO_4)_3$ (HMPA = hexamethylphosphoramide) [21], $Ln(phen)_3(ClO_4)_3$ [7], and $Ce(TBPO)_4(ClO_4)_3$ (TBPO = tri-n-butylphosphine oxide) [19]. However, all the perchlorate groups are indicated to be ionic in the species $Ln(en)_4(ClO_4)_3$ [5], $Ln(pn)_4(ClO_4)_3$ [8], and $Ln(phen)_4$ $(ClO_4)_3$ [7]. Similar results have been obtained for the thiocyanate group, where both coordinated and ionic thiocyanate is found for the species $Yb(phen)_2(SCN)_3$ but only the ionic group for $Ln(phen)_3(SCN)_3$ [30].

Where both coordinated and ionic nitrate, perchlorate, or thiocyanate groups are present, relative intensities of appropriate bands allow rough assignments of the proportions of each, e.g., as $[Yb(phen)_2(SCN)_2](SCN)$ [30]. The coordinated nitrate group can be either unidentate, bidentate, or bridging, but i.r. data alone cannot distinguish among these possibilities. Raman polarisation studies can distinguish between unidentate nitrate and the other two, e.g., in showing that in $Ln(TBP)_3(NO_3)_3$ complexes, the nitrate group is bidentate or bridging [31]. Vibrational studies also suggest the presence of bidentate formate and acetate groups in the anhydrous species [32].

Vibrational data cannot alone establish coordination numbers, but they can suggest at least minimal values or reasonable values when combined with a knowledge of stoichiometry and the behaviour of the other groups present. Thus, such assignments as $[Ln(HMPA)_4(ClO_4)_2]ClO_4$ [21], $[Ln(en)_4$ $(NO_3)](NO_3)_2$ (Ln = La–Sm) [5], $[Ln(en)_3(NO_3)_2](NO_3)$ [5], $[Ln(dien)_2(NO_3)_2]$ (NO_3) [9], and $[Yb(phen)_2(SCN)_2](SCN)$ [30] have been made.

7.2.1.2 Electronic spectra

The line-like electronic absorption spectra of the Ln^{3+} ions (Ln ≠ Y, La, Ce, Yb, Lu) result from inner 4f transitions of a predominantly electric dipole character[33] that for a free ion are Laporte forbidden but are allowed by interactions produced by external ligand fields that mix in states of opposite parity. When these spectra for systems containing complexing ligands are compared with those of the aquated cations, three general changes, all of which are related to alterations in the strength and symmetry of the ligand fields, are observed[3].

These changes are small shifts toward longer (occasionally shorter) wavelengths, splitting of certain bands into several small maxima, and alteration of specific absorptivity of individual bands. Those bands that are particularly susceptible to splitting and intensity changes are termed hypersensitive. Although these effects are useful in determining the symmetries of complex species in solution and thus suggesting coordination numbers and molecular structures, they are far less definitive for the lanthanide ions than for the d-type ions since the order of perturbation for the former is crystal field < spin-orbit coupling < inter-electron repulsion. However, the hypersensitive transitions are consistent with the selection rules for quadrupole radiation within the 4f shell, and the intensities of these transitions are probably related to asymmetrical distribution of electromagnetically induced dipoles surrounding the lanthanide ion.

Definitive evaluations of coordination number by means of electronic spectra are limited. Studies of the 5780 Å hypersensitive band of the Nd^{3+} ion suggest that in aqueous solution this ion is 9-coordinate [34], thus supporting a similar conclusion based upon partial molal volume studies[35]. Thus the band shape is remarkably similar to that for crystalline $[Nd(H_2O)_9]$ $(BrO_3)_3$, where 9-coordination has been established by x-ray techniques[36], but different from that of $[Nd(H_2O)_6Cl_2]Cl$, where 8-coordination has been confirmed[37]. Hypersensitive transitions have also been correlated with coordination numbers 6, 7, and 8 for the β-diketonates of the ions Nd^{3+}, Ho^{3+}, and Er^{3+} [38], and coordination numbers of 8–9 for Ln^{3+} in $Ln(CH_3 CO_2)_3 \cdot 1.5\ H_2O$ (Ln = La, Ce), 8 in $Ln(CH_3CO_2)_3 \cdot 1.5\ H_2O$ (Ln = Pr, Nd), 6–7 in $Ln(CH_3CO_2)_3 \cdot 4\ H_2O$ (Ln = Sm–Lu), 9 in $Ln(CH_3CO_2)_3$ (Ln = La–Pr), 8–9 in $Nd(CH_3CO_2)_3$, 7–8 in $Ln(CH_3CO_2)_3$ (Ln = Gd–Ho), and 6–7 in $Yb(CH_3CO_2)_3$ have been similarly assigned[39]. Octahedral symmetry and 6-coordination are established by the electronic spectra for the species $[LnX_6]^{3-}$ (X = Cl, Br, I) [40, 41].

7.2.1.3 Fluorescence spectra

Complexes of the Ln^{3+} ions (except Y^{3+}, La^{3+}, Gd^{3+}, Lu^{3+}) exhibit fluorescence if the ligand has in its molecular structure a bond system that can absorb sufficient energy to be raised to an appropriate singlet state that can revert to a triplet state and thence either lose this energy directly or transfer it to appropriate excited Ln^{3+} ionic states[42]. Aromatic and conjugated systems are particularly suitable, especially in combination with Eu^{3+} or Tb^{3+} ion. The fluorescence spectra of the tetrakis dibenzoylmethanide and tetrakis benzoylacetonate chelates of europium(III) in ethanol have been interpreted in terms of 8-coordination in a triangular dodecahedral geometry[43]. The fluorescence spectrum of the compound $[Eu(terpy)_3](ClO_4)_3$ (2,2',6',2''-terpyridine) is taken to indicate 9-coordination in a distorted D_3 site symmetry[44].

7.2.1.4 X-ray crystallography

X-ray diffraction data for single crystals allow the definitive establishment of both coordination numbers and molecular geometries for solid complexes. To the extent that data of this type establish the arrangements of nearest neighbour atoms about a central Ln atom, all types of compounds are similarly treated, and a decision as to what is a complex and what is not is somewhat arbitrary. Herein, the term complex is restricted to a species, containing an Ln^{n+} ion and a recognised ligand, which can be regarded as an entity that could exist in solution and could undergo reactions as such. In this sense, an anhydrous chloride of the YCl_3 type structure would not qualify as a complex even though in its crystal structure there exist octahedral arrangements of six chloride ions around each lanthanide ion, which are linked by shared edges[45]. However, a hydrated chloride, containing the group $[Ln(H_2O)_6Cl_2]^+$ would so qualify[37].

Coordination numbers 6, 7, 8, 9, 10, and 12 have been defined by x-ray methods. Inasmuch as these methods have given to us most of our definitive structural data, discussion of specific results is deferred to the subsequent section on molecular geometry.

7.2.1.5 Miscellaneous methodology

Mass spectral data are more indicative of coordination number than absolutely definitive. Thus the mass spectrum of the compound $Cs[Y(hfac)_4]$ (hfac = hexafluoroacetylacetonate) contains peaks assignable to the species $Cs[Y(hfac)_4]^+$ and $Cs[Y(hfac)_3]^+$, suggestive, respectively, of coordination numbers 8 and 6 on the assumption that the ligand is bidentate[46]. Similarly, peaks for the ions $[Ho(dipav)_3]^+$, $[Ho(dipav)_2]^+$, and $[Ho(dipav)]^+$ (dipav = dipivaloylmethane) suggest coordination numbers 6, 4, and 2, and peaks for the ions $Na[Ho(TDH)_4]^+$ and $Na[Ho(TDH)_3]^+$ (TDH = trifluorodipivaloylmethane) suggest coordination numbers 8 and 6 [47].

Changes in relaxation frequency with increasing atomic number of the cation in the ultrasonic absorption spectroscopy of aqueous nitrates have been associated with changes in coordination number[48]. Proton magnetic resonance spectra have given some data on hydration numbers in aqueous solution, e.g., 2.4 for the Y^{3+} ion in nitrate and perchlorate solutions containing acetone[49]. Such studies are restricted to the diamagnetic ions Y^{3+}, La^{3+}, and Lu^{3+}.

7.2.2 Coordination polyhedra

Irrespective of whether the bonding in a complex species is electrostatic or covalent, the most probable spatial positions of the donor atoms with respect to the central atom are predictable in terms of electron-pair, ligand, or Coulombic repulsions or the requirements of maximum symmetry[13, 14, 50–53]. The polyhedra, in order of preference for a given coordination number, and their site symmetries, so predicted are:

$$
\begin{array}{lll}
\text{C.N.} = 6 & 1 & \text{octahedron } (O_h) \\
 & 2 & \text{trigonal prism } (D_{3h}) \\
= 7 & 1 & \text{pentagonal bipyramid } (D_{5h}) \\
 & 2 & \text{monocapped octahedron } (C_{3v}) \\
 & 3 & \text{monocapped trigonal prism } (C_{2v}) \\
= 8 & 1 & \text{triangular faced dodecahedron } (D_{2d}) \\
 & 2 & \text{square antiprism } (D_{4d}) \\
= 9 & 1 & \text{tricapped trigonal prism } (D_{3h}) \\
= 10 & 1 & \text{4,4-bicapped square antiprism } (D_{4d}) \\
= 11 & 1 & \text{pentacapped trigonal prism} \\
= 12 & 1 & \text{icosahedron } (I_h)
\end{array}
$$

For coordination number 6, experience shows that the octahedral arrangement is preferred by a wide margin, with distortion to the trigonal prism occurring only with a ligand of fixed geometry. In other instances where there is more than one possibility, the choice is less definite. This is particularly true of coordination number 8, where the dodecahedral and square antiprismatic arrangements are so close to each other that the choice may depend on measured angles and may not always have been correctly made[13, 14, 54, 55].

A point of particular importance with regard to complexes of the lanthanide ions is that the energy of reorganisation in going from coordination number 7 to 8, 8 to 9, or 9 to 10 is sufficiently small that solvation upon dissolution may readily change the coordination number[13].

7.2.3   Molecular structures

7.2.3.1   Coordination number 6

6-Coordination in an octahedral geometry has been established spectroscopically for the ions $[LnX_6]^{3-}$ ($X = Cl, Br, I$) in their triphenylphosphonium or pyridinium salts[40, 41]. A single-crystal x-ray diffraction study has given the same results for the ion $[Er(NCS)_6]^{3-}$ in its tetrabutylammonium salt[56]. It is probable that octahedral 6-coordinate structures are characteristic of the thermally stable tris(β-diketone) chelates where the ligand is very bulky (e.g., 2,2,6,6-tetramethyl-3,5-heptanedione [57]) and the species $Cs_2Na[LnCl_6]$ ($Ln = La-Lu$, Y) [58]. The tris(tropolonates) of the Ln^{3+} ions are largely obtained anhydrous, suggesting 6-coordination [59]. However, these compounds are so intractable that they are believed to be polymeric with oxygen bridges and to represent 7- or 8-coordination [59].

7.2.3.2   Coordination number 7

Two and possibly all three of the preferred coordination polyhedra for this coordination number have been observed. The ion $[TbF_7]^{3-}$, as the salt $Cs_3[TbF_7]$, has been reported as isostructural[60] with the ion $[ZrF_7]^{3-}$, which in turn may have pentagonal bipyramidal geometry[61]. Monocapped octahedral geometry is characteristic of the tris(benzoylacetonato)aquoyttrium(III) molecule, $[Y(C_6H_5COCHCOCH_3)_3(H_2O)]$[62], and the analogous tris(dibenzoylmethanide)aquoholmium(III) molecule, $[Ho(C_6H_5COCHCOC_6H_5)_3(H_2O)]$ [63]. In each instance, the octahedron is substantially distorted. In the former molecular structure, the β-diketonate ligands wrap asymmetrically about the coordination polyhedron, with two phenyl groups 'up' and one 'down'. In the latter structure, the arrangement is that of a three-bladed propellor, each blade of which is a planar six-membered β-diketonate ring with the phenyl groups twisted 10 degrees and 19 degrees to the plane. In each molecule, the water molecule lies above a triangular octahedral face. Details of these and other molecular structures are best found in the original articles.

Monocapped trigonal prismatic geometry is characteristic of the tris(acetylacetonato)aquoyttrium(III) molecule, $[Y(CH_3COCHCOCH_3)_3(H_2O)]$ [64]; the analogous tris(acetylacetonato)aquoytterbium(III) molecule in the compound $[Yb(CH_3COCHCOCH_3)_3(H_2O)]\cdot0.5$ C_6H_6 [65]; and the molecule of the acetylacetonimine adduct of tris(acetylacetonato)ytterbium

(III), $[Yb(CH_3COCHCOCH_3)_3(CH_3COCH=C(NH_2)CH_3)_3$ [66]. In each instance the Ln^{3+} ion is bonded to seven oxygen atoms, six from the chelating ligand and one from the adduct molecule. The last of these occupies a position above a rectangular face. Individual trigonal prismatic groups are hydrogen-bonded to each other through the adduct molecules. The chelating ligands span between the triangular faces.

7.2.3.3 Coordination number 8

Both the square-antiprism and the triangularly faced dodecahedron are characteristic polyhedra for this very common coordination number, although, as previously mentioned, the choice between the two for a specific compound is not necessarily unambiguous[13, 14, 54, 55]. Examples of distorted square antiprismatic geometry are found for the following molecular species $[Ln(H_2O)_6Cl_2]^+$ (Ln = Nd,   Sm,   Eu,   Gd,   Er)[37, 67]; $[Ce(CH_3COCHCOCH_3)_4]$ [68]; $[Y(CH_3COCHCOCH_3)_3(H_2O)_2]$ [69]; and $[La(CH_3COCHCOCH_3)_3(H_2O)_2]$ [70]. The same geometry is found in crystals of the complex compounds $(NH_4)_2[CeF_6]$ [71], $Li[LnF_4]$ (Ln = Eu–Lu) [72], and $[Er(C_2O_4)(HC_2O_4)(H_2O)_3]$ [73], but bridging precludes the delineation of discrete 8-coordinate molecular groups.

Examples of triangular dodecahedral geometry include the following molecular species $[Ce(C_6H_5COCHCOC_6H_5)_4]$ [74], $[Pr(C_4H_3SCOCHCOCF_3)_4]^-$ in its monohydrated ammonium salt[75], and $[Y(CF_3COCHCOCF_3)_4]^-$ in its caesium salt[76]. By contrast with molecular structures reported for other 8-coordinate dodecahedral species, in each of the latter two structures each chelating ligand spans adjacent vertices between the two intersecting bisphenoids that constitute the dodecahedron. The result is idealised overall D_2 symmetry.

It is immediately obvious that the entire problem of 8-coordination geometry can be solved only through the determination of the molecular structures of such apparently 8-coordinate species as $[Ln(DMA)_8](ClO_4)_3$ (DMA = N,N-dimethylacetamide; Ln = La–Nd) [24]; $[Ln(BuL)_8](ClO_4)_3$ (BuL = γ-butyrolactam; Ln = La, Pr–Sm, Gd, Dy, Er, Yb, Y) [77]; $[Ln(pd)_4]$ $[B(C_6H_5)_4]_3$ (pd = pyramidone; Ln = La, Nd, Gd, Er, Yb, Lu, Y)[23], $Na[LnT_4]$ (T = tropolone; Ln = Ce–Lu) [55, 74]; $[Ln(en)_4]$ $(ClO_4)_3$ [5]; $[Ln(pn)_4](ClO_4)_3$ [8]; and $[Ln(diket)_4]^-$ (diket = β-diketonate) [80, 81] – to cite only a few of the many species that have been described.

7.2.3.4 Coordination number 9

The tricapped trigonal prism is the coordination polyhedron characteristic of the following complex species: $[Ln(H_2O)_9]^{3+}$ in the compounds $[Nd(H_2O)_9]X_3$ (X = Cl, BrO_3) [36] and $[Ln(H_2O)_9](C_2H_5SO_4)_3$ (Ln = Pr, Er, Y) [82]; $[Ln(OH)_9]^{6-}$ in the crystalline trihydroxides (Ln = La, Pr–Sm, Gd, Dy) [83]; and $[La(H_2O)_6(SO_4)_3]^{3-}$ in the compound $La_2(SO_4)_3 \cdot 9 H_2O$ [84].

A somewhat different coordination polyhedron describes the ions $[Ln(EDTA)(H_2O)_3]^-$ (EDTA = ethylenediamine-N,N,N',N'-tetra-acetate; Ln = La, Tb) in salts of the type $M^I[Ln(EDTA)(H_2O)_3]\cdot 5\,H_2O$ [85]. The polyhedron is defined by four oxygen atoms and two nitrogen atoms from the $EDTA^{4-}$ ion and three oxygen atoms from the coordinated water molecules, but it is *quasi* D_2 dodecahedral with two of the coordinated water molecules in roughly the same site.

7.2.3.5 Coordination number 10

Detailed molecular structure data are available for only a few species, each of which has its own geometry. The molecular structure of $[La(EDTA\cdot H)(H_2O)_4]$ (EDTA·H = monoprotonated EDTA ion) is comparable with that of the $[La(EDTA)(H_2O)_3]^-$ ion, except that three water molecules are close to one dodecahedral site [86]. Molecules of the species $[La(bipy)_2(NO_3)_3]$ have a rough bicapped dodecahedral geometry (D_2 symmetry) with four nitrogen atoms from the bipyridine groups and six oxygen atoms from the bidentate nitrate groups occupying the polyhedral sites [87]. The coordination polyhedron of oxygen donors in the ion $[Ce(NO_3)_5]^{2-}$, in its triphenylethyl phosphonium salt, again more closely approximates the bicapped dodecahedron than the energetically more favoured bicapped square antiprism [88], but the midpoints of the bidentate nitrate groups define almost exactly a regular trigonal bipyramid around the central Ce^{3+} ion. The coordination polyhedron characteristic of the molecule $[Ce(NO_3)_4\{OP(C_6H_5)_3\}_2]$ is based upon oxygen atoms from four bidentate nitrate groups and two unidentate phosphine oxide molecules but does not correspond to a regular coordination polyhedron [89]. The best arrangement appears to be a distorted octahedron with two phosphonyl oxygen atoms in *trans* positions and each nitrate group in a single coordination position. In crystals of the compound $La_2(CO_3)_3\cdot 8\,H_2O$, two different 10-coordinate groups based upon oxygen donor atoms from water molecules and both carbonate and hydrogen carbonate groups can be distinguished [90]. The coordination polyhedron is a dodecahedron with two sites occupied by bidentate carbonate groups. Observed geometry is apparently determined by differences in the steric requirements of the several groups present. Other studies with but a single kind of donor group will be important supplements to that reported for the $[Ce(NO_3)_5]^{2-}$ ion.

7.2.3.6 Coordination number 12

This high coordination number appears to be limited to the bidentate nitrate group, which has an extremely short 'bite'. In crystals of the double nitrate $Ce_2Mg_3(NO_3)_{12}\cdot 24\,H_2O$, each Ce^{3+} ion is at the centre of a regular icosahedron provided by six bidentate nitrate groups [91]. The same arrangement is found for the $[Ce(NO_3)_6]^{2-}$ ion in crystals of the salt $(NH_4)_2[Ce(NO_3)_6]$ [92]. In each instance, the site symmetry is tetrahedral, and the midpoints of the six nitrate groups describe a rough octahedron around the central cerium ion. The same molecular structure is observed for the $[Ce(NO_3)_6]^{2-}$ ion in a concentrated aqueous solution of its ammonium salt[93].

7.2.4 Constancy of coordination number

There is evidence that for a given ligand or combination of ligands the co-ordination number of an Ln^{3+} ion decreases as its crystal radius decreases. The change is seldom abrupt and may not appear at the same Ln^{3+} ion as the ligand changes. Detailed molecular structure data to identify such changes are lacking. The series $[Ln(en)_4(NO_3)](NO_3)_2$ (Ln = La–Sm)—$[Ln(en)_4](NO_3)_3$ (Ln = Eu–Yb); $[Ln(en)_4Cl]Cl_2$ (Ln = La, Nd)—$[Ln(en)_4]Cl_3$ (Ln = Sm, Gd, Er); and $[La(en)_4Br]Br_2$—$[Ln(en)_4]Br_3$ (Ln = Nd, Gd) illustrate both this type of change and the effect of ligand size on where the change occurs[5]. Alterations in thermodynamic stability in solution and in kinetics of ligand exchange (Sections 7.3, 7.4) can be accounted for in terms of change in coordination number.

7.2.5 Isomerism

The possibilities of geometrical isomerism increase substantially with increasing coordination number, assuming that ligands occupy fixed positions and do not readily exchange positions[13]. If geometrical isomers do exist, they are most likely to be found in crystals, where ligand mobilities are reduced. Although cases of polymorphism, that could arise from geo-metrical isomerism, are known, no crystal-structure determination has yet indicated isomerism. A report ascribing differences in the fluorescence spectra of samples of the piperidinium salt of the tetrakis(dibenzoylmethanide)-europium(III) ion to the presence of stereoisomers of different symmetry[81] has been questioned and the phenomenon ascribed to changes in the environment of the Eu^{3+} ion caused by solvation by water or dimethyl-formamide molecules[94]. In 8-coordinated β-diketone complexes of dode-cahedral geometry, the edges spanned by the ligand may differ[75, 76] but the arrangement for a given ligand appears to be fixed. Crystal geometry may be expected to persist for the rather ionic lanthanide complexes only in solvents of low polarity. In this connection, it is of interest that chroma-tographic studies with tris(acetylacetonato) complexes in benzene with d-lactose as adsorbant indicate partial resolution into optical isomers[95], whereas crystallisation of alkaloid salts of ethylenediaminetetraacetato chelates from an aqueous medium gives no indication of comparable reso-lution[96].

7.3 THERMODYNAMIC STABILITY IN SOLUTION

The fundamental aspects of thermodynamic stability in solution have been summarised in earlier reviews[1, 3, 4]. In brief, it is measured by the free-energy change (ΔG) for the overall complexation process, which is in turn related to the enthalpy change (ΔH) and the entropy change (ΔS) and to the overall formation constant (β_n) as

$$\Delta G = \Delta H - T\Delta S = -RT \ln\beta_n$$

Of these quantities, ΔH and β_n are measurable experimentally; ΔG and ΔS are then calculable from experimental data. Formation constants have been evaluated for many species in aqueous systems by application of the standard techniques. Enthalpy changes have been determined both from the temperature coefficients of formation constants and by direct calorimetric measurement. Few, if any data, are available at the standard state of infinite dilution, but many have been determined at constant ionic strength. Almost all data are concentration based and thus not absolutely thermodynamic in character. Where complexation proceeds in a stepwise fashion, stepwise formation constants ($K_{1,2}...u$) and thermodynamic functions have often been evaluated. The extensive tabulations of data[1, 3, 4, 97] are being supplemented at a rapid rate. Typical of more current research are papers dealing with complexes derived from acetate[98]; β-hydroxypropionate[99]; isobutyrate and α-hydroxyisobutyrate [100]; 2-hydroxy-2-methylbutyrate [101]; tropolonate [78, 79]; 1,1,1,2,2,3,3-heptafluoro-7,7-dimethyl-octanedionate [102]; fluoride, oxalate, and sulphate[103] ions. Data for mixed chelates involving more than one type of ligand are appearing also[104].

Comparisons for a particular ligand show that either in terms of ΔG or β_n the lanthanide complexes are less stable than those for d-transition metal ions of the same charge. Comparisons of thermodynamic stabilities among the various lanthanide ions (i.e., Ln^{3+}) show no common trend and no single type of variation. The first formation constants (and $-\Delta G$ values) for complexes of the lighter lanthanide ions (La^{3+}–Eu^{3+} or Gd^{3+}) increase quite uniformly with decreasing crystal radius, irrespective of the ligand used, as might be expected if the interaction is predominantly electrostatic. For the heavier Ln^{3+} ions, however, the variations are more complex and are dependent upon the ligand present. At least the following broad trends can be distinguished in this region: (1) a more or less regular increase with decreasing crystal radius, (2) essential constancy for the region Gd^{3+}–Lu^{3+}, and (3) an increase to a maximum in the vicinity of the Dy^{3+} ion. Trend (1) has been explained in terms of increasing Coulombic attraction with decreasing crystal radius. On this basis, the stability of the Y^{3+} complex of a given ligand should be comparable with that of the Ho^{3+} or Dy^{3+} complex. This is commonly the case. Trends (2) and (3) cannot be explained simply. They probably reflect combinations of such factors as steric effects, changes in coordination number of the cation, and differences in the coordinating abilities of the various donor atoms of the ligand. With trend (2) or (3) ligands, the yttrium ion appears with the lighter lanthanides rather than where its crystal radius would predict. It is important, however, to realise that in solution one deals with solvated ions and that solvation may smooth differences in size that characterise the unsolvated ions in crystals.

For nearly every ligand that has been investigated, there exists a discontinuity or irregularity in log K, log β_n, or ΔG values at or close to the Gd^{3+} ion. However, this *gadolinium break* [105] is not distinguishable in the log K_1

and log K_2 values for the tropolonate chelates[78, 79] and is probably absent in other systems. Rather than associate it with a particular structural factor, it is probably more reasonable to associate it with a fortuitous balance of factors, among which change in coordination number or hydration number may be significant. In addition to the evidences from crystal structure determinations previously noted, variations in the enthalpy and entropy contributions to the free energy of complexation also lead to this conclusion[106–109].

The enthalpy change accompanying complexation of a lanthanide ion in aqueous solution is a measure of the difference in energy between the ligand-cation bond and the coordinated water molecule-cation bond. Inasmuch as coordinated water molecules may be within or outside the primary coordination sphere of the lanthanide ion, the enthalpy change really expresses a difference in energy between the ligand bond and the hydration sphere bond. Included also is whatever energy is needed to rearrange hydrogen bonds in the vicinity of the complex species. For a given ligand, ΔH values seldom change monotonically from the La^{3+} to the Lu^{3+} ion[106]. Rather the variation is somewhat sinusoidal, suggesting that no simple bonding process is involved. Experimentally, the change in enthalpy is always small (1–5 kcal mol^{-1}), and it may either favour $(-\Delta H)$ or oppose $(+\Delta H)$ complexation. The entropy change accompanying complexation is related to alteration in order in the system, both in terms of changing the number of particles and the modes of vibration of these particles. In aqueous systems, complexation by chelation releases bound water molecules, thereby increasing the entropy change. Experimentally, the entropy change is the major contribution to the free energy change, and the resulting chelates are entropy stabilised[1, 3, 4]. Although partial molal entropies for various chelates suggest that the Ln^{3+} ions fall into two groups with the Gd^{3+} ion as the dividing species[110–114], it is incorrect to ascribe this division to a change in coordination requirements[110] since both configurational and translational contributions to entropy are important[111].

For complexation in aqueous solution, commonly the more negative the ΔH contribution the less positive the ΔS contribution[115]. For 1:1 non-chelated species, it has been concluded that when the enthalpy change favours complexation and the entropy change opposes it (e.g., NO_3^-, SCN^-), the primary hydration sphere of the Ln^{3+} ion is essentially retained, and the resulting species is an outer-sphere, ion-pair type[115]. When the reverse is true (e.g., with F^-, RCO_2^-), the primary hydration sphere is ruptured, and the resulting species is an inner-sphere type[106, 107, 108, 116]. An evaluation of the entropies of hydration of the Ln^{3+} ions from the solubilities and enthalpies of solution of the iodates in water at 25 °C, supports a change in hydration number somewhere near the centre of the series and suggests that changes in the thermodynamic parameters of complexation are commonly consequences of dehydration rather than ligation[117]. Change in hydration number is supported by conductance[108] and molar volume[118] data. On the other hand, an exacting evaluation of the entropies of the Ln^{3+} ions in aqueous solution indicates a regular increase with decreasing crystal radius and gives no indication of change in hydration number anywhere in the series[119]. The conclusions of Bertha and Choppin[117] are believed to represent an artifact in the treatment of their data. It is apparent that the problem is not yet solved. It is significant that $-\Delta H_1$ and ΔS_1 for the formation of 1:1 tropolonate complexes increase almost linearly with decreasing crystal radius[79]. It has been pointed out also that linear correlations of ΔH or ΔS for Ln^{3+} ions with a common ligand should exist if ΔG for the complexation process is zero or constant[120].

A number of the properties of the Ln^{3+} ions that relate to the thermo-dynamic stabilities of their complexes show a more subtle 'tetrad' effect when related to atomic number[121]. This effect is generalised as 'In systems involving all 15 lanthanides(III), the points on a plot of the logarithm of a suitable numerical measure of a given property of these elements $v.$ Z(atomic number) may be grouped, through the use of four smooth curves without inflections,

Table 7.1   Stepwise thermodynamic functions for formation of [Ln(en)n](ClO$_4$)$_3$ complexes in acetonitrile

Function	Ln^{3+}			
	La^{3+}	Tb^{3+}	Yb^{3+}	Y^{3+}
Formation constant				
$\log K_1$	9.5	10.4	11.5	10.4
$\log K_2$	7.5	8.4	9.3	8.2
$\log K_3$	6.2	6.2	6.5	5.6
$\log K_4$	3.3	3.2	3.8	3.1
Enthalpy change, kcal mol^{-1}				
ΔH_1	-17.3	-19.9	-20.1	-20.0
ΔH_2	-15.5	-18.6	-18.8	-18.7
ΔH_3	-13.8	-13.1	-14.4	-12.9
ΔH_4	-11.0	-9.0	-12.8	-10.5
Entropy change, cal mol^{-1} deg^{-1}				
ΔS_1	-15.1	-19.8	-15.4	-17.6
ΔS_2	-18.2	-24.5	-21.1	-23.2
ΔS_3	-18.4	-16.0	-19.0	-15.2
ΔS_4	-22.1	-15.8	-25.9	-18.5

into four tetrads with the gadolinium point being common to the second and third tetrads and the extended smooth curves intersecting, additionally, in the 60–61 amd 67–68 regions[122]'. This tetrad effect has been ascribed either to a variation in the nephelauxetic ratio[123] in the third decimal place[124] or the dependence of the coefficients of the Racah interelectronic repulsion parameters E^1 and E^3 on the quantum numbers L and S for the electronic ground state of the Ln^{3+} ion[125]. Similarly, most complexes of the ion pairs $Ce^{3+}-Pr^{3+}$, $Sm^{3+}-Eu^{3+}$, $Tb^{3+}-Dy^{3+}$, and $Tm^{3+}-Yb^{3+}$ show extra thermodynamic stability in comparison with linear interpolations between $La^{3+}-Nd^{3+}$, $Pm^{3+}-Gd^{3+}$, $Gd^{3+}-Ho^{3+}$, and $Er^{3+}-Lu^{3+}$ [126].

Thermodynamic considerations are somewhat simplified by elimination of the competitive effects of water through the use of anhydrous conditions and less strongly coordinating solvents. For example, in anhydrous aceto-nitrile, the anhydrous lanthanide(III) perchlorates react with ethylenediamine in distinguishable steps to form the species $Ln(en)_n^{3+}$ $(n = 1,2,3,4)$[5]. The large negative enthalpy changes (Table 7.1) indicate strong Ln—N bonds in the complexes. In combination with negative entropy changes, these values indicate that the complexes are substantially enthalpy stabilised. Although stability quite generally increases with decreasing crystal radius, with the Y^{3+} ion occupying its expected position, no absolute trends or regularities are distinguishable. The stepwise enthalpies of formation of the ions $[Ln(dien)_n]^{3+}$ $(n = 1,2,3)$ in perchlorate solutions in acetonitrile[9] increase generally with decreasing crystal radius of the Ln^{3+} ion for $n = 1$ or 2 but decrease significantly for $n = 3$. Steric accommodation of the third large diethylenetriamine group is more difficult for the smaller Ln^{3+} ions, with a substantial change occurring between the Gd^{3+} and Dy^{3+} ions. Indeed when nitrate is present the most stable species in solution is the ion $[Ln(dien)_2(NO_3)_2]^+$. For each ion, the total enthalpy of complexation is less for diethylenetriamine than for ethylenediamine[5] even though the nitrogen atoms present are of comparable basicity, thus indicating again the restrictions of steric accommodation. The order of stability of the ytterbium(III) isobutyrate and α-hydroxyisobutyrate complexes increases with the polarity and coordinating ability of the solvent as[100]

ethanol > dimethylsulphoxide > ethyleneglycol >
40% dimethylsulphoxide − water ~ formamide > water.

Many additional investigations that minimise the effects of water are in order.

7.4 KINETICS AND COMPLEXATION

This area of lanthanide chemistry has been essentially unexplored, although important contributions are beginning to appear. In the sense that complexation reactions involve competition between ligands, these are reactions of ligand exchange. Very broadly, ligand exchange reactions of the lanthanide ions are either (1) those with relatively simple ligands that are so fast that relaxation techniques are essential in their investigation or (2) those with more complicated ligands that are slow enough that they can be followed by stop-flow or analysis-of-aliquots techniques.

The kinetics of the formation and dissociation of the 1:1 oxalato complexes, $Ln(C_2O_4)^+$, in aqueous solution have been evaluated by the pressure-jump procedure[127]. The specific rate constants for formation of the species are essentially constant at 7.7–$8.6 \, mol \, l^{-1} \, s^{-1}$ for the series La^{3+}–Eu^{3+} and at $0.63 \, mol \, l^{-1} \, s^{-1}$ for Er^{3+} and Tm^{3+}, with a rapid drop in the Gd^{3+}–Ho^{3+} region. They are of the same orders of magnitude with the same changes as those for murexide complexation[128]. Both are in accord with the Eigen–Tamm mechanism[129] for other cations, which for the lanthanides (without attention to ionic charges) and a ligand L can be formulated as

$$Ln(aq) + L(aq) \rightleftharpoons Ln(aq)(aq)L \rightleftharpoons Ln(aq)L(aq) \rightleftharpoons LnL(aq) \rightleftharpoons LnL(aq)'$$

Loss of water from the inner hydration sphere (Step 3) is the rate-determining step. The implication is that the primary hydration sphere remains unaltered in the region La^{3+}–Eu^{3+} but is decreased in the region Eu^{3+}–Ho^{3+} to a new size.

The temperature-jump technique has been used to study the bimolecular complexation of Ln^{3+} ions with o-aminobenzoate ion in aqueous systems[130, 131]. A rapid decrease in specific rate constant occurs in the region Eu^{3+}–Dy^{3+}, and values in the La^{3+}–Eu^{3+} and Dy^{3+}–Lu^{3+} regions differ in magnitude. A change in coordination number again appears reasonable. Some comparable preliminary data for the Ln^{3+}-acetate system are reported also[131]. A flash photolysis investigation of the bimolecular electron transfer process

$$Ce^{III} + NO_3^- \rightarrow Ce^{IV}(NO_3)^-$$

has shown that at least five different nitratocerium(III) species are involved and thus suggests the significance of ligand exchange in this reaction[132]. Ultrasonic absorption measurements indicate a marked decrease in relaxation frequency for the formation of $Ln(NO_3)^{2+}$ species between the ions Sm^{3+} and Er^{3+}, consistent with a change in average cation coordination number[133].

The isotopic exchange of radioactive $^{144}Ce^{3+}$ ion with inactive cerium(III) in the species CeL^{n-} ($n = 0$, L = N-hydroxyethylethylenediamine-N,N',N'-triacetate (HEDTA); $n = 1$, L = ethylenediamine-N,N,N',N'-tetraacetate (EDTA), trans-1,2-hexanediamine-N,N,N',N'-tetraacetate (DCTA), ethyleneglycol-bis(aminoethyl)-N,N,N',N'-tetraacetate (EGTA); $n = 2$, L = diethylenetriamine-N,N,N',N'',N''-pentaacetate (DTPA)), evaluated by removal of ionic cerium(III) from aliquots and measurement of the radiocerium content of the remaining chelate, indicates initial slow protonation of the ligand in the complex, followed by more rapid further protonation, removal of the ligand, and ultimate rapid complexation of radiocerium(III) by the protonated ligand[134, 135]. Similar conclusions have been reached in independent studies[136, 137] with the radionuclides $^{152}Eu^{3+}$ and $^{154}Eu^{3+}$ and the species $[Ln(EDTA)]^-$ where the kinetics of exchange agree with the rate equation

$$\text{rate} = kc_H\, c_{Eu^{3+}}\, c_{(LnEDTA)}/c_{Ln^{3+}}$$

A comparable study involving the exchange of radio $^{147}Nd^{3+}$ with the inactive ion $[Nd(DTPA)]^{2-}$ indicates that under acidic conditions the process proceeds by protonation and bimolecular collision, whereas under neutral conditions a bimolecular collision in which protons participate occurs[138]. The general mechanism is in all of these cases that were proposed by Betts, Dahlinger, and Munro[139]. A stopped-flow technique indicates that the rate of transfer of ligand from $[Ln(DCTA)]^-$ to Cu^{2+} ion in aqueous solution is dependent on proton concentration and independent of copper ion concentration[140], with acid dissociation rate constants at 25 °C decreasing regularly with decreasing crystal radius (e.g., La^{3+}, 129; Y^{3+}, 0.36; Lu^{3+}, 0.017, each in l mol^{-1} s^{-1}). The formation of the species $[Ln(DCTA)]^-$ proceeds via the rapid formation of the protonated species $H(DCTA)^{3-}$, followed by a slower first-order reaction in which the Ln^{3+} ion is incorporated into the coordination cage of the ligand and displaces the proton.

These formation reactions are slower than expected from water exchange[141] or murexide complexation[128].

7.5 BONDING IN COMPLEX SPECIES

Many of the properties of the lanthanide complexes, as already described, support the argument that the ligands are held by substantially electrostatic interaction. To the evidences for at least minor covalent interaction may be added the existence of a derived nephelauxetic series of ligands for the Pr^{3+} and Nd^{3+} ions closely comparable with that for d-transition metal ions[142, 143]. Furthermore, by adding to the enthalpy changes measured for 1:3 mol ratio complexing in aqueous solution by diglycollate and dipicolinate ions the integral heats of solution of the nonahydrated solid bromates or ethyl sulphates, the enthalpy change for the process ($X = BrO_3$ or $C_2H_5SO_4$)

$$LnX_3 \cdot 9H_2O(c) + aL^{n-}(aq) \rightleftharpoons LnY_n^{(an-3)}(aq) + 3X^-(aq) + 9H_2O(aq)$$

so calculated, thus eliminating the complication induced by the ion $Ln(aq)^{3+}$, decreases quite regularly with increasing atomic number, and relative to the values for the ions La^{3+} ($4f^0$), Gd^{3+} ($4f^7$), and Lu^{3+} ($4f^{14}$), the enthalpy changes for the other cations are lower by up to a few tenths of a kilocalorie per mole[144]. These observations may indicate minor ligand field interaction, or minor covalent bonding.

It might be expected, by analogy to behaviour of the d-transition metal species, that isotropic shifts in the proton magnetic resonance spectra of suitable ligands bonded to paramagnetic lanthanide ions could provide some indication of the degree of covalency in the bond in question[4]. The observed isotropic shift is a combination of a contact term[145], which is a measure of covalency, and a pseudo-contact term[146], which is determined by the geometrical arrangement of ligands about the central atom and interactions between the cation and molecules of the solvent. For d-transition metal complexes, the geometries are known and essentially constant and ion–solvent molecule interactions are minimal or known so that correction for the pseudo-contact contribution is easily made, and the degree of covalency can be established reasonably. For a lanthanide ion, evaluation of the pseudo-contact term, particularly in aqueous systems, is very difficult as a consequence of lack of exact definition of either geometry or composition of the coordination sphere and the presence of substantial and variable interactions between cation and solvent molecules. Data for the interaction of substituted pyridine molecules of known and rigid geometries with the Pr^{3+} and Nd^{3+} ions, as perchlorates, in acetonitrile under strictly anhydrous conditions have been interpreted in terms of contact and pseudo-contact contributions of similar magnitude but different sign[6]. Although the contact shifts are large enough to indicate substantial covalent interaction, the g and J values of the two cations are such that they may be more significant in determining the magnitude of the contact term than the nature of the bond itself. Isotropic paramagnetic shifts in proton resonance, ascribable to combinations of contact and pseudo-contact terms, are measures of thermodynamic parameters in ligand exchange reactions of alkyl-substituted bipyridine complexes of lanthanide ions, which may relate to bonding[147].

It is reasonable to ask what orbitals can be involved if covalent inter-actions do exist. The distinguishing electrons in the 4f orbitals are comparable in a formal sense to the electrons in nd orbitals of the ordinary transition metal ions. Although these orbitals are spatially oriented toward favourable donor sites for the high coordination numbers observed[148, 149] and poten-tially capable of forming appropriately directed hybrids[150], shielding by the $5s^2 5p^6$ octet effectively screens these orbitals from interaction with ligand orbitals. Evidences for their participation in bonding are limited[151]. As already stated, perturbations in electronic spectra upon complexing are accounted for in terms of alteration of crystal field symmetry and do not imply covalent interactions[152, 153]. Interpretations in terms of even weak 4f orbital involvement[154] are not convincing. Magnetic data, which are also dependent upon 4f electrons[155], are limited for complexes of the lanthanides, but those data that have been published give no definitive indication that these electrons are involved in bonding[156–158]. Furthermore, magnetic data for the species $Ln(Cp)_3$ (Cp = cyclopentadienide)[159], $Ln(Cp)_2Cl$[160], $Ln(Cp)_2$ (Ln = Eu, Yb)[161], and $Ln(Cot)$ (Ln = Eu, Yb; Cot = cyclo–octatraenide)[162] indicate neither involvement of 4f electrons nor covalent interaction by these potentially π-bonding ligands. The same is true of the species $K[Ln(Cot)_2]$ (Ln = Ce, Pr, Nd, Sm, Tb)[163].

It is inferred then that any covalent contribution to a bond must involve 5d or higher orbitals, that are normally unoccupied in the lanthanide ions but are potentially energetically available. With the Y^{3+} ion, of course, the 4d orbitals are analogous.

7.6 COMPLEXES IN SEPARATIONS

Except in those instances where a lanthanide can appear in a different oxidation state, and thereby give compounds that differ substantially in properties, separations of the lanthanides are fractional in character. The tediousness of operation and the slowness of effecting complete separations by classical techniques of precipitation, dissolution, or crystallisation have been substantially obviated by the enhancement of slight differences between adjacent lanthanide ions and within the series as a whole by complexation. Three types of separation based upon these effects are ion exchange, solvent extraction, and volatilisation.

The principles and practice of ion exchange separations have been ade-quately reviewed[3, 164]. Recent publications are concerned mainly with refine-ments of procedure using chelating agents, such as EDTA, HEDTA, NTA, DTPA, and α-hydroxyisobutyrate, that were developed earlier. Few other complexing agents have shown real promise.

The principles and practice of solvent extraction separations have also been reviewed adequately[3, 165]. The most successful extractants appear to be esters of orthophosphoric acid, such as tri-n-butylphosphate (TBP) and di(-2-ethylhexyl) phosphoric acid (HDEHP), both of which effectively extract the Ln^{3+} ions from nitric acid solutions without undergoing ex-cessive hydrolytic decomposition. Tri-n-butylphosphate extracts the species

$Ln(NO_3)_3 \cdot 3$ TBP into itself or an inert, non-aqueous medium such as kerosene. Extractability increases with increasing atomic number (and decreasing crystal radius) at a given aqueous nitric acid concentration and, for a given Ln^{3+} ion, with increasing nitric acid concentration. The reagent HDEHP gives better separations, although in the same general order, probably because its behaviour as a chelating agent enhances property differences between adjacent Ln^{3+} ions, and is the preferred reagent for large-scale separations[165]. A mixture of TBP and HDEHP is a better extractant than pure HDEHP for the ions La^{3+}–Gd^{3+} but shows no difference for the heavier Ln^{3+} ions, presumably because the larger cations can also accommodate TBP molecules in their coordination spheres, whereas the smaller ones cannot[166]. Similar instances of synergistic behaviour are reported when an extractant such as thenoyltrifluoroacetone is used.

The lack of volatility of simple anhydrous lanthanide salts except at very high temperatures has precluded their use in extensive separations. The majority of the reported complexes decompose thermally before they volatilise. Important exceptions are found in tris(β-diketonates), $[Ln(diket)_3]$, where the organic groups are large and/or contain fluorine. Both the tris 2,2,6,6-tetramethyl-3,5-heptanedione (or dipivaloylmethane, thd) and the 1,1,1,2,2,3,3-heptafluoro-7,7-dimethyl-4,6-octanedione (fod) chelates are volatile, and the individual lanthanides can be separated chromatographically as these chelates[167, 168]. Volatility is also characteristic of the tetrakis 1,1,1,5,5,5-hexafluoro-2,4-pentanedione (hfac) salt, $Cs[Y(hfac)_4]$ [46], and other comparable fluorinated species derived from the alkali metal ions[47].

7.7 EMISSION OF RADIANT ENERGY

The early observation that excitation of an appropriate organic ligand bonded to a lanthanide ion can result in emission of energy from that ion[169] and subsequent observations that on occasion and particularly if the ion is Eu^{3+} laser behaviour can occur[4, 12, 170] have directed particular attention to complex species. The non-radiative intramolecular transfer of energy absorbed by the ligand to the cation, followed by radiative emission from the latter, is the significant process (p. 280)[42]. If radiation released in the drop to the ground state in the last of these processes is coherent, laser action results. For example, the europium(III) tetrakisbenzoylacetone chelates loses energy coherently at $c.$ $1634\ cm^{-1}$ (6120 Å) largely as a consequence of an internal ionic $^5D_0 \rightarrow {}^7F_2$ transition. However, not all chelates that fluoresce behave as lasers[171], since energy from the triplet state can be lost as fluorescence by direct change to the ground state.

Lanthanide complexes have been classified in terms of their fluorescence characteristics as[42]

(1) derivatives of the ions La^{3+} ($4f^0$), Gd^{3+} ($4f^7$), and Lu^{3+} ($4f^{14}$) that give no ion fluorescence. For the La^{3+} and Lu^{3+} ions no intra-4f transitions are possible. For the Gd^{3+} ion, the lowest-lying excited term lies at a higher level than the triplet term for each ligand thus far investigated, thus forbidding any metal ion energy transfer.

(2) complexes of the ions Sm^{3+}, Eu^{3+}, Tb^{3+}, and Dy^{3+}, which exhibit

strong ion fluorescence. For each of these ions, there exists an excited term lying close to the ligand triplet level.

(3) complex derivatives of the ions Pr^{3+}, Nd^{3+}, Ho^{3+}, Er^{3+}, Tm^{3+}, and Yb^{3+} which show weak ion fluorescence. For each of these ions, only small energy differences between terms exist, increasing the probability of non-radiative energy transfer with the dissipation of smaller quantities of energy.

The intensity of fluorescence depends upon the quantity of energy available in the triplet state, the efficiency of the transfer of energy to the metal ion, and the probability that ionic emission, rather than non-radiative deactivation, will occur. The efficiency of transfer is related both to the energy difference between the triplet state and the resonance level of the cation and to the nature of the bond between cation and ligand. Coupling between solvent and cation can result in non-radiative loss of energy and decrease in the intensity of ion fluorescence. Changes in the donor ability of oxygen atoms of a β-diketonate toward the Eu^{3+} ion apparently account for increased fluorescence intensity when electron donating methoxy groups are substituted in *meta* positions in the dibenzoylmethanide chelate as opposed to decreased intensity when electron withdrawing groups are substituted in meta or para positions[172]. Fluorescence intensity increases for europium(III) β-diketonates with increase in dipole moment of the solvent and decreases with increasing temperature[173]. Increase in intensity by formation of a protective sheath has been noted with europium(III) tris(β-diketonates) when donor ligands such as trioctylphosphine or a fourth molecule of diketone[175] are present.

References

1. Moeller, T. (1967). *Werner Centennial,* 306. (Washington, D.C.: American Chemical Society)
2. Moeller, T. (1953). *Rec. Chem. Progr.,* **14,** 69
3. Moeller, T., Martin, D. F., Thompson, L. C., Ferrús, R., Feistel, G. R. and Randall, W. J. (1965). *Chem. Rev.,* **65,** 1
4. Moeller, T., Birnbaum, E. R., Forsberg, J. H. and Gayhart, R. B. (1968). *Progress in the Science and Technology of the Rare Earths.* Vol. 3, 61. (London: Pergamon Press)
5. Forsberg, J. H. and Moeller, T. (1969). *Inorg. Chem.,* **8,** 883, 889
6. Birnbaum, E. R. and Moeller, T. (1969). *J. Amer. Chem. Soc.,* **91,** 7274
7. Grandey, R. C. and Moeller, T. (1970). *J. Inorg. Nucl. Chem.,* **32,** 333
8. Charpentier, L. J. and Moeller, T. (1970). *J. Inorg. Nucl. Chem.,* **32,** 3575
9. Forsberg, J. H. and Wathen, C. A. (1971). *Inorg. Chem.,* **10,**
10. Williams, R. J. P. (1970). *Quart. Rev. Chem. Soc.,* **24,** 331
11. Templeton, D. H. and Dauben, C. H. (1954). *J. Amer. Chem. Soc.,* **76,** 5237
12. Sinha, S. P. (1966). *Complexes of the Rare Earths.* (New York: Pergamon Press)
13. Muetterties, E. L. and Wright, C. M. (1967). *Quart. Rev. Chem. Soc.,* **21,** 109
14. Lippard, S. J. (1967). *Progress in Inorganic Chemistry.* Vol. 8, 109. (New York: Interscience Publishers)
15. Cousins, D. R. and Hart, F. A. (1967). *J. Inorg. Nucl. Chem.,* **29,** 1745, 2965, 3009
16. Graham, P. and Joesten, M. (1970). *J. Inorg. Nucl. Chem.,* **32,** 531
17. Karayannis, N. M., Bradshaw, E. E., Pytlewski, L. L. and Labes, M. M. (1970). *J. Inorg. Nucl. Chem.,* **32,** 1079
18. Stewart, W. E. and Siddall III, T. H. (1970). *J. Inorg. Nucl. Chem.,* **32,** 3599
19. Karayannis, N. M., Mikulski, C. M., Pytlewski, L. L. and Labes, M. M. (1970). *Inorg. Chem.,* **9,** 582
20. Giesbrecht, E. and Zinner, L. B. (1969). *Inorg. Nucl. Chem. Lett.,* **5,** 575

21. Durney, M. T. and Marianelli, R. S. (1970). *Inorg. Nucl. Chem. Lett.*, **6**, 895
22. Parris, G. E. and Long, G. G. (1970). *J. Inorg. Nucl. Chem.*, **32**, 1593
23. Sauro, L. J. and Moeller, T. (1968). *J. Inorg. Nucl. Chem.*, **30**, 953
24. Moeller, T. and Vicentini, G. (1965). *J. Inorg. Nucl. Chem.*, **27**, 1477
25. Vicentini, G. and Najjar, R. (1968). *J. Inorg. Nucl. Chem.*, **30**, 2771
26. Vicentini, G., Perrier, M. and Prado, J. C. (1969). *J. Inorg. Nucl. Chem.*, **31**, 825
27. Gatehouse, B. M., Livingstone, S. E. and Nyholm, R. S. (1957). *J. Chem. Soc.*, 4222
28. Hart, F. A. and Laming, F. P. (1965). *J. Inorg. Nucl. Chem.*, **27**, 1605, 1825
29. Sinha, S. P. (1965). *Z. Naturforsch. A.*, **20**, 552, 1661
30. Hart, F. A. and Laming, F. P. (1964). *J. Inorg. Nucl. Chem.*, **26**, 579
31. Ferraro, J. R., Walker, A. and Cristallini, C. (1965). *Inorg. Nucl. Chem. Lett.*, **1**, 25
32. Ferraro, J. R. and Becker, M. (1970). *J. Inorg. Nucl. Chem.*, **32**, 1495
33. Wybourne, B. G. (1965). *Spectroscopic Properties of Rare Earths*, Ch. 6. (New York: Interscience Publishers)
34. Karraker, D. G. (1968). *Inorg. Chem.*, **7**, 473
35. Spedding, F. H., Pikal, M. J. and Ayers, B. O. (1966). *J. Phys. Chem.*, **70**, 2440
36. Helmholz, L. (1939). *J. Amer. Chem. Soc.*, **61**, 1544
37. Marezio, M., Plettinger, H. A. and Zachariasen, W. H. (1961). *Acta Crystallogr.*, **14**, 234
38. Karraker, D. G. (1967). *Inorg. Chem.*, **6**, 1863
39. Karraker, D. G. (1969). *J. Inorg. Nucl. Chem.*, **31**, 2815
40. Ryan, J. L. and Jørgensen, C. K. (1966). *J. Phys. Chem.*, **70**, 2845
41. Ryan, J. L. (1969). *Inorg. Chem.*, **8**, 2053
42. Whan, R. E. and Crosby, G. A. (1962). *J. Mol. Spectrosc.*, **8**, 315
43. Brecher, C., Lempicki, A. and Samelson, H. (1964). *J. Chem. Phys.*, **41**, 279
44. Durham, D. A., Frost, G. H. and Hart, F. A. (1969). *J. Inorg. Nucl. Chem.*, **31**, 833
45. Templeton, D. H. and Carter, G. F. (1954). *J. Phys. Chem.*, **58**, 940
46. Lippard, S. J. (1966). *J. Amer. Chem. Soc.*, **88**, 4300
47. Belcher, R., Majer, J., Perry, R. and Stephen, W. I. (1969). *J. Inorg. Nucl. Chem.*, **31**, 471
48. Garnsey, R. and Ebdon, D. W. (1969). *J. Amer. Chem. Soc.*, **91**, 50
49. Frattiello, A., Lee, R. E. and Schuster, R. E. (1970). *Inorg. Chem.*, **9**, 391
50. Gillespie, R. J. (1967). *Angew. Chem., Int. Ed. Engl.*, **6**, 819
51. Gillespie, R. J. (1970). *J. Chem. Educ.*, **47**, 18
52. King, R. B. (1969). *J. Amer. Chem. Soc.*, **91**, 7211, 7217
53. King, R. B. (1970). *J. Amer. Chem. Soc.*, **92**, 6455, 6460
54. Hoard, J. L. and Silverton, J. V. (1963). *Inorg. Chem.*, **2**, 235
55. Lippard, S. J. (1968). *Inorg. Chem.*, **7**, 1686
56. Martin, J. L., Thompson, L. C., Radanovich, L. J. and Glick, M. D. (1968). *J. Amer. Chem. Soc.*, **90**, 4493
57. Sicre, J. E., Dubois, J. T., Eisentraut, K. J. and Sievers, R. E. (1969). *J. Amer. Chem. Soc.*, **91**, 3476
58. Morss, L. R., Siegal, M., Stenger, L. and Edelstein, N. (1970). *Inorg. Chem.*, **9**, 1771
59. Muetterties, E. L. and Wright, C. M. (1965). *J. Amer. Chem. Soc.*, **87**, 4706
60. Hoppe, R. and Rödder, K.-M. (1961). *Z. Anorg. Allg. Chem.*, **313**, 154
61. Zachariasen, W. H. (1954). *Acta Crystallogr.*, **7**, 792
62. Cotton, F. A. and Legzdins, P. (1968). *Inorg. Chem.*, **7**, 1777
63. Zalkin, A., Templeton, D. H. and Karraker, D. G. (1969). *Inorg. Chem.*, **8**, 2680
64. Cunningham, J. A., Sands, D. E., Wagner, W. F. and Richardson, M. F. (1969). *Inorg. Chem.*, **8**, 22
65. Watkins II, E. D., Cunningham, J. A., Phillips II, Theodore, Sands, D. E. and Wagner, W. F. (1969). *Inorg. Chem.*, **8**, 29
66. Richardson, M. F., Corfield, P. W. R., Sands, D. E. and Sievers, R. E. (1970). *Inorg. Chem.*, **9**, 1632
67. Bel'skii, N. K. and Struchkov, Yu T. (1965). *Kristallografiya*, **10**, 16
68. Matkovic, B. and Grdenic, D. (1963). *Acta Crystallogr.*, **16**, 456
69. Cunningham, J. A., Sands, D. E. and Wagner, W. F. (1967). *Inorg. Chem.*, **6**, 499
70. Phillips II, T., Sands, D. E. and Wagner, W. F. (1968). *Inorg. Chem.*, **7**, 2295
71. Ryan, R. R., Larson, A. C. and Kruse, F. H. (1969). *Inorg. Chem.*, **8**, 33
72. Thoma, R. E., Brunton, G. D., Penneman, R. A. and Keenan, T. K. (1970). *Inorg. Chem.*, **9**, 1096

73. Steinfink, H. and Brunton, G. D. (1970). *Inorg. Chem.*, **9**, 2112
74. Wolf, L. and Bärnighausen, H. (1960). *Acta Crystallogr.*, **13**, 778
75. Lalancette, R. A., Cefola, M., Hamilton, W. C. and LaPlaca, S. J. (1967). *Inorg. Chem.*, **6**, 2127
76. Bennett, M. J., Cotton, F. A., Legzdins, P. and Lippard, S. J. (1968). *Inorg. Chem.*, **7**, 1770
77. Miller, W. V. and Madan, S. K. (1968). *J. Inorg. Nucl. Chem.*, **30**, 3287
78. Campbell, D. L. and Moeller, T. (1969). *J. Inorg. Nucl. Chem.*, **31**, 1077
79. Campbell, D. L. and Moeller, T. (1970). *J. Inorg. Nucl. Chem.*, **32**, 945
80. Melby, L. R., Rose, N. J., Abramson, E. and Cavis, J. C. (1964). *J. Amer. Chem. Soc.*, **86**, 5117
81. Bauer, H., Blanc, J. and Ross, D. L. (1964). *J. Amer. Chem. Soc.*, **86**, 5125
82. Fitzwater, D. R. and Rundle, R. E. (1959). *Z. Kristallog.*, **112**, 362
83. Schubert, K. and Seitz, A. (1947). *Z. Anorg. Allg. Chem.*, **254**, 116
84. Iverovna, V. I., Tarasova, V. P. and Umanskii, M. M. (1951). *Izvest. Akad. Nauk SSSR Ser. Fiz.*, **15**, 164
85. Hoard, J. L., Lee, B. and Lind, M. D. (1965). *J. Amer. Chem. Soc.*, **87**, 1612
86. Lind, M. D., Lee, B. K. and Hoard, J. L. (1965). *J. Amer. Chem. Soc.*, **87**, 1611
87. Al-Karaghouli, A. R. and Wood, J. S. (1968). *J. Amer. Chem. Soc.*, **90**, 6548
88. Al-Karaghouli, A. R. and Wood, J. S. (1970). *Chem. Commun.*, 135
89. Mazhar-ul-Haque, Caughlin, C. N., Hart, F. A. and Van Nice, R. (1971). *Inorg. Chem.*, **10**, 115
90. Shinn, D. B. and Eick, H. A. (1968). *J. Amer. Chem. Soc.*, **7**, 1340
91. Zalkin, A., Forrester, J. D. and Templeton, D. H. (1963). *J. Chem. Phys.*, **39**, 2881
92. Beineke, T. A. and Delgaudio, J. (1968). *Inorg. Chem.*, **7**, 715
93. Larsen, R. D. and Brown, G. H. (1964). *J. Phys. Chem.*, **68**, 3060
94. Workman, M. O. and Burus, J. H. (1969). *Inorg. Chem.*, **8**, 1542
95. Moeller, T., Gulyas, E. and Marshall, R. H. (1959). *J. Inorg. Nucl. Chem.*, **9**, 82
96. Moeller, T., Moss, F. A. J. and Marshall, R. H. (1955). *J. Amer. Chem. Soc.*, **77**, 3182
97. Sillén, L. G. and Martell, A. E. (1965). *Stability Constants of Metal Ion Complexes.* Special Publication 17. (London: The Chemical Society)
98. Choppin, G. R. and Schneider, J. K. (1970). *J. Inorg. Nucl. Chem.*, **32**, 3283
99. Jones, A. D. and Choppin, G. R. (1969). *J. Inorg. Nucl. Chem.*, **31**, 3523
100. Clark, M. E. and Bear, J. L. (1970). *J. Inorg. Nucl. Chem.*, **32**, 3569
101. Powell, J. E., Chughtai, A. R. and Ingemanson, J. W. (1969). *Inorg. Chem.*, **8**, 2216
102. Sweet, T. R. and Brengartner, D. (1970). *Anal. Chim. Acta*, **52**, 173
103. Aziz, A. and Lyle, S. J. (1970). *J. Inorg. Nucl. Chem.*, **32**, 1925
104. Özer, U. Y. (1970). *J. Inorg. Nucl. Chem.*, **32**, 1279
105. Schwarzenbach, G. and Gut, R. (1956). *Helv. Chim. Acta*, **39**, 1589
106. Grenthe, I. (1964). *Acta Chem. Scand.*, **18**, 293
107. de la Praudiere, P. L. E. and Staveley, L. A. K. (1964). *J. Inorg. Nucl. Chem.*, **26**, 1713
108. Choppin, G. R. and Graffeo, A. J. (1965). *Inorg. Chem.*, **4**, 1254
109. Bertha, S. L. and Choppin, G. R. (1969). *Inorg. Chem.*, **8**, 613
110. Betts, R. H. and Dahlinger, O. F. (1959). *Can. J. Chem.*, **37**, 91
111. Moeller, T. and Ferrús, R. (1961). *J. Inorg. Nucl. Chem.*, **20**, 261
112. Moeller, T. and Ferrús, R. (1962). *Inorg. Chem.*, **1**, 49
113. Moeller, T. and Thompson, L. C. (1962). *J. Inorg. Nucl. Chem.*, **24**, 499
114. Moeller, T. and Hseu, T.-M. (1962). *J. Inorg. Nucl. Chem.*, **24**, 1635
115. Choppin, G. R. and Strazik, W. F. (1965). *Inorg. Chem.*, **4**, 1250
116. Walker, J. B. and Choppin, G. R. (1967). *Lanthanide/Actinide Chemistry*, 127. (Washington, D.C.: American Chemical Society)
117. Bertha, S. L. and Choppin, G. R. (1969). *Inorg. Chem.*, **8**, 613
118. Padova, J. (1967). *J. Phys. Chem.*, **71**, 2347
119. Hinchey, R. J. and Cobble, J. W. (1970). *Inorg. Chem.*, **9**, 917
120. Fay, D. P. and Purdie, N. (1970). *Inorg. Chem.*, **9**, 195
121. Peppard, D. F., Mason, G. W. and Lewey, S. (1969). *J. Inorg. Nucl. Chem.*, **31**, 2271
122. Peppard, D. F., Bloomquist, C. A. A., Horwitz, E. P., Lewey, S. and Mason, G. W. (1970). *J. Inorg. Nucl. Chem.*, **32**, 339
123. Schäffer, C. E. and Jørgensen, C. K. (1958). *J. Inorg. Nucl. Chem.*, **8**, 143
124. Jørgensen, C. K. (1970). *J. Inorg. Nucl. Chem.*, **32**, 3127

125. Nugent, L. J. (1970). *J. Inorg. Nucl. Chem.*, **32**, 3485
126. Siekierski, S. (1970). *J. Inorg. Nucl. Chem.*, **32,** 519
127. Graffeo, A. J. and Bear, J. L. (1968). *J. Inorg. Nucl. Chem.*, **30**, 1577
128. Geier, G. (1965). *Ber. Bunsenges. Phys. Chem.*, **69**, 617
129. Eigen, M. and Tamm, K. (1962). *Z. Elektrochem.*, **66**, 93, 107
130. Silber, H. and Swinehart, J. H. (1967). *J. Phys. Chem.*, **71**, 4344
131. Silber, H., Farina, R. D. and Swinehart, J. H. (1969). *Inorg. Chem.*, **8**, 819
132. Martin, T. W. and Glass, R. W. (1970). *J. Amer. Chem. Soc.*, **92**, 5075
133. Garnsey, R. and Ebdon, D. W. (1969). *J. Amer. Chem. Soc.*, **91**, 50
134. Glentworth, P., Wiseall, B., Wright, C. L. and Mahmood, A. J. (1968). *J. Inorg. Nucl. Chem.*, **30**, 967
135. Wiseall, B. and Balcombe, C. (1970). *J. Inorg. Nucl. Chem.*, **32**, 1751
136. D'Olieslager, W. and Choppin, G. R. (1971). *J. Inorg. Nucl. Chem.*, **33**, 127
137. D'Olieslager, W., Choppin, G. R. and Williams, K. R. (1970). *J. Inorg. Nucl. Chem.*, **32**, 3605
138. Asano, T., Okada, S. and Taniguchi, S. (1970). *J. Inorg. Nucl. Chem.*, **32**, 1287
139. Betts, R. H., Dahlinger, O. F. and Munro, D. M. (1958). *Radioisotopes in Scientific Research*, Vol. 2, 326. (Oxford: Pergamon Press)
140. Nyssen, G. A. and Margerum, D. W. (1970). *Inorg. Chem.*, **9**, 1814
141. Geier, G. (1968). *Helv. Chim. Acta*, **51**, 94
142. Sinha, S. P. and Schmidtke, H. H. (1965). *Mol. Phys.*, **10**, 7
143. Yatsimirskii, K. B. and Kostromina, N. A. (1964). *Russ. J. Inorg. Chem.*, **9**, 971
144. Staveley, L. A. K., Markham, D. R. and Jones, M. R. (1968). *J. Inorg. Nucl. Chem.*, **30**, 231
145. McConnell, H. M. and Chestnut, D. B. (1958). *J. Chem. Phys.*, **28**, 107
146. McConnell, H. M. and Robertson, R. E. (1958). *J. Chem. Phys.*, **29**, 1361
147. Hart, F. A., Newberry, J. E. and Shaw, D. (1970). *J. Inorg. Nucl. Chem.*, **32**, 3585
148. Friedman, Jr., H. G., Choppin, G. R. and Feuerbacher, D. G. (1964). *J. Chem. Educ.*, **41**, 354
149. Becker, C. (1964). *J. Chem. Educ.*, **41**, 358
150. Eisenstein, J. C. (1956). *J. Chem. Phys.*, **25**, 142
151. Johnson, O. (1970). *J. Chem. Educ.*, **47**, 431
152. Sen Gupta, S. K. and Rohatgi, K. K. (1970). *J. Inorg. Nucl. Chem.*, **32**, 2247
153. Choppin, G. R., Henrie, D. E. and Buijs, K. (1966). *Inorg. Chem.*, **5**, 1743
154. Dutt, N. K. and Rahut, S. (1970). *J. Inorg. Nucl. Chem.*, **32**, 2905
155. Van Vleck, J. H. (1932). *The Theory of Electric and Magnetic Susceptibilities*, Chapter 9. (Oxford: University Press)
156. Moeller, T. and Horwitz, E. P. (1959). *J. Inorg. Nucl. Chem.*, **12**, 49
157. Fritz, J. J., Grenthe, I., Field, P. E. and Fernelius, W. C. (1960). *J. Amer. Chem. Soc.*, **82**, 6200
158. Fritz, J. J., Field, P. E. and Grenthe, I. (1961). *J. Phys. Chem.*, **65**, 2070
159. Birmingham, J. M. and Wilkinson, G. (1956). *J. Amer. Chem. Soc.*, **78**, 42
160. Maginn, R. E., Manastyrskyj, S. and Dubeck, M. (1963). *J. Amer. Chem. Soc.*, **85**, 627
161. Fischer, E. O. and Fischer, H. (1965). *J. Organometal. Chem.*, **3**, 181
162. Hayes, R. G. and Thomas, J. L. (1969). *J. Amer. Chem. Soc.*, **91**, 6876
163. Mares, F., Hodgson, K. and Streitwieser, Jr., A. (1970). *J. Organometal. Chem.*, **24**, C68
164. Powell, J. E. (1964). *Progress in the Science and Technology of the Rare Earths*, Vol. 1, 62. (London: Pergamon Press)
165. Peppard, D. F. (1964). *Progress in the Science and Technology of the Rare Earths*, Vol. 1, 89. (London: Pergamon Press)
166. Zangen, M. (1963). *J. Inorg. Nucl. Chem.*, **25**, 1051
167. Eisentraut, K. J. and Sievers, R. E. (1965). *J. Amer. Chem. Soc.*, **87**, 5254
168. Sievers, R. E., Eisentraut, K. J. and Springer, Jr., C. S. (1967). *Lanthanide/Actinide Chemistry*, 141. (Washington, D.C.: American Chemical Society)
169. Weisman, S. I. (1942). *J. Chem. Phys.*, **10**, 214
170. Lempicki, A. and Samelson, H. (1963). *Phys. Lett.*, **4**, 133
171. Samelson, H., Lempicki, A. and Brecker, C. (1964). *J. Chem. Phys.*, **40**, 2553
172. Filipescu, N., Sager, W. F. and Serafin, F. A. (1964). *J. Phys. Chem.*, **68**, 3324
173. Filipescu, N. and McAvoy, N. (1966). *J. Inorg. Nucl. Chem.*, **28**, 253

174. Halverson, F., Brinen, J. S. and Leto, J. R. (1964). *J. Chem. Phys.,* **41,** 157, 2752
175. Metlay, M. (1963). *J. Chem. Phys.,* **39,** 491

8
The Organometallic Chemistry of the Lanthanides and Actinides

BASIL KANELLAKOPULOS
Institut für Heisse Chemie, Karlsruhe

and

K. W. BAGNALL
University of Manchester

8.1 INTRODUCTION

Organometallic compounds of the 4f and 5f elements are a relatively recent development in the chemistry of these elements. The work which provided the basis for this development was the synthesis of dicyclopentadienyl iron (ferrocene, $Fe(C_5H_5)_2$) in 1951, a discovery[1,2] which soon led to the opening of a new and interesting field of coordination chemistry in both the d- and f- transition elements.

Reynolds and Wilkinson[3] reported the synthesis of the first organometallic compound of a 5f element, a cyclopentadienyl complex, in 1956 and in the same year analogous complexes of the 4f elements were described by Birmingham and Wilkinson[4]. Since then a large number of analogous organometallic compounds of the lanthanides and actinides have been synthesised and characterised, a rapid expansion in knowledge which has, to some extent, resulted from the availability of new physical and physico-chemical methods which can be applied to the study of structures and reaction mechanisms. Thus the main features of interest in these compounds, their structures, the nature of the metal–carbon bond and the possibility of the participation of f electrons in the metal–ligand bond, can now be fully investigated.

The known organometallic compounds of the lanthanides and actinides can be classified conveniently in groups based on the parent organic ligands[5] as under:

Number of ligand electrons	Class	Example
3	enyl	π-allyl
5	dienyl	π-cyclopentadienyl, π-indenyl
10	tetraenyl	π-cyclo-octatetraenyl

The cyclopentadienyl compounds are the best known and most extensively investigated of these compounds, largely because the allene, indene and cyclo-octatetraene complexes have only been known for a relatively short time. Simple alkyl and aryl compounds cannot, however, be prepared[73].

8.2   CYCLOPENTADIENYL COMPOUNDS

The known di- and tri-cyclopentadienyl compounds of the lanthanide elements are listed in Table 8.1, in which preparative yields and some physical data are included; a similar survey of the known cyclopentadienyl compounds of the actinide elements is presented in Table 8.2.

8.2.1   Preparative methods

(a) *Preparation in organic solvents*—Cyclopentadiene is a weak acid (pK_a about 17) so that ionic alkali metal cyclopentadienyls are easily formed by reaction with an alkali metal. Thus Thiele[6] obtained the white salt, KC_5H_5, the first metal derivative of a five-membered ring system, by reaction of potassium metal with freshly prepared cyclopentadiene in benzene solution.

The general method for the preparation of both lanthanide and actinide cyclopentadienyl complexes is simply by reaction of the anhydrous metal halide with cyclopentadienyl potassium or sodium in an organic solvent, such as benzene, ether or tetrahydrofuran:

$$MX_n + nNaC_5H_5 \rightarrow M(C_5H_5)_n + nNaX \tag{8.1}$$

$$MX_n + (n-1)KC_5H_5 \rightarrow M(C_5H_5)_{n-1}X + (n-1)KX \tag{8.2}$$

Table 8.1
A   Tricyclopentadienyl compounds of the lanthanide elements, $M(C_5H_5)_3$

Metal M	Colour	Sublimation temperature (°C) (10^{-3}–10^{-4} mmHg)	Preparative yield (%)	m.p. (°C)	μ_{eff} (BM)	Reference
La	Colourless	260	25	295*	Diamagnetic	4
Ce	Orange-yellow	230	72	435*	2.46	4
Pr	Pale green	220	83	415*	3.61	4, 16
Pr	Pale green	160–180	—	418	—	22
Nd	Reddish-blue	220	78	380	3.62	4
Pm	Yellow-orange	145–260	—	—	—	16†, 22‡, 23‡, 31‡
Sm	Orange	220	75	365	1.54	4, 16†, 40
Eu	Brown	—	84	Stable up to 700	—	24
Gd	Yellow	220	84	350	7.95–7.98	4
Tb	Colourless	230	34	316	8.9	20, 16†
Dy	Yellow	220	85	302	10.0	4
Ho	Yellow	230	75	295	10.2	20
Er	Pink	200	88	285	9.44	4
Tm	Yellow-green	220	60	278	7.1	20
Yb	Dark green	150	82	273*	4.00	4
Lu	Colourless	180–210	66	264	diamagnetic	20

B   Dicyclopentadienyl compounds, $M(C_5H_5)_2$

Compound						
$(C_5H_5)_2Eu$	Yellow	400–420	20	—	7.63	20
$(C_5H_5)_2Yb$	Red	400–420	11	—	Diamagnetic	20
$(C_5H_5)_2Sm·OC_4H_8$	Purple	—	—	—	3.6	40

*Slight decomposition
†Microchemical preparations
‡Carrier method

(X = halogen). The metal cyclopentadienyl halides, produced according to equation (8.2), react with a variety of alkali metal salts in organic solvents with replacement of halide ion by other ligands:

$$U(C_5H_5)_3Cl + NaBH_4 \rightarrow U(C_5H_5)_3BH_4 + NaCl \tag{8.3}$$

$$U(C_5H_5)_3Br + NaOR \rightarrow U(C_5H_5)_3OR + NaBr \tag{8.4}$$

$$U(C_5H_5)_3I + NaSCN \rightarrow U(C_5H_5)_3SCN + NaI \tag{8.5}$$

Table 8.2 Cyclopentadienyl complexes of the actinide elements

Metal	Compound	Colour	Sublimation temperature (°C)	Preparative yield (%)	m.p. (°C)	Reference
Th	$Th(C_5H_5)_4$	Colourless	250–290	40.7	no m.p.	25
	$Th(C_5H_5)_3F$	Pale yellow	200	—	—	26
	$Th(C_5H_5)_3Cl$	Colourless	200–	44	—	28
	$Th(C_5H_5)_3Br$	Pale yellow	180	70	—	26, 27
	$Th(C_5H_5)_3I$	Pale yellow	190	—	—	26
	$Th(C_5H_5)_3OCH_3$	Colourless	—	—	—	28
	$Th(C_5H_5)_3OC_2H_5$	Colourless	—	—	—	—
	$Th(C_5H_5)_3n—OC_4H_9$	Colourless	135	38	148–150	28
	$Th(C_5H_5)_3t—OC_4H_9$	Colourless	110	92	—	27
Pa	$Pa(C_5H_5)_4$	Orange	220 dec.	54	—	13
U	$U(C_5H_5)_4$	Red	200–220 dec.	$\begin{cases} 6 \\ 99 \end{cases}$	—	$\begin{cases} 29, 30 \\ 27 \end{cases}$
	$U(C_5H_5)_3F$	Green	170	80	—	31, 32
	$U(C_5H_5)_3Cl$	Pale brown	120–130	82–85	—	3, 30
	$U(C_5H_5)_3Br$	Dark brown	160	80	—	32
	$U(C_5H_5)_3I$	Brown	170	80	—	32
	$U(C_5H_5)_3OCH_3$	Green	120	74	299–302	28, 33
	$U(C_5H_5)_3OC_2H_5$	Green	120	—	210–213	30
				54	210–213	33
	$U(C_5H_5)_3i—OC_3H_7$	Green	130	90	200–201	33
	$U(C_5H_5)_3n—OC_4H_9$	Green	120	15/83	149–151	28, 33
	$U(C_5H_5)_3t—OC_4H_9$	Green	120–130	78	—	33
	$U(C_5H_5)_3n—OC_6H_{13}$	Green	120	90	76	27
	$U(C_5H_5)_3n—OC_8H_{17}$	Green	120	25	38–40	27
	$U(C_5H_5)_3OC_6H_{11}(cyclo)$	Green	—	80	247–248	27
	$U(C_5H_5)_3BH_4$	Red	170	95	—	30, 34, 27
	$U(C_5H_5)_3SCN$	Green	180 dec.	95	—	27
	$U(C_5H_5)_3OCN$	Green	—	56	—	27
	$U(C_5H_5)_3$	Bronze	—	10–40	—	30, 35
	$U(C_5H_5)_3THF$	Brown	—	95	—	35
	$U(C_5H_5)_3nic$	Brown	—	75	—	35
	$U(C_5H_5)_3CNC_6H_{11}$	Pale brown	—	82	—	35
Np	$Np(C_5H_5)_4$	Brown-red	200–220 dec.	72	—	36
	$Np(C_5H_5)_3F$	Green	170	—	—	12, 31, 37
	$Np(C_5H_5)_3Cl$	Brown	100	45	—	12, 31, 38
Pu	$Pu(C_5H_5)_3$	Moss green	140–165	60	—	10, 31
	$Pu(C_5H_5)_3nic$	Green	—	70	—	27, 39
Am	$Am(C_5H_5)_3$	Rose	160–200	50	—	11, 31
Cm	$Cm(C_5H_5)_3$	Colourless	180	—	—	14, 15
Bk	$Bk(C_5H_5)_3$	Amber	135–165	—	—	16*
	$(Bk(C_5H_5)_2Cl)_2$	—	220–300	—	—	17*
Cf	$Cf(C_5H_5)_3$	Ruby red	135–320	—	—	16*

dec: decomposes   nic: l-nicotine
*microchemical preparation

The products are usually isolated by evaporating the solvent and subliming them from the residue at higher temperatures under vacuum or, if the compound is not stable to heat, by extracting the residue with a suitable organic solvent, such as benzene, diethyl ether, pentane or tetrahydrofuran, from which the product can be recovered by evaporation of the solvent.

(b) *Synthesis using molten ionic cyclopentadienyl compounds* — It is essential to adopt a working procedure which minimises the fire risk in the glove-box when working on the synthesis of compounds of highly α-radioactive elements such as protactinium and the transuranium actinides. With such elements there is also the problem of a α-radiolysis of the solvent, reagents and product if the reaction time is prolonged, so that a rapid preparative method which does not involve an organic solvent is extremely useful. Apart from these considerations, the reaction of metal chlorides with a melt of dicyclopentadienyl beryllium or magnesium is of general utility, for the beryllium compound melts at 59–60 °C and has a low vapour pressure[7], while the magnesium compound melts at 177 °C and has a high vapour pressure at this temperature[8], although this is appreciably lower than that of the beryllium compound at the same temperature.

Reid and Wailes[9] have used molten dicyclopentadienylmagnesium extensively for such preparations, while Baumgärtner *et al.*[10-14] have used the beryllium compound for the preparation of the cyclopentadienyl complexes of the transuranium elements. This preparative method has also been used for microscale preparation of cyclopentadienyl complexes of very highly α-radioactive elements by Baumgärtner *et al.*[14] and by Laubereau and Burns[15-17].

(c) *Synthesis from the metal and cyclopentadiene* — E. O. Fischer and co-workers[18] have found that alkali metal (Li,Na,K,Rb,Cs) cyclopentadienyls can be prepared by reaction of the metal with freshly prepared cyclopentadiene in anhydrous liquid ammonia, and have used this method to prepare the dicyclopentadienyl complexes of europium and ytterbium[19, 20]. The ytterbium compound can also be prepared by reduction of $Yb(C_5H_5)_2Cl$ with finely dispersed sodium metal in tetrahydrofuran[42]. A similar procedure is used to prepare[40] the dicyclopentadienyl samarium complex, $Sm(C_5H_5)_2 \cdot$ THF, by reduction of $Sm(C_5H_5)_3$ with potassium naphthalene in tetrahydrofuran.

The yields from the various preparative reactions discussed in this and preceding sections depend on the reaction times and sublimation conditions; with well dried halides and long reaction times (about 200 h) one of us (B.K.) has found that yields exceeding 95% are quite usual.

(d) *Other preparative methods* — A number of special methods have been developed for the preparation of cyclopentadienyl halides and related compounds[27]. For example, $U(C_5H_5)_3F$ is obtained in good (>90%) yield by heating $U(C_5H_5)_3Br$ with sodium fluoride, and the base, $U(C_5H_5)_3OH$, can be synthesised in the same way by treating $U(C_5H_5)_3Br$ with alkali hydroxide; this base can be obtained pure by extraction into benzene.

Other simple preparative procedures included the isolation of pure $Th(C_5H_5)_3Cl$ by subliming $Th(C_5H_5)_4$ through a layer of thorium tetrachloride[26] and the preparation of $U(C_5H_5)_3BF_4$ by passing a stream of boron trifluoride through a solution of $U(C_5H_5)_3F$ in benzene[27].

8.2.2   Physical and chemical properties

All of the tricyclopentadienyl complexes of the actinide elements, except for $U(C_5H_5)_3$, are very stable to heat, as are the corresponding lanthanide complexes, which have well-defined melting points (Table 8.1). These tricyclopentadienyl complexes are quite volatile and many of them sublime under reduced pressure $(10^{-3}-10^{-4}Torr)$ at temperatures below 200 °C. The lanthanide complexes, apart from the cerium compound, are relatively stable in dry air, especially in large crystals, but $U(C_5H_5)_3$ and $Pu(C_5H_5)_3$ are pyrophoric in air. All of these compounds react with water to form cyclopentadiene and the metal hydroxide.

The tetracyclopentadienyl complexes, $M(C_5H_5)_4$, (Th,Pa,U,Np) are thermally stable below 200–250°C, but they are not very volatile and, except for the thorium compound, they decompose in this temperature range. The tricyclopentadienyl metal halides, $M(C_5H_5)_3X$, are much more stable to

Table 8.3   Dicyclopentadienyllanthanide- and di(methylcyclopentadienyl) lanthanide- halides (41)

Compound	Colour	m.p. (°C)	μ_{eff} (exptl.) (BM)
$(C_5H_5)_2SmCl$	Yellow	No m.p.; dec. > 200°	1.62–1.94
$(C_5H_5)_2GdCl$	Colourless	No m.p.; dec. > 140°	8.86
$(CH_3C_5H_4)_2GdCl$	Colourless	188–197	—
$(C_5H_5)_2DyCl$	Yellow	343–346 dec.	10.6
$(C_5H_5)_2HoCl$	Yellow-orange	340–343 dec.	10.3
$(C_5H_5)_3ErCl$	Pink	No mp; dec. > 200°	9.79
$(C_5H_5)_2ErI$	Pink	270	—
$(CH_3C_5H_4)ErCl$	Pink	119–122	—
$(C_5H_5)_2YbCl$	Orange-red	No m.p.; dec. > 240°	4.81
$(CH_3C_5H_4)_2YbCl$	Red	115–120	—

dec. = decomposes

heat than the tetracyclopentadienyl complexes, and can be sublimed quantitatively without decomposition. All of the above complexes are soluble in a variety of organic solvents, such as tetrahydrofuran and ethylene glycol dimethyl ether.

The tri-and tetracyclopentadienyl complexes react with halogens to form the corresponding cyclopentadienyl metal halides. For example, $U(C_5H_5)_3X$ (X = Cl,Br,I) can be prepared by treating $U(C_5H_5)_4$ with chlorine, bromine or iodine[27] and $Er(C_5H_5)_3$ reacts[41] with iodine to form $Er(C_5H_5)_2I$. The lanthanum, praseodymium and neodymium analogues do not react under the same conditions[41].

$U(C_5H_5)_4$ is reduced quantitatively to the metal by reaction with potassium in benzene[27] and the finely divided, highly reactive uranium reduces any excess $U(C_5H_5)_4$ to the tricyclopentadienyl complex, $U(C_5H_5)_3$. $Pu(C_5H_5)_3$ is reduced to plutonium metal under similar conditions; the metals prepared in this way are pryphoric in air.

Maginn et al.[41] have prepared the dicyclopentadienyl lanthanide chlorides by treating the trichlorides with two equivalents of cyclopentadienyl sodium in tetrahydrofuran:

$$MCl_3 + 2NaC_5H_5 \xrightarrow{\text{THF}} (C_5H_5)_2MCl + 2NaCl \qquad (8.6)$$

these compounds are also formed [41] by reaction of the tricyclopentadienyl complexes with the corresponding trichlorides:

$$2M(C_5H_5)_3 + MCl_3 \xrightarrow{\text{THF}} 3(C_5H_5)_2MCl \qquad (8.7)$$

An alternative method of preparing them is by treating the tricyclopentadienyl complexes with a solution of hydrogen chloride in tetrahydrofuran or by passing dry hydrogen chloride through a solution of the tricyclopentadienyl complex[27].

The complexes $(C_5H_5)_2MCl$ are dimeric in non-polar solvents, the chlorine atom acting as a bridge:

The bonding in the bridge is evidently sufficiently strong to allow sublimation of the complexes as dimers, as shown by mass spectrometry[68], but the bonding must really be quite weak because tetrahydrofuran is a sufficiently strong base to dissociate the dimeric unit by solvation; the complexes then behave as monomers in that solvent[41]. All the compounds of this type sublime at 150–200 °C under reduced pressure (about 10^{-5} Torr).

Laubereau[17] has recently reported the preparation of an analogous berkelium compound $[(C_5H_5)_2BkCl]_2$, which is isostructural with the corresponding samarium complex. The chloride ion in the complexes can be replaced by other anions, such as amide, acetate, methoxide or phenoxide[41]; bis(methylcyclopentadienyl) erbium acetate is also dimeric in benzene[41]. Many of these substitution products are remarkably stable in air; some data on these compounds are given in Table 18.4.

Fischer et al.[32] have found that $(C_5H_5)_3UF$ is dimeric, the fluorine atom forming a bridge as in the complexes $(C_5H_5)_2MCl$; this fluoride appears to be the only halide complex of this type to exist in the dimeric form in benzene. The Lewis acid character of $(C_5H_5)_3UF$ with respect to bases such as pyridine and tetrahydrofuran has also been investigated[32]. Heteroassociates of this compound with the trincyclopentadienyl compounds of the lanthanides and actinides have recently been reported by Kanellakopulos et al.[43]. These are 1:1 adducts of the type $(C_5H_5)_3UF \cdot M(C_5H_5)_3$, obtained by mixing equimolar amounts of $(C_5H_5)_3UF$ and $Yb(C_5H_5)_3$ or $U(C_5H_5)_3 \cdot THF$

in solution in benzene. $Tm(C_5H_5)_3$ does not yield a stable adduct[32] whereas $Pu(C_5H_5)_3$ yields a crystalline product[27].

Table 8.4   Substitution products of the dicyclopentadienyllanthanide chlorides $(C_5H_5)_2MX$[41]

Compound	Preparative yield (%)	Colour	m.p. (°C)
$(C_5H_5)_2Er\ O\!-\!\overset{O}{\overset{\|}{C}}CH_3$	45	Pink	331–335 dec.
$(CH_3C_5H_4)_2Er\ O\!-\!\overset{O}{\overset{\|}{C}}CH_3$	55	Pink	199–201
$(C_5H_5)_2Er\ O\!-\!\overset{O}{\overset{\|}{C}}H$	27	Pink	dec. > 270
$(C_5H_5)_2Er\ OCH_3$	52	Pink	236–240
$(C_5H_5)_2ErNH_2$	33	Pink	330–334
$(CH_3C_5H_4)_2Gd\!-\!\overset{O}{\overset{\|}{O}}CCH_3$	65	White	207–209
$(C_5H_5)_2Dy\!-\!OCH_3$	5	Yellow	dec. > 235
$(C_5H_5)_2Yb\!-\!\overset{O}{\overset{\|}{O}}CCH_3$	65	Orange	325–329
$(C_5H_5)_2Yb\!-\!\overset{O}{\overset{\|}{O}}CC_6H_5$	54	Orange	350–375
$(C_5H_5)_2YbOCH_3$	60	Orange	290–305
$(C_5H_5)_2Yb\!-\!OC_6H_5$	64	Red	382–386

dec. = decomposes

The $(C_5H_5)_3U^+$ cation is precipitated by silicotungstic acid, chloroplatinic acid, Reinecke's salt and potassium tri-iodide when an aqueous solution of $(C_5H_5)_3UCl$ is treated with the appropriate reagent[3].

Lanthanide complexes of the type $(C_5H_5)MCl_2\cdot3THF$ have been prepared[44] by treating the trichlorides with one equivalent of cyclopentadienyl sodium in tetrahydrofuran:

$$C_5H_5Na + MCl_3 \xrightarrow{THF} (C_5H_5)MCl_2\cdot3THF + NaCl \qquad (8.8)$$

and by reaction of the tricyclopentadienyl complex with two moles of the trichloride, or by treating a dicyclopentadienyl lanthanide chloride with one equivalent of hydrogen chloride in tetrahydrofuran[44]:

$$(C_5H_5)_3M + 2MCl_3 \xrightarrow{THF} 3(C_5H_5)MCl_2\cdot3THF \qquad (8.9)$$

$$(C_5H_5)_2MCl + HCl \xrightarrow{THF} (C_5H_5)MCl_2\cdot3THF + C_5H_6 \qquad (8.10)$$

the lanthanum, praseodymium and neodymium complexes of this type could not be prepared.

The complexes are isolated[44] by evaporating the filtrate from (8.8), or the solutions obtained by reactions 8.9 and 8.10, until crystallisation occurs at 0 °C, for the compounds of this type are not volatile and decompose when heated under reduced pressure. Some data for the known complexes are given in Table 8.5.

Table 8.5   Tris(tetrahydrofuran)cyclopentadienyllanthanide dichlorides $(C_5H_5)MCl_2(C_4H_8O)_3$[44]

Element M	Colour	m.p. (°C)	θ (K)	μ_{eff} (BM)
Sm	Beige	No m.p.	—	—
Eu	Purple	No m.p.	157	4.24
Gd	Lavender	82–86 dec.	—	—
Dy	Colourless	85–90 dec.	90.9	11.81
Ho	Yellow	84–92	—	—
Er	Pink	91–94	24.2	9.68
Yb	Orange	78–81	7.52	4.33
Lu	Colourless	76–78	—	—

dec. = decomposes

The tricyclopentadienyl lanthanide and actinide complexes have a marked acidic character and yield 1:1 complexes with neutral basic molecules such as tetrahydrofuran, 1-nicotine, cyclohexylisonitrile, ammonia or triphenyl-phosphine[4, 35, 42, 45, 46] in organic solvents such as benzene or pentane. The 1:1 tetrahydrofuran adducts are always obtained when the tricyclopen-tadienyl complexes are prepared in tetrahydrofuran. These are more deeply coloured than the parent tricyclopentadienyl compound, a difference which is due to the bathochromic shift through the metal oxygen bonding. Some of the THF adducts are thermally stable with well defined melting points, such as the emerald green ytterbium complex[42], $(C_5H_5)_3Yb\cdot THF$, which sublimes at 100–120 °C/10^{-2} Torr and melts at 223–226 °C. In other cases, such as the mahogany brown europium[51] complex, $(C_5H_5)_3Eu\cdot THF$, the compounds are thermally unstable and cannot be sublimed. They all appear to be sensitive to atmospheric oxidation. The cyclohexylisonitrile complexes (Table 8.6) sublime *in vacuo* at 150–160 °C and have definite melting points but are also air sensitive.

Table 8.6   Properties of some cyclohexylisonitrile complexes of the tri-cyclopentadienyl lanthanides[46]

Compound	Colour	m.p. (°C)	Preparative yield (%)	μ_{eff} (BM)
$(C_5H_5)_3Nd\cdot CNC_6H_{11}$	Violet	147	80	3.4
$(C_5H_5)_3Tb\cdot CNC_6H_{11}$	Colourless	162	62	10.1
$(C_5H_5)_3Ho\cdot CNC_6H_{11}$	Yellow	165	80	10.6
$(C_5H_5)_3Yb\cdot CNC_6H_{11}$	Dark green	167	78	4.4

The tricyclopentadienyl lanthanide complexes with ammonia are conveniently prepared by treating the parent compounds with anhydrous liquid ammonia, in which they are sparingly soluble[46, 42, 46]. The dark-green ytterbium complex is thermally stable and can be sublimed[42, 46] at 120–160 °C, but the very pale green, almost white, praseodymium complex and the yellow samarium complex lose ammonia[46] when heated at 100–150 °C. Dicyclopentadienylytterbium amide is formed[46] when the tricyclopentadienyl ammine is heated above 200 °C:

$$(C_5H_5)_3Yb \cdot NH_3 \xrightarrow{200-250\ ^\circ C} (C_5H_5)_2Yb \cdot NH_2 + C_5H_6 \qquad (8.11)$$

This is a volatile, yellow crystalline solid which melts at 345 °C.

8.2.3   Infrared spectra

A theoretical treatment of the infrared and Raman spectra of π-complexes formed between metals and C_nH_n rings has been given by Fritz[47]. Following this, it is possible to assign the bands in the infrared spectra of the lanthanide and actinide compounds. The following modes are characteristic of π-bonded cyclopentadienyl groups:

(a) a C—H stretching frequency at about 3100 cm^{-1} (v_{C-H})
(b) a C—C stretching frequency at 1410–1440 cm^{-1} (ω_{C-C})
(c) an antisymmetric ring stretching frequency at 1100–1150 cm^{-1} (v_{C-C})
(d) a C—H frequency at 1000 cm^{-1} (δ_{C-H})
(e) a C—H frequency at 770–830 cm^{-1} (γ_{C-H})

The infrared spectra[82] of the complexes $M(C_5H_5)_4$ (M=Th,Pa,U,Np) are very similar (Figure 8.1) and it can be assumed that these compounds are isostructural, very probably tetrahedral with the four rings at the apices. In view of the failure[30] of attempts to prepare a phenyl derivative $(C_5H_5)_3UC_6H_5$, it would be very unlikely for one cyclopentadienyl ring to be σ-bonded, and the spectra show no bands characteristic of the σ-bonded ligand, with diene character. Moreover, ^{1}H-n.m.r. spectra show only one proton signal, indicating that all the ring protons are equivalent. The dipole moment of $U(C_5H_5)_4$ is zero[29, 30] unlike that of the analogous niobium, tantalum[48] and molybdenum[49] compounds.

The tricyclopentadienyl lanthanide and actinide complexes have similar structures (see 8.2.5) and infrared spectra, and the tricyclopentadienyl actinide halides (Cl,Br,I) must likewise form an isostructural series since they have very similar electronic absorption and infrared spectra. Bands characteristic of bridging fluorine appear at 466 cm^{-1} in the infrared spectrum[32] of $(C_5H_5)_3UF$ and at 432 and 423 cm^{-1} in the spectra of the heteroassociates[43] $(C_5H_5)_3UF \cdot Yb(C_5H_5)_3$ and $(C_5H_5)_3UF \cdot U(C_5H_5)_3$ respectively.

The infrared spectra of the cyclohexylisonitrile complexes have been recorded[27, 35, 46]; information concerning the metal—ligand bonding can be obtained from the shifts in the C—N band on complexation. In the lanthanide complexes (Table 8.7) the C—N band, which appears at 2203–2208 cm^{-1},

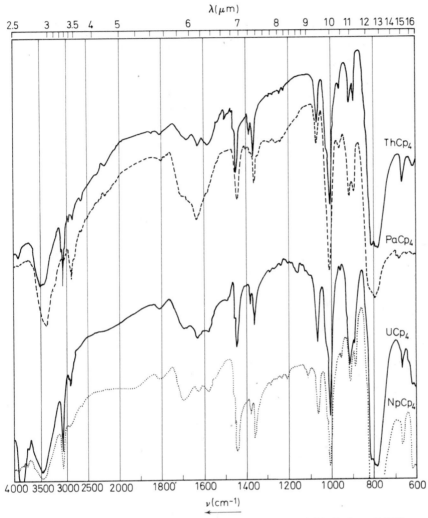

Figure 8.1 Infrared spectra of the tetracyclopentadienyl actinides[82] (Cp $= C_5H_5^-$)

Table 8.7 C≡N frequencies of some tricyclopentadienylmetal–cyclohexylisonitrile compounds, $(C_5H_5)_3M \cdot CNC_6H_{11}$

Metal	$\nu_{C\equiv N}(cm^{-1})$	Reference
Y	2208	46
Nd	2207	46
Pr	2203	27
Tb	2205	46
Ho	2205	46
Yb	2203	46
U	2160	35
Pu	2190	27

is shifted towards higher frequency as compared with the free ligand, in which the band appears at $2130 \ cm^{-1}$, and it is considered that the carbon — metal bond in these compounds has σ-donor character[46], the degree of which depends on the oxidation state and the electron density at the central metal atom. The shift in the actinide complexes is smaller and in the uranium(III) complex a somewhat stronger σ-donor character, with some back donation, is suggested[35], for the ligand has a strong donor–acceptor character[50]. The plutonium compound seems to be intermediate between the uranium(III) complex and the lanthanide complexes (Table 8.7), while the americium complex should be bonded in a similar manner to the lanthanide compounds.

8.2.4 Absorption spectra

Systematic studies of the u.v./visible spectra of the lanthanide and actinide cyclopentadienyl complexes are of value in assessing the influence of the bonding on the f-levels, the participation of f-electrons in the chemical bonding and for the comparison of the influence of the ligand on the 4f and 5f electrons. In addition, such studies would be useful in investigating whether the bonding to the cyclopentadienyl ligand has any favourable effect on the fluorescent properties of such systems[52] and it would also be valuable to determine excited levels by obtaining spectra recorded for samples cooled to low temperatures and then to combine the results with magnetic data.

Pappalardo[52] has investigated the absorption spectrum of $Nd(C_5H_5)_3$ in solution in 2-methyltetrahydrofuran, finding four excited levels at 70, 310, 570 $670 \ cm^{-1}$, while from the spectra[4] of $Pr(C_5H_5)_3$ and $Er(C_5H_5)_3$, recorded at room temperature, it appears that the cyclopentadienyl ligand in these two complexes does not markedly affect the positions of the 4f levels, in contrast to the pronounced alteration of the 3d levels in the sandwich compounds of the first transition series[52].

Fluorescence studies have shown that $Gd(C_5H_5)_3$ shows an intense green fluorescence under excitation with 3660 Å radiation which shifts to the blue[53] at 78 K. $La(C_5H_5)_3$ shows a weaker, yellow fluorescence with a maximum at $19\ 300 \ cm^{-1}$, the more intense emission from gadolinium being favoured by the presence of the paramagnetic Gd^{3+} ion[53]. The curium complex, $Cm(C_5H_5)_3$, exhibits a bright red fluorescence on excitation with ultraviolet light of 3600 Å wavelength[15]. Fluorescence data such as these can be interpreted as transitions from the essentially triplet states of the cyclopentadienyl ion to the ground state, essentially a singlet state[53].

The spectra of di- and trivalent ytterbium cyclopentadienyl complexes have also been reported[42, 54, 55]. The group of transitions observed at about $10\ 000 \ cm^{-1}$ for the ytterbium(III) compounds is useful for identification purposes and has been interpreted as the $^2F_{7/2}$ to $^2F_{5/2}$ transition in the $4f^{13}$ configuration of Yb^{3+}. The broad bands around $16\ 000 \ cm^{-1}$ are due to 5p–4f excitations[54].

The cyclopentadienyl ligand is stronger than other ligands in the lanthanide complexes[54] and tends to shift the $4f \rightarrow 5d$ transitions to lower frequencies.

Pappalardo and Jørgensen[55] have calculated the frequencies of the first electron transfer band of the compounds $M(C_5H_5)_3$ (M=Sm,Eu,Tm,Yb) and have compared these with the experimental results for these compounds and the lanthanide halide complexes, the agreement with the predicted values being fair. These authors also reported a molecular orbital treatment of $M(C_5H_5)_3$, assuming D_{3h} symmetry in which the centres of the three aromatic ligands form an equilateral triangle with the metal atom at its centre; Fischer[56] has suggested that there is some mixing of the filled π-orbitals of the cyclopentadienyl anion with empty 5d orbitals of the central metal atom.

Pappalardo[57] has reported the appearance of hypersensitive transitions at 19 400 and 26 900 cm^{-1} in the absorption spectrum of $Er(C_5H_5)_3$ in methyltetrahydrofuran; these very intense transitions should essentially lead to $^4G_{11/2}$ states of the Er^{3+} ion and are less conspicuous in the spectrum of unsolvated, solid $Er(C_5H_5)_3$. $Nd(C_5H_5)_3$ also exhibits hypersensitivity and in both cases this is accompanied by large splittings of the $J = 3/2$ levels. Fischer and Fischer[58] have observed hypersensitive transitions in the spectrum of $Tm(C_5H_5)_3$, which was recorded over the range 5000–29 000 cm^{-1}; in this instance an unusually large ligand field splitting of the $4f^n$ multiplet terms was also observed and the observed shifts in the bands above 20 000 cm^{-1} are ascribed to electron transfers from non-bonding ligand orbitals into appropriate 4f orbitals. A large nephelauxetic effect would also be expected from the estimated Racah parameter. Pappalardo[59] has also studied the absorption spectrum of this compound, as well as those of tri(methylcyclopentadienyl) thulium and ytterbium, finding no marked changes in the fine structure of the spectrum of thulium when the cyclopentadiene ring is replaced by its methyl derivative.

Reynolds and Wilkinson[3] have reported the visible spectra of $(C_5H_5)_3UCl$ and $U(C_5H_5)_3^+$ for the range 400–800 nm; Anderson and Crisler[34] have also studied these spectra, as well as that of $(C_5H_5)_3UBH_4$. The extinction coefficients of the last are about double those of the corresponding chloride; term assignments for spectral bands from 4252 to 21 584 cm^{-1} were also reported. The spectra of $U(C_5H_5)_3X$ (X=Cl,Br,I) have been found by Fischer et al.[32] to be independent of the solvent used and of the temperature, whereas that of $U(C_5H_5)_3F$ depends on both. The same authors have reported the spectra of several alkoxides, as well as the halides and the borohydride in benzene solution[33]; a bathochromic shift, which depends on the electronegativity of the group X in these compounds $(C_5H_5)_3UX$, has been observed. The absorption spectra of $U(C_5H_5)_3$ and its derivatives have been reported by Kanellakopulos et al.[35] and it has been shown that 5f electrons are more sensitive to variation of the ligands than are the 4f electrons in analogous lanthanide complexes.

The spectrum of $Am(C_5H_5)_3$ has been compared with the spectra of the americium(III) halides[60]; the marked overlaps of contiguous absorption bands, especially in the visible region, makes it difficult to make the upper level assignments. Similar difficulties have been encountered with the spectra[37] of $Pu(C_5H_5)_3$ and $(C_5H_5)_3NpX(X=halide)$. The absorption spectra of $Bk(C_5H_5)_3$ and $(C_5H_5)_2BkCl$, recorded by Laubereau and Burns[16, 17] by the use of microtechniques, are very similar over the range 400–720 nm.

8.2.5 X-ray investigations

Wong et al.[61] have determined the crystal structure of tricyclopentadienyl samarium, $Sm(C_5H_5)_3$; the crystal symmetry is orthorhombic, space group *Pbcm*, with eight formula units per unit cell, of which the dimensions are

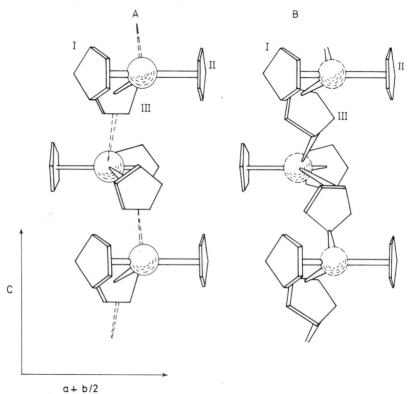

Figure 8.2 The crystal structure of tricyclopentadienylsamarium, $(C_5H_5)_3Sm$ (From Wong et al.[61], by courtesy of Munksgaard)

$a = 14.23$, $b = 17.40$ and $c = 9.37$ Å. The eight $Sm(C_5H_5)_3$ molecules in the unit cell are divided into two symmetrically independent and structurally different groups, A and B. The A and B groups form close-packed, infinite zig-zag chains along the *c*-axis and these chains alternate in layers parallel to (100) with an average spacing of $a/2$; the arrangement around the samarium atom in both chains is a distorted tetrahedron (Figure 8.2). In the A group the π-bonded cyclopentadienyl rings are almost equidistant from the samarium atom (A_I—Sm, A_{II}—Sm, 2.49 Å; A_{III}—Sm, 2.55 Å). A_{III}, however, is rotated upwards (or downwards) by about 30.8 degrees from the

$A_I - Sm - A_{II}$ plane, so that the samarium atoms in the A chain can bond to a fourth ring, the A_{III} ring half a period away. The angle $A_I - Sm - A_{II}$ is

Table 8.8 Unit-cell dimensions of some tricyclopentadienyl lanthanide and actinide compounds, $M(C_5H_5)_3$ [15, 16]

Metal (M)	Radius (M^{3+}), Å	a, Å	b, Å	c, Å	Vol, Å³	d calc. g cm⁻³
Pr	1.103	14.20	17.62	9.79	2449	1.82
Cm	0.985	14.16	17.66	9.69	—	—
Pm	0.979	14.12	17.60	9.76	2425	1.88
Sm	0.964	14.15	17.52	9.77	2422	1.90
Bk	0.954	14.11	17.55	9.63	2385	2.47
Cf	0.944	14.10	17.50	9.69	2391	2.47
Gd	0.938	14.09	17.52	9.65	2382	1.97
Tb	0.923	14.20	17.28	9.65	2368	1.99

123.5 ± 0.4 degree. In the B group, two rings, B_I, B_{II}, are bonded as in the A group, the ring–samarium distances and the angle $B_I - Sm - B_{II}$ not differing significantly from the corresponding A_I, A_{II} values. B_{III} has nearly the same inclination with respect to (001) as A_{III}, but is 0.6 Å further along the plane

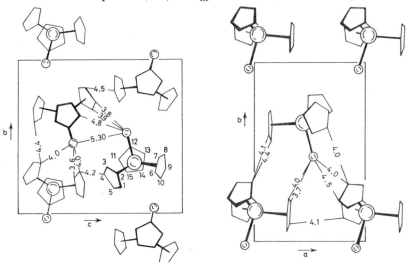

Figure 8.3 The crystal structure of tricyclopentadienyluranium(IV) chloride, $(C_5H_5)_3UCl$ (From Wong et al.[62], by courtesy of Munskgaard)

of the ring from the A_{III} position. In addition, the observed temperature factor for Sm(B) is about four times that of Sm(A), so that Sm(B) is evidently disordered. There is about 37% ionic character in the Sm—C bond.

Laubereau and Burns[15, 16] have shown that other tricyclopentadienyl complexes, $M(C_5H_5)_3$ (M = Pr,Pm,Gd,Tb,Bk,Cf), have the same structure as the samarium compound; crystallographic data for the compounds are summarised in Table 8.8. Single crystal diffraction studies of $Tm(C_5H_5)_3$

indicate that this has a different structure, the compound being orthorhombic with unit cell dimensions $a = 19.98$, $b = 13.82$ and $c = 8.59$ Å. The structural change[16] probably occurs earlier in the lanthanide series, but it is not certain at which element this first becomes apparent. It has also been shown[17] that $[(C_5H_5)_2BkCl]_2$ is isomorphous with $[(C_5H_5)_2SmCl]_2$.

The crystal structure of tricyclopentadienyl uranium chloride, $(C_5H_5)_3UCl$, has been reported by Wong et al.[62]; the crystal symmetry is monoclinic, space group $P2_1/n$, with lattice parameters $a = 8.26$, $b = 12.50$ and $c = 13.81$ Å; $\beta = 90.6$ degrees. There are four formula units per unit cell. The cyclopentadienyl rings and the chlorine atom form a distorted tetrahedron around the uranium atom, the ring—uranium—ring angles being 115, 120 and 115 degrees, and the corresponding ring—uranium—chlorine angles are 101, 100 and 101 degrees. The U – Cl and U – C bond lengths are 2.559 and 2.74 Å respectively; the U – Cl bond is essentially ionic. The structure of the compound, viewed down the a and c axis, is shown in Figure 8.3. X-ray powder photographs of $(C_5H_5)_3UBr$ and $(C_5H_5)_3UI$ are similar to that of the chloride analogue, but that of $(C_5H_5)_3UF$ is not[27].

8.2.6 Nuclear magnetic resonance spectra

The n.m.r. spectra of a large number of lanthanide and actinide cyclopentadienyl complexes have been investigated by Ammon et al.[32, 33, 63–67]; the temperature dependence of the paramagnetic p.m.r. shifts of several of these compounds were also studied[32, 66, 67]. Some n.m.r. data for the ring protons are given in Table 8.9. $^1H - ^{11}B$ nuclear spin coupling has been

Table 8.9 ^{1}H-n.m.r.-shifts of several cyclopentadienyl lanthanide and actinide complexes [3, 64]

f^n-System	Compound	Solvent *	Shift (p.p.m.)	Line-width (c s$^-$)
f⁰	Th(C₅H₅)₄	C₆D₆	1.1	—
	Th(C₅H₅)₄	THF	0.93	—
	Th(C₅H₅)₃BH₄	C₆D₆	1.12	—
f²	Pr(C₅H₅)₃	THF	−7.73	4.3
	Pr(C₅H₅)₃	2-Me-THF	−7.88	3.0
	Pr(C₅H₅)₃nic	C₆H₆	−7.58	6.0
	U(C₅H₅)₄	C₆H₆	20.36	—
	U(C₅H₅)₄	C₆D₆	20.42	—
	U(C₅H₅)₄	THF	20.20	—
	U(C₅H₅)₃F	C₆D₆	12.6	3.2
	U(C₅H₅)₃Cl	C₆D₆	9.56	1.5
	U(C₅H₅)₃Br	C₆D₆	9.79	1.1
	U(C₅H₅)₃I	C₆D₆	10.4	1.3
	[U(C₅H₅)₃]⁺	H₂O	6.40	1.0
	U(C₅H₅)₃BH₄	C₆D₆	13.77	2.4
	U(C₅H₅)₃BH₄	THF	13.41	2.9
f³	Np(C₅H₅)₃Cl	C₆D₆	27.4	30
	Nd(C₅H₅)₃	THF-d₈	3.22	27
	Nd(C₅H₅)₃N-Me-Pyrr	C₆H₆	3.22	34

f^n-System	Compound	Solvent *	Shift (p.p.m.)	Line width (c s^{-1})
f^3	Nd(C$_5$H$_5$)$_3$·CNC$_6$H$_{11}$	C$_6$H$_6$	3.32	27
	Nd(C$_5$H$_5$)$_3$·nic	C$_6$H$_6$	3.33	26
f^5	Pu(C$_5$H$_5$)$_3$	THF	-5.24	66
	Pu(C$_5$H$_5$)$_3$nic	C$_6$H$_6$	-5.61	76
f^{10}	Ho(C$_5$H$_5$)$_3$	C$_6$D$_6$	-197	200
	Ho(C$_5$H$_5$)$_3$CNC$_6$H$_{11}$	C$_6$D$_6$	-137	145
	Ho(C$_5$H$_5$)$_3$CNC$_6$H$_{11}$	THF	-134	130
f^{12}	Tm(C$_5$H$_5$)$_3$	C$_6$H$_6$	164	220
	Tm(C$_5$H$_5$)$_3$	THF-d$_8$	69	160
	Tm(C$_5$H$_5$)$_3$·nic	C$_6$D$_6$	60.5	65
	Tm(C$_5$H$_5$)$_3$·CNC$_6$H$_{11}$	C$_6$D$_6$	55.1	65
f^{13}	Yb(C$_5$H$_5$)$_3$	C$_6$D$_{12}$	59	300
	Yb(C$_5$H$_5$)$_3$	C$_6$D$_6$	56	290
	Yb(C$_5$H$_5$)$_3$	THF	54	280
	Yb(C$_5$H$_5$)$_3$·Pyr	C$_6$D$_6$	52	270
	Yb(C$_5$H$_5$)$_3$·CNC$_6$H$_{11}$	C$_6$D$_6$	51	290
	(Yb(C$_5$H$_5$)$_2$Cl)$_2$	C$_6$D$_6$	74	260
	(Yb(C$_5$H$_5$)$_2$Cl)$_2$	THF	47.8	245
	(Yb(C$_5$H$_5$)$_2$NH$_2$)$_2$	C$_6$D$_6$	24.0	45
	(Yb(C$_5$H$_5$)$_2$NH$_2$)$_2$	THF	21.4	50

*THF = tetrahydrofuran
2-Me-THF = 2-methyl-tetrahydrofuran
nic = l-nicotine
Pyr = pyridine
N-Me-Pyrr = N-methyl pyrrolidine

observed for (C$_5$H$_5$)$_3$UBH$_4$ at low temperatures[67] and from a systematic investigation of the alkoxides (C$_5$H$_5$)$_3$UOR, it has been shown that it is possible to estimate the pseudocontact and Fermi contact contributions to the overall isotropic shift of the ring and alkyl protons[33].

The main factor determining the magnitude of the paramagnetic shift in the compounds (C$_5$H$_5$)$_3$MX is probably the number of bonds between the ligand X and the central metal atom; there is only a single bond when X is a halide (including fluoride when reactants prevent dimerization of the molecule), whereas there are two bonds in (C$_5$H$_5$)$_3$UBH$_4$, and there are three bonds in the case of (C$_5$H$_5$)$_4$U (three π bonds per ring).

8.2.7 Mass spectra

Müller[68] has investigated the mass spectra of several lanthanide and actinide cyclopentadienyl complexes and has calculated the dissociation energies for the C$_5$H$_5$ ring removed according to the equations:

$$M(C_5H_5)_n^+ \rightarrow M(C_5H_5)_{n-1}^+ + C_5H_5 \qquad (8.12)$$

$$M(C_5H_5)_{n-1}^+ \rightarrow M(C_5H_5)_{n-2}^+ + C_5H_5 \qquad (8.13)$$

The fragmentation behaviour of the tricyclopentadienyl halides depends on the halogen present; for example $(C_5H_5)_3ThI$ yields mainly $(C_5H_5)_3Th^+$ wheareas $(C_5H_5)_3ThF$ yields mainly $(C_5H_5)_2ThF^+$, a reflection of the comparative strength of the $Th-I$ and $Th-F$ bonds. The mass spectrum of $U(C_5H_5)_3$ differs[35] from those of the corresponding lanthanide complexes, while $Cm(C_5H_5)_3$ behaves[14] like the latter.

8.2.8 Magnetic susceptibility data

The magnetic moments of all the cyclopentadienyl lanthanide complexes are essentially those of the corresponding free ions. This behaviour is not unexpected, for the 4f orbitals are too well shielded by the outer 5s and 5p orbitals for them to be markedly influenced by ligand fields. In this respect the lanthanides differ from the d transition metals in which the d-orbitals are strongly affected by the surrounding ligands. Some magnetic data for the lanthanide complexes are given in Tables 8.1, 8.3 and 8.5.

The magnetic moments of the actinide complexes are appreciably smaller than those of the free ions, but are not adequately explicable on present theories. In addition to the magnetic data for the actinide complexes presented

Table 8.10 Magnetic susceptibility data for some cyclopentadienyl actinide complexes

Central metal ion	Number of 5f electrons	Compound	θ (K)	μ_{eff}*(BM) (exptl.)	Reference
Th^{4+}	0	$(C_5H_5)_4Th$	—	diamag. $\chi_m = -133 \times 10^{-6}$ cm^3 mol^{-1}	25
		$(C_5H_5)_3ThCl$	—	diamag. $\chi_m = -122 \times 10^{-6}$ cm^3 mol^{-1}	28
U^{4+}	2	$(C_5H_5)_4U$	− 82	2.78†	—
		$(C_5H_5)_3UBH_4$	—	2.82	30
		$(C_5H_5)_3UBH_4$	− 87	2.85	27
		$(C_5H_5)_3UOCH_3$	− 93	2.87	27
		$(C_5H_5)_3UOC_2H_5$	—	2.80	30
		$(C_5H_5)_3UOC_4H_9$	− 82	2.68	28
		$(C_5H_5)_3USCN$	−164	3.39	27
		$(C_5H_5)_3UBF_4$	−160	3.30	27
		$(C_5H_5)_3UF$	−158	3.35	32
		$(C_5H_5)_3UCl$	−138	3.16	28
		$(C_5H_5)_3UBr$	−130	3.37†	—
U^{3+}	3	$(C_5H_5)_3U \cdot OC_4H_8$	−140	2.33	35
Np^{4+}	3	$(C_5H_5)_4Np$	−152	2.43	69
		$(C_5H_5)_3NpCl$	− 45	2.20	69
Pu^{3+}	5	$(C_5H_5)_3Pu$	− 49	1.12	39
Am^{3+}	6	$(C_5H_5)_3Am$	−138	1.74†	—

*Calculated according to the equation $\mu_{eff} = 2.84\sqrt{\chi_m(T-\theta)}$
†Measurements by P. Laubereau (Ref. 26)

in Table 8.10, Fischer et al.[39] have reported the magnetic susceptibility of nicotine and its adduct with $Pu(C_5H_5)_3$.

8.2.9 Thermodynamic and other data

Duncan and Thomas[70] reported the vapour pressure of $Nd(C_5H_5)_3$ to be 3.93×10^{-3} Torr at 200 °C, and found the heat of sublimation to be 23.5 kcal mol^{-1}. These values seem to be low, for Haug[71] has reported that the enthalpies of sublimation of $(C_5H_5)_3UCl$, $(C_5H_5)_2GdCl$ and the tricyclopentadienyl complexes of praseodymium, neodymium, holmium and thulium are 27.7, 33.7, 31.7, 32.2, 28.6 and 26.1 kcal mol^{-1} respectively.

The Mössbauer spectrum of $Eu(C_5H_5)_2$ has been investigated by Hüfner et al.[72], using the 21.7 keV γ transition of ^{151}Eu. They found an isomer shift

Table 8.11   Dipole moments of some actinide and lanthanide cyclopentadienyl compounds

Compound	Dipole-moment (Debye Units)	Reference
$(C_5H_5)_4U$	0	29, 30
$(C_5H_5)_3UF$	2.33	32
$(C_5H_5)_3UCl$	3.88	30
$(C_5H_5)_3UBH_4$	3.51	30
$(C_5H_5)_3UOCH_3$	1.58	33
$(C_5H_5)_3OC_2H_5$	1.06	30
$(C_5H_5)_3UO(i)C_3H_7$	1.28	33
$(C_5H_5)_3UO(n)C_4H_9$	1.23	33
$(C_5H_5)_3Lu$	0.85	46

of -1.32 cm s^{-1} (with EuF_3 as standard), a value close to those of $EuSe(-1.32$ cm s$^{-1})$ and $EuCl_2(-1.34$ cm s$^{-1})$, demonstrating the ionic character of the compound, as postulated by Fischer and Fischer[20].

Measurement of the dipole moments of the cyclopentadienyl complexes can also be of value in the assignment of the structure and type of bonding in the compounds; some results are summarised in Table 8.11.

8.3 ALLYL COMPLEXES

Wilke et al.[83] have obtained tetra-allyl thorium, $Th(C_3H_5)_4$, a dark yellow solid which is stable below 0 °C, by reaction of a thorium tetrahalide with an allyl magnesium halide in ether at low temperature:

$$4C_3H_5MgBr + MX_4 \rightarrow (C_3H_5)_4M + 4MgBrX \qquad (8.14)$$

The compound is a classical π allyl complex, as shown by its ^{1}H n.m.r. spectrum, which is temperature dependent. The spectrum in C_6D_6 at 10 °C (H_a, τ, 3.97; H_b, τ, 6.46; H_c, τ, 7.61, with tetramethylsilane as internal standard) changes at 80 °C to a spectrum characteristic of a dynamic π allyl system (H_a, τ, 3.95; $H_{b,c}$, τ, 7.09). The $C=C$ mode in the infrared spectrum does not change with temperature.

The same method has been used to prepare tetra-allyl uranium[74], a dark red compound which, like the thorium compound, is purified by extraction with a hydrocarbon and recrystallised from pentane at low temperatures. The ^1H n.m.r. spectrum[75] of the uranium compound shows that this also is a π allyl complex. The uranium compound[74] burns spontaneously in air, but is stable in an inert atmosphere; it is thermally stable up to -20 °C and decomposes above this temperature to give propylene (81.5%) and propane (18.5%). The compound is paramagnetic with a Weiss constant of 100 K and the magnetic moment is 2.6 BM.

8.4 INDENYL COMPOUNDS

Indenyl complexes of the lanthanides have been prepared by Tsutsui and Gysling[76] by reaction of the anhydrous lanthanide trichlorides with sodium indenyl in tetrahydrofuran solution. The products are tetrahydrofuran adducts, $M(C_9H_7)_3.THF$ (M = La, Sm, Gd, Tb, Dy and Yb); their magnetic moments correspond to those of the cyclopentadienyl complexes (Table (8.12). The ^1H n.m.r. spectra of the lanthanum compound is very similar to

Table 8.12 Properties of some trisindenyllanthanide tetrahydrofuran complexes $(C_9H_7)_3MOC_4H_8$ [76]

Metal	Colour	μ_{eff}(BM)
La	Pale tan	0
Sm	Deep red	1.55
Gd	Pale green	7.89
Tb	Pale yellow	9.43
Dy	Pale tan	9.95
Yb	Dark green	4.10

that of the ionic sodium compound, wheareas that of the samarium compound indicates appreciably less ionic character in the bonding, a difference which may be due to the smaller radius of the samarium ion, resulting from the lanthanide contraction.

8.5 CYCLO-OCTATETRAENYL COMPOUNDS

Streitwieser and Müller–Westerhoff[77] prepared the first cyclo-octatetraenyl (COT) complex of a f-transition element, $U(COT)_2$ (uranocene), which has a D_{8h} sandwich structure[81], like ferrocene. It was obtained by reaction of K_2COT with uranium tetrachloride in dry, oxygen free tetrahydrofuran at 0 °C. After stirring overnight, degassed water is added and the green, crystalline product is separated and extracted with benzene or toluene.

Uranocene inflames in air, but is stable to water, acetic acid and aqueous sodium hydroxide; it sublimes at 180 °C/0.03 Torr. Analogous compounds of thorium, neptunium and plutonium have been prepared in a similar manner by Streitwieser and Yoshida[78] and by Karracker et al.[79], using $ThCl_4$, $NpCl_4$ and $((C_2H_5)_4N)_2PuCl_6$ respectively.

The cyclo-octatetraenyl complexes of cerium, praseodymium, neodymium, samarium and terbium, all of the form $K[M(COT)_2]$, were prepared in much the same way as the actinide complexes by Mares et al.[80]. In these preparations a suspension of the anhydrous lanthanide trichloride in tetrahydrofuran was added to a mixture of 1.5 equivalents of K_2COT and 0.5 equivalent of COT, also in tetrahydrofuran. Europium and ytterbium trichlorides did not yield complexes under these conditions, evidently because of reduction to the divalent state.

However, Hayes and Thomas[21] have synthesised the bivalent europium and ytterbium COT complexes by reaction of a solution of the lanthanide metal in anhydrous liquid ammonia with cyclo-octatetraene. The orange Eu(COT) and pink Yb(COT) are stable to 500 °C but do not sublime at that temperature even at 10^{-3} Torr. Both compounds are almost explosively oxidised in air and their colours are markedly affected by interaction with solvents. The lanthanide complexes, $KM(COT)_2$, decompose without melting at about 160 °C and cannot be sublimed. They are soluble in tetrahydrofuran but virtually insoluble in solvents such as hexane, ether and methyl cyanide; they ignite on exposure to air and react with protic solvents such as alcohols, amines or water. The rate of methanolysis of the complexes increases from cerium to samarium, and yields a mixture of cyclo-octatetraene and -triene. Their infrared spectra in the 600–1200 cm^{-1} region are identical with that of uranocene, indicating the anions also have a D_{8h} sandwich structure.

Uranocene is monoclinic[81], space group $P 2_1/n$, with lattice parameters $a = 7.084$, $b = 8.701$, $c = 10.631$ Å and $\beta = 98.75$ degrees. The calculated density (for two formula units in the unit cell) is 2.29 g cm^{-3}. The average U—C bond length is 2.648 Å. The structure is shown in Figure 8.4. The infrared spectra of the actinide compounds are essentially identical, as also are their x-ray diffraction patterns, so that these compounds evidently have the same structure[78]. The bis(cyclo-octatetraenyl) compounds of the actinides

Table 8.13 Properties of cyclo-octatetraenyl complexes of the actinides

Compound	Colour	Sublimation temperature (°C)	μ_{eff}(BM)	Reference
$Th(C_8H_8)_2$	Bright yellow	160	—	78
$U(C_8H_8)_2$	Green	180	2.43	77
$Np(C_8H_8)_2$	Yellow red	—	1.81	79
$Pu(C_8H_8)_2$	Cherry red	—	diamag. [$(\chi m = -218\pm92)$ $\times 10^{-6}$ cm^3 mol^{-1}]	79

(Table 8.13) have similar chemical properties; they react rapidly with air to form oxides and are sparingly soluble in organic solvents.

The Mössbauer spectrum[79] of $Np(COT)_2$ at 4.2 K is similar to that of neptunium tetrachloride apart from the isomer shift, which is $+2.5 \, \mathrm{cm \, s^{-1}}$, relative to NpO_2. This large positive isomer shift is indicative of an unusual

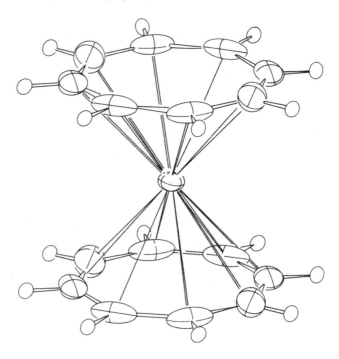

Figure 8.4   The structure of bis-(cyclo-octatetraenyl)uranium(IV), 'uranocene', $U(C_8H_8)_2$
(From Zalkin and Raymond[81], by courtesy of The American Chemical Society)

additional shielding of the 6s shell which suggests a strong electron contribution from the ligand to the metal orbitals. It is interesting to note also that the intensities of the absorption bands in the visible spectra of the uranium, neptunium and plutonium complexes are about an order of magnitude greater than the intensities of bands resulting from normal 5f–5f transitions, which may indicate an unusual degree of mixing between the 5f and 6d orbitals[79]. The metal–ligand bond in the uranium complex can be discussed[77] in terms of a sharing of the 20 π electrons of the two COT rings with vacant uranium orbitals, including the f_{xyz} and $f_{z(x^2-y^2)}$ orbitals (E_{2u}). In addition, the two electrons originally associated with the U^{4+} ion can be placed in the degenerate backbonding combination, $E_{3u} \leftarrow f_{x(x^2-3y^2)}, f_{y(3x^2-y^2)}$.

The magnetic susceptibilities of the actinide compounds have been measured over the temperature range 2.4–45 K. $Pu(COT)_2$ is diamagnetic, whereas the uranium and neptunium compounds exhibit Curie–Weiss dependence. The calculated values of the paramagnetic susceptibilities are larger than the observed values, indicating that there are covalent contributions to the metal – ligand bonding[79].

References

1. Kealy, T. J. and Pauson, P. L. (1951). *Nature*, **168**, 1039
2. Miller, S. A., Tebboth, J. A. and Tremaine, J. F. (1952). *J. Chem. Soc.*, 632
3. Reynolds, L. T. and Wilkinson, G. (1956). *J. Inorg. Nucl. Chem.*, **2**, 246
4. Birmingham, J. M. and Wilkinson, G. (a) (1954). *J. Am. Chem. Soc.*, **76**, 6210; (b) (1956). *J. Am. Chem. Soc.*, **78**, 42
5. Green, M. L. (1968). *Organometallic Compounds*, Vol. II, 'The Transition Elements' (Ed. by G. E. Coates, M. L. Green and K. Wade). (London: Methuen)
6. Thiele, J. (1901). *Ber.*, **34**, 71
7. Fischer, E. O. and Hoffmann, H. P. (1959). *Chem. Ber.*, **92**, 482
8. Hull, H. S., Reid, A. F. and Turnbull, A. G. (1965). *Australian J. Chem.*, **18**, 249
9. Reid, A. F. and Wailes, P. C. (1965). *Inorg. Chem.*, **5**, 1213
10. Baumgärtner, F., Fischer, E. O., Kanellakopulos, B. and Laubereau, P. (1965). *Angew. Chem.*, **77**, 866; *Int. Ed.*, **4**, 878
11. Baumgärtner, F., Fischer, E. O., Kanellakopulos, B. and Laubereau, P. (1966). *Angew. Chem.*, **77**, 112; *Int. Ed.*, **5**, 134
12. Fischer, E. O., Laubereau, P., Baumgärtner, F. and Kanellakopulos, B. (1966). *J. Organometal. Chem.*, **5**, 583
13. Baumgärtner, F., Fischer, E. O., Kanellakopulos, B. and Laubereau, P. (1969). *Angew. Chem.*, **81**, 182; *Int. Ed.*, **8**, 202
14. Baumgärtner, F., Fischer, E. O., Billich, H., Dornberger, E., Kanellakopulos, B., Roth, W. and Stieglitz, L. (1970). *J. Organometal. Chem.*, **22**, C 17
15. Laubereau, P. G. and Burns, J. H. (1970). *Inorg. Nucl. Chem. Letters*, **6**, 59
16. Laubereau, P. G. and Burns, J. H. (1970). *Inorg. Chem.*, **9**, 1091
17. Laubereau, P. G. (1970). *Inorg. Nucl. Chem. Letters*, **6**, 611
18. Fischer, E. O. and Fritz, H. P. (1959). In *Adv. Inorg. Chem. and Radiochem.*, Vol. 1, 55 (Ed. by H. J. Emeléus and A. G. Sharpe)
19. Fischer, H. (1965). *Thesis*, Tech. University of Munich
20. Fischer, E. O. and Fischer, H. (1964). *Angew. Chem.*, **76**, 52; (1965). *J. Organometal. Chem.*, **3**, 181
21. Hayes, R. G. and Thomas, J. L. (1969). *J. Am. Chem. Soc.*, **91**, 6876
22. Kopunec, R., Macašek, F., Mikulaj, V. and Drienovsky, P. (1969). *Radiochim. Radioanal. Letters*, **1**, 117
23. Baumgärtner, F., Fischer, E. O. and Laubereau, P. (1967). *Radiochim Acta*, **7**, 188
24. Tsutsui, M., Takino, T. and Lorenz, D. (1966). *Z. Naturforsch.*, **21b**, 1
25. Fischer, E. O. and Treiber, A. (1962). *Z. Naturforsch.*, **17b**, 276
26. Laubereau, P. (1968). (Private communication)
27. Kanellakopulos, B., Baumgärtner, F. and Dornberger, E. (Unpublished results)
28. Ter Haar, G. L. and Dubeck, M. (1964). *Inorg. Chem.*, **3**, 1648
29. Fischer, E. O. and Hristidu, Y. (1962). *Z. Naturforsch.*, **17b**, 275
30. Hristidu, Y. (1962). Thesis, Ludwig-Maximilian-University of Munich
31. Laubereau, P. G. (1966). *Thesis*. Technical University of Munich
32. Fischer, R. D., Von Ammon, R. and Kanellakopulos, B. (1970). *J. Organometal. Chem.*, **25**, 123
33. Von Ammon, R., Kanellakopulos, B. and Fischer, R. D. (1969). *Radiochim. Acta*, **11**, 162
34. Anderson, M. L. and Crisler, L. R. (1968). CRDL-940327-2; (1969). *J. Organometal. Chem.*, **17**, 345
35. Kanellakopulos, B., Fischer, E. O., Dornberger, E. and Baumgärtner, F. (1970). *J. Organometal. Chem.*, **24**, 507
36. Baumgärtner, F., Fischer, E. O., Kanellakopulos, B. and Laubereau, P. (1968). *Angew. Chem.*, **80**, 661; *Int. Ed.*, **7**, 634
37. Carnall, W. T., Fields, P. R. and Pappalardo, R. G. (1968). *Progress in Coord. Chem.*, Haifa, Jerusalem (Ed. by M. Cais), pp. 411–413 Elsevier
38. Baumgärtner, F., Fischer, E. O. and Laubereau, P. (1965). *Naturwiss.*, **52**, 560

39. Fischer, R. D., Laubereau, P. and Kanellakopulos, B. (1968). *Westdeutsche Chemie-dozententagung*, Hamburg
40. Watt, G. W. and Gillow, E. W. (1969). *J. Am. Chem. Soc.*, **91**, 775
41. Maginn, R. E., Manastyrskyj, S. and Dubeck, M. (1963). *J. Am. Chem. Soc.*, **85**, 672
42. Calderazzo, F., Pappalardo, R. and Losi, S. (1966). *J. Inorg. Nucl. Chem.*, **28**, 987
43. Kanellakopulos, B., Dornberger, E., Von Ammon, R. and Fischer, R. D. (1970). *Angew. Chem.*, **82**, 956
44. Manastyrskyj, S., Maginn, R. E. and Dubeck, M. (1963). *Inorg. Chem.*, **5**, 904
45. Fischer, E. O. and Fischer, H. (1965). *Angew. Chem.*, **77**, 261
46. Fischer, E. O. and Fischer, H. (1966). *J. Organometal. Chem.*, **6**, 141
47. Fritz, H. P. (1964). In *Adv. Organometal. Chem.* (Ed. by F. G. A. Stone and R. West), pp. 239–316, Academic Press
48. Fischer, E. O. and Treiber, A. (1961). *Chem. Ber.*, **94**, 2193
49. Fischer, E. O. and Hristidu, Y. (1962). *Chem. Ber.*, **95**, 253
50. Cotton, F. A. and Zingales, F. (1961). *J. Am. Chem. Soc.*, **83**, 351
51. Manastyrskyj, S. and Dubeck, M. (1964). *Inorg. Chem.*, **3**, 1647
52. Pappalardo, R. (1964). *Helv. Phys. Acta*, **38**, 178
53. Pappalardo, R. and Losi, S. (1965). *J. Inorg. Nucl. Chem.*, **27**, 733
54. Fischer, R. D. and Fischer, H. (1965). *J. Organometal. Chem.*, **4**, 412
55. Pappalardo, R. and Jørgensen, C. K. (1967). *J. Chem. Phys.*, **46**, 632
56. Fischer, R. D. (1965). *Angew. Chem.*, **77**, 1019; *Int. Ed.*, **4**, 972
57. Pappalardo, R. (1968). *J. Chem. Phys.*, **49**, 1545
58. Fischer, R. D. and Fischer, H. (1967). *J. Organometal. Chem.*, **8**, 155
59. Pappalardo, R. (1969). *J. Mol. Spectr.*, **29**, 13
60. Pappalardo, R., Carnall, W. T. and Fields, P. R. (1969). *J. Chem. Phys.*, **51**, 842
61. Wong, C.-H., Lee, T.-Y. and Lee, Y.-T. (1969). *Acta Cryst.*, **25B**, 2580
62. Wong, C.-H., Yen, T.-U. and Lee, T.-Y. (1965). *Acta Cryst.*, **18**, 340
63. Von Ammon, R., Kanellakopulos, B. and Fischer, R. D. (1968). *Chem. Phys. Letters*, **2**, 513
64. Von Ammon, R., Kanellakopulos, B., Fischer, R. D. and Laubereau, P. (1969). *Inorg. Nucl. Chem. Letters*, **5**, 219
65. Von Ammon, R., Kanellakopulos, B., Fischer, R. D. and Laubereau, P. (1969). *Inorg. Nucl. Chem. Letters*, **5**, 315
66. Von Ammon, R., Kanellakopulos, B. and Fischer, R. D. (1969). *Chem. Phys. Letters*, **4**, 553
67. Von Ammon, R., Kanellakopulos, B., Schmid, G. and Fischer, R. D. (1970). *J. Organometal. Chem.*, **25**, C 1
68. Müller, J. (1969). *Chem. Ber.*, **102**, 152
69. Fischer, R. D., Laubereau, P. and Kanellakopulos, B. (1969). *Z. Naturforsch.*, **24a**, 616
70. Duncan, J. F. and Thomas, F. G. (1964). *J. Chem. Soc.*, 360
71. Haug, H. O., *J. Organometal. Chem.*, (to be published)
72. Hüfner, S., Kienle, P., Quitmann, D. and Brix, P. (1965). *Z. Physik.*, **187**, 67
73. Gilman, H. and Jones, R. G. (1945). *J. Org. Chem.*, **10**, 505
74. Lugli, G., Marconi, W., Mazzei, A., Paladino, N. and Pedretti, U. (1969). *Inorg. Chim. Acta*, **3**, 253
75. Paladino, N., Lugli, G., Pedretti, U. and Brunelli, M. (1970). *Chem. Phys. Letters*, **5**, 15
76. Tsutsui, M. and Gysling, H. J. (1969). *J. Am. Chem. Soc.*, **91**, 3175
77. Streitwieser, A., Jr. and Müller-Westerhoff, U. (1968). *J. Am. Chem. Soc.*, **90**, 7364
78. Streitwieser, A., Jr. and Yoshida, N. (1969). *J. Am. Chem. Soc.*, **91**, 7528
79. Karraker, D. G., Stone, J. A., Jones, E. R., Jr. and Edelstein, N. (1970). *J. Am. Chem. Soc.*, **92**, 4841
80. Mares, F., Hodgson, K. and Streitwieser, A., Jr. (1970). *J. Organometal. Chem.*, **24**, C 68
81. Zalkin, A. and Raymond, K. N. (1969). *J. Am. Chem. Soc.*, **91**, 5667
82. Kanellakopulos, B. (1969). III. *Internat. Protactinium Conf.*, Schloss Elmau, Germany
83. Wilke, G., Bogdanović, B., Hardt, P., Heimbach, P., Keim, W., Körner, M., Oberkirch, W., Tanaka, K., Steinrücke, E., Walter, D. and Zimmermann, H. (1966). *Angew. Chem.*, **78**, 157

9
Absorption Spectra of Actinide Compounds*

J. L. RYAN

Pacific Northwest Laboratories, Battelle Memorial Institute, Washington

* This paper is based on work performed under United States Atomic Energy Commission Contract AT(45-1)-1830.

9.1 INTRODUCTION

The actinide hypothesis[1-4] contains as its most important assumption, the concept that the actinides and lanthanides are similar in that their electronic configurations include 5f- and 4f-electrons, respectively. Because of this, close similarities in the spectra of the compounds of these two series of elements are observed. Jørgensen has discussed the various types of transitions for transition-group complexes which give rise to absorption spectra in the accessible u.v., visible, and near infrared regions[5]. Of these, the following have been observed for stoichiometric actinide and lanthanide complexes and compounds: (1) Internal $nf^2 \rightarrow nf^2$-transitions which are Laporte-forbidden and cause the relatively very weak and narrow bands commonly associated with the actinides and lanthanides. (2) Allowed $nf^2 \rightarrow nf^{2-1} (n+1)$d-transitions which cause intense and relatively broad bands usually, but not necessarily, observed in the ultraviolet region. (3) Broad and usually intense $nf^2 \rightarrow \lambda^{-1}$ nf^{2+1} electron-transfer bands (where λ^{-1} denotes a hole in the orbitals concentrated mainly on the ligands) which occur most commonly in the u.v. region but are often present in the visible region where they produce the rather intense yellow, red, brown, or black colours of many actinide compounds. In addition, metal–ligand frequencies are observed in the infrared and Raman spectra.

The presence of quite detailed internal f-electron transition spectra in easily accessible spectral regions for most of the actinides, and the sensitivity of their spectra to the environment around the metal ion (which is intermediate between d-group and lanthanide elements, particularly for higher oxidation state actinides), makes absorption spectrophotometry an extremely useful tool for studying actinide chemistry, particularly that in solution. In fact, this and the variability of the coordination numbers of the actinide elements makes the actinide elements a very suitable subject for the general study of both aqueous and non-aqueous solution chemistry and particularly coordination chemistry. Considerable progress has been made in the last five to ten years in understanding actinide absorption spectra. This has been due, to a large extent, to advances in knowledge of the chemistry of the lighter actinides which has made available discrete complex species, particularly halides, of known structure. The recent availability of macroscopic quantities of the heavy actinides has also contributed significantly in this regard.

This review will attempt to cover the aspects of the spectra of greatest interest to the practising actinide chemist. It will be concerned directly only with the spectra of molecular species: solid compounds, complexes in aqueous or non-aqueous media and in ion-exchange resins, molecular gases, etc. Methods of obtaining information of direct chemical interest such as coordination number, structure, nature of complex ion species in solution etc., and the results of these methods will be emphasised, but the theory of actinide spectra, distribution of energy levels, etc. will only be covered to the extent necessary to achieve this goal. Lanthanide spectra will only be reviewed to the extent that they are needed to correlate properties of the actinide and lanthanide series or to substantiate interpretations of actinide spectra.

Actinide spectra can be used in two basic ways. First the energies, intensities, and structure of the spectrum itself can, in many cases, give information regarding symmetry, coordination number, the nature and strength of bonding, relative oxidising tendency of the actinide ion, etc. Often this approach yields information on bonding and actinide systematics that applies to an overall understanding beyond that of the particular actinide in question. The second method is to use the spectrum as a tool or a fingerprint to identify complexes in solutions, molten salts, ion-exchange resins, etc. and to measure complex formation constants in the more or less classical manner. In this type of usage the actinide spectra are not unique, but because of their nature they are often more useful in this regard than d-group transition element absorption spectra. Thus actinide spectra can be used as a powerful tool to study reactions in aqueous and non-aqueous media, for example, in solvent extraction, and in similar systems such as ion-exchange resins.

9.2 INTERNAL $5f^q \rightarrow 5f^q$-TRANSITIONS

9.2.1 Ligand effects on transition energies

One of the most characteristic properties of actinide and lanthanide ions (except those containing no f-electrons) is the presence of a large number of relatively weak and narrow absorption bands in the u.v., visible, and near infrared regions of the spectrum. These are due to Laporte-forbidden internal transitions within the 5f- or 4f-shells. The complexity of these spectra is demonstrated by the fact that the number of energy states expected, neglecting crystal or ligand field splitting, and spin–orbit coupling effects, is (in parenthesis) f^1 and f^{13} (1), f^2 and f^{12} (7), f^3 and f^{11} (17), f^4 and f^{10} (47), f^5 and f^9 (73), f^6 and f^8 (119), and f^7 (119) (Reference 6, p. 15). These are further split by spin–orbit interaction and ligand-field effects. As an example, in the relatively simple f^1-system the single energy state is split into two by spin–orbit interaction and these are in turn split into a total of five energy levels (giving four transitions) by an octahedral field, or into seven levels (six transitions) by lower symmetries. In addition to these splittings, the complexity of the spectra is increased by the coupling of metal–ligand and other vibrational transitions to the various f-levels. Thus, in the example of an f^1-system, it is found experimentally for UCl_6^- that about 12 transitions are easily observed in the solution spectrum at 25 °C, about 20 are easily detected in solid salts at 25 °C, and at low temperature the number of easily-resolved transitions increases markedly[7].

The crystal field splittings of the trivalent lanthanides are small (a few hundred wave numbers) so that the spectra closely resemble those of the free ions (Reference 6, p.190). The crystal field splittings of the trivalent actinides are apparently only slightly greater than those of the lanthanides[8, 9] (Reference 6, p.192), but those of the higher oxidation state light actinides in octahedral complexes are often much larger being of the same order of magnitude as the spin–orbit interaction[10–19]. The crystal field splitting is reported to be small for U^{IV} in CaF_2[20], relatively small for Np^{IV} in ThO_2[21], and small for tetravalent Np, Pu, Am, and Cm in various media[22]. The splittings for U^{IV} in UCl_4 solid[23], and matrix-isolated[24] UCl_4, and for Np^{IV} in $PbMoO_4$[25] are moderately large, and it has been postulated[23] that it may vary widely with the nature of the compound in question. The splitting in UX_6^- complexes appears, from the data of Reference 7, to vary rather markedly with halide. It is tempting to propose[15] that the splitting might be much higher in the more strongly bonded, low coordination number hexa-halides than in high coordination number species. It should be noted, however, from the close similarity, except for intensities, of the spectra of $PuCl_6^{3-}$ and the Pu^{III} aquo ion[26] that the splitting cannot be appreciably larger in $PuCl_6^{3-}$ than in the higher coordination number aquo ion. This is also true for Am^{III}[27] and is certainly true for the trivalent lanthanide hexa-halides[28, 29]. The large splittings and the relatively large nephelauxetic effects in the higher oxidation state actinide ions render their $5f \rightarrow 5f$ spectra quite sensitive to environment in comparison to the trivalent actinide and lan-thanide ions. In fact, change in the ligand environment often changes the spectrum so drastically that transitions recognised in one environment are not readily recognised in the new environment, and in some cases casual examination of the absorption spectrum is not sufficient to convince one that the same actinide oxidation state is involved.

Since splitting by the crystal (ligand) field varies in a known manner with the symmetry of the metal ion environment (Reference 6, p.179), it is in theory possible to use the spectra to partially determine the symmetry. Because of the complexity of the spectra of actinide ions, particularly those which contain several f-electrons, which require rather detailed low-tem-perature spectral measurements and detailed interpretation, and because of the difficulty, discussed above, of interpreting the splittings which occur with higher oxidation state actinides, this is at best an inconvenient and often (particularly in the case of solution spectra) an unusable method of deter-mining the symmetry of actinide compounds and complexes. Wybourne (Reference 6, p.202) has also noted the difficulty of interpreting actinide crystal field splittings. In the relatively simple $5f^1$-system the problem is not as great, and Selbin et al.[17, 18] have proposed that the degree of splitting of the $\Gamma_7 \rightarrow \Gamma_{8'}$-transition in (commonly 6-coordinate) U^V complexes is a measure of the degree of distortion from octahedral symmetry (see also Karracker[19] and the discussion of the spectra of UOX_5^{2-} complexes[7]). Because crystal field splittings do not provide a convenient measure of symmetry and coordination number in actinide complexes, the theory of the detailed energy level schemes is not discussed further, and the reader interested in this is referred to books on the subject[6, 30-32].

It has been pointed out (see for example, Reference 6, p.216–219) that the electrostatic crystal field model, although providing much useful information, cannot lead to a complete understanding of lanthanide spectra, and does not take into account the degree of covalency in the bonding as exemplified by the nephelauxetic effect[33-35]. This will be even more true of actinide spectra, particularly those of the higher oxidation state ions. The nephel-auxetic effect has been studied in a variety of lanthanide ions[36, 37, 38]; the maximum value of the relative decrease in interelectronic repulsion parameters reported for a lanthanide ion is $d\beta = 4.2\%$ for Nd in $BaNd_2S_4$ compared with the aquo ion, the values for oxides being somewhat smaller except for Pr in one form of Pr_2O_3 [33], and those for lanthanides in $LaCl_3$ being typically $<1\%$. For the Pr^{III} aquo ion $d\beta\approx4\%$ versus gaseous Pr^{3+} [39] indicating overall $d\beta$ values of $<10\%$ in lanthanide compounds. The effect of bond length on the nephelauxetic effect has been emphasised, and it has been pointed out that $d\beta$ is 2.5–4.5 times larger in the hexahalides than in the M^{III} ions substituted in $LaCl_3$ [28]. The nephelauxetic effect also decreases with increase in atomic number across the 4f-group [28, 36] and no doubt does also for the actinides. An increase in covalent bonding effects on the f-orbitals with decrease in bond length would be expected. Similar effects of bond length in the actinide series should also occur, and comparison of the actinides and lanthanides should be, as far as possible, limited to compounds of similar composition and symmetry.

The nephelauxetic effect is much larger for the tetravalent actinides [15]. Although the spectrum of the gaseous U^{IV} ion and the energy level scheme and symmetry for the aquo ion are not known, the shifts of about 4% between the MCl_6^{2-} and MBr_6^{2-} (M = U, Np, Pu) and between UBr_6^{2-} and UI_6^{2-} [15] against $\sim0.4\%$ between $PrCl_6^{3-}$ and $PrBr_6^{3-}$ [28] indicate about ten times as large a nephelauxetic effect for the tetravalent actinides as for the trivalent lanthanides. This indicates a value of $d\beta$ of up to 20% relative to the aquo ion [40]. Preliminary analysis [41] of the absorption spectrum of what appears almost certain to be octahedral UF_6^{2-} indicates a very large negative nephelauxetic shift (of perhaps 20%) versus UCl_6^{2-}. This indicates that the f-orbitals have an appreciable amount of covalent character in the MX_6^{2-} (X = Cl, Br, I) complexes. Johnston et al. [42] have discussed the effect of the covalency change between UCl_6^{2-} and UBr_6^{2-} on the crystal field and free ion parameters.

The author noted [43] that $d\beta$ for PuX_6^{3-} and AmX_6^{3-} (X = Cl, Br) relative to the aquo ions was somewhat less than twice that for the lanthanides. The nephelauxetic effect has also been found to be about twice as large for the trivalent actinide halides [44] and somewhat less than twice as large for the cyclopentadienides [45] as for the corresponding lanthanide compounds. Nugent et al. [45] have discussed in some detail the nephelauxetic effect in Am and Cm cyclopentadienides, which should be among the more covalent trivalent actinide compounds. They attributed about 2.8% and 2.5% covalent character to the f-orbitals of Am and Cm respectively, as compared to about 1.9% and 1.6% for the lanthanides of the same ionic radius.

The fact that the f→f spectra of the actinides, particularly in their higher oxidation states, are relatively sensitive to their environment makes them extremely useful for the study of complex ion species by the more conventional techniques. These include comparative or 'fingerprint' methods as described earlier. The variability of actinide spectra with environment is also useful for the determination of stability constants; in the case of relatively stable complexes, these can often be identified by Job's method of continuous variations.

9.2.2 Ligand effects on intensities

The intensity of spectra can in some cases give direct information concerning the coordination number and symmetry of actinide complexes. The internal

f^q-transitions are forbidden as electric dipole radiation if the metal ion is at a site of inversion symmetry but this restriction is relaxed to some extent when the centre of symmetry is destroyed. Satten et al.[11, 12, 46, 47] and Pappalardo and Jørgensen[14] have shown in the case of the octahedral UCl_6^{2-} and UBr_6^{2-} ions in solid salts that pure electronic transitions are missing (except as weak magnetic dipole radiation in certain cases), and that the weak vibronic spectrum that results is due to the coupling of vibrational frequencies to the missing pure electronic levels. The frequencies observed are those which destroy the centre of symmetry of the complex during the vibration. The most important of these are the odd fundamental modes, v_3, v_4, and v_6 in the terminology of Adams[48]. Other vibrational frequencies such as anion complex to cation vibrations are observed as weaker contributions. Since these vibrations couple to both the ground and excited electronic states, the actual spectrum consists of groups of bands essentially equally spaced above and below the positions of the weak or missing pure electronic transitions, the spacing from the centre position corresponding to the vibrational frequencies. The vibrationally excited levels of the ground state are depopulated by cooling so that at low temperatures the lower halves of these vibronic groups decrease in intensity and eventually disappear. In hexahalide complexes the vibrational frequencies vary with halide mass in the manner expected and the infrared active modes v_3 and v_4 are also observed in the infrared[7, 17, 49, 50].

The absorption spectra of the hexahalide complexes in solutions in which they are stable are very similar to those in solid salts[7, 28, 50, 51], with those in the salts being slightly better resolved[7, 28]. The weak vibronic transitions due to crystal lattice vibrations are, of course, missing in the solution spectra. At 25 °C in either solids or solutions, only a few of the internal f^q-transitions have the vibronic transitions sufficiently resolved and are sufficiently free of overlap with other groups to allow easy recognition of these frequencies. Even in these cases v_4 and v_6 are sometimes not resolved[28, 51] (see for example the group centred at 598 nm in Figure 9.1).

The coordination numbers of actinide ions (particularly the non-oxygenated ions) are typically > 6 (see the excellent review of high coordination numbers by Muetterties and Wright[52]). The relatively small U^V ion may be, to some extent, an exception to this and may commonly exhibit 6-coordination. Although the coordination number eight appears to be fairly common, discrete 8-coordinate cubic symmetry is known only in Na_3PaF_8[53] and is not known in solution for any actinide ions having f-electrons. As a result, a centre of symmetry in actinide complexes is rare, the most common examples being the hexahalide complexes. Thus octahedral or very slightly distorted octahedral complexes have 5f→5f spectra that are generally an order of magnitude less intense than those of typical actinide complexes (such as the aquo ions)[15, 26–28, 50]. The situation is comparable to the well-known case of the d→d-transitions of Co^{II} in which pink octahedral Co^{II} species have very low molar absorptivities and blue tetrahedral complexes have high molar absorptivities. Although intensities are more easily measured in solutions than in solids, salts of the hexahalide complexes are very pale coloured except when allowed (f→d or electron-transfer) transitions contribute to absorption in the visible region. The intensities of the hexahalide spectra are found experimentally always to increase with increase in halide atomic number[7, 15, 28].

Gruber and Menzel[21] have shown that the spectrum of Np^{IV} in single crystals of 8-coordinate cubic ThO_2 is about tenfold weaker than that of more typical Np^{IV} spectra. They do observe the pure electronic transitions

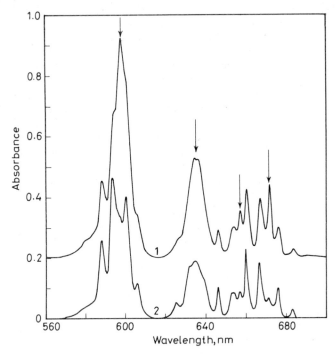

Figure 9.1 The effect of hydrogen bonding on the vibronic absorption spectrum of UCl_6^{2-}. (1) 3.9×10^{-2} M $[(C_2H_5)_4N]_2UCl_6 + 0.3$ M H_2O in HCl-saturated acetonitrile (absorbance scale displaced 0.2) and (2) 3.9×10^{-2} M $[(C_2H_5)_4N]_2UCl_6$ in acetonitrile; 2.00 cm cells. Arrows mark positions of pure electronic transitions. (From Ryan[51] by courtesy of the American Chemical Society).

and conclude[54] that this is due to dynamic distortion resulting from the fact that the Np^{IV} ion is in a lattice cage that is too large for it. It has been presumed[53] that Na_3UF_8 and Na_3NpF_8, which are isostructural with Na_3PaF_8, are also cubic with a centre of symmetry, but no spectral studies of these, particularly with regard to intensities, are available. In 12-coordinate icosahedral symmetry the metal ion is at a centre of symmetry and the $Th(NO_3)_6^{2-}$ anion has been shown[55] to have slightly distorted icosahedral symmetry. The spectra of the U, Np, and Pu analogues are known[56, 57]. The overall intensity of the visible–near infrared absorption spectrum of the $U(NO_3)_6^{2-}$ ion is markedly less than that of the U^{IV} aquo ion and other typical U^{IV} complexes. The decrease in intensity is much less in the case of Np^{IV} and the intensity is only slightly smaller for $Pu(NO_3)_6^{2-}$ than for the aquo ion. It would appear that either the symmetry of the $M(NO_3)_6^{2-}$ ion decreases rather markedly with decrease in metal radius or that the symmetry of the aquo ion increases. The fact that the intensity of the $PuCl_6^{2-}$ spectrum is lower than that for the aquo ion by almost as large a factor as in the U^{IV} case would support the first hypothesis. The reported complete non-absorbability of Bk^{IV} into anion-exchange resins from HNO_3 [58], if correct, also supports such a conclusion, and indicates that the decrease in metal ionic radius is sufficient to prevent formation of the 12-coordinate hexanitrato complex in HNO_3 with Bk^{IV}. In this regard, the relatively large radii of the actinides and the tendency of nitrate to form bidentate complexes is probably the main reason for the marked tendency of the actinides and lanthanides, in contrast to other transition elements, to form nitrato complexes in solution.

Small distortions from centrosymmetric in the octahedral UCl_6^{2-} complex cause the pure electronic transitions to appear to a small extent, and in the solid salts this has been observed to be caused by impurities[46] or by melting and rapidly freezing the salt thereby inducing physical strain[59]. For vibronic groups that are sufficiently resolved, the pure electronic transitions can be observed to a very small extent in solution spectra in cases in which they are completely absent in the crystal spectra[51] (Figure 9.1). This is presumably due to slight distortion of the complex by weak solvent interaction. Stronger interactions cause the pure electronic transitions to become considerably more intense but do not affect the vibronic transitions. Hydrogen bonding to the UCl_6^{2-} and UBr_6^{2-} ions by primary, secondary, and tertiary ammonium ions and the hydronium ion has been demonstrated in this manner[51] as shown in Figure 9.1.

Jørgensen[60] has prepared mixed hexahalide complexes of U^{IV} and has published the spectra of $UBrCl_5^{2-}$ and UBr_5Cl^{2-}. Although the vibronic structure is more complex than for UCl_6^{2-} and UBr_6^{2-}, the overall intensities are not appreciably higher. Gans et al.[61] compared the reflection spectra of several U^{IV} compounds to those of various UCl_6^{2-} salts. They found that $UCl_4 \cdot 2Et_3PO$, $UCl_4 \cdot 2Et_2PhPO$, and $UCl_4 \cdot 2Ph_3PO$ have weak spectra which are similar to and only slightly more intense than those of UCl_6^{2-} salts and concluded that these are octahedral trans phosphine oxide complexes. They categorise other U^{IV} spectra as 'strong' or 'medium' in intensity. In this writer's opinion, the differences they observed between these 'strong' and 'medium' spectra are within the error of determination of intensities from reflection spectra, and $UCl_4 \cdot 2MeCN$, which they conclude on this basis to be cis-octahedral, apparently does not even have this stoichiometry[62, 63] and is very likely not to be 6-coordinate at all. In view of the small differences in intensity of the mixed halide UX_6^{2-} complexes just discussed, it would appear that the conclusion that weak spectra of UCl_4 addition compounds are due to a trans-octahedral configuration is not necessarily valid. Only if the U–ligand bond strength is appreciably different from the U–Cl bond strength, will the cis configuration distort the field sufficiently to cause a large intensity increase. The spectrum of $U(OH)_2Cl_4^{2-}$ is reported[64] to be similar to and slightly more intense than that of UCl_6^{2-}.

The intensities of the internal $5f^q$-transitions of Am^{III} are about 20-fold lower in $AmCl_6^{3-}$ than in the aquo ion[26]. Marcus and Bomse[27] have shown that dissolution of $[(C_6H_5)_3PH]_3AmCl_6$ in succinonitrile–acetonitrile solvent produces a pentachloro complex having a spectrum similar to and not more than twice as intense as that of $AmCl_6^{3-}$. Since this spectrum is quite different from and still about tenfold less intense than that of the aquo ion, it is no doubt due to an octahedral $AmCl_5(RCN)^{2-}$ complex. A spectrum of similar intensity was obtained for Am^{III} in HCl-saturated ethanol[27]. The twofold increase in intensity is due to distortion of the octahedral field because of non-equivalence of the ligands. Similar behaviour occurs with U^V; the UCl_6^- complex is stable in HCl-saturated ethanol[7] and at somewhat lower HCl concentration a presumably 6-coordinate U^V species is obtained the spectrum of which, while similar to that of UCl_6^-, shows distinct differences, with about twice the intensity of the UCl_6^- spectrum[41]. In the UOX_5^{2-}, the strong U—O bond distorts the octahedral field sufficiently to increase the intensities about fivefold over those for the UX_6^- [7]. Thus some care must be exercised in relating intensity directly to coordination number in mixed ligand complexes. The scarcity of actinide spectra of low intensity and vibronic character, particularly for complexes in water, is evidence of the rarity of a centre of symmetry and 6-coordination in actinide complexes.

It has been generally assumed[65-70] that formation of outer-sphere actinide complexes has little or no affect on the internal $5f^q$ spectra of actinide ions whereas formation of inner-sphere complexes does (see also discussion of 4f- and d-group complexes in Reference 71, pp. 44–45). This is a valid and very logical conclusion based on the expected small perturbation of the field around the metal ion by second-sphere effects relative to first sphere. It is a very useful correlation, but some care and understanding of the nature of the spectra should be applied in its use, and it is probable that no fixed rules can be made as to the extent to which each type of complex will affect the spectrum. As discussed earlier, the ligand field splittings and nephelauxetic effects are greater for the higher oxidation state than for the trivalent actinides. For this reason the effect of change of ligand in the first coordination sphere in a trivalent actinide will be small compared to that in tetravalent actinides. The degree of the effect on the spectrum caused by a change in the first sphere ligand will depend on the actinide ion oxidation state, the relative ligand-field strengths of the ligand entering the coordination sphere and the ligand it is replacing, and on whether the coordination number and/or symmetry is changed simultaneously. It is completely reasonable to expect that relatively strong second-sphere effects may, at least in some special cases, produce as much spectral change with a higher oxidation state actinide as first-sphere effects do in some cases with trivalent actinides. Thus, with the octahedral UX_6^{2-} the relatively strong second-sphere effects of hydrogen bonding[51] or the inclusion of these ions in different solid salts produce spectral changes that are as large as those observed[67, 70] on forming inner-sphere chloride complexes of trivalent actinides. These in turn are small compared to the well-known[72] effects of forming inner-sphere complexes of the higher oxidation state actinides.

Because of the relatively large effect of environment on the internal $5f^q$ spectra of higher oxidation state actinides, there has been some tendency[73] to attribute them to hypersensitive transitions. These are transitions observed with the trivalent lanthanides[74] which show very large intensity variations with change in environment and are found to obey certain selection rules[44]. Legitimate hypersensitive transitions have apparently been observed for trivalent $Am^{44, 67, 68}$, but assignment of transitions of higher valence state actinides as hypersensitive transitions simply on the basis of appreciable intensity change alone is not valid.

9.2.3 Individual spectra

In this section the individual internal f-electron transition spectra of the actinides are reviewed briefly, covering work published within the last five years except in cases of particular actinide valence states for which data are scarce and for mention of older work of somewhat greater interest to actinide chemists such as the spectra of discrete halide, nitrate, etc. complexes. Compounds or complexes mentioned specifically in Sections 9.2.1 and 9.2.2 will generally not be covered here. It should be remembered that there is a large amount of older spectral work for light actinides in their common oxidation states.

9.2.3.1 5f¹ Systems

The only detailed study of Pa^{IV} spectra appears to be that of Axe[10], on $PaCl_6^{2-}$ in Cs_2ZrCl_6. Asprey et al.[75] have, however, measured the spectrum of $Rb_7Pa_6F_{31}$.

Uranium(V) systems have been quite extensively investigated recently; the review by Selbin and Ortego[18] covers the results, including spectral data, up to about 1968 and most of the work reviewed is not discussed here. Gritzner and Selbin[76] measured the spectrum of the UO_2^+ ion in dimethyl-sulphoxide and Cohen[77] has measured the spectrum in aqueous $CaCl_2$ and K_2CO_3 solutions; Kemmler-Sack[78] has also discussed the spectra of various U^V-containing ternary oxides. Frlec and Hyman[79] published part of the spectrum of $N_2H_6(UF_6)_2$, $N_2H_6UF_7$, and $(NH_3OH)UF_6$; only the first of these appears to contain the UF_6^- ion, the others having spectra similar to those of non-octrahedral fluoro complexes published earlier[80]. Gruen and McBeth[81] studied the spectrum of gaseous uranium pentachloride and interpreted it on the basis of the dimer U_2Cl_{10}, consisting of two octahedra sharing edges. They also demonstrated the formation of a gas phase complex, $UCl_5 \cdot AlCl_3$ and obtained its spectrum. Lux et al.[82] measured the spectrum of UBr_5 as a solid and in methylene dichloride and showed that it reacted with HBr in methylene dichloride to produce the UBr_6^- complex.

Selbin et al.[17] re-measured the spectrum of UCl_6^- in $SOCl_2$, assigned the various energy levels, and identified the v_3, v_4, and v_6 metal–chlorine vibrational frequencies in the vibronic spectrum. Ryan[7] measured the spectra of the UF_6^-, UCl_6^-, and UBr_6^- complexes in both solid salts and solutions. This work identified v_3, v_4, and v_6 for all three hexahalides (v_3 and v_4 were also measured in the infrared in all cases) and generally confirmed the assignments made by Selbin et al. for UCl_6^-. Values of the vibrational frequencies are shown in Table 9.1. This work disagrees with much of the previous[13] interpretation of the UF_6^- spectrum, and indicates that the pure electronic

Table 9.1   Frequencies in cm^{-1} of the fundamental modes[7] of UX$_6^-$

	UF$_6^-$	UCl$_6^-$	UBr$_6^-$
v_3	525	309	217
v_4	164	122	86
v_6	125	95	61

$\Gamma_7 \rightarrow \Gamma_7$-transition is present in all cases, probably as a magnetic dipole transition. The preparation and spectra of the rather uncommon (for actinides) mono-oxo complexes, UOF$_5^{2-}$, UOCl$_5^{2-}$, and UOBr$_5^{2-}$, were also reported, and spectral methods were used to study the stability of U^V relative to U^{IV} plus U^{VI} in a variety of non-aqueous media.

Hagan and Cleveland[83] published a very carefully measured spectrum of the NpO$_2^{2+}$ aquo ion. The spectra of the NpO$_2$(NO$_3$)$_3^-$ [57] and NpO$_2$Cl$_4^{2-}$ [84] ions have been published and used to identify species involved in solvent extraction; distortion of the NpO$_2$Cl$_4^{2-}$ ion by hydrogen bonding changes its spectrum. Stafsudd et al.[85] have studied the low-temperature spectrum of NpO$_2$Cl$_4^{2-}$ in Cs$_2$UO$_2$Cl$_4$ and conclude that not all of the spectrum can be attributed to the 5f^1 configuration (see also Sections 9.3 and 9.4.1). Np^{VI} oxalate complexes have been studied[86] by spectrophotometric methods, and the spectra of Np^{VI} in hydroxide[87] and concentrated fluoride solutions[88] have been reported. A spectrum of NpF$_6$ has been published[89] which is perhaps of somewhat better quality than that used for the original interpretation[16] of its spectrum.

9.2.3.2 5f^2 Systems

Spectra of several U^{IV} compounds, including the octahedral halide complexes, have been discussed in Sections 9.2.1. and 9.2.2. Zeeman effects in the spectrum of UBr$_6^{2-}$ have also been measured[90].

Morrey et al.[91] measured and discussed the spectra of gaseous and solid UCl$_4$ at various temperatures. Gruen and McBeth[92] also measured the spectrum of UCl$_4$ gas and demonstrated the existence of a gas phase UCl$_4$ complex with aluminium chloride. They later[81] showed this to be UCl$_2$·(AlCl$_4$)$_2$. The spectrum of U^{IV} in molten fluorides has been measured[93]. Earlier work[59] on the spectra of U^{IV} in molten chlorides concluded that species of coordination number lower than six were obtained in NaCl, for instance, as against predominantly UCl$_6^{2-}$ in CsCl. This seems unlikely and it is probable that the coordination numbers of the non-octahedral species observed were >6. The higher charge density of Na$^+$ as compared with Cs$^+$ probably lowers the effective Cl$^-$ activity, but in this condensed system the higher charge density of small alkali ions continues to act on the Cl$^-$ ions attached to the U, weakening, and thereby lengthening, the U—Cl bond and thus producing higher coordination numbers.

Bagnall *et al.*[94] have measured the spectrum of $U(N,N\text{-diethyldithio-carbamate})_4$ in which they conclude U^{IV} is probably 8-coordinate. Selbin and Ortego[95] have measured the absorption spectra of several U^{IV} chelates, all of which were thought to be 8-coordinate including one with tetra-dentate ligands, and have reported the band positions, noting that at least some of the bands are at considerably different energies from those of UX_6^{2-}. They[96] have also reported spectra of $U(\text{tropolonate})_4$ and $LiU(\text{tropolonate})_5$ and there is reason to believe the latter may be 10-coordinate. The spectrum of $U(C_5H_5)_3F$ at both 77 K and 298 K have recently been published[97], and spectrophotometric methods has recently been used[98] to study mixed oxalate-fluoride complexes of U^{IV}. Spectral measurements were used[82] to show that the compound $UOBr_3 \cdot 2.5$ DMA (DMA = N,N-dimethylacetamide) contains U^{IV} and U^{VI}. Spectral measurements in glacial acetic acid[99] indicate formation of mixed acetato–perchlorato complexes of U^{IV}.

The spectra of Np^V in aqueous hydroxide[87] and fluoride[88] solutions have been reported, and the spectrum in $HClO_4$ has been carefully remeasured[83]. The detailed interpretation of the absorption spectra of both the NpO_2^+ and PuO_2^{2+} ions in $HClO_4$ has been discussed by Eisenstein and Pryce[100]. Diethylaminetriaminepenta-acetic acid complexes of U^{IV}, Np^V, and Pu^{VI} have been studied by spectrophotometric methods[101], as have plutonyl acetate complexes in aqueous solution[102], and the spectrum of Pu^{VI} in glacial acetic acid has also been published[99]. The absorption spectra of the $PuO_2(C_2H_3O_2)_3^-$ and $PuO_2(C_2H_3O_2)_4^{2-}$ ions have been used to identify species in anion-exchange resins and amine extracts[103]. Comparison of these spectra with the spectra of $PuO_2Cl_4^{2-}$ [94], $PuO_2(NO_3)_3^-$ [57], and various other Pu^{VI} complexes, including the aquo ion, indicates that most of the transitions vary markedly in intensity with change in the PuO_2^{2+} ion environment. Thus for PuO_2^{2+} in $1MHClO_4$, $\varepsilon_{831nm} = 550$ [104] whereas in $PuO_2Cl_4^{2-}$ the most likely corresponding transition is a doublet at $\sim 880\text{–}890$ nm with $\varepsilon = 4.3$. In $PuO_2Cl_4^{2-}$, $PuO_2(NO_3)_3^-$ and $PuO_2(C_2H_3O_2)_3^-$ this particular transition is in all cases weak while in $PuO_2(C_2H_3O_2)_4^{2-}$, $PuO_2(NO_3)_4^{2-}$ [105], and the aquo ion it is very strong. Whether this is connected with the fact that all the ligand atoms are in a plane perpendicular to the $O = Pu = O$ group in the former is not certain at this point but is an interesting speculation. Sutton[106] has proposed, on the basis of Raman data, that the UO_2^{2+} aquo ion is $UO_2(H_2O)_6^{2+}$ in which the six water molecules form a puckered ring around the UO_2 axis, but the structures of $PuO_2(C_2H_3O_2)_4^{2-}$ and $PuO_2(NO_3)_4^{2-}$ are not known.

Varga *et al.*[107, 108] have reported the spectrum of $CsNpF_6$, the NpF_6^- ion being the only non-oxo complex of Np^V for which f→f-transition spectra are at present available. They have correlated essentially all the lines observed with calculated levels assuming small crystal field splitting by an octahedral field and have correlated the number of lines with those predicted by such splitting. They neglected vibronic coupling, and it is well known[11, 12, 46, 47] that the spectra of the octahedral isoelectronic UX_6^{2-} and the chemically similar octahedral UF_6^- ions (including $CsUF_6$)[7, 13] consist almost entirely of vibronic transitions with only a very few electronic lines observed as magnetic dipole transitions. Because of this, it is not apparent to the writer how such a correlation can be correct.

9.2.3.3 5f³ Systems

The absorption spectrum of aqueous U^{III} has been compared to a calculated energy level scheme and an attempt to correlate observed and calculated intensities was also made[109]. Good quality spectra of UCl_3, UF_3, and $U(C_5H_5)_3$ at 25 °C have been published[97]. The spectra of U^{III} in molten 60 mol % $AlCl_3$–40 mol % $NaCl$[110], in molten fluorides[93], and in glacial acetic acid[99] have been reported.

The energy shifts between the spectrum of $NpCl_6^{2-}$ and that of $NpBr_6^{2-}$ have been discussed[15], and these spectra are shown in Figure 9.2. The spectrum of $Np(NO_3)_6^{2-}$ is also available[57]. Detailed studies of the absorption spectra of Np^{IV} doped in ThO_2 [21, 54], and in $PbMoO_4$ [25], have recently been published. The spectrum of presumably 8-coordinate, $Np(diethyldithio-carbamate)_4$ has been reported[94] and the low-temperature spectrum of $Np(C_5H_5)_3F$ has been measured[111]. The spectrum of the Np^{IV} aquo ion has been re-measured[83], the spectrum of Np^{IV} in saturated KF reported[88], and Np^{IV}–diethylaminetriaminepenta-acetic acid complexes have been studied by spectrophotometric methods[101].

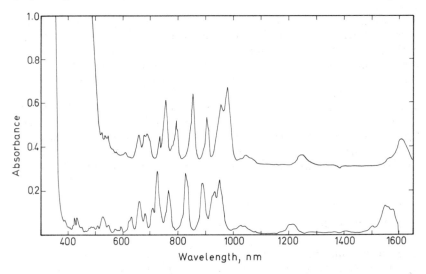

Figure 9.2 Absorption spectra of Np^{IV} hexahalides: 2.4×10^{-2} M $[(C_2H_5)_4N]_2NpCl_6$ (bottom) and 1.2×10^{-2} M $[(C_2H_5)_4N]_2NpBr_6$ (top) in acetonitrile, 1.00 cm cells

Varga et al.[112] have compared the room-temperature absorption spectrum of Np^{IV} in CsF·2HF with a calculated energy level scheme. They made a similar comparison with some of the many absorption peaks in a rather poorly resolved room-temperature spectrum of Pu^V in solid Rb_2PuF_7 and also with the aqueous solution spectrum of AmO_2^{2+}. It appears to this writer that assignments made in this manner with room-temperature spectra (with extra peaks not assigned) of compounds of unknown symmetry and unknown extent of ligand-field splitting must be considered as quite tentative.

9.2.3.4 5f⁴ Systems

A detailed study[9], including polarisation effects and Zeeman effects, of the low-temperature absorption and fluorescence spectrum of Np^{III} in $LaBr_3$ has been carried out. The spectrum of aqueous Np^{III} has been re-measured[83] and has been compared to a calculated energy level scheme[109]. Brown et al.[113] have published the spectrum of $(C_2H_5)_4NNp(dtc)_4$, where dtc = diethyl-dithiocarbamate. The Np^{III} ion is coordinated to eight sulphur atoms in a grossly distorted dodecahedral arrangement. Shiloh and Marcus[114] have shown the effect of Cl^- and Br^- complexing in LiX solutions on the spectrum of Np^{III}, and spectral methods have been used[101] to study Np^{III} and Pu^{IV} complexes of diethylaminetriaminepenta-acetic acid.

The absorption spectra of $PuCl_6^{2-}$, $PuBr_6^{2-}$ [15] and $Pu(NO_3)_6^{2-}$ [56] are well-known, and the chloro and nitrato complexes have been identified by spectral methods in anion-exchange resins and amine extracts. Cleveland et al.[115] have shown, on the basis of the absorption spectra, that Pu metal dissolves readily with excess Cl_2 in the high dielectric, relatively non-complexing, solvent propane-1,2-diol carbonate (propylene carbonate), to produce a solution which always contained 60% Pu^{IV} as $PuCl_6^{2-}$ and 40% cationic Pu^{III}, from which a compound of formula $[Pu^{III}(PDC)_7]_2(Pu^{IV}Cl_6)_3$ was isolated; this had the same absorption spectrum as the solution. Apparently, in the absence of a better complexing agent for Pu^{IV} than the solvent, Pu^{III} is not oxidised by Cl_2. If excess Cl^- is present, oxidation to $PuCl_6^{2-}$ occurs very readily as has been noted previously[26] in other relatively non-complexing solvents. McLaughlin et al.[116, 117] studied Pu^{III} in CaF_2. They found that an absorption spectrum due to Pu^{IV}, rather than the expected Pu^{II}, grew into the original Pu^{III} spectrum as a function of time due to radiation damage. This conversion to Pu^{IV} was found to be temperature-sensitive. The spectrum of the presumably 8-coordinate Pu(N,N-diethyldithiocarbamate)$_4$ has been published[94]. There does not appear to be any recent work on the spectrum of Am^V.

9.2.3.5 5f⁵ Systems

The $PuCl_6^{3-}$ and $PuBr_6^{3-}$ ions have been prepared in non-aqueous solvents and in solid triphenylphosphonium salts and the very weak absorption spectrum of the former has been published[26]. The absorption spectrum of $PuBr_6^{3-}$ (as the solid salt) is almost identical to that of $PuCl_6^{3-}$ except for a very slight red-shift. Conway and Rajnak[118] made a detailed study of the low-temperature spectrum of Pu^{III} in $LaCl_3$, including polarisation and Zeeman effects, and compared the results with a calculated energy level scheme. The spectrum of the aquo ion has also been compared with calculated energy levels[109]. The spectra of Pu^{III} in CaF_2[116, 117], and (at low temperature) of $Pu(C_5H_5)_3$ have been obtained[111]; the spectrum of Pu^{III} in the grossly distorted dodecahedral $(C_2H_5)_4NPu(dtc)_4$, where dtc = diethyl-dithiocarbamate, has been reported[113]. Chloride and bromide complexing in aqueous LiX have a fairly small effect on the internal 5f⁵ spectrum of Pu^{III}[114].

Spectra of Am^{IV} have been reported only in 15M NH_4F[119] and 10M H_3PO_4[120] solutions.

9.2.3.6 5f⁶ Systems

The author[26] has prepared both salts and solutions of $AmCl_6^{3-}$ and $AmBr_6^{3-}$ complexes and has obtained the relatively very weak absorption spectra of both of these. These spectra were not published[26] but they were very similar, except for a very slight red-shift of most of the $AmBr_6^{3-}$ peaks relative to $AmCl_6^{3-}$ and somewhat greater intensities in $AmBr_6^{3-}$ consistent with the order of intensities in lanthanide[28] and other actinide hexahalides[7, 15]. Some of the band groups are wider for $AmCl_6^{3-}$ than for $AmBr_6^{3-}$ indicating a vibronic nature, and in particular the band group at about 9.6 kK for $AmCl_6^{3-}$ appears to consist of a central peak with two side peaks or shoulders equally spaced above and below. For $AmCl_6^{3-}$ the outer pair are about 210 cm^{-1} and for $AmBr_6^{3-}$ about 145 cm^{-1} above and below the central peak. The transitions on the low energy side of the central peak are less intense as expected for vibronic transitions at room temperature. The centre peak is probably a pure electronic transition, and the outer pair are probably due to coupling of the v_3 vibrational modes, the decrease with halide mass being that expected. These vibrational frequency values are about the same as those for $PrCl_6^{3-}$ in solution[28] and, since the mass of Am is greater than that of Pr, indicate stronger bonding in the actinide halide complexes as expected[121, 122]. Marcus and Bomse[27] have recently published the $AmCl_6^{3-}$ spectrum.

Pappalardo et al.[44] have recently carried out a detailed analysis of the low-temperature absorption spectra of $AmCl_3$, $AmBr_3$, and AmI_3. Energy levels were assigned, vibronic transitions and hypersensitive transitions identified, and nephelauxetic effects discussed. They[123] also studied the spectrum of americium tricyclopentadienide and concluded that the bonding in it is somewhat more covalent than that in AmI_3. The absorption spectrum of the Am^{III} aquo ion has been compared to a calculated energy level scheme[109]. Marcus and Shiloh reported the spectra of Am^{III} in various chloride environments[68].

Delle Site and Baybarz[124] have published an excellent and very thorough spectrophotometric study of Am^{III} complexes with eight different aminopolyacetic acids. The stoichiometries and ranges of stability of 22 complexes were determined and instability constants were measured for 13 of these. The changes in the Am^{III} spectrum were small but distinct in all cases, amounting to energy shifts of up to slightly greater than 1 % and intensity changes of up to ±40% against the aquo ion. The portion of the spectrum studied was red-shifted relative to the aquo ion in all cases. Hafez[101] has also studied the Am^{III} complex with diethylaminetriaminepenta-acetic acid by spectrophotometric methods. Shiloh et al.[67] have made a spectrophotometric study of Am^{III} complexes in aqueous lithium bromide, iodide, and nitrate; magnesium iodide; potassium carbonate; and nitric acid. Barbanel[1] and Mikhailova[70] have studied the chloro complexes in aqueous HCl.

Keenan[125] published the spectrum of Cm^{IV} in 15M CsF solution. More recently Edelstein *et al.*[117] observed lines due to Cm^{IV} in crystals of Cm-doped CaF_2 in which radiation damage had occurred.

9.2.3.7 5f⁷ Systems

The $CmCl_6^{3-}$ complex has recently been prepared[27] in solution in the same way as the lanthanide, Pu^{III} and Am^{III} hexachloro complexes[26, 28]. The spectrum is drastically (up to 60-fold for one transition) reduced in intensity relative to the aquo ion because of the octahedral symmetry. All the transitions are red-shifted relative to the aquo ion as expected.

The absorption spectrum of the Cm^{III} aquo ion has been compared to a calculated energy level scheme[109, 126] and the fit is quite good. Gruber *et al.*[127] have studied the low-temperature polarised absorption and emission spectra of Cm in $LaCl_3$, while Edelstein *et al.*[117] have measured the spectrum in CaF_2.

A few transitions of Bk^{IV} in 5M $HClO_4$ have been reported[128] and Baybarz *et al.*[129] have recently measured the Bk^{IV} spectrum in four different solutions including 1M $HClO_4$.

9.2.3.8 5f⁸ Systems

The absorption spectrum of Bk^{III} has been measured by Gutmacher *et al.*[128] in several aqueous acid solutions and is reported for HCl solution in the range 3200 Å to 15 000 Å. It has also been measured[129] in 1 M $DClO_4$ solution to 17 000 Å. Calculated energy levels for Bk^{III} are available[109].

9.2.3.9 5f⁹ Systems

The absorption spectra of Cf^{III} as a single crystal of $CfCl_3$ and as Cf^{III} loaded on to a Dowex-50 cation exchange resin bead have been measured by Green and Cunningham[130], and calculated energy levels have been reported[109].

9.2.3.10 5f¹⁰ Systems

Cunningham *et al.*[131] and Fujita *et al.*[132] measured the absorption spectrum of Es^{III} from 3500 Å to 10 600 Å in 3–6M HCl. Later measurements by Nugent *et al.*[133] in 1M $HClO_4$ extend the range measured and confirm most of the previous data. They also studied the luminescence of an einsteinium β-diketone chelate. Carnall and Fields[109] have compared calculated and measured energy levels for Es^{III}.

9.3 URANYL SPECTRA

The spectrum of the uranyl, UO_2^{2+}, ion is treated separately here since in many of its properties which are of greatest use to the actinide chemist it is similar to the internal f^q-transition spectra, yet in its origins it does not fit in that category. The uranyl spectrum in the visible region, like the internal f^q-transition spectra, has molar absorptivities, ε_{max}, generally but not always less than 50, often has sharp fine structure, and is overall rather sensitive to ligand environment. This combination makes the visible spectrum quite useful for the study and identification of uranyl complex ion species in aqueous and non-aqueous solvents, ion-exchange resins, solid salts, etc. The strong bands in the u.v. are the typical electron-transfer bands associated with the ligands attached to the UO_2^{2+} ion and are discussed in Section 9.4.3.

Although the uranyl spectrum has been known and studied in more detail and for a longer period than other actinide ion spectra, it is not, as pointed out by McGlynn and Smith[134] and by Rabinowitch and Belford[135], in many respects well understood. Basically, it appears that the weak uranyl spectrum is due to the coupling of several vibrational frequencies of the complex, principally the symmetrical O—U—O frequencies[135], to several fairly closely spaced very weak or forbidden electron-transfer transitions[134, 136]. Series of up to about eight members of the symmetrical vibrational frequency are coupled to each electronic level. The electron-transfer transition character is discussed further in Section 9.4.1. Transitions similar to those of the uranyl ion may be expected to occur for the 5f-electron-containing MO_2^{2+}- and MO_2^+-species. Although the spectra of these species were discussed in the preceding section, their spectra in the region of weak narrow transitions cannot be expected to be due solely to typical internal $5f^q$-transitions. Detailed discussion of the study and theory of uranyl spectra is beyond the scope of this review, but information, including both theory and experimental data, on uranyl spectra up to about 1960 is covered in the book by Rabinowitch and Belford[135]. More recent theoretical interpretations of uranyl spectra are also available[134, 137–140]. The principal effect of ligand environment on the uranyl spectrum appears to be on the relative intensities and to a smaller extent on the energies of the series of vibrational peaks. There are however, several empirical or semi-empirical correlations of uranyl complex spectra to structure that are of interest here.

Ryan and Keder[103] have pointed out the very close similarity of the spectra of $UO_2(NO_3)_3^-$, $UO_2(SO_4)_3^{4-}$, $UO_2(ClO_4)_3^-$, $UO_2(CO_3)_3^{4-}$, and UO_2 $(C_2H_3O_2)_3^-$ ions, this last being shown in Figure 9.3. The spectrum of the tris-(propionato) ion is also very similar to these[141]. The general appearance of the spectra of these complexes is almost identical but overall intensities vary between the different complexes. It appears that spectra of this type are peculiar to the uranyl ion bonded to six oxygen atoms of three 'short-bite' bidentate ligands in a plane perpendicular to the uranyl oxygen atoms. The spectra of the tetranitrato-[105] and tetra-acetato-uranyl[103] ions differ markedly from those of the trinitrato- and triacetato-complexes (Figure 9.3). Both are more intense, and both, but particularly the tetranitrato complex, show much less vibrational structure than most other known discrete uranyl complexes. Crystal structure data for $[(C_2H_5)_4N]_2 \, UO_2(NO_3)_4$ would be of considerable interest because neither monodentate nitrato complexes of the actinides nor discrete 10-coordinate complexes of U^{VI} are well-established. (Evidence[142] showing that the compound of composition Cs_2UO_2

$(NO_3)_4$ does not contain $UO_2(NO_3)_4^{2-}$ and thus purporting to demonstrate the non-existence of the $UO_2(NO_3)_4^{2-}$ complex actually supports the original work[105] on the $UO_2(NO_3)_4^{2-}$ ion, since in that work alkali metal salts containing the $UO_2(NO_3)_4^{2-}$ ion also could not be obtained.)

Hydrogen bonding to the $UO_2Cl_4^{2-}$ ion[84] distorts the ion in such a way as to markedly increase the intensity of two of the vibrational series. Although the intensities of the other vibrational series and the energies were almost

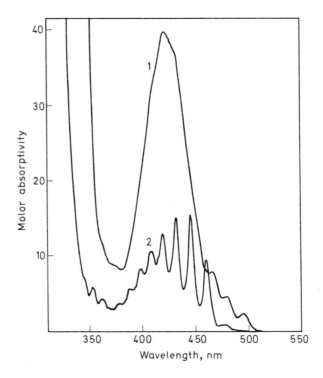

Figure 9.3 Absorption spectra of: (1) $UO_2(C_2H_3O_2)_4^{2-}$ in liquid $(C_2H_5)_4NC_2H_3O_2 \cdot H_2O$ and (2) $UO_2(C_2H_3O_2)_3^-$ in acetonitrile. (From Ryan and Keder[103] by courtesy of the American Chemical Society.)

unchanged, an appreciable change in the appearance of the overall spectrum resulted. The author[41] has prepared by fusion techniques a compound of stoichiometry $(Pr_4N)_3[(UO_2)_2Cl_7]$. The spectrum of this material differs from that of $UO_2Cl_4^{2-}$ only in an even larger change in intensity of these same two vibrational series, along with a small shift in energy of the entire spectrum. The spectrum of a fusion preparation of composition $Pr_4NUO_2Cl_3$ indicated an obvious mixture, but Vdovenko et al.[143] have prepared the composition $R_4NUO_2Cl_3$ in benzene and the published spectrum appears (within the limits of the quality of the journal reproduction) to be almost the same as that of the $(Pr_4N)_3[(UO_2)_2Cl_7]$. Vdovenko et al.[143, 144] conclude that the $R_4NUO_2Cl_3$ solutions contain

$$\begin{array}{c}\text{Cl}\diagdown \\ \text{Cl}\diagup\end{array}\text{UO}_2^{}\begin{array}{c}\diagdown\text{Cl}\diagdown \\ \diagup\text{Cl}\diagup\end{array}\text{UO}_2^{}\begin{array}{c}\diagdown\text{Cl} \\ \diagdown\text{Cl}\end{array}\quad\text{or}\quad -\text{Cl}-\overset{\overset{\displaystyle\text{Cl}}{|}}{\underset{\underset{\displaystyle\text{Cl}}{|}}{\text{UO}_2}}-\text{Cl}-\overset{\overset{\displaystyle\text{Cl}}{|}}{\underset{\underset{\displaystyle\text{Cl}}{|}}{\text{UO}_2}}-\text{Cl}-$$

chains. It would appear that the $(Pr_4N)_3[(UO_2)_2Cl_7]$ probably contains

$$\text{Cl}-\overset{\overset{\displaystyle\text{Cl}}{|}}{\underset{\underset{\displaystyle\text{Cl}}{|}}{\text{UO}_2}}-\text{Cl}-\overset{\overset{\displaystyle\text{Cl}}{|}}{\underset{\underset{\displaystyle\text{Cl}}{|}}{\text{UO}_2}}-\text{Cl}$$

All of these chloro complexes, including $UO_2Cl_4^{2-}$, appear to have essentially the same spectrum, except that the intensities of two of the vibrational series are very sensitive to distortion of the square-planar chloride array. The distorting effect of the hydrogen bonding is small compared to that of chlorine bridging. The intensities of these two series, relative to the rest of the $UO_2Cl_4^{2-}$ spectrum, change with the solvent but to a much smaller extent than the change produced by hydrogen bonding[84]. The only difference between the spectra of $UO_2Cl_4^{2-}$ in solution and in solid salts is again a small change in the intensities of these two series relative to the others[84]. The marked sensitivity of these particular vibrational series to environment was noted in the original work[84]. Vdovenko et al.[145] have studied the effect on the UO_2^{2+}-stretching vibrations in the infrared of various distorting interactions on the $UO_2Cl_4^{2-}$ ion, but unfortunately no visible spectra were reported.

Vdovenko et al.[144] have made a very interesting correlation of the energy of the first intense vibrational band in the uranyl spectrum to coordination number. Based on the examples of known structure shown and on examples known to the writer, this appears to be valid and this writer has observed that uranyl complexes known by him to be 6-coordinate (total 8-coordination for the U^{VI} including the oxide ligands) generally do not absorb above 500 nm whereas those with known 4-coordination number do. It should be noted, however, that the types of examples of known coordination number studied[144] are rather limited, all the 6-coordinate examples being of the four-membered ring bidentate ligand structure discussed earlier, and all the 4-coordinate examples (except the somewhat questionable $UO_2(ClO_4)_4^{2-}$ of unknown structure) being chloro or bromo complexes. The proposed 5-coordinate compounds overlap the 6-coordinate species and their co-ordination numbers are not well-established; in particular $M_3[UO_2(SO_4)_2 NO_3]$ is probably 6-coordinate. This is not meant as a criticism of this very good observation but merely as a caution that further confirmation with complexes of greater chemical and structural variability, such as chelating groups other than the 'short-bite' four-membered ring-forming type, would be very desirable. The position of the first intense vibrational band in mixed halide complexes of 4-coordination number has also been quantitatively correlated[144] with the ligand basicity and bond distance. The correlation by Vdovenko et al.[144] of structure with the relative intensities of vibrational bands of different electronic transitions does not appear completely valid since all the intense bands of $UO_2(NO_3)_3$ and $UO_2(C_2H_3O_2)_3^-$ seem to belong to the same vibrational series. The problem here is that there are not just two but several (at least six[84]) overlapping series of vibrational peaks in some of these complexes.

Absorption spectra of U^{VI} in LiOH have recently been published by Tomažič et al.[146]. The amphoteric nature of the plutonyl ion was studied[147, 148] much earlier in NH_4OH and KOH, and the present author several years ago studied, but did not publish, the spectra of bright yellow solutions of U^{VI} in Pr_4NOH. The spectra in LiOH and Pr_4NOH are similar and indicate the presence of at least three soluble species as a function of hydroxide concentration. The molar extinction coefficients decrease with increasing hydroxide concentration and a terminal species is formed at about 0.15 M OH^- (~ 0.01 M U^{VI}) with $\varepsilon_{max.400nm}$ 23. Solid uranyl salts obtained from the alkaline solutions had spectra distinctly different from those of the solution species and the identity of the latter are not known.

Bell and Biggers[149, 150] have applied curve-resolution techniques to the uranyl spectrum in 1 M $HClO_4$ and have discussed in detail the proposed energy levels involved. Even with the best techniques, the resolution of the 25 °C spectrum of the uranyl aquo complex of unproven symmetry will be extremely poor compared to that of many other uranyl complexes for which measurements are available, and for which the symmetry is known; a greater understanding of the uranyl spectrum would be obtained by study of the latter. The paper by Vdovenko et al.[144], and references therein, give the absorption spectra of a large number of uranyl complexes including those with mixed ligands. Spectra of uranyl oxalate complexes have also appeared[151] and recent spectral studies of solid uranyl compounds have been reviewed[152].

9.4   ELECTRON-TRANSFER SPECTRA

9.4.1   Optical electronegativities

Jørgensen[136] has recently reviewed the entire field of electron-transfer spectra, including a section on the 4f and 5f elements. In his review, he considers that electron-transfer spectra are of even greater importance to the chemist than those due to the internal d^q or f^q-transitions. This statement is probably correct if it is qualified to limit it to the information obtained directly from the spectrum as such (band positions, etc.), rather than to apply it to the use of spectra as a tool to identify species in solutions, etc. by comparative methods, equilibrium studies, etc. In the latter case, the electron-transfer spectra are too broad and nondescript to serve as good 'fingerprints' to identify complex ion species as compared with the internal f^q spectra or even perhaps the f→d spectra, and the only appreciable use of actinide electron-transfer spectra in this way has been with Pa^V (see References 153, 154, and earlier work referred to therein). The energies of electron-transfer spectra (redox spectra) are a measure of the difference in central metal ion and ligand electronegativity (and are thus related closely to redox potentials), increasing oxidising strength of the metal ions and reducing strength of the ligands decreasing the electron-transfer band energy. A simple knowledge of electron-transfer band systematics often allows the oxidation state or the presence or absence of a particular ligand in the co-ordination sphere to be determined by colour alone. This is particularly valuable in work with the relatively unstable higher oxidation state actinides and lanthanides. Thus, the reported[155] preparation of salts of $PrCl_6^{2-}$ can be rejected simply on the basis of the reported pale-yellow colour, since $CeCl_6^{2-}$ is yellow and $PrCl_6^{2-}$, having a much more strongly oxidising central metal ion, must be intense red, brown, or black.

Jørgensen[5, 71, 156] has discussed the concept of electronegativity and its relationship to electron-transfer spectra. He developed the concept of *optical electronegativities*, χ_{opt}, which vary strongly with oxidation number of a given atom. χ_{opt} is defined by the relation:

$$\sigma_{corr} = 30 \text{ kK} [\chi_{opt}(X) - \chi_{opt}(M)] \qquad (9.1)$$

where σ_{corr} is the wave number (in units of $1 \text{ kK} = 1000 \text{ cm}^{-1}$) of the first strong electron-transfer band, corrected for the effects of spin-pairing energy and other forms of interelectronic repulsion on the partly filled shell, and (X) and (M) refer to ligand and metal respectively. This allows the fixing of values for χ_{opt} using Pauling's electronegativity values for the halogens. Actually the variations of χ_{opt} across the 4f- or 5f-series reflect essentially only the monotonic effect of increasing nuclear charge on each f-electron.

Spin-pairing energy and other interelectronic repulsion and relativistic effects are major factors in the variation of the electron-transfer band wave number with the number of electrons in the 4f or 5f shells[157]. Since these affect the oxidation potentials as well, the *uncorrected* electronegativity values, χ_{uncorr}, not taking spin-pairing or other interelectronic repulsion or relativistic effects into account[28, 136], are more closely related to chemical criteria such as standard oxidation potentials[136]. Using this concept Equation (9.1) becomes[28, 136]:

$$\sigma_{obs} = 30 \text{ kK} [\chi_{opt}(X) - \chi_{uncorr}(M)] \qquad (9.2)$$

where σ_{obs} is the observed frequency of the lowest energy electron-transfer band.

Although strong absorption in the u.v. region by 4f- and 5f-group elements was occasionally attributed to electron transfer, the first real study was made by Jørgensen, on the lanthanide bromides in nearly anhydrous ethanol[157]. At this time very few hexahalides were known for the actinides and lanthanides, but these possess definite advantages for the study of electron-transfer spectra[50]. The electronegativities of the halide ions are known; the hexahalides provide a constant known coordination number, N = 6, of simple symmetry unlike the usual values of N = 7 to 12 for actinides and lanthanides[15, 136] and they provide a convenient base for comparison with d-group elements. The study of the systematics of the electron-transfer spectra and, to a lesser extent, the internal f^n and 5f→6d spectra, has provided the major

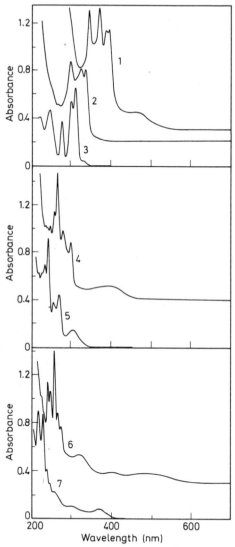

Figure 9.4 Absorption spectra of acetonitrile solutions of M^{IV} hexahalides in the region of The Laporte-allowed transitions. (1) 4.6×10^{-2} M UI_6^{2-} (U^{IV} iodide in concentrated tributyl-propylammonium iodide) in 0.0019 cm cell (absorbance scale displaced 0.3), (2) 1.1×10^{-4} M $[(C_2H_5)_4N]_2UBr_6$ (absorbance scale displaced 0.2), (3) 1.4×10^{-4} M $[(C_2H_5)_4N]_2UCl_6$, (4) 5.5×10^{-5} M $[(C_2H_5)_4N]_2NpBr_6$ (absorbance scale displaced 0.4), (5) 6.7×10^{-5} M $[(C_2H_5)_4N]_2NpCl_6$, (6) 5.0×10^{-5} M $[(C_2H_5)_4N]_2PuBr_6$ (absorbance displaced 0.3), and (7) 4.4×10^{-5} M $[(C_2H_5)_4N]_2PuCl_6$. All cell thicknesses 1.00 cm except in (1). (From Ryan and Jørgensen[15] by courtesy of Taylor and Francis Ltd.)

incentive for the preparation of many of the actinide and lanthanide hexahalide complexes; these are often difficult to obtain because of the strong A-group (hard acid) behaviour of actinides and lanthanides[158-160].

The actinide (and lanthanide) hexahalides exhibit two types of allowed transitions (Figure 9.4), strong bands of half-widths toward lower energy $\sigma(-) \sim 0.6$ kK and broad, generally weaker, bands of $\sigma(\frac{1}{2}) \sim 2$ kK. The former are the f$\rightarrow$d-bands to be discussed later, and the latter are the electron-transfer bands discussed here. The two types can usually be distinguished by the larger half-width and relatively weak intensity of the electron-transfer bands compared to the f$\rightarrow$d-bands[40], and the change in energy with variation in halide in accordance with Equation (9.2), the f$\rightarrow$d-transitions shifting less with change in halogen. The setting of arbitrary limits of half-width and, particularly, intensity to distinguish these[161] is not completely valid however. In general, it is not possible to distinguish electron-transfer bands occurring at energies higher than those of the first f$\rightarrow$d-transitions.

The first (lowest energy) electron-transfer band of the f-group hexahalides has been assigned[136] as the Laporte-allowed transition $\pi t_{1g} \rightarrow$ f. Higher-energy transitions are presumably due to transitions both from other sets of molecular orbitals and to other f levels and will not be considered further here.

The electron-transfer spectra of the known actinide hexahalides, or the colour due to electron-transfer bands in cases where measurements were not made, are listed in Table 9.2. Bagnall and Brown[162] have reported pale-yellow salts of $PaCl_6^-$, while Brown and Jones[163] have reported salts of $PaBr_6^-$ as orange or orange-red in colour, which would imply that the first electron-transfer band is in the region 21–24 kK. The preparation of a salt of PaI_6^-, and thus its colour, has not been unambiguously established[164]. The data for UCl_6 are based on somewhat incomplete work[165] which gives values of 17 and 20.8 kK for the solid and 21 and 27 kK for the gas. The

Table 9.2   Electron-transfer bands (or colours due to electron-transfer bands) of actinide hexahalides*

$5f^0$		$PaCl_6^-$ (pale-yellow)	$PaBr_6^-$ (orange-red)	PaI_6^-? (dark brown)
		UCl_6 (17, 20, 27?)		
$5f^1$		UCl_6^- 25.3, (29.4)	UBr_6^- (17.4), (20.5), (23.3), (26.0)	UI_6^- (black, salts unstable above $-35\,°C$.)
	NpF_6 38.8, 45.2			
$5f^2$				UI_6^{2-} 20.9
		$NpCl_6^-$ ~ 18.5		
	PuF_6 31.7, 40.0, 44.5			
$5f^3$		$NpCl_6^{2-}$ 33.1	$NpBr_6^{2-}$ 25.0, 27.4	
$5f^4$		$PuCl_6^{2-}$ 27.2, 33.2, 38.9	$PuBr_6^{2-}$ (19.6), 20.8, 24.9, 31.4	
$5f^7$		$BkCl_6^{2-}$ 21.1		

*Values are in kK with shoulders in parenthesis. Other known actinide hexahalides not having electron-transfer bands in a measurable range or having such bands obscured by 5f$\rightarrow$6d-transitions are listed in Table 9.4. References are in the text.

spectrum of UCl_6 in perfluoroheptane solution is similar to that of the gas except for a further, weaker band at 15.6 kK[166]. The values for NpF_6[89] and PuF_6[167] are for the gases. The data for UCl_6^-, UBr_6^-, and the very unstable UI_6^- are from reference 7 and the data for UI_6^{2-}, $NpCl_6^{2-}$, $NpBr_6^{2-}$, $PuCl_6^{2-}$, and $PuBr_6^{2-}$ are from reference 15. The result for $NpCl_6^-$ is from unpublished data of the author, based on preliminary measurements on the quite impure, deep red-brown salt $(C_6H_5)_4AsNpCl_6$ prepared[168] from anhydrous HCl solution. The first electron-transfer band of red-orange $BkCl_6^{2-}$ has been measured only recently[169]. The absorption spectrum of $AmBr_6^-$ has been reported[170]; although it has been proposed[136] that a shoulder at 37.0 kK might be an electron-transfer band, the band is sufficiently well resolved to indicate $\sigma(-) < 0.7$ kK and thus it is no doubt not due to electron transfer. Electron-transfer bands of the reducible lanthanide (Ce^{IV}, Sm^{III}, Eu^{III}, Yb^{III}, and Tm^{III}) hexachloro- and hexabromo-complexes[28] and, except for Ce^{IV}, the hexaiodo-complexes[50], have been measured and the results are also tabulated in Jørgensen's review[136].

Values of χ_{uncorr}, calculated from the data of Table 9.2 using Equation (9.2), for actinide ions of various oxidation states are tabulated in Table 9.3. In cases where hexahalides cannot be prepared, a measurement of χ_{uncorr} can sometimes be made using other compounds. The measured value for Am^{IV} in Table 9.3 was determined from the reported[169] 8.6 kK (0.29 electronegativity units) difference in the absorption cut-off due to electron-transfer transitions for Am- and Pu^{IV}-hydrous oxides. The measured value for Cm^{IV} was obtained from the cut-off in the absorption spectrum of the Cm^{IV} fluoride complex in aqueous solution[125] using Equation (9.2) and assuming[169] that the wave number difference between the $\varepsilon = 100$ point and the position of the first transition is the same as for typical actinide hexahalide complexes. The values of χ_{uncorr} observed[28] for trivalent lanthanide bromide complexes of somewhat uncertain stoichiometry in ethanol[157] are up to 0.2 electronegativity unit lower than those for the corresponding hexachlorides and hexabromides. Thus the values[171] for Cf and Es^{III} in Table 9.3 are for bromo complexes of unknown stoichiometry. These preliminary results[171] should not, therefore, be directly compared with the rest of Table 9.3 until further work now in progress, which includes results for the heavy actinide hexahalides, is complete. In this regard, the value for Cm^{IV} may also not compare as

Table 9.3   Uncorrected optical electronegativities, χ_{uncorr}, for actinides of various oxidation states*

	Th	Pa	U	Np	Pu	Am	Cm	Bk	Cf	Es	Fm	Md	No	Lr
(VI)			2.4	2.6	2.85									
(V)		2.0–2.1‡	2.2	2.4										
(IV)	1.0†	1.4†	1.8	1.95	2.1	2.4§	2.7§	2.3	2.8†	3.2†	3.2†	3.4†	3.8†	4.3†
(III)						<1.55‖			1.6¶	1.7¶				

*Values determined from spectra of hexahalides in Table 9.1 except: †Calculated values from Reference 169, ‡value estimated from the colour of $PaBr_6^-$ (Table 9.2), §values measured from species other than hexahalides from data of Reference 169, ‖limit imposed by the spectrum of $AmBr_6^{3-}$ as discussed in text, and ¶values calculated from data from tribromides in ethanol from Reference 171.

closely to the other values since the nature of the Cm^{IV} fluoro complex in 15 M CsF is not known.

The characteristic absorption spectrum of the UO_2^{2+} ion at about 24 kK has been interpreted by Jørgensen[15, 40, 136, 172] as due to electron transfer from an occupied molecular orbital, localised mainly on the two oxygen ligands, to the empty 5f-orbital of the central atom. The low intensity was explained as being due to the extraordinarily low oscillator strength of $\pi_u \rightarrow 5f$ in linear molecules of symmetry $D_{\infty h}$[136]. (Other interpretations of this[85, 135] are also available). The value of $\chi_{(opt)}(X)$ for O^{2-} of 3.15 obtained from Equation (9.2) and χ_{uncorr} for U^{VI} (Table 9.3) is identical to that obtained from the spectrum of Eu in Y_2O_3[36]. Bands at 20.8 kK for NpO_2^{2+}, 19 kK for PuO_2^{2+}, and possibly at 18 kK for AmO_2^{2+} have been assigned to similar transitions[136]. It must be remembered that these ions have internal f-electron transitions of similar intensities and since these electron-transfer transitions have metal–oxygen vibrational frequencies superimposed on them, some care must be exercised in making the correct assignments.

9.4.2 Actinide electron-transfer band systematics and actinide oxidation-reduction potentials

Jørgensen[15, 157] has developed a refined spin-pairing energy treatment to explain the variation of the energy of electron-transfer bands across the 4f- and 5f-groups and has successfully applied this to the interpretation of the electron-transfer spectra of the lanthanide bromides in ethanol[157]. His treatment consists of equations[15, 136, 157] containing two unknown parameters, W and $(E-A)$, the parameters E^1, E^3 and ξ_{4f} or $_{5f}$ being determined or estimated from the internal nf^q-transitions. The unknown parameters can be obtained by fitting to known data points which span the centre of the 4f or 5f series. These equations have been collected together[169] into:

$$\sigma_{obs} = W - q(E-A) + (9/104)N(S)E^1 + M(L)E^3 + P(S,L,J)\xi_{4f \text{ or } 5f} \quad (9.3)$$

where σ_{obs} is the same as in Equation (9.2), W is the constant energy the electron-transfer band would have if all f-electron shielding were perfect, and q is the number of f-electrons. The treatment takes into account the three main factors which cause the variation in the band energies across the series. These are the regular Z- or q-dependent increase in the energy of complexation due to the well-known lanthanide or actinide contraction of the ionic radii, and the regular Z- or q-dependent increase in the effective nuclear charge felt by each f-electron and arising from ineffective infra-f-electron screening. These are contained in the combined parameter $(E—A)$, an empirical constant for a given f-group series. The rest (last three terms) of Equation (9.3) accounts for the irregular q-dependent variation in the interelectronic repulsion energy of the q electrons in each ground state f^q electronic configuration. The latter factor reflects the difference in χ_{opt} and χ_{uncorr} and contains the effect of spin-pairing energy, special stabilisation effects on H and I ground terms, and relativistic effects[15, 136, 157]. The meaning of the quantities in Equation (9.3) are discussed in greater detail, and values for known parameters are tabulated in Reference 169. Using values of the electron-transfer bands of the known tetravalent actinide hexahalide complexes, Nugent et al.[169] have used this treatment to obtain values of χ_{uncorr} (M) for all the tetravalent actinides and these are given in Table 9.3.

Barnes and Day[173] have shown linear correlations between the energy of the first electron-transfer bands of various trivalent lanthanide complexes and the standard (II–III) lanthanide potentials. Miles[174] made similar plots for hexavalent, U, Np, Pu, and Am and for tetravalent U, Np, and Pu in 1M $HClO_4$ using the wave number on the u.v. cut-off where $\varepsilon = 50$ for the hexavalent and $\varepsilon = 400$ for the tetravalent actinides. The lack of fit and deviation from the expected slope for his values for the tetravalent actinides is due to the cut-off for the U^{IV} aquo ion, unlike that for Np^{IV} and Pu^{IV}, being caused by 5f→6d-transitions instead of electron transfer[15]. He used this plot to predict a value of the thorium (III–IV) potential based on the very weak absorption ($A = 0.07$) at 195 nm of a 10^{-3} M Th solution. Considering the difficulty of measurement at this wavelength, the result is very questionable.

Nugent et al.[171] have published a preliminary report in which they made a linear unit slope correlation of standard (II–III) potentials of the lanthanides with the energies (in electron volts) of the first electron-transfer bands of the trivalent lanthanide bromides in ethanol. They assumed a similar unit slope relationship for the trivalent actinides which allowed them to calculate the standard (II–III) potentials of Am, Cm, Bk, Cf, Es, and Fm using the measured potentials of Md and No. The results agreed well with the measured difference between the first electron-transfer bands of Cf and Es bromides in ethanol.

A linear unit slope correlation between the energy (in electron volts) of electron-transfer bands of the tetravalent hexahalide complexes given in Table 9.2 and the standard (III–IV) potentials has been demonstrated by Nugent et al.[169]. The values of the (III–IV) potentials of Am and Cm were determined from the electron-transfer spectra of Am^{IV}-hydrous oxide and of Cm^{IV} in 15 M CsF already mentioned, and corrected to those of the hexachlorides. The value for Am, 2.1 V, is thought[169] to be much closer to the true value than that of 2.8 V previously reported[175]. The (III–IV) potentials of all the actinides were calculated[169] in the same way as the electronegativities using Jørgensen's treatment (Equation (9.3)), and the linear relationship. Good agreement between measured (in the cases available) and calculated results was obtained. The lanthanides were treated similarly, and the results are shown in Figure 9.5. The fundamental assumption of this treatment is that the relative bond strength of the tetravalent and trivalent actinide or lanthanide hexahalide complexes changes in going across the series in the same manner as the relative bond strength of the tetravalent and trivalent metal aquo complexes, which are of course the basis for the potentials. The validity of the potential values in Figure 9.5 are, of course, subject to the limitations of this assumption and others made[169] in the determination of the parameters used in Equation (9.3). The limitations involved in making such a correlation have also been discussed by Jørgensen (Reference 40, pp. 329–330). Since the variation in the chemical properties of an ion in a particular oxidation state are small across an f-group series, this assumption should hold reasonably well for such series. The close fit to the unit slope line indicates this to be essentially true, but a 'calibration shift' is observed between the actinides and lanthanides[169, 171]. Care must be exercised in making such correlations however, and the direct correlation of electron-transfer band energy and potential between elements in different oxidation states and between different transition-group series is not valid.

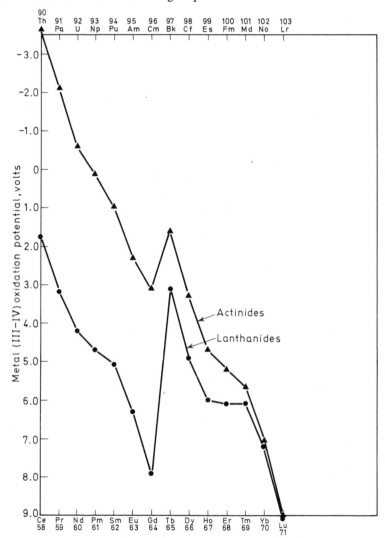

Figure 9.5 Metal (III–IV) oxidation potentials v atomic number for the lanthanides and actinides. (From Nugent, Baybarz, Burnett and Ryan[169] by courtesy of Pergamon Press Ltd.)

The greater overall variation of the actinide (III–IV) potentials (Figure 9.5) relative to the lanthanides is due to a 1.7-fold larger value[169] of the parameter $(E - A)$, which accounts for the increase in the effective nuclear charge felt by each f-electron with increase in Z or q. Similarly, the values for the trivalent actinides[171] are larger than those for the trivalent lanthanides[157, 176].

This difference in $(E-A)$, less effective mutual screening of the 5f- than 4f-electrons from the increasing nuclear charge[176], is then the reason for the greater oxidation potential variability in the actinides[176, 177]. The smaller interelectronic repulsion parameters (smaller effect of spin-pairing) for the actinides as compared with the lanthanides[169] causes the half-filled shell effect to be smaller for the actinides than for the lanthanides (Figure 9.5), as predicted by Jørgensen[157]. It is interesting to note that values of $(E-A)$ for both the actinides and lanthanides[169, 176] are still much less than those[176] for the 3d- and 4d-group elements, indicating that the d-electrons are less effective in mutual screening than the 5f- and 4f-electrons[176]. Other reports[178, 179] propose less effective mutual shielding by f- than d-electrons in relation to the lanthanide contraction. These are based on the statement that the lanthanide contraction is more spectacular than the contractions which occur in the d-block transition series[178], whereas actually the contraction per unit increase in Z is much greater for the d-group than for the f-group series, in agreement with the spectral data. In fact Hf would be much smaller than Zr if the lanthanide contraction per unit increase in Z was as large as in any of the d-groups (see Figure 4–3 of Reference 180).

Parameters occurring in Jørgensen's refined spin-pairing energy treatment of electron-transfer spectra, terms E^1 and E^3 of Equation (9.3), have been successfully used[181, 182] to explain the tetrad effect[183, 184]. The tetrad effect is the cyclic variation in various chemical properties of actinide and lanthanide ions, such as solvent extraction and ion-exchange distribution coefficients, with minima (or maxima) occurring at approximately the $\frac{1}{4}$, $\frac{1}{2}$, and $\frac{3}{4}$ filled shells.

9.4.3   Charge-separation effects in actinide complexes

Jørgensen[185] has discussed the reasons for a *charge-separation effect* in electron-transfer spectra, and has expanded this idea to the concept of *anisotropic* complexes[172]. This effect is important for the explanation of the electron-transfer spectra of many actinide and lanthanide complexes and should be of interest to actinide chemists. Basically the charge-separation effect is the increase in photon energy required for electron transfer with increase in metal–ligand distance. This is caused by the fact that the hole created in the molecular orbital, located mainly on the ligands, by the optical transition attracts the excited electron and the greater the distance involved in the transition the greater will be the total energy required to overcome this attraction. This causes the energy of the electron-transfer bands to depend to a varying extent on whether more than one ligand is attached to the metal with different bonding strengths (anisotropic complexes) and on coordination number. An outstanding example of an anisotropic complex is $UO_2Cl_4^{2-}$, an ion with very strong metal–oxygen bonds in which the U—Cl distances are abnormally long, the metal–halogen electron-transfer transitions being shifted drastically to higher wave number compared with those for hexahalides[15, 136, 172]. Jørgensen[172] indicates that $\sigma_1-\sigma_2$ is about $1.15 \times 10^{-3} \left(\frac{1}{r2} - \frac{1}{r1}\right)$ where σ_1 and σ_2 are the band energies and r_1 and r_2 are the distances between the hole and the excited electron for two complexes at the same metal ligand. He has also applied the charge-separation concept to the even longer distances involved in second-sphere complexes[172].

It should be apparent from this that in many cases the position of an electron-transfer band can be used to qualitatively (if $\chi_{uncorr}(M)$ is known) predict whether a ligand is actually in the metal coordination sphere, to estimate relative ligand bond strengths, and to make some estimate of coordination numbers. Thus, the presumably monochloro-[173, 186] and mono-bromo-complexes[157] of the trivalent lanthanides, with electron-transfer bands at 3–6 kK higher energy than the respective MX_6^{3-} [28], probably have co-ordination numbers > 6. The $UO_2X_4^{2-}$ (and other $MO_2X_4^{2-}$) have electron-transfer bands due to the X ligands at almost 20 kK lower energies than would be expected from the data for UCl_6 (Table 9.2) resulting in a value of 1.8 for χ_{uncorr} for U^{VI} in the uranyl ion[28, 136].

The first electron-transfer band for the $UOBr_5^{2-}$ complex is at 9.3 kK higher energy[7] than that for UBr_6^-. This results in $\chi_{\pi ncorr}(M) = 1.9$ in $UOBr_5^{2-}$ against 2.2 for U^V in UX_6^-, a result due to the abnormally weak U—X bonds in the anisotropic UOX_5^{2-} complexes, as also confirmed by an approximately 15% decrease in frequency of the highest energy strong U—X peak in the far infrared in going from UX_6^- to UOX_5^{2-} (see Section 9.6). The decrease in $\chi_{uncorr}(M)$ of 0.3 units between UX_6^- and UOX_5^{2-}, compared to the decrease of 0.6 units between UCl_6 and $UO_2Cl_4^{2-}$, indicates that the U—O bond is much weaker in UOX_5^{2-} than in $UO_2X_4^{2-}$. This is confirmed by the fact that UOX_5^{2-} is readily de-oxygenated by HX in non-aqueous solvents whereas $UO_2X_4^{2-}$ is not. The fact that $\chi_{uncorr}(M)$ in $UOBr_5^{2-}$ is 1.9 against values of 1.8 for both U^{VI} in $UO_2Br_4^{2-}$ and U^{IV} in UBr_6^{2-} indicates that $UOBr_5^{2-}$ is unstable towards disproportionation, as is observed[7]. The colours of most of the known UX_5L ($X = Cl^-$ or Br^-) addition-compounds are[18, 187] very similar to those of the corresponding UX_6^- compound indicating that the ligand, L, is not bound much more strongly than the halogens in any of these cases (except when $L = O^{2-}$ just discussed and possibly a few other[187] somewhat less well-defined compounds).

Gainar and Sykes[65] have attributed the u.v. cut-off of Np^{IV} in 1 M $HClO_4$ to an electron-transfer transition from the water ligand at 48 kK. Sykes and Taylor[66, 188] have observed a decrease in the wave number of this cut-off with increasing Cl^- concentration and attributed this to electron transfer. This writer has examined more dilute solutions of Np^{IV} in aqueous HCl, and no maxima or distinct shoulders below ~ 50 kK were observed. This makes it somewhat difficult to say with certainty that 5f→6d-bands are not involved, but the cut-offs are not sharp indicating broad bands. The large shift, > 8 kK, which occurs between the first band of $NpCl_6^{2-}$ and $NpCl^{3+}$ in water, together with the lack of change of the internal f-electron spectrum of $NpCl^{3+}$ from that of the aquo ion, strongly supports the conclusion[66] that $NpCl^{3+}$ is a second-sphere complex. The lack of discrete peaks in the spectrum probably also supports this since the Np—H_2O—Cl bond will be less rigid than the Np—Cl bond.

Np^{VII}[189, 190] and Pu^{VII}[191] have intense absorption bands in the visible region and these are undoubtedly due to electron-transfer bands, as also pointed out by Jørgensen[176], who concluded from the high energy of the strong bands for M^{VII} in alkaline solutions that the MO_5^{3-} species was highly anisotropic and contained the strongly bonded MO_2^{3+}-entity. This conclusion has also been reached by Krot et al. from other evidence[190]. Actually the lowest energy transition for a NpO_2^+ species might be expected to be a weak oxygen-to-metal electron-transfer band, corresponding to the well-known UO_2^{2+}-transition at about 420 nm already discussed. The published spectrum of Np^{VII} in alkaline solution[189] indicates that there may be a band at 925 nm (10.8 kK) with $\varepsilon \sim 20$, and the spectrum of NpO_2^{3+} in acid solutions[190] indicates appreciable absorption above 700 nm. The 10.8 kK value in combination with $\chi_{opt}(O^{2-}) = 3.15$ would give a reasonable value of $\chi_{uncorr} = 2.8$ for Np^{VII}. Further spectral measurements of more concentrated Np^{VIII} solutions should be made to determine if this band exists.

9.5 5f$^q \to$ 5f^{q-1} 6d-TRANSITIONS

9.5.1 Effect of variation of atomic number

Jørgensen[15, 157] has discussed the theory of f→d-transitions and has pointed out that they can be treated in the same manner (using Equation (9.3)) as electron-transfer spectra. He has[157] treated the 4f→5d-transitions of the lanthanide tribromides in ethanol in this manner. The energies of the first 4f→5d-transitions of the lanthanides in CaF_2 were found to agree very well with those calculated using Equation (9.3), being 18 kK below those for the free ions[192]. Jørgensen[136, 157] has also commented that the term $(E-A)$ in Equation (9.3) is somewhat larger for f→d than for electron-transfer bands, but the difference is small (~ 0.9 kK) for trivalent lanthanides[157]. Because of this, $nf^q \to nf^{q-1} (n+1)$ d actinide- and lanthanide-transitions show almost exactly the opposite variation with Z or q as the electron-transfer $nf^{q-1} \to \lambda^{-1} nf^q$ transitions, and for a constant chemical species, the f→d-bands behave much like inverted electron-transfer bands[136]. These transitions have, in fact, been interpreted[174] as being due to inverted electron transfer. In view of the width of these strong bands compared to electron-transfer bands (as discussed in the previous section) and the fact that f→d-transitions are also seen in the gaseous ions[193], shifted in accordance with the nephelauxetic effect[5], this seems unlikely. The shift from the free ion values is of a very reasonable magnitude considering the effects of the metal ion environment on the f→d-band energies, as discussed in 9.5.2. One can attempt to make a rough estimate of the magnitude expected for inverted electron transfer. The effective value of the optical electronegativities of the first acceptor orbitals of the halides, $\chi_{accept} (X^-)$, can be determined using Equation (9.2) where σ_{obs} is the lowest energy absorption band of the halide ion (Reference 40, p.284) and $\chi_{uncorr} (M)$ in the equation becomes $\chi_{accept} (X^-)$. Using values so obtained and converting Equation (9.2) to: $\sigma_{calc} (MX_6^{-n}) = 30$ kK $[\chi_{uncorr} (M') - \chi_{accept} (X^-)]$ where $\chi_{uncorr} (M')$ is the value obtained from normal electron transfer (Table 9.2) for the metal in one oxidation state higher than that for (M), one obtains values for $\sigma_{calc} (MX_6^{-n})$ in fair agreement with those measured for the lowest energy 5f→6d-bands of the actinide hexachloro- and hexabromo-complexes. The fit is poor for UI_6^{2-}, $CeCl_6^{3-}$, and $CeBr_6^{3-}$, and the calculated variation with halide is distinctly the reverse of that observed. A further very strong argument against this interpretation was kindly pointed out to the author by Dr. C. K. Jørgensen. This is that acetylacetonate, dipyridyl, phenanthroline, picolinate, pyridine-N-oxide, and cyclopentadienide have absorption bands similar to, and at much lower energy than, halides, yet their complexes with 4f- and 5f-group elements do not show inverted electron-transfer bands at appreciably lower energy than the halide f→d bands. If the calculation just described is carried out for acetylacetonate using the reported[136] χ_{opt} (2.7) and applied to U^{IV}, the result indicates that $U(acac)_4$ would have the most intense allowed transitions throughout the visible region, whereas the compound is actually known[194, 195] to be the green colour typical of U^{IV} compounds having allowed transitions only in the u.v. Thus, it appears that these bands are definitely due to f→d-transitions rather than to inverse electron transfer, but it should be realised from the extent of d-orbital participation in bonding that these transitions in compounds and complexes are not pure f→d-transitions in quite the same context that they are in the gaseous ions.

Miles[174] plotted the energy of the cut-off ($\varepsilon = 500$) in the u.v. spectrum of trivalent U, Np, Pu, and Am aquo ions against potential, and obtained a reasonably good straight line. His slope varied somewhat from that predicted by equivalence in band energy and oxidation potential. The fact that the slope was not that predicted appears to be at least partially due to the use of 2.44 V for the (III–IV) oxidation potential of Am which now appears to be in error[169]. Miles predicted 2300 Å for the u.v. edge ($\varepsilon = 500$) for the Bk^{III} aquo ion. How he arrived at this value, which is the same as that later determined experimentally[169], is not clear since his plot distinctly gives a value of 2470 Å. Nugent et al.[169] have re-plotted these data, excluding Am, and using the measured value for Bk^{III}, and get a better fit to a line of unit slope (where the band energy, hc/λ, is in volts). This yields a value for the Americium (III–IV) potential of 2.0 V, in fair agreement with that obtained from Am^{IV} electron-transfer spectra and with that calculated using Equation (9.3) (see Figure 9.5). The value obtained for the curium (III–IV) potential from this plot (3.2 V) was also in agreement with that obtained from electron-transfer spectra. The reason for the use of the frequency where $\varepsilon = 500$ for this plot instead of the frequency of the first transition is that the u.v. spectra of the trivalent U, Np, and Pu aquo ions[196] are not sufficiently resolved, and that the value for Cm^{197} has to be extrapolated somewhat beyond the measurable range. Because the intensity of the first transition is not the same for each element, some error will be introduced, and possible variation in hydration number across the series would introduce further uncertainty. A similar unit slope plot was made[169] for the u.v. cut-offs due to 4f→5d-transitions for the trivalent lanthanides; use of the energy of the first f→d-transitions of the actinide and lanthanide tribromides and the lanthanide trichlorides in ethanol gives unit slope lines parallel to those for the aquo ions[169].

The fact that the trivalent actinide and lanthanide f→d-transitions follow the potentials in a manner analogous to the tetravalent actinide and lanthanide electron-transfer bands requires, for a specific complex, that the separation between the highest energy level of the ligand and the normally vacant d-level of the metal cation remain constant across each series[169]. This does not mean that the d-level energy must remain constant across the series, but in fact, it is generally concluded[193] (also p.465, Reference 4) that the variation of d-level energy will be much less than the variation of the f-level energy. Thus, at least a major portion of the variation of the f→d-transition energies is due, as in the electron-transfer spectra, to variation of the energy of the f-orbital ground state. It should be emphasised that Equation (9.3) basically treats only the energy variation of the f ground state as a function of q and Z. For electron-transfer transitions in a given chemical species, the donor orbital energy is essentially constant, and the variation of electron transfer band energy with q is essentially due entirely to variation of f ground-state energy. The transitions treated here will show the behaviour of inverse electron-transfer bands with variation in q, regardless of whether they are transitions to a relatively constant energy d-level or to a relatively constant energy ligand-acceptor orbital.

Divalent Am has recently been prepared[117, 198] both by radiolytic and electrolytic reduction in crystalline CaF_2. The spectrum exhibits f→d-bands across the entire visible region as would be expected for this highly reducing species.

9.5.2 Halide complexes

As in the case of the electron-transfer spectra, it appears that the hexahalide complexes offer certain distinct advantages over the aquo ions for the study of the systematics of the f→d-transitions. These transitions are moved to lower energy by as much as 9–23 kK compared with the aquo ions and thus in many cases are more accessible to measurement; the spectra also often appear to be sharper and better resolved in the hexahalides. With the aquo ions, it is uncertain whether constant complex species are being considered, and, in fact, the coordination number of the trivalent lanthanide does vary across the series [199-207].

Measured values of hexahalide complex 5f→6d-transitions are given in Table 9.4. $PaCl_6^{2-}$, $PaBr_6^{2-}$ and PaI_6^{2-} complexes have been prepared[205], and although the spectra have apparently not been measured, it is expected that the lowest energy allowed transitions will be 5f→6d. $PuCl_6^{3-}$, $PuBr_6^{3-}$, $AmCl_6^{3-}$, and $AmBr_6^{3-}$ have also been prepared[26]; of these only $AmBr_6^{3-}$ has been measured in the u.v.[170], but the u.v. cut-off of $PuCl_6^{3-}$ is shifted ~ 7 kK to lower energy from that of Pu^{III} in 1 M $HClO_4$[26]. Although $CmCl_6^{3-}$ is known[27], its u.v. spectrum has not been measured. All these actinide MX_6^{3-} compounds are expected to have 5f→6d-transitions as their lowest energy allowed bands and several of them are being measured in work currently under way[206]. Reisfeld and Crosby measured the u.v. spectrum of the UF_6^- ion and proposed that the bands observed were due to 5f→6d-transitions[13]; recent measurements of the electron-transfer spectra of UCl_6^- and UBr_6^- confirm this assignment[7]. The data for UCl_6^{2-}, UBr_6^{2-}, UI_6^{2-}, $NpCl_6^{2-}$, $NpBr_6^{2-}$, $PuCl_6^{2-}$, and $PuBr_6^{2-}$ are from Ryan and Jørgensen[15], and these transitions, along with the electron-transfer bands, are shown in Figure 9.4. The first strong transitions of the UX_6^{2-} are shifted by about

17 kK for UCl_6^{2-} and about 23 kK for UI_6^{2-} from the first transition for the aquo ion at approximately 48 kK[196, 207]. In addition to the hexahalides in Table 9.4, Morrey has found the first intense 5f→6d-transition at 15.6 kK for UCl_3 in a frozen CsCl melt which appears to contain the UCl_6^{3-} ion[208].

Bagnall and Brown[209] have observed a strong peak for Pa^{IV} in 12 M HCl at 24.3 kK which is not present in more dilute HCl. Since the rest of the spectrum changes very little from that in dilute HCl, and decreases only slightly in intensity with increase in HCl concentration, it appears that the lowest energy f→d-transition for the Pa^{IV} chloro complex formed in concentrated HCl is quite intense and is probably widely separated from the next transition, as in the isoelectronic $CeCl_6^{3-}$ [28]. Jørgensen[136] has noted this shift from about 35 kK for the first transition of other Pa^{IV} complexes, including the 36 kK first transition for Pa^{IV} in $HClO_4$[210], but not that for $Rb_7Pa_6F_{31}$ of 28.6 kK[75], and has mildly implied that the 24.3 kK peak is due to $PaCl_6^{2-}$.

Table 9.4 **5f^q → 5f^{q-1}6d-Transitions of actinide hexahalides, including those expected to show 5f^q → 5f^{q-1}6d-transitions***

	$PaCl_6^{2-}$ (yellow)	$PaBr_6^{2-}$ (orange-red)	PaI_6^{2-} (dark blue)
$5f^1$			
UF_6^- 36.2, 40.0, 44.2			
$5f^2$	UCl_6^{2-} (29.8), 31.8, 33.0, 35.7, 40.0	UBr_6^{2-} (27.4), 29.4, 30.4, (31.5), 33.1, 37.0?, (39.0?)	UI_6^{2-} 24.5, 24.8, 25.9, 27.9
$5f^3$	$NpCl_6^{2-}$ 36.9, 39.1, 41.2, 42.5, 44.0, 45.6	$NpBr_6^{2-}$ 33.4, 35.6, 37.6, 38.6, 40.4, 41.4, 42.2	
$5f^4$	$PuCl_6^{2-}$ 40.9, 42.1, 43.7, 45.9	$PuBr_6^{2-}$ 36.4, 37.6, 38.8, 40.4, 40.9, 41.7, (43.5)	
$5f^5$	$PuCl_6^{3-}$	$PuBr_6^{3-}$	
$5f^6$	$AmCl_6^{3-}$	$AmBr_6^{3-}$ (37.0), 39.65, 42.1	
$5f^7$	$CmCl_6^{3-}$		

*Wave numbers in kK with shoulders in parenthesis. Hexahalides without transitions listed have been prepared, but have not been measured, and the lowest energy Laporte-allowed bands are expected to be due to 5f→6d-transitions. References in text.

This is somewhat doubtful since it is known that concentrated HCl solutions of U^{IV} contain practically no UCl_6^{2-} (in fact, the f→f spectra[211] indicate a coordination number > 6) and chloro complexing appears to increase as Z increases for tetravalent actinides[211]. This writer has found that the first strong transition for U^{IV} in 12 M HCl is 3 kK above the first strong (neglecting the relatively very weak shoulder) transition for UCl_6^{2-}. Thus one might expect (taking the colour into consideration also) that the first transition for $PaCl_6^{2-}$ might be closer to 22 kK. The orange-red and dark blue colours of $PaBr_6^{2-}$ and PaI_6^{2-} are readily explained by shifting a single transition across the visible region by the amounts shown in Table 9.3 for the first strong 5f→6d-transitions for UCl_6^{2-}, UBr_6^{2-}, and UI_6^{2-}. The apparent shift of 12–14 kK between the first 5f→6d-transitions of $PaCl_6^{2-}$ and the Pa^{IV} aquo ion is somewhat less than the 17 kK in the case of U^{IV}, and measurement of the spectra of the PaX_6^{2-} ions in acetonitrile would be of value both to the study of the systematics of the tetravalent actinide $5f \rightarrow 6d$-transitions and to the further understanding of the extent of chloride-complexing of the tetravalent actinides. Poturaj-Gutniak and Taube[161] have measured the u.v. spectrum of U^{IV} in methanol–HCl, and at high HCl concentration complete conversion to UCl_6^{2-} occurs. At much lower HCl concentration, the spectrum (curve A, Figure 1 of their paper) is identical to that observed by this writer for U^{IV} in 12 M aqueous HCl, confirming the weaker coordinating power of methanol as compared with water.

9.5.3   Effects of metal ion environment

Although the actinide and lanthanide $f \rightarrow d$-transitions do not appear to be as well understood in this regard as the electron-transfer spectra, the $f \rightarrow d$-transitions appear to be qualitatively, but not quantitatively, affected in the same manner by variation in ligand electronegativity and bond strength (coordination number, etc.) as the electron-transfer transitions. Thus, $f \rightarrow d$-transitions are shifted to lower energy with ligand in the order H_2O, Cl^-, Br^-, I^-. For a constant known coordination number this shift with ligand is much less (Table 9.4) than that observed in the case of electron-transfer bands[15, 28] (see Table 9.2), and is in the reverse direction, in the case of halides, to that expected for inverted electron transfer. The energy of the $f \rightarrow d$-transitions does, however, appear to be as strongly or more strongly affected by bond length and/or coordination number as are the electron-transfer band energies. Thus the first $f \rightarrow d$-bands[157, 169] of Am, Ce, and Tb tribromides in ethanol (symmetry unknown) are shifted about 2–3 kK to lower energy from those of $AmBr_6^{3-}$ (Table 9.4), $CeBr_6^{3-}$ and $TbBr_6^{3-}$[28] respectively. Similar shifts occur between Ce, Tb, and Pr chlorides in ethanol[169] and $CeCl_6^{3-}$, $TbCl_6^{3-}$[28] and $PrCl_6^{3-}$[199]. These shifts are about the same as the shifts in the first electron-transfer transitions of trivalent Sm, Yb, Eu, and Tm halides in ethanol and the respective hexahalides[28].

Large differences in the $f \rightarrow d$-transition energies of hexahalides and aquo ions have already been mentioned; since the actinide and lanthanide aquo ions apparently always have coordination numbers > 6, this effect is due to both change in ligand and coordination number. Jørgensen and Brinen[212] studied the u.v. spectrum of Ce^{III} in 0.1 M $HClO_4$ and concluded that since six bands are observed, where only five $f \rightarrow d$-transitions are possible for a single chemical species, the first (temperature-sensitive) weak band is due to a small amount of a Ce^{III} aquo ion of different, and probably lower, coordination number. Spectral work currently in progress[199] confirms this, indicates that the first weak $f \rightarrow d$-transition for aqueous Pr is caused by a similar effect, and indicates, in agreement with physical measurements of the bulk solution properties by Spedding et al.[200] (see also References 201–204), that the Tb^{III} aquo ion has a different coordination number from that for the principal Ce and Pr species. Similar studies are under way with the trivalent actinides[199]. Thus the 6 kK difference[212] between the first weak and first strong bands of Ce^{III} in $HClO_4$ represents a shift caused by change in coordination number. The principal Ce species is probably[213, 214] $Ce(H_2O)_9^{3+}$, and it would thus be expected that the other species would be $Ce(H_2O)_8^{3+}$, but the fact that this is not cubic with a centre of symmetry (or octahedral 6-coordinate) is demonstrated by the fact that the conversion is complete with the heaviest lanthanides[199-204], the intensities of the internal f→f spectra of which indicate that their aquo ions are not centrosymmetric and have a coordination number >6[28].

The energy difference between the first f→d-transitions at 39.7 kK for $Ce(H_2O)_x^{3+}$ and at 33.7 kK $Ce(H_2O)_{x-n}^{3+}$ where x is probably 9[213, 214] and $x-n>6$, is 6 kK; between $Ce(H_2O)_{x-n}^{3+}$ and $CeCl_6^{3-}$ (at 30.3 kK)[28], where there must be a further decrease in coordination number, the difference is only 3.4 kK. In concentrated HCl, Ce[III], presumably as a mixed chloro–aquo complex of coordination number eight as concluded for Nd[III] [214], has its first band at 32.5 kK [215], only 1.2 kK from that for $Ce(H_2O)_{x-n}^{3+}$. The first band for cerium trichloride in ethanol is at 33.0 kK [169], only 0.7 kK from that of $Ce(H_2O)_{x-n}^{3+}$. This, and the shifts of 2–3 kK between trichlorides or tribromides in ethanol and the hexahalides, compared to only about 1 kK between the MCl_6^{3-} and MBr_6^{3-}, makes it apparent that coordination number is the most important factor influencing the position of f→d-transitions for a given lanthanide or actinide of constant valence but varying environment.

This hypothesis is also supported by the somewhat more limited data for tetravalent uranium discussed earlier, for which the energy differences are similar, but larger, than for the trivalent lanthanides and actinides. A shift of 2.4 kK between UBr_6^{2-} and UCl_6^{2-} (Table 9.4) may be compared to 3 kK between UCl_6^{2-} and the U[IV] chloro complexes of N>6 in aq. 12 M HCl, and to 17 kK between UCl_6^{2-} and the aquo ion. The spectrum of UCl_4 in acetonitrile differs from that of UCl_6^{2-}, but the 5f→6d-transitions start at about the same energy. Although the coordination number is probably eight[62, 63], it is known that acetonitrile is very weakly coordinating towards actinides[15, 27, 28, 63] and the complex would be expected to be anisotropic with strong U—Cl bands. This is confirmed by infrared measurements[216] in which it was found that the U—Cl stretching frequencies are very slightly higher for $UCl_4.4CH_3CN$ than for the 6-coordinate UCl_6^{2-} salts. The rather poorly resolved spectrum of $UCl_4.2TBP$[161] also appears to have the first 5f→6d-transitions (the assignment[161] as electron-transfer transitions is probably incorrect) at the same energy as in UCl_6^{2-}.

The energy shift of from about 6% for U to 10% for Pu in going from the MCl_6^{2-} to the MBr_6^{2-} (Table 9.4) is about two- to three-times the corresponding shifts in the internal f-electron transitions[15]. The 4f→5d-band shifts between the lanthanide MCl_6^{3-} and MBr_6^{3-} also appear to be approximately two- to three-times as large as for the internal f-electron transitions[28]. If the f→d-energy variation is attributed to a nephelauxetic effect, the apparent effect of coordination number on f→d-transition energies can be understood in terms of a pronounced increase in nephelauxetic effect (or increase in covalency) with decrease in coordination number and thus also bond distances. It has been noted in the case of 4f→4f-transitions that variation in M—Cl bond distances can produce up to fourfold variation in the nephelauxetic parameter, $d\beta$; because of the pronounced variability of coordination number in the lanthanides (and actinides), a unique nephelauxetic series of ligands cannot be established as can be done for the d-group series[28]. It is not surprising that the effect on the f→d-transitions is larger than that on the f→f in view of the greater effect of bonding on the d-orbitals.

Pa^{IV} in 15 M NH_4F has the first 5f→6d-transition at 28.5 kK[217], compared with 36.0 kK for the Pa^{IV} aquo ion and 24.3 for the Pa^{IV} species in 12 M HCl. It seems likely that the species present in such a solution is PaF_8^{4-} since Penneman et al.[218] have reported that solutions of U^{IV} in >10 M NH_4F have internal f^1 spectra very similar to that of $(NH_4)_4 UF_8$, now known[219] to contain discrete UF_8^{4-} units as distorted tetragonal antiprisms. Since neither the nephelauxetic effect (or the electronegativity) of F^- are expected to be greater than that of H_2O[33], it appears that the 7.5 kK decrease (comparable to the difference in the two Ce^{III} aquo ions) is due mainly to a decrease in coordination number on forming the fluoro complex. It is tempting to predict on this basis that the Pa^{IV} aquo ion has a coordination number greater than eight.

Shiloh, Marcus and Givon have measured the 5f→6d-transitions of U^{III}[220], Np^{III}, Pu^{III}[114], and Am^{III}[67, 68, 221] in concentrated LiCl and LiBr solutions. They find, in each case, that intense 5f→6d-bands grow in with increasing halide concentration at lower energy than those present in dilute halide solution. They attribute these to formation of inner-sphere halide complexes. The positions of these transitions in LiBr are only about 0.3 kK below those in LiCl and appear to be about the same as those observed for the tribromides in ethanol[169], but some of them have shoulders at somewhat lower energy. That these bands are due to species having coordination number >6 is apparent from the intensities of the f→f spectra. On the basis of stability constant calculations, they concluded that MX^{2+} and MX_2^+, where X = Cl and Br, are both formed, and that the latter are primarily responsible for the strong f→d-absorption. They neglected water activity in these very concentrated LiX solutions and water must be involved in the equilibrium constant expressions to at least as high a power as the halide activity, due to its replacement in the coordination sphere (and quite possibly to a higher power due to a possible simultaneous decrease in coordination number). As a result, it is difficult to see how such stability constants have any meaning. If water activity is included, it appears that the data would fit a loss of two water molecules and addition of one Cl^- to the coordination sphere for a net coordination number decrease of one. The first transition of U^{III} in concentrated LiCl is at 18.3 kK[114, 220] compared to that of the aquo ion at about 25 kK[5]. The first strong transition for U^{III} in frozen KCl[208], or in LiCl—KCl eutectic[222], is also at 18 kK, where the f→f spectra indicate that the symmetry is not octahedral and the coordination number is undoubtedly greater than six, but where the number of chloride ions in the first coordination sphere no doubt differs from that in concentrated LiCl. In CsCl, where the f→f-transition intensities indicate the presence of UCl_6^{3-}, the energy of the first strong transition is decreased to 15.6 kK, and the energy increases with increase in temperature[208] and thus bond length. In frozen NaCl the first U^{III} f→d-transition is at about 21.5 kK and the f→f-transition intensities indicate low symmetry[208]. Thus, it definitely appears that the positions of the f→d-bands are influenced more by bond length and the coordination number than by the nature of the ligand. Further studies of these transitions with species of known structure are needed.

9.6 INFRARED AND RAMAN SPECTRA

Since metal–ligand vibrational frequencies are by no means unique to actinide elements, and since several current books and review articles[48, 223–225] treat the subject in detail, the theory is not discussed here, except to re-emphasise that many of the metal–ligand vibrational frequencies couple to many of the electronic transitions responsible for other types of actinide spectra. These are notably the internal f^q-transition spectra (particularly when there is a centre of symmetry and the pure electronic transitions are very weak or missing) and the uranyl (actinyl) ion spectra. As a result the far infrared and Raman spectra can be very helpful in understanding the internal $5f^q$ spectra, particularly of hexahalides, and the spectra of uranyl compounds. In addition, the number of transitions observed can often be used directly to determine symmetry, but in most reports of actinide ion–ligand frequencies, infrared and Raman data are not available over a sufficiently complete range. In fact, only the highest energy infrared-allowed metal–ligand frequency is usually available for actinide compounds.

Most of the actinide ion–ligand vibrational frequencies measured have been for halide ligand or, in the case of the actinyl ions, oxide ligand. The infrared and Raman spectra of the actinyl ions have been reviewed[48, 135, 224], as have the metal–halogen and metal–oxygen vibrational spectra of halide and oxo- and dioxo-halide compounds[226, 227]. Only the more recent work will be covered here.

Claasen et al.[228] have recently measured the Raman spectra of gaseous UF_6, obtaining $v_1 = 654$ cm^{-1}, $v_2 = 532.5$ cm^{-1}, and $v_5 = 202$ cm^{-1}. They have also tabulated the best values available for all the fundamental modes of UF_6, NpF_6, and PuF_6. Gasner and Frlec[229] measured the Raman spectra of NpF_6, obtaining for the gas $v_1 = 654$ cm^{-1}, $v_2 = 535$ cm^{-1}, and $v_5 = 208$ cm^{-1}, and for the liquid v_1 (polarised) $= 651$ cm^{-1}, v_2 (depolarised) $= 524$ cm^{-1}, and v_5 (depolarised) $= 218$ cm^{-1}. Frlec and Claasen[230] measured the low frequency infrared-active mode, v_4, for UF_6, NpF_6, and PuF_6, the values being 186.2, 198.6, and 206.0 cm^{-1} respectively. Bougon[231] studied the infrared spectrum of gaseous UF_6, and both the infrared and Raman spectrum of solid UF_6, confirming O_h symmetry but indicating tetragonal distortion in the solid, in agreement with earlier x-ray and n.m.r. data.

Ryan[7] reported the infrared spectra of the U^V hexahalide complexes, UX_6^-. The values obtained for the salts were: $(C_6H_5)_4AsUF_6 : v_3 = 525$ cm^{-1}, $v_4 = 173$ cm^{-1}; $(C_2H_5)_4NUCl_6 : v_3 = 310$ cm^{-1}, $v_4 = 122$ cm^{-1}; $(C_2H_5)_4NUBr_6 : v_3 = 214$ cm^{-1} and $v_4 = 87$ cm^{-1} (see also 9.2.3.1). Frlec and Hyman[79] reported $v_3 = 526$ cm^{-1} in $N_2H_6(UF_6)_2$ and Brown et al.[232] also reported $v_3 = 214$ cm^{-1} in Et_4NUBr_6. MacCordick et al.[233] have reported comparable values for UCl_6^-, and also report the Raman active modes, $v_1 = 343$ cm^{-1}, $v_2 = 277$ cm^{-1}, and $v_5 = 136$ cm^{-1} for UCl_6^-, in agreement with other, unpublished[234] data for several UCl_6^- salts. MacCordick et al.[233] also reported the Raman spectrum of $(NO)_3PaCl_8$, which is very similar to the spectrum of UCl_6^-, and this writer would seriously question whether the compound really contained the $PaCl_8^{3-}$ ion as proposed (see also comments regarding the existence of UCl_8^{3-} in Reference 7). The short wavelength portion of the infrared spectrum of $N_2H_6UF_7$ (no UF_6^- ion present) exhibits[79] a very strong band at 435 cm^{-1} with weak bands at 590 and 800 cm^{-1}.

Ryan[7] has reported the infrared spectra of UOF_5^{2-}, $UOCl_5^{2-}$ and $UOBr_5^{2-}$ in which the U—O frequencies were at 760 and 853 cm^{-1} (UOF_5^{2-}), 813 and 913 cm^{-1} ($UOCl_5^{2-}$), and 817 and 919 cm^{-1} ($UOBr_5^{2-}$). The U—Cl and U—Br frequencies were: $UOCl_5^{2-}$, 120 and 253 cm^{-1}, with weak side bands at 296 and 197 cm^{-1}, and $UOBr_5^{2-}$, 80 and 190 cm^{-1}, with a weak side band at 250 cm^{-1}. The lower U—X frequencies in these complexes, as compared with the UX_6^- complexes, was attributed to weakening of the U—X bonds due to the strong U—O bond in the MOX_5^{2-} (anisotropic complex effect as discussed in 9.4.3). Similarly, the decrease in the U—O frequencies in UOF_5^{2-}, compared to $UOCl_5^{2-}$ or $UOBr_5^{2-}$, was attributed to stronger U—F than U—Cl or U—Br bonds.

Woodward and Ware[235] have measured Raman and infrared spectra of UCl_6^{2-} and $ThCl_6^{2-}$. They did not measure v_4, and in the UCl_6^{2-} case obtained it and v_6 (Raman and infrared inactive) from the vibronic data reported by Satten et al. (see 9.2.2). The results were for UCl_6^{2-}: $v_1 = 299$ cm^{-1}, $v_2 = 237$ cm^{-1}, $v_3 = 262$ cm^{-1}, $v_4 = 114$ cm^{-1}, $v_5 = 121$ cm^{-1}, and $v_6 = 80$ cm^{-1}, and for $ThCl_6^{2-}$: $v_1 = 294$ cm^{-1}, $v_2 = 255$ cm^{-1}, $v_3 = 259$ cm^{-1}, and $v_5 = 114$ cm^{-1}. They calculated M—Cl bond-stretching force constants and compared them to those for $PbCl_6^{2-}$. The results indicated stronger bonding in the actinide hexachloro complexes. Satten and Stafsudd[49, 236] have reported v_3 (258 cm^{-1}) and v_4 (121 cm^{-1}) for $[(CH_3)_4N]_2UCl_6$, and Stafsudd[236] has obtained v_3 (195 cm^{-1}) and v_4 (84 cm^{-1}) for Cs_2UBr_6. Pandey et al.[237] have calculated force constants for UCl_6^{2-}.

Berringer et al.[238] have recently measured the infrared and Raman spectra of several $NpCl_6^{2-}$ salts, and find for $[(C_2H_5)_4N]_2NpCl_6$: $v_1 = 301$ cm^{-1}, $v_3 = 258$ cm^{-1}, $v_4 = 112$ cm^{-1}, and $v_5 = 123$ cm^{-1}. Krasser and Nürnberg[239] measured the infrared and Raman spectra of UF_4, and interpreted the results on the assumption of a tetrahedral structure. The U^{IV} atom in UF_4 is, in fact, 8-coordinate with the F$^-$ atoms arranged in a slightly distorted antiprism[240].

Kharitonov et al.[241] made a detailed infrared study of $K_3UO_2F_5$ including the effects of ^{18}O substitution. They calculated force constants for the uranyl ion, taking into account the effects of the F$^-$ ligands, and showed that the error caused by neglecting these effects is 8–9%. Other[242] spectral data and uranyl force constant calculations for the K$^+$ and NH_4^+ salts of $UO_2F_5^{3-}$ are also available. Ohwada[243] measured the symmetric and asymmetric uranyl frequencies in a variety of compounds, calculated uranyl ion force constants for these, and estimated force constants for the NpO_2^{2+}, PuO_2^{2+}, and AmO_2^{2+} analogues of the uranyl compounds. Vdovenko et al.[244] have discussed bond strength in actinyl ions in terms of bond multiplicities, and have recorded[245] the infrared spectra of various ^{18}O-substituted uranyl complexes.

Vdovenko et al.[145] and Sergienko and Davidovich[246] have observed regular decreases in the asymmetric uranyl frequency with increase in cation radius in salts of $UO_2Cl_4^{2-}$ and $UO_2F_5^{3-}$ respectively. This was attributed[145] to greater interaction of the small cations with the halide ligands. Thus, it was proposed that a decrease in U–halogen bond strength causes an increase in U—O bond strength, as exemplified by the asymmetric stretching frequency. Other second-sphere interactions with the chloride ligands in $UO_2Cl_4^{2-}$ produced similar effects[145]. Tertiary and quaternary ammonium salts of $UO_2F_5^{3-}$ and $UO_2F_3^-$ were also studied[247]. The higher asymmetric uranyl frequency and the lower U—F frequency in the tertiary ammonium salt of $UO_2F_5^{3-}$ were attributed to hydrogen bonding to the cation. Both bridging and terminal U—F frequencies were observed in $UO_2F_3^-$. It should be noted that hydrogen bonding was not found to produce an appreciable change in energy of the asymmetric uranyl frequency in $UO_2Cl_4^{2-}$, but the band was broadened considerably[84].

Belyaev et al.[248] and Vdovenko et al.[249] measured the far infrared spectra of several uranyl complexes and some of their results are shown in Table 9.5. The infrared results of Newbery[250], who also measured the Raman spectra, for $UO_2Cl_4^{2-}$ and $UO_2Br_4^{2-}$ agree reasonably well with the data

Table 9.5 Infrared frequencies in the range 75–500 cm^{-1} for anionic uranyl complexes*

Compound	ν_2 (UO_2^{2+}) (cm^{-1})	ν (U—L) (cm^{-1})	Unassigned (cm^{-1})
$Li_2[UO_2Cl_4]$	270 s, (260)	236 m broad	†
$Cs_2[UO_2Cl_4]$	(267), 259 s	(240)	138w, 111m
$[(C_2H_5)_4N]_2[UO_2Cl_4]$	259 s	238 m	125m, 100m
$(R_4N)_2[UO_2Cl_3Br]$	255 s	236 s, 165	125m, 85w
$(R_4N)_2[UO_2Cl_3I]$	261 s	236 m, 135	120m, 95w, 85w
$(R_4N)_2[UO_2Cl_3NO_3]$	258 s	233 s, (219)	†
$R_4N[UO_2Cl_3]$	270 s	243 s, 216 m	†
$(R_4N)_2[UO_2Br_4]$	251 s	160	
$R_4N[UO_2(ClO_4)_3]$	256 s	177	
$R_4N[UO_2(NO_3)_3]$	267 s	233 m, 219 m	†
$R_4N[UO_2(CH_3COO)_3]$	263 s	232 m, 217 m	†
$(NH_4)_4[UO_2(C_2O_4)_3]$	262 s	240 m, 211 m	†
$(NH_4)_4[UO_2(CO_3)_3]$	277 s, (295)	237 m, 214 m	†
$K_3[UO_2F_5]$	286 s	390 s, 370 s (232)‡, 220‡m	†

*Data from References 248 and 249; s = strong; m = medium; w = weak; shoulders in parenthesis; R = $C_{10}H_{21}$; R_4N^+ salts were as viscous liquids or solutions in CCl_4 or C_6H_6
†Not measured below 190 cm^{-1}.
‡Bands assigned to δ(OUF) and δ(FUF).

shown in Table 9.5. Bukalov et al.[251] have measured the Raman spectra of $UO_2Cl_4^{2-}$, $UO_2Br_4^{2-}$, $UO_2(NO_3)_3^-$, $UO_2F_5^{3-}$ and $UO_2(CO_3)_3^{4-}$ salts. Garg and Narasimham[252] have measured both the infrared and Raman spectra of $Zn[UO_2(C_2H_3O_2)_3]_2$, assigned the frequencies, and used the information in the analysis of the luminescence spectrum. Bullock[253] has studied the infrared spectra of several uranyl nitrate complexes and concluded that nitrate was bidentate in all of them; he found no correlation between the uranyl stretching frequencies and the position of the ligands in the spectrochemical series.

References

1. Seaborg, G. T. (1963). *Man Made Transuranium Elements*, Chap. 3. (Englewood Cliffs, N.J.: Prentice-Hall)
2. Seaborg, G. T., Katz, J. J. and Manning, W. M. (1949). *The Transuranium Elements*, *Nat. Nucl. Energy Series IV*, 14B. (New York: McGraw-Hill)
3. Seaborg, G. T. and Katz, J. J. (1954). *The Actinide Elements*, *Nat. Nucl. Energy Series IV*, 14A. (New York: McGraw-Hill)
4. Katz, J. J. and Seaborg, G. T. (1957). *The Chemistry of the Actinide Elements*. (London: Methuen & Co. Ltd.)
5. Jørgensen, C. K. (1963). *Advan. Chem. Phys.*, **5**, 33
6. Wybourne, B. G. (1965). *Spectroscopic Properties of Rare Earths*. (New York: Interscience Publishers)
7. Ryan, J. L. (1971). *J. Inorg. Nucl. Chem.*, **33**, 153
8. Carnall, W. T. and Wybourne, B. G. (1964). *J. Chem. Phys.*, **40**, 3428
9. Krupke, W. F. and Gruber, J. B. (1967). *J. Chem. Phys.*, **46**, 542
10. Axe, J. D. (1960). *U.S. At. Energy Comm.*, UCRL-9293
11. Satten, R. A., Young, D. J., and Gruen, D. M. (1960). *J. Chem. Phys.*, **33**, 1140
12. Satten, R. A., Schreiber, C. L., and Wong, E. Y. (1965). *J. Chem. Phys.*, **42**, 162
13. Reisfeld, M. J. and Grosby, G. A. (1965). *Inorg. Chem.*, **4**, 65
14. Pappalardo, R. and Jørgensen, C. K. (1964). *Helv. Phys. Acta.*, **37**, 79
15. Ryan, J. L. and Jørgensen, C. K. (1963). *Mol. Phys.*, **7**, 17
16. Eisenstein, J. C. and Pryce, M. H. L. (1960). *Proc. Roy. Soc. (A)*, **255**, 181
17. Selbin, J., Ortego, J. D. and Gritzner, G. (1968). *Inorg. Chem.*, **7**, 976
18. Selbin, J. and Ortego, J. D. (1969). *Chem. Rev.*, **69**, 657
19. Karraker, D. G. (1964). *Inorg. Chem.*, **3**, 1618
20. Conway, J. G. (1959). *J. Chem. Phys.*, **31**, 1002
21. Gruber, J. B. and Menzel, E. R. (1969). *J. Chem. Phys.*, **50**, 3772
22. Conway, J. G. (1964). *J. Chem. Phys.*, **41**, 904
23. McLaughlin, R. (1962). *J. Chem. Phys.* **36**, 2699
24. Clifton, J. R., Gruen, D. M., and Ron, A. (1969). *J. Chem. Phys.*, **51**, 224
25. Sharma, K. K. and Artman, J. O. (1969). *J. Chem. Phys.*, **50**, 1241
26. Ryan, J. L. (1967). *Advan. Chem. Ser.*, **71**, 331
27. Marcus, Y. and Bomse, M. (In press). *Israel J. Chem.*
28. Ryan, J. L. and Jørgensen, C. K. (1966). *J. Phys. Chem.*, **70**, 2845
29. Gruber, J. B., Menzel, E. R., and Ryan, J. L. (1969). *J. Chem. Phys.*, **51**, 3816
30. Judd, B. R. (1963). *Operator Techniques in Atomic Spectroscopy*. (New York: McGraw-Hill)
31. Jørgensen, C. K. (1962). *Orbitals in Atoms and Molecules*, Chapt. 11. (London: Academic Press)
32. Jørgensen, C. K. (In press). *Lanthanides and 5f Elements*. (London: Academic Press)
33. Jørgensen, C. K. (1962). *Absorption Spectra and Chemical Bonding in Complexes*, 138. (Oxford: Pergamon Press)
34. Jørgensen, C. K. (1962). *Progr. Inorg. Chem.*, **4**, 73
35. Jørgensen, C. K. (1969). *Oxidation Numbers and Oxidation States*. (New York: Springer-Verlag)
36. Jørgensen, C. K. and Rittershaus, E. (1967). *Mat. Fys. Med. Dan. Vid. Selsk.*, **35**, No. 15
37. Tandon, S. P. and Mehta, P. C. (1970). *J. Chem. Phys.*, **52**, 4896
38. Tandon, S. P. and Mehta, P. C. (1970). *J. Chem. Phys.*, **52**, 5417
39. Sinha, S. P. and Schmidtke, H. H. (1965). *Mol. Phys.*, **10**, 7
40. Jørgensen, C. K. (1967). *Intern. Rev. Halogen Chem.* (V. Gutmann, Ed.) **1**, 265. (London: Academic Press)
41. Ryan, J. L. Unpublished results
42. Johnston, D. R., Satten, R. A., Schreiber, C. L., and Wong, E. Y. (1966). *J. Chem. Phys.*, **44**, 3141
43. Ryan, J. L. (1966). Presented to the Symposium on Lanthanide and Actinide Chemistry, 152nd Meeting of the American Chemical Society (New York). See also Reference 26
44. Pappalardo, R. G., Carnall, W. T., and Fields, P. R. (1969). *J. Chem. Phys.*, **51**, 1182

45. Nugent, L. J., Lauberau, P. G., Werner, G. K. and Vander Sluis, K. L. (1971). *J. Organometal. Chem., 27,* 365
46. Pollack, S. A. and Satten, R. A. (1962). *J. Chem. Phys., 36,* 804
47. Satten, R. A. (1958). *J. Chem. Phys., 29,* 658
48. Adams, D. A. (1968). *Metal-Ligand and Related Vibrations.* (New York: St. Martin's Press)
49. Satten, R. A. and Stafsudd, O. M. (1967). *Opt. Properties Ions Cryst. Conf. Baltimore* 1966, 423
50. Ryan, J. L. (1969). *Inorg. Chem., 8,* 2053
51. Ryan, J. L. (1964). *Inorg. Chem., 3,* 211
52. Muetterties, E. L. and Wright, C. M. (1967). *Quart. Rev. Chem. Soc., 21,* 109
53. Brown, D., Easey, J. F., and Rickard, C. E. F. (1969). *J. Chem. Soc. (A),* 1161
54. Menzel, E. R. and Gruber, J. B. (1970). *J. Chem. Phys., 52,* 4830
55. Šćavničar, S. and Prodić, B. (1965). *Acta Crystallogr., 18,* 698
56. Ryan, J. L. (1960). *J. Phys. Chem., 64,* 1375
57. Keder, W. E., Ryan, J. L., and Wilson, A. S. (1961). *J. Inorg. Nucl. Chem., 20,* 131
58. Moore, F. L. (1967). *Anal. Chem., 39,* 1874
59. Morrey, J. R. (1963). *Inorg. Chem., 2,* 163
60. Jørgensen, C. K. (1963). *Acta Chem. Scand., 17,* 251
61. Gans, P., Hathaway, B. J., and Smith, B. C. (1965). *Spectrochim. Acta., 21,* 1589
62. Bagnall, K. W., Brown, D., and Jones, P. J. (1966). *J. Chem. Soc. A.,* 1763
63. Vdovenko, V. M., Volkov, V. A., Suglobova, I. G., and Suglabov, D. N. (1969). *Radiokhimiya, 11,* 26
64. Hayton, B. and Smith, B. C. (1970). *J. Inorg. Nucl. Chem., 32,* 1219
65. Gainar, I. and Sykes, K. W. (1964). *J. Chem. Soc.,* 4452
66. Sykes, K. W. and Taylor, B. L. (1962). *Proc. 7th ICCC.,* 31. (Stockholm: Almqvist and Wixsell, A. B.)
67. Shiloh, M., Givon, M., and Marcus, Y. (1969). *J. Inorg. Nucl. Chem., 31,* 1807
68. Marcus, Y., and Shiloh, M. (1969). *Israel J. Chem., 7,* 31
69. Jones, A. D. and Choppin, G. R. (1969). *Actinides Rev., 1,* 311
70. Barbanel', Yu. A. and Mikhailova, N. K. (1969). *Radiokhimiya, 11,* 595
71. Jørgensen, C. K. (1963). *Inorganic Complexes,* 1–10. (London: Academic Press)
72. See for example papers in Reference 2
73. Ekstron, A. E., Farrell, M. S., and Lawrence, J. J. (1968). *J. Inorg. Nucl. Chem., 30,* 660
74. Henrie, D. E. and Choppin, G. R. (1968). *J. Chem. Phys., 49,* 477 and references given therein
75. Asprey, L. B., Kruse, F. H., and Penneman, R. A. (1967). *Inorg. Chem., 6,* 544
76. Gritzner, G. and Selbin, J. (1968). *J. Inorg. Nucl. Chem., 30,* 1799
77. Cohen, D. (1970). *J. Inorg. Nucl. Chem., 32,* 3525
78. Kemmler-Sack, S. (1968). *Z. Anorg. Allgem. Chem., 363,* 282 and 295
79. Frlec, B. and Hyman, H. H. (1967). *Inorg. Chem. 6,* 2233
80. Penneman, R. A., Sturgeon, G. D., and Asprey, L. B. (1964). *Inorg. Chem., 3,* 126
81. Gruen, D. M. and McBeth, R. L. (1969). *Inorg. Chem., 8,* 2625
82. Lux, F., Wirth, G., and Bagnall, K. W. (1970). *Chem. Ber., 103,* 2807
83. Hagan, P. G. and Cleveland, J. M. (1966). *J. Inorg. Nucl. Chem., 28,* 2905
84. Ryan, J. L. (1963). *Inorg. Chem., 2,* 348
85. Stafsudd, O. M., Leung, A. F., and Wong, E. Y. (1969). *Phys. Rev., 180,* 339
86. Mefod'eva, M. P., Krot, N. N., Smirnova, T. V., and Gel'man, A. D. (1969). *Radiokhimiya, 11,* 193
87. Cohen, D. and Fried, S. (1969). *Inorg. Nucl. Chem. Lett., 5,* 653
88. Thalmayer, C. E. and Cohen, D. (1967). *Advan. Chem. Ser., 71,* 256
89. Steindler, M. J. and Gerding, T. J. (1966). *Spectrochim. Acta., 22,* 1197
90. Satten, R. A., Johnston, D. R., and Wong, E. Y. (1968). *Phys. Rev., 171,* 370
91. Morrey, J. R., Carter, D. G., and Gruber, J. B. (1967). *J. Chem. Phys., 46,* 804
92. Gruen, D. M. and McBeth, R. L. (1968). *Inorg. Nucl. Chem. Letters., 4,* 299
93. Young, J. P. (1967). *Inorg. Chem., 6,* 1486
94. Bagnall, K. W., Brown, D., and Holah, D. G. (1968). *J. Chem. Soc. (A),* 1149
95. Selbin, J. and Ortego, J. D. (1967). *J. Inorg. Nucl. Chem., 29,* 1449
96. Selbin, J. and Ortego, J. D. (1968). *J. Inorg. Nucl. Chem., 30,* 317
97. Schmieder, H., Dornberger, E., and Kanellakopulos, B. (1970). *Appl. Spectrosc., 24,* 499

98. Nikitina, S. A. and Lipovskii (1969). *Radiokhimiya*, **11**, 187
99. Alei (Jr.), M., Johnson, Q. C., Cowan, H. D., and Lemons, J. F. (1967). *J. Inorg. Nucl. Chem.*, **29**, 2327
100. Eisenstein, J. C. and Pryce, M. H. L. (1966). *J. Res. Natl. Bur. Std. A.*, **70**, 165
101. Hafez, M. B. (1968) *Comm. Energ. At.* [*Fr.*], *Rap. CEA-R-3521*
102. Eberle, S. H., Schaefer, J. B., and Brandau, E. (1968). *Radiochim. Acta.*, **10**, 91
103. Ryan, J. L. and Keder, W. E. (1967). *Advan. Chem. Ser.*, **71**, 335
104. Cohen, D. (1961). *J. Inorg. Nucl. Chem.*, **18**, 217
105. Ryan, J. L. (1961). *J. Phys. Chem.*, **65**, 1099
106. Sutton, J. (1952). *Nature*, **169**, 235
107. Varga, L. P., Asprey, L. B., Keenan, T. K., and Penneman, R. A. (1970). *J. Chem. Phys.*, **52**, 1664
108. Varga, L. P., Brown, J. D., Reisfeld, M. J. and Cowan, R. D. (1970). *J. Chem. Phys.*, **52**, 4233
109. Carnall, W. T. and Fields, P. R. (1967). *Advan. Chem. Ser.*, **71**, 86
110. Poturaj-Gutniak, S. (1969). *Nukleonika*, **14**, 269
111. Carnall, W. T., Fields, P. R., and Pappalardo, R. G. (1968). 11*th International Conference on Coordination Chemistry*, 411. (New York: Elsevier Publishing Co.)
112. Varga, L. P., Reisfeld, M. J., and Asprey, L. B. (1970). *J. Chem. Phys.*, **53**, 250
113. Brown, D., Holah, D. G., and Rickard, C. E. F. (1970). *J. Chem. Soc. (A)*, 786
114. Shiloh, M. and Marcus, Y. (1966). *J. Inorg. Nucl. Chem.*, **28**, 2725
115. Cleveland, J. M., Bryan, G. H., and Eggerman, W. G. (1970). *Inorg. Chem.*, **9**, 964
116. McLaughlin, R., White, R., Edelstein, N., and Conway, J. G. (1968). *J. Chem. Phys.*, **48**, 967
117. Edelstein, N., Easley, W.; and McLaughlin, R. (1967). *Advan. Chem. Ser.*, **71**, 203
118. Conway, J. G. and Rajnak, K. (1966). *J. Chem. Phys.*, **44**, 348
119. Asprey, L. B. and Penneman, R. A. (1962). *Inorg. Chem.*, **1**, 134
120. Yanir, E., Givon, M., and Marcus, Y. (1969). *Inorg. Nucl. Chem. Letters*, **5**, 369
121. Thompson, S. G., Harvey, B. G., Choppin, G. T., and Seaborg, G. T. (1954). *J. Amer. Chem. Soc.*, **76**, 6229
122. Hulet, E. K., Gutmacher, R. G., and Coops, M. S. (1961). *J. Inorg. Nucl. Chem.*, **17**, 350
123. Pappalardo, R., Carnall, W. T., and Fields, P. R. (1969). *J. Chem. Phys.*, **51**, 842
124. Delle Site, A. and Baybarz, R. D. (1969). *J. Inorg. Nucl. Chem.*, **31**, 2201
125. Keenan, T. K. (1961). *J. Amer. Chem. Soc.*, **83**, 3719
126. Carnall, W. T. (1967). *J. Chem. Phys.*, **47**, 3081
127. Gruber, J. B., Cochran, W. R., Conway, J. G., and Nicol, A. (1966). *J. Chem. Phys.*, **45**, 1423
128. Gutmacher, R. G., Hulet, E. K., Lougheed, R., Conway, J. G., Carnall, W. T., Cohen, D., Keenan, T. K., and Baybarz, R. D. (1967). *J. Inorg. Nucl. Chem.*, **29**, 2341. See also Peterson, J. R. (1967). *U.S. At. Energy Comm.*, *UCRL-17875*
129. Baybarz, R. D., Stokely, J. R., and Peterson, J. R. To be submitted to *J. Inorg. Nucl. Chem.*
130. Green, J. L. and Cunningham, B. B. (1966). *Inorg. Nucl. Chem. Letters*, **2**, 365. See also Green, J. L. (1965). *U.S. At. Energy Comm.*, *UCRL-16516*
131. Cunningham, B. B., Peterson, J. R., Baybarz, R. D., and Parsons, T. C. (1967). *Inorg. Nucl. Chem. Letters*, **3**, 519
132. Fujita, D. K., Cunningham, B. B., Parsons, T. C., and Peterson, J. R. (1969). *Inorg. Nucl. Chem. Letters*, **5**, 245
133. Nugent, L. J., Baybarz, R. D., Werner, G. K., and Friedman, H. A. (1970). *Chem. Phys. Letters*, **7**, 179
134. McGlynn, S. P. and Smith, J. K. (1961). *J. Mol. Spectrosc.*, **6**, 164
135. Rabinowitch, E. and Belford, R. L. (1964). *Spectroscopy and Photochemistry of Uranyl Compounds.* (Oxford: Pergamon Press)
136. Jørgensen, C. K. (1970). *Prog. Inorg. Chem.*, **12**, 101
137. Belford, R. L. and Belford, G. (1961). *J. Chem. Phys.*, **34**, 1330
138. Newman, J. B. (1967). *J. Chem. Phys.*, **47**, 85
139. Jatkar, S. K. K., Khedekar, A. V., Mukhedkar, A. J. (1968). *J. Univ. Poona Sci. Technol.*, No. 34, 67
140. Umrieko, D. S., Sevchenko, A. N., Novitskii, G. G. (1968). *Dokl. Akad. Nauk Beloruss., SSSR*, **12**, 884

141. Alikhanova, Z. M., Burkov, V. I., Kizel, V. A., Krasilov, Yu. I., and Safronov, G. M. (1969). *Zh. Prikl. Spektrosk.*, **10**, 134
142. Lipovskii, A. A. and Kuzina, M. G. (1963). *Radiokhimiya*, **5**, 268
143. Vdovenko, V. M., Skoblo, A. I., and Suglobov, D. N. (1966). *Radiokhimiya*, **8**, 651
144. Vdovenko, V. M., Skoblo, A. I., and Suglobov, D. N. (1969). *Radiokhimiya*, **11**, 30
145. Vdovenko, V. M., Mashirov, L. G., Skoblo, A. I. and Suglobov, D. N. (1967). *Zhur. Neorg. Khim.*, **12**, 2914
146. Tomažič, B., Žutić, V., and Branica, M. (1969). *Inorg. Nucl. Chem. Letters*, **5**, 271
147. Moskvin, A. I. and Zaitseva, V. P. (1962). *Radiokhimiya*, **4**, 63
148. Pérez-Bustamante, J. A. (1965). *Radiochimica Acta.*, **4**, 67
149. Bell, J. T. and Biggers, R. E. (1967). *J. Mol. Spectr.*, **22**, 262 and (1968) ibid., **25**, 312
150. Bell, J. T. (1969). *J. Inorg. Nucl. Chem.*, **31**, 703
151. Kuzina, M. G. and Lipovskii, A. A. (1969). *Radiokhymiya*, **11**, 91
152. Narasimham, K. V. and Garg, C. L. (1969). *Bhagavantam Vol.*, 84 (see *CA* **83**, 50175)
153. Davydov, A. V., Pal'shin, E. S., and Palei, P. N. (1969). *Radiokhimiya*, **11**, 110
154. Plaisance, M. and Guillaumont, R. (1969). *Radiochim. Acta.* **12**, 32
155. Pajakoff, S. (1963). *Monatsh*, **94**, 482
156. Jørgensen, C. K. (1962). *Solid State Physics*, **13**, 375
157. Jørgensen, C. K. (1962). *Mol. Phys.*, **5**, 271
158. Ahrland, S., Chatt, J. and Davies, N. R. (1958). *Quart. Rev. (London)*, **12**, 265
159. Ahrland, S. (1966). *Struct. Bonding (Berlin)*, **1**, 207
160. Pearson, R. G. (1966). *Science*, **151**, 172
161. Poturaj-Gutniak, S. and Taube, M. (1968). *J. Inorg. Nucl. Chem.*, **30**, 1005
162. Bagnall, K. W. and Brown, D. (1964). *J. Chem. Soc.*, 2031
163. Brown, D. and Jones, P. J. (1967). *J. Chem. Soc. (A)*, 247
164. Brown, D., Easey, J. F. and Jones, P. J. (1967). *J. Chem. Soc. (A)*, 1698
165. Urey, H. (1943). *U.S. At. Energy Comm. A-750*
166. Katz. J. J. and Rabinowitch, E. (1951). *The Chemistry of Uranium. Nat. Nucl. Energy Series VII, 5*, 504. (New York: McGraw-Hill)
167. Steindler, M. J. and Gunther, W. H. (1964). *Spectrochim. Acta*, **20**, 1319
168. Ryan, J. L. (1968). *11th International Conference on Coordination Chemistry*, 220. (New York: Elsevier Publishing Co.)
169. Nugent, L. J., Baybarz, R. D., Burnett, J. L. and Ryan, J. L. (In Press). *J. Inorg. Nucl. Chem.*
170. Ryan, J. L. (Unpublished results quoted by C. K. Jørgensen in Reference 136)
171. Nugent, L. J., Baybarz, R. D. and Burnett, J. L. (1969). *J. Phys. Chem.*, **73**, 1177
172. Jørgensen, C. K. (1965). *Proceedings of the Symposium on Coordination Chemistry, Tihany, Hungary, 1964*, 11. (Budapest: Publishing House of the Hungarian Academy of Sciences)
173. Barnes, J. C. and Day, P. (1964). *J. Chem. Soc.*, 3886
174. Miles, J. H. (1965). *J. Inorg. Nucl. Chem.*, **27**, 1595
175. Cunningham, B. B. (1964). *Ann. Rev. Nucl. Sci.*, **14**, 323
176. Jørgensen, C. K. (1968). *Chem. Phys. Letters*, **2**, 549
177. Jørgensen, C. K. (1969). *Chimia*, **23**, 292
178. Cotton, F. A. and Wilkinson, G. (1966). *Advanced Inorganic Chemistry*, 1055 (see also 681–682). (New York: Interscience)
179. Johnson, O. (1970). *J. Chem. Ed.*, **47**, 431
180. Day, M. C. and Selbin, J. (1969). *Theoretical Inorganic Chemistry*, 119. (New York: Reinhold Book Corp.)
181. Jørgensen, C. K. (1970). *J. Inorg. Nucl. Chem.*, **32**, 3127
182. Nugent, L. J. (1970). *J. Inorg. Nucl. Chem.*, **32**, 3485
183. Peppard, D. F., Mason, G. W. and Lewey, S. (1969). *J. Inorg. Nucl. Chem.*, **31**, 2271
184. Peppard, D. F., Bloomquist, C. A. A., Horwitz, E. P., Lewey, S. and Mason, G. W. (1970). *J. Inorg. Nucl. Chem.*, **32**, 339
185. Jørgensen, C. K. (1962). *Orbitals in Atoms and Molecules*, 95–97. (London: Academic Press)
186. Barnes, J. C. (1964). *J. Chem. Soc.*, 3880
187. Selbin, J., Ahmad, N. and Pribble, M. J. (1970). *J. Inorg. Nucl. Chem.*, **32**, 3249
188. Taylor, B. L. (1959). *The Solution Chemistry of Neptunium*, Ph.D. Thesis, Queen Mary College, University of London

189. Krot, N. N., Mefod'eva, M. P., Smirnova, T. V., and Gelman, A. D. (1968). *Radiokhimiya,* **10,** 412
190. Krot, N. N., Mefod'eva, M. P. and Gelman, A. D. (1968). *Radiokhimiya,* **10,** 634
191. Komkov, Yu. A., Krot, N. N., and Gelman, A. D. (1968). *Radiokhimiya,* **10,** 625
192. Loh, E. (1966). *Phys. Rev.,* **147,** 332
193. Dieke, G. H., Crosswhite, H. M. and Dunn, B. (1961). *J. Opt. Soc. Amer.,* **51,** 820
194. Biltz, W. and Clinch, J. A. (1904). *Z Anorg. Chem.,* **40,** 218
195. Albers, H., Deutsch, M., Krastinat, W., and von Orsten, H. (1952). *Chem. Ber.,* **85,** 267
196. Cohen, D. and Carnall, W. T. (1960). *J. Phys. Chem.,* **64,** 1933
197. Carnall, W. T., Fields, P. R., Stewart, D. C. and Keenan, T. K. (1958). *J. Inorg. Nucl. Chem.,* **6,** 213
198. Edelstein, N., Easley, W. and McLaughlin, R. (1966). *J. Chem. Phys.,* **44,** 3130
199. Ryan, J. L., Nugent, L. J. and Baybarz, R. D. (To be published)
200. Spedding, F. H., et al. (1966). *J. Phys. Chem.,* **70,** 2423, 2440, and 2450 and Spedding, F. H., et al. (1954). *J. Amer. Chem. Soc.,* **76,** 879, 882, and 884
201. Graffeo, A. J. and Bear, J. L. (1968). *J. Inorg. Nucl. Chem.,* **30,** 1577
202. Geier, G. (1965). *Ber. Bunsenges, Phys. Chem.,* **69,** 617
203. Silber, H. B. and Swinehart, J. H. (1967). *J. Phys. Chem.,* **71,** 4344
204. Garnsey, R. and Ebdon, D. W. (1969). *J. Amer. Chem. Soc.,* **91,** 50
205. Brown, D. and Jones, P. J. (1967). *J. Chem. Soc. (A),* 243
206. Nugent, L. J., Baybarz, R. D., Burnett, J. L., and Ryan, J. L. (To be published)
207. Stewart, D. C. (1952). *U.S. At. Energy Comm., AECD-3351*
208. Morrey, J. R. (Private communication)
209. Bagnall, K. W. and Brown, D. (1967). *J. Chem. Soc. (A),* 275
210. Brown, D. and Wilkins, R. G. (1961). *J. Chem. Soc.,* 3804
211. Ryan, J. L. (1961). *J. Phys. Chem.,* **65,** 1856
212. Jørgensen, C. K. and Brinen, J. S. (1963). *Mol. Phys.,* **6,** 629
213. Krumholz, P. (1958). *Spectrochim. Acta,* **10,** 274
214. Karraker, D. G. (1970). *J. Chem. Ed.,* **47,** 424
215. Mitsuji, T. (1967). *Bull. Chem. Soc. Japan,* **40,** 2091
216. Brown, D. (1966). *J. Chem. Soc. (A),* 766
217. Haissinsky, M., Muxart, R. and Arapaki, H. (1961). *Bull. Soc. Chim. France,* 2248
218. Penneman, R. A., Kruse, F. H., George, R. S. and Coleman, J. S. (1964). *Inorg. Chem.,* **3,** 309
219. Rozenzweig, A. and Cromer, D. T. (1970). *Acta Crystallogr.,* **B26,** 38
220. Shiloh, M. and Marcus, Y. (1965). *Israel J. Chem.,* **3,** 123
221. Shiloh, M., Givon, M. and Marcus, Y. (1965). *Proc. 3rd U.N. Int. Conf. Peaceful Uses Atom. Energy, Geneva,* 1964, **10,** 585
222. Gruen, D. M., Fried, S., Graf, P. and McBeth, R. L. (1958). *Proc. 2nd U.N. Int. Conf. Peaceful Uses Atom. Energy, Geneva,* 1958, **28,** 112
223. Nakamoto, K. (1963). *Infrared Spectra of Inorganic and Coordination Compounds.* (New York: John Wiley & Sons)
224. Ferraro, J. R. (1970). *Low Frequency Vibrations of Inorganic Compounds.* (New York: Plenum Press)
225. Clark, R. J. H. (1967). *Intern. Rev. Halogen Chem.* (V. Gutmann, Ed.), **3,** 85. (London: Academic Press)
226. Bagnall, K. W. (1967). *Coordin. Chem. Rev.,* **2,** 145
227. Brown, D. (1968). *Halides of the Lanthanides and Actinides,* 248–253. (London: John Wiley and Sons, Ltd.)
228. Claasen, H. H., Goodman, G. L., Holloway, J. H., and Selig, H. (1970). *J. Chem. Phys.,* **53,** 341
229. Gasner, E. I. and Frlec, B. (1968). *J. Chem. Phys.,* **49,** 5135
230. Frlec, B. and Claasen, H. H. (1967). *J. Chem. Phys.,* **46,** 4603
231. Bougon, R. (1967). *Commis. Energ. At. (Fr.) Rapp.,* No. 3235
232. Brown, D., Hill, J., and Rickard, C. E. F. (1970). *J. Chem. Soc. (A),* 476
233. MacCordick, J., Kaufmann, G., and Rohmer, R. (1969). *J. Inorg. Nucl. Chem.,* **31,** 3059
234. Morgan, L. G. and O'Leary, G. P. (Unpublished data).
235. Woodward, L. A. and Ware, M. J. (1968). *Spectrochim. Acta,* **24A,** 921
236. Stafsudd, O. M. (1967). *U.S. At. Energy Comm., UCLA-34P103-3*

237. Pandey, A. N., Singh, H. S., and Sanyal, N. K. (1969). *Curr. Sci.,* **38,** 108
238. Berringer, B. W., Gruber, J. B., Loehr, T. C., and O'Leary, G. P. (1970). *Bull. Am. Phys. Soc.,* **15,** 1602
239. Krasser, W. and Nürnberg, H. W. (1970). *Spectrochim. Acta,* **26A,** 1059
240. Larson, A. C., Roof (Jr.), R. B., and Cromer, D. T. (1964). *Acta. Crystallogr.,* **17,** 555
241. Kharitonov, Yu. Ya., Knyazeva, N. A., and Buslaev, Yu. A. (1969). *Zh. Neorg. Khim,* **14,** 1034
242. Nguyem-Quy-Dao (1968). *Bull. Soc. Chim. Fr.,* 3976
243. Ohwada, K. (1968). *Spectrochim. Acta,* **24A,** 97
244. Vdovenko, V. M., Mashirov, L. G., and Suglobov, D. N. (1968). *Radiokhimiya,* **10,** 587
245. Vdovenko, V. M., Mashirov, L. G. and Suglobov, D. N. (1969). *Dokl. Akad. Nauk. SSSR,* **185,** 824
246. Sergienko, V. I. and Davidovich, R. L. (1970). *Spectrosc. Lett.,* **3,** 27
247. Vdovenko, V. M., Ladygin, I. N., and Suglobov, D. N. (1970). *Zh. Neorg. Khim.,* **15,** 265
248. Belyaev, Yu. I., Vdovenko, V. M., Ladygin, I. N., and Suglobov, D. N. (1967). *Zh. Neorg. Khim.,* **12,** 3222
249. Vdovenko, V. M., Ladygin, I. N., and Suglobov, D. N. (1968). *Zh. Neorg. Khim.,* **13,** 297
250. Newbery, J. E. (1969). *Spectrochim. Acta,* **25A,** 1699
251. Bukolov, S. S., Vdovenko, V. M., Ladygin, I. N., and Suglobov, D. N. (1970). *Zh. Prikl. Spectrosk.,* **12,** 341
252. Garg, C. L. and Narasimham, K. V. (1970). *Spectrochim. Acta.,* **26A,** 627
253. Bullock, J. I. (1967). *J. Inorg. Nucl. Chem.,* **29,** 2257